CLIMATE CHANGE

AND

ENERGY POLICY

CLIMATE CHANGE

AND

ENERGY POLICY

Proceedings of the
International Conference
on

Global Climate Change: Its Mitigation
Through Improved Production and Use of Energy

Los Alamos National Laboratory

October 21–24, 1991
Los Alamos, New Mexico, USA

Editors-in-Chief

Louis Rosen
Robert Glasser

Los Alamos National Laboratory document LA-UR-92-502.

L.C. Catalog Card No. 92-72649
ISBN 1-56396-017-6
DOE CONF-9110127

Organizing Committee

Sumner Barr
Los Alamos National Laboratory

Ralph Cicerone
University of California, Irvine

Robert Glasser
Los Alamos National Laboratory

George Hidy
Electric Power Research Institute

Charles Keller
Los Alamos National Laboratory

Louis Rosen
Los Alamos National Laboratory

Charles Teclaw
U.S. Department of Energy/Los Alamos National Laboratory

Linda Trocki
Los Alamos National Laboratory

William Zagotta
Lawrence Livermore National Laboratory

Editors
John C. Allred
Kay Harper
Arthur S. Nichols
Beverly Talley

Contents

PART THREE: CLIMATE-CHANGE POLICY AND DECISION MAKING

Introduction

Layout of the Book

Robert Glasser

Los Alamos National Laboratory

As we approach the end of the twentieth century, we find ourselves at a crossroads. Until now, many have assumed that our technology and industry will enable us to sustain economic and social progress indefinitely. This assumption is currently being tested as never before. The population of the planet is approaching 5.5 billion and could reach 14 billion by the year 2100. All around us we see evidence of unsustainable activity: desertification, deforestation, urban pollution, damaged watersheds, loss of biological diversity, and, in many areas, a widening gap between the rich and poor. Today almost one-fourth of mankind lives in abject poverty. All this fosters instability, which is exacerbated by the revolution in communications.

Feeding and caring for the inhabitants of our world is a daunting technological, economic, and sociological challenge. At the very heart of satisfying their needs will be decisions involving the production, distribution, and use of energy. Coal, oil, and gas, which fueled the industrial revolution, currently account for about 90 per cent of the world's production of commercial energy. It is not at all clear, however, that the energy requirements of the next century can be met in the same way they were met in previous times. Proven oil and natural gas reserves will run out in less than 50 years (though coal reserves could last many decades longer) and these three sources of energy have disturbing byproducts, such as smog and acid rain. Most importantly, many leading climate-change scientists suspect that the use of fossil fuels is contributing to significant changes in the global climate that could have severe ecological and sociological ramifications. If the most disturbing predictions of these scientists are true, global warming is among the gravest threats facing humanity.

The "Greenhouse Effect" is a well-understood natural phenomenon. A number of gases occur in the atmosphere in small quantities. Together with water vapor they serve to warm the surface of the planet by absorbing infrared radiation that would otherwise escape into space. Without these gases, the planet would become uninhabitably cold — indeed the temperature would drop to a point similar to that on the surface of the moon. Human activity since the industrial revolution — particularly the emission of CO_2 from the burning of fossil fuels — has added significantly to the quantities of greenhouse gases in the atmosphere. Less well understood is the impact these added, anthropogenic gases will have on the temperature, although the evidence suggests that warming is likely.

Even a small increase in global temperature could cause severe dislocations, particularly in the less-developed countries. The potential dislocations are sufficiently serious to warrant national and international attention. Moreover, the posited close connections between global warming and our production and use of energy suggest a need for integrated and interdisciplinary approaches to understanding this problem. The situation is particularly acute because the world-wide demand for energy will increase dramatically in the coming decades due to unprecedented population growth. It is, therefore, important for us, at this critical juncture, to focus our efforts on understanding better the science and energy policy implications of climate change.

This book presents the proceedings of an international conference which was held 21–24 October 1991 at the Los Alamos National Laboratory. The Conference, which was attended by experts from more than 15 countries, was explicitly designed to explore the connections between climate change and energy policy, and to foster dialogue between these disciplines.

Part One of the book presents the latest information available on the status of the science of climate change. Leading climate-change scientists describe the reliability of climate-change models, the observational evidence for global warming, and the extent to which the results of the models match the observations. Of critical importance to policy makers is understanding when we are likely to have better evidence on global warming. Equally pressing is the need for data on the regional impacts of climate change. A number of papers in this part of the book address these issues. Part One concludes by presenting the results of some specific research efforts involving climate modeling, including some interesting preliminary data suggesting a connection between the smoke from Kuwaiti Oil wells (which were set afire in the recent Persian Gulf War) and local and regional climate change.

Part Two of the book is entitled "Climate Change, Energy Technologies, and Economics." Among other things, it explores the economics of energy alternatives in response to global warming. In addition, it presents the latest information on the status of technologies for mitigating emissions of CO_2. The contributions of energy efficiency and renewable forms of energy are elucidated, as are some exciting new research efforts, such as "Hot Dry Rock," and technologies in support of nuclear energy, such as "Accelerator-Driven Transmutation of Nuclear Waste."

Part Three of the book focuses on climate-change policy and decision making. Scientists and policy makers have different constituents, agendas, and constraints. As a result, the decisions policy makers take in response to national problems are often inadequate reflections of the scientific information available. Part Three of the book explores the constraints on decision making in light of scientific uncertainty. It also provides an update on how countries such as Britain, Canada, and the Netherlands are developing their national strategies in response to climate change.

The fourth and final part of the book contains the results of panel discussions on the subjects of integrating climate change and energy policy. It is becoming increasingly apparent that it is inappropriate and artificial to treat energy and environment as separate issues. They are two sides of the same coin. A main goal of the Conference was to encourage energy and environmental experts to integrate. The panel sessions provide some interesting insights into the roadblocks and opportunities for this integration. Part Four of the book also presents some observations on the progress that was made in the

Conference's energy workshops and in the workshops on climate-change science and energy policy. The final chapter of the book is a summary of the Conference proceedings and some concluding remarks by Louis Rosen.

Our world faces some daunting challenges in the coming decades. As the human population and demands for improved quality of life increase, the stresses and strains on our environment will also increase. Global climate change is simply the latest—albeit extremely worrisome—manifestation of a crowded planet. The environmental problems we are witnessing have arisen in the context of a system in which nations have pursued their interests more or less independently. The solutions to these problems, however, will clearly require international technological and political cooperation and collaboration on an unprecedented scale. The alternative may very well be global disaster.

Opening Remarks

John Whetten
Associate Director
Los Alamos National Laboratory

A little over two years ago Los Alamos National Laboratory co-sponsored an international conference in Santa Fe, New Mexico, on technology-based confidence building focused on energy and environment. Among the conclusions that were reached at the conference was the need for further research into global warming and, in particular, further discussion on the connection between climate change and energy policy. As a means of describing some of the goals for this conference, I would like to say a little bit about our activities at Los Alamos.

My directorate at Los Alamos has responsibility for the energy and technology programs, which include environmental R&D. We have been increasing our portfolio in these programs for the past couple of years and have come to realize the expediential linkage between energy and environment. In fact, in many of our programs it's difficult to tell whether it's an energy program or environment program; it's very difficult to separate the two. Let me cite an example.

About a year ago, under the sponsorship of the Department of Energy and the Mexican Petroleum Institute, we began a program jointly with the Mexican Petroleum Institute to look for options for improving the air quality in Mexico City. We have just had our first review of that program, at some of the data to come out of the experiments, and at some of the first models and some of the first options selected for improving the air quality of the Mexico City basin. We were somewhat startled to see the extraordinary relationship between energy in that particular basin and the deterioration of the air quality there. In fact, most of the options being explored for improving the air quality really have to do with energy, relocating and improving refining processes, substituting natural gas for fuel oil, removing the worst polluting automobiles from the roads, and so forth. In fact, we're finding this true with many of our energy programs—that they are intertwined with environmental R&D.

As most of you are probably aware, Los Alamos and other national laboratories have ongoing programs in climate change. At Los Alamos we make use of some of the world's most powerful computers and emphasize the modeling in our activities, and, at the same time, we have significant programs to study possible alternative energy sources and means to improve energy efficiency, such as cleaner combustion technology, enhanced oil-recovery methods, high-temperature superconductors, and advanced energy-efficient materials. It's my belief that for the most part scientists involved in global climate change at Los Alamos, and probably elsewhere, have had relatively little interaction with energy policy makers and energy researchers. Yet at the heart of the issue of global warming is the production and use of energy.

One of our main objectives in convening this meeting is to encourage a closer integration of these two communities — energy policy researchers and makers and climate-change scientists. If nothing else, this conference should provide an occasion for people from these widely different disciplines to compare notes, establish relationships, and perhaps begin collaborations. We would like to encourage you to seize this opportunity. How do these two communities, the climate-change people and the energy people think about global warming? Do they share the same assumptions concerning the present and future, the potential responses of society, and expectations with regard to national priorities?

We think that the past may yield some useful insights for the future. What has been the climate-change scientists' previous experience with the policy-making community and how has their experience affected progress on particular issues? The acid-rain and nuclear-winter examples may be instructive. When, if ever, are the uncertainties about global warming likely to disappear and what are the implications for energy policy making of the timing of the disappearance of uncertainty? The answer to these and many other questions will not only help us to understand better the pressures and problems associated with climate change, but also to address the larger issue of the connection between science and policy making.

We have some additional objectives for the conference that I will discuss only briefly. Obviously, with such a distinguished group of participants, we have an opportunity here to review the present status and future directions of climate-change studies, especially climate modeling. How good are the models that we're using? How well do the results match what we are observing in nature? Are they likely to improve significantly and if so, how soon? A similar opportunity exists to explore some of the relevant cutting-edge technologies that relate to global warming and the economic and political trade-offs associated with energy alternatives. The possibilities range from continuing business-as-usual to drastic cuts in CO_2 emissions with the resulting high economic costs.

Finally we hope the conference will enable us to investigate how we might enhance ongoing initiatives in the areas of climate change and energy and identify and evaluate the potential for new measures and ideas.

We think we've given you a set of ambitious goals for this conference. We may not be able to achieve them all in four days of deliberations, but your attendance here today suggests that you share with us a willingness to make that attempt. We welcome you to the laboratory and look for success in this conference. Thank you very much.

Introduction of Keynote Speaker

Siegfried S. Hecker
Director
Los Alamos National Laboratory

I would like to add my welcome to all of you to sunny New Mexico. It's my pleasure to make a few opening remarks and introduce the keynote speaker this morning.

National security today, now more than ever, is not just military strength, and in fact the dizzying pace of global, geopolitical change in the past few weeks has brought that home. We have for a number of years moved in the direction of applying some of the very important technical strengths that we had the good fortune to build up as part of our defense mission to our civilian missions. This is especially true in the field of global climate-change research. There are a couple of what we call our "core technical competencies" that are especially pertinent. I'll just mention two: high-performance computing and modeling and dynamic experimentation and diagnostics (the ability to measure).

We're very fortunate, because of the strong defense mission, to have been able, over the years, to build up at Los Alamos the world's most powerful scientific computational facility. In fact, we expect that it will become even more powerful, perhaps in the next few months' time. We are especially proud of what we've been able to do, not only with conventional supercomputers, but in the area of massively parallel processing. We have two Thinking Machines' computers (CM2s), the connection machine models that are massively parallel processing machines, and we're in the process of finalizing an arrangement to bring the first true production machine version of Thinking Machines' latest models, called the CM5, here to Los Alamos. The CM5 will be placed in an "open partition," which means it will be available to do work of an unclassified nature, such as the enormous challenges associated with global climate-change research.

So we have both an interest and some sophisticated capabilities to apply to this important subject.

I am pleased to introduce the keynote speaker this morning, Serguei Kapitsa, who will talk about global change. It was my pleasure to meet Professor Kapitsa for the first time about four years ago when he was here at Los Alamos to give a lecture on the subject of arms control. At that time, the INF Treaty (Intermediate Nuclear Forces Treaty) was still under discussion. Professor Kapitsa predicted that the treaty would be signed and enter into effect, and that it was to be the first treaty that would actually reduce the nuclear arsenals of both countries. He indeed was correct. Looking back now, of course, INF was a rather small step compared to what's happened in recent times. No one would have been able to predict the incredible pace of events in the areas of arms control and arms reductions.

We had a very interesting time with Professor Kapitsa the last time he was here. Let me just tell you a little bit about his truly exceptional background.

He was born in England when his father was at the Rutherford Laboratory. He holds a PhD in aeronautical engineering as well as in physics. The latter was earned at the Joint Institute for Nuclear Research in Dubna. He holds a chair in physics at the Moscow Physical Technical Institute and is on the Board of Editors of the Soviet equivalent of *Nature*. He's also the editor of the Russian edition of *Scientific American*. He is broadly schooled and has a remarkably diverse set of interests.

Professor Kapitsa has been a member of the Soviet Committee of UNESCO and Vice President of the European Physical Society; today he is Chairman of the Soviet Physical Society, Vice Chairman of the Soviet Scientists Against Nuclear War, and Deputy to Academician Velikhov on arms control issues. I think all of you know that Velikhov is Scientific Advisor to President Gorbachev as well as Vice President of the Soviet Academy of Sciences.

Professor Kapitsa is the recipient of numerous prizes and honorary degrees, among them the UNESCO prize, which probably came to him because of his frequent and outstanding television appearances aimed at educating the public in matters of science and technology. That's something that we would very much like to import to this country.

We are very pleased to have you with us, Professor Kapitsa, to present the keynote address.

Keynote Address

Professor Serguei Kapitsa
Institute for Physical Problems
Academy of Science
Moscow 117334 Russia

Thank you, Dr. Hecker, for your kind introduction. It is not easy to be the speaker who opens this conference because there are so many things we have to think about. But I will single out a few things that seem most important and should be treated differently on our agenda.

I will begin by indicating how I would like to see this conference develop and what I believe we should be really engaged in. What is lacking most is not so much the detailed assessments of various technologies and ideas that are so well represented here, but a general overview of the whole situation. What are the really important factors and how do they interact with each other? We try to understand this through global modeling. I think it is not so true in the world today, as it was eight or ten years ago when the *Limits to Growth* was published, that people believe that through using powerful computer data we can understand the future. We have to explain what happened in the past—and perhaps there we could test some of our ideas. You know, in our country we are now engaged in rewriting our history to a great extent. And history is said to be the way of predicting the past. So, that could be an exercise in reverse global modeling, rather than looking at what is to be in the future. However, it seems that we are not always good at predicting what happened in the past and even poorer at predicting what can happen in the future. I could remind you that in the days of the Queen of Sheba, Arabia was a very flourishing and green part of the world. On the other hand, Europe was just recovering from the last ice age and the conditions there were far from favorable. Now things have slightly changed.

In our country we have seen major changes in the environment, especially concerning the Caspian Sea. Its level went down by a few meters over the observable decades, but now the level has risen. I would like our climatologists to explain how it happened. On the other hand, we have the Aral Sea, half of which has vanished, although we know why it happened—the water was used for irrigating the very fertile regions of Central Asia. In fact, I think that the very painful experience of the Soviet Union on these matters can be instructive because our country is so large that it can serve as a subglobal model. We can see many events that certainly will have a different magnitude and a different mechanism on a global scale, but you can already see them happening in our country. I should say that with respect to the environment, the damage to the environment, and the difference between the South and the North, tensions are building and have now become politically manifest. In fact, within your country there are large regional economic differences that generate strains of their own and pressures on the environment that have to be taken into account.

The outcome of what we are doing profoundly influences the whole issue of interdependence and security. Today, certainly, security is a key word, and national security is defined more in terms of energy or of the economy or of the stability of the whole system than in terms of military power. In fact, I think we've been seeing a rather powerful demonstration during the past decade of the futility of solving and resolving major social conflicts and events through military action. Even in the action in the Gulf, however spectacular it was, one can question to what extent it has resolved the issues of the world and of the countries we're facing there. I don't know to what extent the political developments have resolved those differences, but it seems that the military arm could do much, but it couldn't really get down to the region's basic issues. In our country also, we have had demonstrated that the most powerful army in the world, when it had to face its own citizens, had to face social issues, and was futile as an instrument of resolving those difficulties. These lessons are now of immediate importance. I think they teach us once again that technological solutions to these problems are only part of the whole picture. One of the issues that is very difficult to resolve is the idea of stability. Again, we must think of stability not only in terms of technology, of energy, but also in the broader social sense. Stability and security, in a certain way, are interdependent.

This conference can deliver to industry, to the large laboratories, to the public, and to the media a cogent message that can have remarkable influence on the way the public thinks. We can do more than educational establishments, more than the arms of political institutions. I would like to remind you that in less than a year, in June 1992, there will be a large United Nations conference where thousands of people will gather to discuss issues of the global change of the climate, of the environment, and of development. I don't know what the outcome of that meeting will be. Its efficiency I think will be very poor. Will the scientific establishment of the world have a chance of being heard and have an effect on what's going on there? I don't know. But perhaps this meeting of ours could generate some sort of signal that, in one way or the other, could be broadcast to that conference.

In fact, Professor Golitsyn and perhaps others in this hall are advisors to a committee of the International Council of Scientific Unions that has been commissioned by the organizers of this UN conference to generate a certain message from the scientific community of the world. Perhaps the expertise and authority that have been gathered here could help in working out a message that could be delivered.

I think we are somehow not concerned to the extent that we should be with the only independent variable in this picture: population: It is an independent variable in the sense that it is the force that is generating the pressures on the climate, and, at the same time, experiencing the pressures from the lack of energy. We do have to think about what is really happening with the world population. It's a global effect that is very poorly understood.

The world's population is growing. I remember that as a small boy I was taught that the world has 2 billion people; now we have passed 5 billion and we're heading to a 37-year doubling time. A decade ago demographers said that the world population will level off below 10 billion. They've already changed their forecasts considerably. Nobody really understands the nature of this remarkable global population explosion through which we are passing. The world's population expanded more or less regularly over hundreds and thousands of years. According to demographers, 2 000 years ago there were 200 million people on the earth. It is a time scale that is very short compared to the history of mankind, which goes back 2 or 3 million years. We are at a very important threshold in human history. It's like a shock wave; it's a discontinuity, it's a singularity that we're entering, and we're well in the middle of it. And nobody can really understand how it is all happening, and I think this is one of the important variables that does have to be taken into account. We have to find some understanding of why it is all happening and what has really changed, because what is happening now is that the time of doubling of the world population is already the same as the length of the human life. In the past hundred years, not to speak of the Middle Ages, things practically didn't change during a human lifetime, including the order of things, the number of people in the world, and everything else. The time scale of change was much longer. The population doubled in a thousand years, in a hundred years, but certainly not in the few decades that is happening now.

This is something that we do have to take into account, one way or the other. In fact, because of the reduced time scale, we see rapid changes not only in the population, but also in energy. And you can see that the climates are changing, maybe in 100 or 500 years, it depends on what you're talking about. The energy buildup, the change in technology, takes about 50 years.* Nuclear energy has already been evolving for 50 years. In that interval the human population has more than doubled. So you can see that the human population is changing more rapidly than even our technology is evolving.

What really drives the increase in human population? We don't even know that. Who controls these things? God or the Pope maybe can tell us — I don't know. We're at a loss, you see. It's a pity that we did not bring population issues into the agenda of our conference. And it seems that the limits of growth that we are now facing in one way or the other are imposed not so much by the environment, but by something in the human race itself. We've yet to see the real pressures of the environment and the climate on the human race. That is an important point. In a certain sense we are dealing without these feedback loops. These things are yet to happen, but the world population is something that is already happening. That is why these time scales really matter. And perhaps we should have this matter in our mind even before it can be done on a more formal basis, using computers, for example. We have simply to bear in mind the phenomena, the connections, and the time scales that are characteristic of the changes that really matter.

---

* A most interesting study of energy and population has been given by John Holdren, "Population and the Energy Problem," *Population and Environment* **12** Spring 1991.

These are the main things that I would like us to keep in mind throughout today's talks. I think that we have to try and integrate these things, and it's very fortunate that the American Institute of Physics will publish the papers of this conference because that certainly is a way of reaching a broader audience. I think that physicists and engineers and people at these large laboratories do have the necessary intellectual resources and, hopefully, an ability to work in an interdisciplinary way where we would all have to multiply our efforts, rather than add them together. Although leaders of research know how difficult it is to generate a team—a collection of experts who will really interact rather than simply sum up each in his own way what is happening. And I do hope that our conference will prove that we can constructively interact in treating these issues, finding what really the priorities are, because I think the priorities are something that are clearly at the heart of any major decisions that have to be taken sooner or later. At present I should say the world is in a certain stage of hesitancy. Major decisions concerning the future of energy have not yet been taken. We're very reluctantly moving away from the primacy of military force and expenditures for the guarantees of security. We have to shift the ideas and the concepts of the ways in which security, both national and global, can be established, and I think these great laboratories, like the one in which we are now present, certainly contributed to one of the mechanisms that generated a certain sense of security in the past.

Can you manage to do it in the future? That is yet to be seen. I must say that in my country the huge laboratories, the research arm of the military and industrial complex, are now also at a loss as to what to do: where to find support for their efforts, what should motivate them, what values should they cherish, and what new social agenda should they follow? This is yet to happen and our country is going through a very complicated turmoil, and perhaps it demonstrates in a peculiar way the very narrow limits of stability that exist in our world. Our country seemed to be the edifice of stability and permanence for many, and suddenly things started to happen. Perhaps the historians of the future will explain to us why the events happened, but that is not my problem. We can only observe these happenings and try to find our way in a very rapidly changing world. Perhaps the rapidity of changes that are imposed on us by the global problems—in the first place, the population problems—lie at the heart of this insecurity that we are all facing today. And I hope we might find a way in which we can somehow mitigate, change, and maybe stabilize the world environmental and energy situation.

I would like to close by once again thanking you for the opportunity to make these remarks.

Part One:

Status of the Science of Climate Change

Assessing Global Climate Change: When Will We Have Better Evidence?

J. D. Mahlman

Geophysical Fluid Dynamics Laboratory/NOAA

PREFACE

The title of this presentation seems to imply that today's evidence on expected climate change is somehow inadequate to begin the serious work of assessing impacts and policy options. ("If only the climate modeling community would give us reliable predictions, then we could get on with the serious work of deciding what to do about it.") I believe that the "we are ready to regulate" versus the "we are not ready to regulate" question is not governed by a simple scientific on/off switch. The simple truth is that, scientifically, much was known about this problem 15 years ago, while much will remain uncertain 15 years from now. The desire to regulate or not to regulate often seems to be governed more by one's personal value system than by the state of the science. As scientists and engineers, it is essential that we recognize this and work to ensure that science is used as a tool for guidance and not as a political bludgeon.

A second introductory point concerns the capability (or lack thereof) of governments to act in response to sharpened scientific information. For the sake of focusing discussion, let me highlight the point by proposing an outrageous sociological thought experiment: Suppose that the infinitely clever climate community, through a stunning breakthrough, announced today the capability to produce a "perfect" climate change prediction (including means and a dazzling array of variability statistics) for all parts of the earth averaged over say, the years 2040 to 2060. Armed with this "perfect" information, how and when and at what capability levels would the impacts analysts and policy makers of the world deal with it? The mind boggles. Clearly, much more research is also going to be required on the energy-use/impacts/economics/policy side of the fence.

Before describing the necessary research to improve scientific confidence, I begin with a listing of those aspects of the climate problem for which we already have high confidence. For example, we are virtually certain that we understand and can quantify, with impressive accuracy, the infrared absorptivity of the so-called greenhouse gases (e.g., CO_2, N_2O, CH_4, and CFCs). Moreover, we are virtually certain that the global concentrations of all these gases are increasing due to emissions related to human activities. These increased concentrations are acting systematically to produce additional heating to the climate system. Finally, we are virtually certain that additional emissions of long-lived greenhouse gases will commit the climate system to additional heating for centuries.

In terms of climate response to this forcing, I am convinced that it is very probable (~9 out of 10 chance) that the global-mean surface temperature will have warmed by ~1–4.5°C by the middle of the next century. I am equally confident that this warming will be accompanied by an increased atmospheric water-vapor loading and an increase in the global-average precipitation rate. I am virtually certain that the middle stratosphere will experience dramatic temperature decreases (>10°C) by that time.

These statements of high confidence evade the issue of the major need for considerably greater confidence in many aspects of the global-warming problem. These other requirements are the dominant theme of this paper. However, the seriousness of these required research advances need not blind us to the inexorable basic truths of the greenhouse problem. Global warming is a real and fundamentally sound issue. It is not a fad that will disappear. Faddish treatment of it will come and go, but the problem will stubbornly remain with us, quite oblivious to all rhetoric, hype (pro and con), and posturing. It is a very safe forecast that impacts analyses and regulatory decisions will continue; it is the pace of those activities and the magnitude of the commitments that will be dependent upon the rate of improvement of scientific understanding.

ASSESSMENT OF THE MAGNITUDES, RATES, AND GEOGRAPHIC PATTERNS OF CLIMATE CHANGE

Impact and policy analyses of greenhouse warming are much affected by the magnitude of warming for a given perturbation as well as the rate at which the response occurs. Since most realized impacts and political perspectives occur at regional scales, the geographical structure of the changing climate has particular significance in determining the kinds of attitudes taken and responses considered by society.

Assuming the usual extrapolations of current anthropogenic trace-gas emissions, the magnitude of predicted change is mainly determined by the degree of effectiveness of various feedback processes in the hydrological cycle. For example, the inclusion of water-vapor feedback very likely will increase the calculated global-mean surface-temperature response by a factor of 1.5 to 2. Quantification of the additional multiplicative effects of cloud-radiative feedback is considerably more difficult, ranging from, say, 0.7 to 2.5 globally, and is even more variable regionally. The magnitude of the high-latitude response is expected to exceed that of the tropics. The rates of response to changing radiative forcing are determined to a large degree by the response of the ocean-ice system and can exhibit considerable geographic variation. Typically, the rates of response appear to be slowest in regions characterized by active exchange to deeper ocean layers and fastest in the tropics.

The rates are also determined by the presence of competing or reinforcing additional radiative forcing mechanisms. Such mechanisms can arise from natural (e.g., volcanic) or anthropogenic causes (e.g., sulfate aerosol, carbonaceous soot, or ozone change). All of these effects can add confusion to the characterization of the system. Moreover, they can, as does natural variability, add changes of either sign depending upon geographical location and altitude.

The problem of predicting geographic patterns is at the mercy of the above issues on the character of the forcing and the magnitudes and rates of response. Also, the smaller the region considered, the more it is affected by natural variability and by local conditions (e.g., changed agricultural or land-use practices). In addition, simulations of climate change in smaller regions are more affected by inadequacies in computer resolution as well as by relatively crude treatment of land-surface processes, precipitation physics, cloud-radiation interaction, topographic influences, etc.

Since cloud-radiation feedback is our best candidate for "Prize Monster" in our uncertainty about the magnitude of the climate system response, progress must be substantial here to effect a relatively quick improvement. I believe this will be found to be as true regionally as it is globally. Significant progress here will require substantial progress on observational, process/diagnostic, and modeling levels. In my own view, intensive cloud-radiation research will begin to yield important reductions of uncertainty in a decade or so.

Understanding the rate of response of the system first requires an improved ability to model the ocean-ice system. More importantly, it may depend upon a far more self-consistent coupling of the atmosphere-ocean system than is currently possible. This will require substantial improvements in modeling accuracy for both atmosphere and ocean. This demands improved physical parameterizations that will depend upon very well-posed field experiments with sharply focused diagnostic analysis. Again, the time scale to yield substantial uncertainty reduction appears to be a decade or so.

Improved prediction of geographic patterns first requires advances on the above magnitude and rate fronts. A number of regional process studies on basic climate mechanisms and balances will be necessary. In addition, substantial increases in computer power will be necessary to develop improved predictive confidence for specific geographic areas. Such computer power must be accompanied by sharply increased physical accuracy in a host of region-specific physical processes that are not well characterized in today's models. The time scale for important advances in predicting geographic patterns, with confidence, likely will be one to two decades.

MECHANISMS AND FEEDBACKS IN THE CLIMATE SYSTEM

The earth's climate is a forced system that, in a bewilderingly complex way, effects a net balance between incoming solar radiation and outgoing infrared radiation. Any process that leads to a change in either of these factors initiates a thermal forcing that inevitably will produce a response in the climate system. For example, the diurnal and annual variations of solar radiation produce large and immediately observable "climate changes" that are central to our daily lives. These are large, rapid responses to large solar radiation changes. On the very long time scales, relatively subtle variations in earth-sun configuration are known to initiate and eliminate major glaciation cycles.

The potential for climate warming due to anthropogenic sources of radiatively active trace species is intermediate in time scale ($\sim$ decades to centuries) to the phenomena described above. Accordingly, some processes that are quantitatively relevant on these time scales may be of lesser relevance to the diurnal/annual responses at one end and the glaciation cycle responses at the other end.

Nevertheless, the capability of climate models to deal with these extreme climate-change problems at the shorter and longer ends of the time scale constitute very important evaluative tests. Current models do a generally good job of simulating the diurnal and annual "climate" changes. However, a close look at specific geographical locations reveals many remaining model deficiencies. These deficiencies become more evident when one looks at more detailed statistics of transient phenomena, such as precipitation. The deficiencies in annual-cycle simulation become even more evident when the model is required to calculate the atmosphere-ocean interactions on the annual time scale in a fully self-consistent manner. The "best" atmosphere and "best" ocean models coupled together still produce significant drifts of the models' local seasonal climates away from the actual climate.

Perhaps surprisingly, the models have lacked the sensitivity required to simulate either the last ice age or the warm climate of 70–100 million years ago. Incorporation of the discovery of low CO_2 and CH_4 amounts in the last ice age, as well as the inference of high CO_2 amounts 70–100 million years ago, have brought the model simulations closer to the reality of those paleoclimates. Intriguingly, models have not yet been able to initiate a glaciation cycle, even given the extremes of solar radiative flux as input. Does this mean that the models are less sensitive than the real climate on the greenhouse warming time scale? Not necessarily. The large ice-age sensitivity may depend upon ice-ocean and/or biogeochemical feedbacks that require centuries or millenia to establish themselves.

All of this shows that there is a daunting amount of research required on a considerable variety of climate-change mechanisms and feedbacks. For this exercise, it is important to note that most of the required research is rather well described in IPCC (1990) and in CEES (1990). Thus, in my view, a detailed summary of the required research projects is unnecessary here.

For the record, however, the kinds of research efforts required to expose mechanisms and feedbacks of climate change are as follows: cloud-radiation feedback; upper-troposphere water vapor; convection-cloud coupling; atmosphere-ocean fluxes; ocean circulation; sea ice; CO_2, CH_4, and N_2O biogeochemistry; radiative-chemical interactions, including ozone; land surface hydrological/biospheric processes; basic budgets and radiative effects of natural and anthropogenic aerosols; atmospheric and oceanic dynamics; turbulence; mathematical modeling techniques; . . . and more.

In each of these research areas, many things are already known. However, accurate predictions of changes in the climate system requires much more quantitative characterization and modeling than currently exists. This will require process studies and field experiments working in association with model simulations in each of these research topics. Moreover, the climate models will have to assimilate this information about physical processes and synthesize it into simulations of the integrated system. This is far more easily said than done. The lesson from our research past is that an indiscriminate conglomeration of "improved" physical processes into an "improved" model usually produces unsatisfactory results. For the foreseeable future, a judicial combination of process-oriented studies evaluated in a model context, performed one-at-a-time, will be the most productive approach. The harsh truth is that this will act to stretch the time scale required for substantial lessening of remaining uncertainties. In my view, it is the requirement for quantitative synthesis of the above-listed research areas that makes the progress time scale stretch to a decade and perhaps well beyond.

THE ROLE OF HUMAN ACTIVITIES IN PRODUCING CLIMATE CHANGE

The human activities that can influence climate are quite varied and can impact the climate system in a variety of ways. A brief listing of each activity and its climate-perturbing effect is given below:

1. Fossil-Fuel Combustion
   a. CO_2 emission (infrared (IR) trapping)
   b. CH_4 emission by natural gas leakage (IR trapping)
   c. NO_x emission alters O_3 (ultraviolet (UV) absorption and IR trapping)
   d. Carbonaceous soot emission (efficient solar absorption)
   e. SO_2-Sulfate emission (solar reflection and IR trapping)
2. Land-Use Changes
   a. Deforestation (releases CO_2 and increases albedo)
   b. Regrowth (absorbs CO_2 and decreases albedo)
   c. Biomass burning (releases CO_2, NO_x, and aerosols)
3. Agricultural Activity
   a. Release of CH_4 (IR trapping and changes O_3)
   b. Release of N_2O (IR trapping and changes O_3)
4. Industrial Activity
   a. Release of CFCs and its substitutes (IR trapping and leads to O_3 destruction)
   b. Release of SF_6, CF_4, and other ultra-long-lived gases (IR trapping virtually forever).

The radiative characterization of the long-lived gases is now rather accurately known, as are the sources and sinks of CFCs. Unfortunately, the chemistry of the uptake and exchanges of CO_2, the sources of N_2O, and the sources/sinks of CH_4 are all insufficiently known. Considerable process-oriented research is required to make the budgets of these greenhouse gases known with sufficient accuracy. Without this knowledge, regulative and mitigative actions cannot be made with high confidence that the desired effect (e.g., decreased rate of CO_2 increase) will occur as anticipated.

The climate impact of NO_x emissions requires knowledge of the global budgets and processes controlling its concentrations. In many instances, the governing chemistry is very poorly understood and quantified. Even the most sophisticated chemical models still disagree with tropospheric observations of the partitioning of reactive nitrogen compounds. Without this, self-consistent modeling of tropospheric and lower stratospheric ozone changes will remain elusive.

The global budgets, distributions, size spectra, and shape factors for natural and anthropogenic aerosols (sulfate and carbonaceous soot) are not nearly well-enough known to make radiative impact calculations with solid confidence. The interactions of aerosols and clouds add further to the confusion. However, reasonable sets of assumptions yield aerosol radiative-forcing magnitudes that are quite competitive with those due to increasing greenhouse gases. Focused field measurements and modeling-research efforts are required to effect meaningful progress toward improved quantification of aerosol impacts on climate change and climate variability, both regionally and globally.

The role of land-surface processes is central to gaining improved ability to predict local and regional climate change near the ground. The influence of climate change on the biosphere and vice-versa will become a subject of greater focus. Accelerated field studies are essential for progress toward quantification. The cultural and spatial-scale gaps between this type of research and the large-scale climate models must be reduced. Because of current immaturity in understanding the coupling aspects of this topic, the difficulty in modeling regional climate change, and the ever-increasing demand for regional improvements, it may take decades before quantitative skill can be claimed for such "earth-system" model predictions.

THE ROLE OF NATURAL VARIABILITY IN DIAGNOSING CLIMATE CHANGE

The presence of natural variability of the climate system has been used and abused in various interpretations of what might happen in the climate. ("We might be in a natural cooling cycle! No problem!" "Natural variability plus global warming could push us over the edge! Act now!")

The simple truth is that the climate-change interpretation problem is more difficult because of the presence of substantial natural variability of the system on the time scale of decades to centuries. Much of this variability occurs due to internal non-linearities of the coupled atmosphere-ocean-ice-biosphere system. In effect, variations that last for one to four human generations are indistinguishable, to the local observer, from that of secular change. Both observational records and coupled atmosphere-ocean models now show that such decade-to-century scale natural variability is ubiquitous in the climate system.

Currently, we have no strong reason to predict that the levels of such natural variability will be either less or more in the next century, or of a particular algebraic sign. Unfortunately, neither climate models nor the current global observational system are sufficient to even characterize where today's climate system sits with respect to natural variations, globally or regionally, let alone predict where it would be going independent of the greenhouse effect. Thus, the observed global-surface warming of a half degree Celsius or so over the last century could conceivably be entirely due to a natural warming cycle. It also could be that a natural cooling cycle is acting to mask a considerably larger greenhouse signal. Like uncertainty, decade-to-century scale natural variability is a double-edged sword that can cut for or against one's preconceived notions about natural versus anthropogenic climate change. Moreover, on a regional basis the likelihood of misinterpreting an apparent observed "climate change" effect of either sign is considerably larger.

Clearly, an improved ability to characterize and understand such low-frequency climate variability is a major research challenge. This is true whether or not such decade-to-century scale natural fluctuations are predictable by climate models of the future. Ability to simulate such fluctuations is quite a different matter than is the ability to predict them. A common assertion is, "I will believe climate model predictions for the future when they show that they can reproduce all the bumps and wiggles of the past 100 years." A superbly capable model may not be able to do that any more than it can predict a detailed cyclone event 3 months in advance. For the foreseeable future, this reduces us to having to deal with the greenhouse warming effect in the context of whatever natural climate variability is playing out over the next decade. However, we must remember that current best predictions indicate that the observed warming will rise well beyond the natural variability by the middle of the next century.

Another important contributor to natural climate variability is the occurrence of episodic volcanic eruptions. A Pinatubo-type eruption can produce a substantial stratospheric warming and a perceptible tropospheric cooling. The forcing time scale is short; about 3 years are required to clear sulfate aerosol out of the stratosphere. The climatic response can be longer than this, depending on how such a cooling signal is sequestered in the upper ocean. The perception of how a Pinatubo type event is felt in the troposphere is confounded by the considerable natural variability on interannual time scales.

It is important to remember that episodic climate variations on the scales of years to decades do nothing to alter the central reality of the greenhouse warming effect. They only act to obscure the issue.

THE OBSERVATIONAL RECORD OF CLIMATE CHANGE: AN EARLY WARNING OR A LONG WAIT?

Ultimately the evidence for global warming must come from the observational data, not from the theoretical predictions themselves. It is popular to assert that the available climate data do not allow one to conclude that the expected climate-change signals are yet evident. That viewpoint may be overly cautious.

In this context, I believe that some important "early-warning" signals are already available. As examples, we have: the infrared absorptivity of greenhouse gases, the systematic increases of a suite of greenhouse gases, the increasing global-mean surface air and ocean temperatures over the last century, warming in the radiosonde record of the past 20 years or so, increasing water vapor amounts in the same radiosonde record, rising global mean sea level during the 20th century, and significant temperature decreases in the stratosphere.

In the deliberately wimpy language of science, individually these phenomena are "not inconsistent" with a hypothesis that the greenhouse warming effect is already underway. Together, they make a rather respectable case. Do they collectively provide the elusive Smoking Gun? My own opinion is, if this were a civil-court case, the preponderance of the evidence would indicate a vote of YES. If it were a criminal case, could we vote for a conviction that is beyond a reasonable doubt? I can visualize a hung jury with a vote of 10 YES and 2 NO.

The problem with any "early warning" pursuits of answers is that the signal is not yet large enough nor long enough to rise clearly out of the "noise" of natural variability. When one focuses on specific regions or short intervals of time, the obscuring effect of natural variability becomes even more vexing. When one appeals to proxy measurements (e.g., indicator species of plants and animals), it gets even worse. In those cases, the baseline of appropriate measurements is much less defined (locally or globally). Moreover, indicator species are intrinsically regional, almost by definition. In addition, such indicator species are frequently affected by regional anthropogenic activities (e.g., land use change) which may have little or no global impact.

This is not to say that the current atmosphere/ocean/chemical/biological monitoring system is in good shape. On the contrary. My personal opinion is that it is by far the weakest link in the entire climate-change research effort. I see little evidence, domestically or internationally, of a program or a commitment to establish a well-designed, long-term global monitoring system. Only tiny parts of the current observational system are dedicated, designed, or directed toward providing the required consistent, stable, and calibrated monitoring of the global system over the next century.

With a genuine awareness of the need and an international commitment, stabilization of current measurement systems and required additions could be achieved. A better system could produce a much-needed synthesis between research and monitoring activities. Such synthesis will be essential for investigation of global change in the context of decade-to-century scale natural variability. (How do you differentiate a decade-scale natural fluctuation from a "trend"?)

From a policymaking perspective, the need for such a monitoring system is virtually self-evident. Almost independent of ideology,* we are going to want to know how the system is changing. We are going to want to evaluate those changes in the context of previous predictions of how the system might change. Moreover, we are going to have

---

* Extreme ideologues may not be interested, because their minds are already made up (a. "Nothing will happen." b. "Global doom will happen."). No uncertainty means new facts are neither necessary nor welcome.

to evaluate such changes and explain them in the context of separating the natural and anthropogenic parts. This is a daunting task; without a well-designed global monitoring system, it is impossible.

The problem with installing such a monitoring system is that it is too difficult. The commitment requires too many people, must be sustained too long, costs too much money, requires too much cooperation, and demands too much leadership. All are excellent excuses to continue to deny the problem. How do we overcome such barriers? I don't know, but ways must be found.

ON SCIENTIFIC UNCERTAINTY AND POLICY DECISIONS

As outlined above, many of the remaining scientific uncertainties are intrinsically difficult and may take decades to unravel. Also, as outlined above, there is much that is known already. It is only the quantitative detail that is significantly uncertain. It is essential to note that none of these uncertainties can make the greenhouse effect go away. They could, however, combine to make the greenhouse warming be of either greater or lesser severity than that of today's consensus best estimates.

I have been on record for over 4 years arguing that the "betting odds" are about 9 out of 10 that the global-mean surface temperature will increase within the range $1.0-4.5°C$ by the middle of the next century. Does the reader think it is "even money" that the warming won't happen or it will be less than $1.0°C$? If so, I have a spare $1000.00 I would be delighted to invest with you in a mutual business transaction.

The point here is not to line up sucker bets, but to identify possible sources of "policy paralysis." Uncertainty is a major issue and will remain so, but to use it as a perpetual excuse for evading responsibility for even addressing the issues is, in my view, indefensible. After all, we know with very high confidence that greenhouse gases are increasing, human activities are causing the increases, these gases are efficient infrared absorbers, and today's emissions are committing the climate system to an altered state of radiative-dynamical balance for centuries.

Indeed, we do admit to substantial uncertainties in the magnitude of the climate response, the rate at which it will happen, and the geographical patterns of the response. These are central to issues of policy response details and decisions on levels of mitigative action. Again, none of these uncertainties can make the greenhouse effect go away. Thus, in my view, the issue is not "Should we respond?" but rather "What are appropriate kinds of responses in the next decade given the scientific uncertainties, the large confusion on the impacts side, and the potentially high social cost of responding?"

Allow me to offer a tutorial example by a comparison of the CFC-ozone issue with that of the greenhouse-gas climate issue. Major international CFC reductions were committed at a time when substantial ozone reductions were still mainly a theoretical projection.

At the time of the Montreal Protocol, the cause of the Antarctic "ozone hole" had not yet been discovered. Its discovery, and the subsequent identification of a chlorine-related

"Smoking Gun," only accelerated the commitment to phase out CFCs. Yet, even today the fundamental uncertainties about CFC-induced ozone depletion remain of comparable magnitude to the uncertainties about the greenhouse warming effect (factor of 2–3 in change predictions of key parameters).

Why, then, has the approach to policymaking been so radically different in the two problems? I submit that the difference has essentially nothing to do with basic scientific uncertainties in the two cases. For the CFC issue, control over global emissions rests with a relatively small number of manufacturers. Moreover, they were nimble-footed enough to invent HCFC substitutes that act to reduce the ability to destroy ozone and trap infrared radiation by about an order of magnitude per unit of emission. For the CO_2 problem no such "quick fixes" are possible. More importantly, there are about, say, 4 billion CO_2 "offenders," namely us. World-wide *per capita* reduction of CO_2 emissions of 60–80% (for today's global population) is apparently needed to stabilize CO_2 concentrations at current levels (IPCC, 1990). This implies a radical global lifestyle switch that is, in all likelihood, unobtainable. In addition, reduced ozone is associated with a perceived bad impact (skin cancer), while climate warming tends to leave a more benign impression.

Thus, to me the issue is not at all one of "regulate vs. don't regulate." Rather, in the near term, it is one of the level of greenhouse radiative forcing we are willing to tolerate in the face of remaining uncertainty. In this context, a policy "non-decision" is indeed a major decision about the future of the radiative forcing for the planet over the next several centuries.

Scientific uncertainty must play an important role in determining the magnitude and character of international policy responses and when such responses should be put into place. Improved scientific knowledge will be invaluable in pinpointing certain details that will help clarify the tradeoffs between the severity of the impacts and the social cost of doing something about it.

This central role for the scientific community should not be construed as an assertion that scientists should play a major role in offering advice as to what policy and regulatory decisions are appropriate. Quite the contrary. In my view the societal decisions as to what should or should not be done to reduce global warming are intrinsically non-scientific. They are decisions based upon expected impacts and how they are perceived through personal and collective value systems. As we are all aware, one's personal values are frequently in severe conflict with others held with equal conviction. In that sense the decisions on what to do appear to be inherently political.

I believe that we scientists must be far more sensitive to the magnitude of the science vs. values cultural gap. It is far too easy for us to tout our personal values in the name of scientific objectivity ("My climate model says that we should/should not regulate now.") Unfortunately, we scientists are frequently asked to answer just that kind of ill-posed question. More unfortunately, we sometimes fail to discipline ourselves to refrain from offering policy advice that is requested from us because we are scientists.

Clearly, scientists, impact analysts, and decision makers need to forge new partnerships to address and deal with the impending global change. When we begin to recognize that all three groups have their unique skills to offer, perhaps we all can get to work and learn from each other.

THE NEED FOR EXPANDED RESEARCH SCOPE AND BREADTH

As the global change questions become increasingly better defined, the need for improved cross-disciplinary perspectives becomes ever more evident. An almost bewildering array of new research interfaces is necessary as we seek greater quantification for an increasing array of potential impacts. How we achieve these new linkages is a daunting challenge.

As the previous discussion makes clear, the problem of predicting climate change is already imposingly cross-disciplinary in scope. Future improvements only will make it more so.

The first category of immediately needed linkages and inputs involves disciplines that are already part of the atmosphere-ocean scientific "culture," at least as seen from the eyes of an outsider. These are: aerosol and cloud physics, moist convection, sea ice, turbulence, and boundary layers. Each of these topics has had trouble matching its small operative spatial scales with the much larger scales associated with the climate system. This scientific and "cultural" gap within our own scientific culture needs the collective attention of more synthesizing research perspectives and strategies. Historically, it is the global modelers who have made the attempt to capture the large-scale essence of intrinsically small-scale physical processes, such as moist convection. Generally speaking, I think it is fair to say that we haven't done a particularly good job. Unfortunately, more qualified specialists have been generally unwilling to volunteer to help do it better. After all, who wants their research "baby" reduced to a model parameterization scheme?

Another category is a nearby discipline that has only been peripherally involved. This is the community of land-surface hydrologists and engineers. Again, their focus has been on scales much smaller than a climate-model grid box. Moreover, they have been inclined to be more interested in the impact of weather and climate on surface and subsurface water than the reverse. That is beginning to change; it is already clear that we have much to learn from each other.

Much more problematic is the interface between climate and the terrestial biosphere. The current efforts are just beginning, few experts are available, perspectives are very different, the scale disparities are large, and quantification of biospheric feedback/response processes is difficult. How long will it be before terrestrial biospheric processes are properly integrated into climate system models? Decades, I bet.

A final category is the interface to disciplines that are relevant to climate changes on 500–50 000 year time scales. Here, improved contacts with the paleoclimatic reconstruction community can be invaluable. On these long time scales, the climate system is known to be quite sensitive to relatively modest changes in radiative forcing. Is this greater climate sensitivity on these time scales solely due to feedbacks that take a thousand years or more to kick in? We don't know, but I suspect that may be the case. To be sure we must know more about the long-term feedbacks in the system.

We know, for example, that ice ages are accompanied by reductions in CO_2 and CH_4, a clear positive (amplifying) feedback in the system. Unfortunately, we do not know yet why this is the case. There are clearly biogeochemical feedbacks in the system; on what time scales do they operate? We do have good reason to believe that such processes are not likely to have much short-term impact on the CO_2 and CH_4 budgets.

In a similar vein, changes in the great continental ice sheets involve processes not well characterized in current models. Will the feedbacks and effects of continental ice-sheet change be significant over the next century? Again, we think not, but there remains an intriguing possibility that we are overlooking something.

In the above examples improved interfaces with both the biogeochemical and continental ice scientists are highly desirable to provide sharper answers. There are likely other such interfaces that I haven't thought about and possibly some that no one has.

Finally, I cannot resist a comment about needed input and linkages from another technical culture, that of computers and computing. To effect substantial progress on the climate modeling front, orders-of-magnitude increases in computing power are required.

The good news is that such increases are forthcoming, thanks to the advent of massively parallel computers. Moreover, large increases in peak computer speed should come at a progressively cheaper relative price. This, of course, assumes that peak speed in a massively parallel system is something that is nearly attainable. The bad news is that such capability produces massive volumes of data, and current and projected systems may not be able to keep up with it. An inevitable truth is that improvements in predictive skill only happen through insightful diagnostic analysis of model output interpreted in the light of theory. It will be a great multicultural computational challenge to overcome these barriers. If they can be overcome, the supercomputer will become an even more powerful tool for helping us penetrate the workings of the climate system.

WHEN WILL WE HAVE BETTER PREDICTIONS?

Many specific issues on mitigation of climate and/or chemical change may depend importantly upon the rate of reduction of current scientific uncertainties. I offer here my own perspectives on the rate of progress expected by given time frames.

a. By 1993?

At the policy-relevant level, the hard answer is practically none.

On this short time scale, certain calculations will be sharpened and some single-model results (e.g., Antarctic Ocean resistance to warming) will be evaluated in other models. It is unlikely that consensus results will shift perceptibly in little more than a year.

b. By 1996?

By this time some higher resolution model calculations should be completed and properly analyzed, giving perhaps some more credible confidence about regional climate-change expectations. With some good luck, characterization of the global-mean cloud-radiation feedback effect could be improved, thus producing an important narrowing of uncertainty. I personally am not confident that regional details of cloud-radiation feedback in climate change will be much further along. The rate of learning about oceanic responses and atmosphere-ocean interaction is currently rather high. It is quite possible that improvements in ocean-atmosphere modeling could yield improved estimates of the rate of climate warming, as well as an improved understanding of the role of natural variability in interpreting observed changes.

By 1996, I believe that useful progress will be made toward understanding and quantifying the impact of other climate-forcing mechanisms that are generally overlooked today. These include tropospheric ozone changes, lower stratospheric ozone changes, volcanic aerosol effects, and aerosol effects due to combustion activities. These effects share some important characteristics. They exert their radiative effects more or less regionally, and they add radiative-forcing effects that are comparable in magnitude and, occasionally, opposite in sign to that of increasing greenhouse gases.

I am hopeful that 1996 will see substantial progress for closing the global budgets of CO_2, CH_4, and N_2O. Although many subtleties are likely to remain unexplained, the basic policymaker's need to understand the global impact of a proposed emissions reduction may very well be under better control.

c. By 2000–2005?

As indicated previously, I expect that, in a decade or so, substantial reduction of uncertainty can be achieved in the problem of quantifying the magnitude and rate of climate change. Perhaps it is relevant to remind ourselves here that new scientific knowledge often does not result in a simple reduction in perceived uncertainty. Occasionally, discoveries are made that give the appearance of figuratively throwing a well-aimed rock at the previous knowledge base. The Antarctic (and now Arctic?) "ozone hole" comes to mind. In this sense, prediction of the rate of progress of scientific improvement is as uncertain as predicting climate change itself. We never know whether or not the next hard-won scientific insight will produce perceptible inching toward "the truth," or if it will inform us that we know much less than we thought we did.

I do expect that we will be looking seriously into meaningful quantification of regional climate-change issues by then. I predict, however, that the social need to know who is going to get hit the most will produce an ever-increasing demand for improved regional detail and accuracy. This is likely to produce a "light at the end of the tunnel" syndrome that will produce the perception of vexingly slow scientific progress, even as substantial improvements are occurring.

Obviously, on this longer time scale, the ability to predict scientific advances can become vulnerable to any number of unforeseen circumstances. Will a viable earth system monitoring program be in place? Will budgets for scientific research increase or decrease? Will the global-warming problem assume a substantially improved position on the funding priority list? Will unfavorable climate-change effects have begun to be noticeable? Will our global social systems remain stable? The simple truth is that such "non-scientific" influences will cloud the scientific crystal ball, perhaps by more than will the inherent unpredictability of scientific advance.

d. By 2025–2030?

Your guess is as good as, or better than, mine. Who knows?

SUMMARY AND DISCUSSION

A summary of the climate issues addressed here leaves some sharp impressions. First, much is already known about the basic science of global warming. Enough, in fact, that no conceivable combination of surprises can make the central fact of the greenhouse effect go away.

This hard truth, however, does not let us evade the conclusion that much remains quite uncertain. Moreover, those uncertainties do provide substantial hurdles to confident policy analysis. Because of the policymaker's need for reliable regional detail in climate-change predictions, satisfying research advances will not come quickly or easily. A decade or more of sustained effort seems to be required in a wide range of scientific questions.

The climate-research efforts in place are generally adequate, although carefully targeted increased efforts should pay large dividends in improved diagnostic and predictive capability. In comparison to the much less mature efforts in impacts (e.g., agriculture, human displacement, ecosystems, economic sensitivity, energy use, and policy analysis), the climate-research efforts appear to be in reasonable shape. A big exception to this characterization is the present state of the required global monitoring system. It is not nearly up to answering the questions that will be asked by the scientific community and the policymakers. I am concerned that needed improvements will require a commitment and a resolve from governments, policymakers, and scientists that may be beyond us. The problem is very big, very hard, very expensive, quite unglamorous, and requires an unprecedented level of global cooperation. Are we up to the challenge? I hope so.

To me it is clear that improved scientific capability is necessary, but far from sufficient, to make sound policy decisions. I believe that sharp increases are also required in the capability to perform analysis of impacts, economic response, energy use, and policy options. How will this happen? Obviously, greater research attention will be required in all of these frontier areas. Clearly more effort will be required to produce the right kinds of talent through our educational institutions.

Can the climate-modeling community help accelerate these research areas towards providing the definitive climate predictions? I think so. For example, already (or soon) we could provide impacts and policy researchers rather detailed model outputs of regional climate-change scenarios. Such scenarios would be credible, but not necessarily accurate. Impacts and policy researchers could study such scenarios in the context of developing and sharpening their own analysis tools. This would allow these researchers to learn more clearly what their conclusions are sensitive to and what information really matters. Moreover, impact researchers could perform blind intercomparison exercises, much as is already done routinely by the climate modeling community. This could yield improved insights on the character and sensitivities of the impact models, long before the climate models have delivered the "perfect" climate forecast.

Are we ready and willing to step up to this challenge?

REFERENCES

Committee on Earth and Environmental Sciences, 1990: Our Changing Planet: The FY 1991 Research Plan. The U.S. Global Change Research Program, 168 pp. (Available U.S. Geological Survey, 104 National Center, Reston, VA 22092).

Intergovernmental Panel on Climate Change, 1990: Climate Change: The IPCC Scientific Assessment. World Meteorological Organization/United Nations Environment Programme, J. T. Houghton, G. J. Jenkins, and J. J. Ephraums, (eds.), 365 pp. (Available Cambridge University Press, 40 West 20th Street, New York, NY 10011).

ACKNOWLEDGMENTS

Much of the text of this paper and the motivation for writing it came from a Workshop sponsored by Science and Policy Associates on "Climate Research Needs." I am grateful to them for giving me permission to use here parts of the text I developed for their workshop. I am indebted to Clinton Andrews, Charles Herrick, Thomas Knutson, Mac McFarland and V. Ramaswamy for helpful and perceptive comments on various aspects of this essay.

When Will We Have Better Evidence for Climate Change Due to Anthropogenic Emissions?

G. S. Golitsyn

Institute of Atmospheric Physics, USSR Academy of Sciences

109017, Moscow, Russia

The changes of environment including climate are abundant everywhere. The record of the globally averaged temperature reveals a statistically significant trend which, for the past 130 years, amounts to about 0.5 K and the mean sea level has risen by some 15–20 cm. The linear trend determination coefficients for these two records are high enough: $r^2 \approx 0.5$, for temperature and higher $r^2 \approx 0.6$ for the mean ocean level. These two values are truly globally averaged variables. Climate for lesser areas is more variable, the increase in variability being proportional roughly to R/a, where R is the Earth radius and a is the characteristic size of the region. Nevertheless, for our country 7 out of the 10 warmest years were in the 1980s and 1989 was 1.4°C warmer than the 1950–1980 average; for 1990 the corresponding number is 1°C. Our winters are much milder now than two or three decades ago. Even in Siberia the monthly mean temperature anomalies reach 5–10°C! Local people say that they forgot bitter frosts of −40°C, or more.

At the same time the atmospheric composition is also changing. We now have instrumental records of the air composition for carbon dioxide (CO_2), methane (CH_4), and nitrous oxide (N_2O), together with temperature (at least, for Antarctica) and aerosol for the last 160 000 years from the analysis of ice cores. These are so-called greenhouse gases, which absorb the solar radiation less than the thermal radiation of the surface and 'ower atmosphere. Therefore, part of the thermal radiation is re-emitted back causing an a'.ditional heating of the surface. Ice cores from Greenland and Antarctica do show that 'he warmer epochs in the Earth's history are accompanied by enhanced concentrations of carbon dioxide and methane (see IPCC I, Lorius et al., 1990 and the literature therein) and vice versa. Unfortunately the time resolution in these records is about 500 years, at least. Nevertheless, Lorius et al. (1990) have estimated that the climate sensitivity for the CO_2 doubling (or equivalent, taking radiative properties of other gg, greenhouse gases) is within 3–4 K.

During the last two hundred years the CO_2 concentration has increased from about 280 ppm to 355 ppm in 1991. The CH_4 concentration rose from 0.7 to 1.7 ppm during the last 300 years. The estimate for NO_2 rise is from about 285 to 310 ppb for the last 200 years. But during the few thousand years before it was level, stratospheric ozone depletions and the ozone hole diverted public attention from a large rise in tropospheric ozone concentration, which is a powerful gg, though it amounts to about 10% of the total ozone content in the atmosphere. According to a model by Crutzen and Zimmermann (1991) since 1800 A.D. it rose by a factor of 3: from about 10 ppb to 30 ppb! Its rise due to the changing tropospheric chemistry (rise of methane and carbon monoxide, decrease of hydroxyl radical OH, the main "cleaner" of atmosphere from VOCs, volatile organic substances) could be a major atmospheric pollution concern in the near future. Even now estimates of losses only in US agriculture due to damage of certain crops and trees due to increase of ozone are from 1 to 5 billion dollars (MacKenzie and El-Ashri, 1989). This is one of the striking examples of the interconnection of environmental problems, and it stresses an urgent necessity to increase greatly scientific efforts to understand much better the atmospheric chemistry.

Stratospheric depletion is due to CFCs, which were practically absent in the atmosphere in 1950. Since that time they are collectively about 1 ppb, increasing annually 5 to 10%. Their radiative forcing is 10 to 20 thousand times greater per molecule than for CO_2 and the life time is 100 to 200 years. Hence the Montreal Protocol of 1987 and the London 1990 amendments to it. Depletion of the total ozone results in an increase of the ultraviolet radiation reaching the surface with a multitude of damaging effects on humans, animals, plants, and materials. This is another example of intricate linkages between human activities and the state of the environment, including climate.

There are opposite effects counteracting the increasing greenhouse effects, and they are related to aerosol, i.e., particulate matter raised from the surface or formed within the atmosphere from gaseous sulphur compounds. Because of the high variability of aerosol content and lifetimes (about 1 week in the lower troposphere to about a month in its upper parts), it is difficult to monitor, and monitoring is done only in a few places on the globe. There are satellite data, but in the global dimension they are only for the last five years or so. Nevertheless there are many indications that the aerosol load is increasing. The aerosol influences climate directly by scattering solar radiation back to space, and also absorbing it, and indirectly increasing the number of cloud condensation nuclei (CCN), thus increasing the cloud amount and therefore, albedo, i.e., the amount of solar radiation reflected back to space. At the recent XX General Assembly of the International Union of Geodesy and Geophysics, held in Vienna in August 1991, during its special Symposium on Aerosol, Clouds and Climate, there were several reports that these effects may cancel from 2/3 to 1 [sic] of radiative forcings of CO_2 which now constitutes just about half of the total greenhouse effect of the increased amount of gg in the atmosphere.

There is an urgent need to understand and monitor the particulate matter and its changes within the atmosphere. If we quantify better these processes and aerosol load changes we could greatly reduce uncertainties in the climate sensitivity estimates to the radiative forcing. Unfortunately, aerosol effects are not at all discussed in Chap. 8 of IPCC-I "Detection of the greenhouse effect in the observations," probably because of large uncertainties in the aerosol trend estimates at the moment. But all the qualitative evidence does point in the right direction: inclusion of them could substantially decrease uncertainties, could about double the empirical sensitivity determined from recent observations, thus narrowing the gap between model estimates of the climate sensitivity in globally averaged temperature rise due to the equivalent doubling of the atmospheric CO_2

concentration, which now lies between 1.5 and 4.5°C, and empirical sensitivity from the last hundred years' record seems to be closer to its lower end (see IPCC-I, Chap. 8).

This is still another example of the close linkages among various environmental problems. The increase in the anthropogenic sulphur emissions, which are now about a half, or more (see IPCC-I) of the total sulphur budget at the Earth, are due to burning of the fossil fuel which produce CO_2 and NO_x. The latter together with SO_2 is the cause for acid rain, which harms lake and forest ecosystems, destroys materials, historical monuments, etc.

The spread in the model estimates of the climate sensitivity to radiative forcing is due to inadequate knowledge of many processes in the climate system consisting of atmosphere, ocean, and land with its biosphere. We discuss only two of them because of their importance. First, there are clouds which the models describe only poorly. My younger colleague from our Institute, my former graduate student I. Mokhov, has done (and is continuing) very important work on the problem. He started with a comparison of several Soviet and American cloud data sets and found a decent correlation between them for monthly mean cloud-cover anomalies for 5° latitude and 10° longitude grid, except for polar regions. This finding supports a notion that we have already cloud data sets which may be used to validate model outputs for clouds. And he has done precisely this, the results of which were striking and now are widely discussed. Looking at the Northern and Southern Hemispheres data he found that the cloud amount is larger in warm seasons than in cold ones, but an analysis of about a dozen model outputs has revealed that they produce exactly the opposite seasonal behaviour, or a level, as few of them (Mokhov, 1991 and several papers by him and M. Schlesinger from UI, Urbana-Champaign, are in preparation). Because cloud cover is a major player in the energy and water budget in the climate system, we see here another urgent need to improve climate models.

The second major source of uncertainly is much more difficult to cure and lies in our ability to understand and model the response of the World Ocean to radiative forcing. The thermal inertia of the ocean is three orders of magnitude larger than that of the atmosphere. Due to currents and overturning processes within the ocean its time delay in thermal adjustment is several decades, not hundreds of years. But our quantitative understanding of the processes within the ocean and of its interaction with the atmosphere are very far from what we desire. Hence, major programs within the World Climate Research Program (WCRP), are Tropical Ocean and Global Atmosphere (TOGA) and the World Ocean Circulation Experiment (WOC), which are intended to improve our knowledge of these processes. It would greatly enhance not only our understanding of the radiative forcing of the climate system but also improve our ability to model the fate of greenhouse gases, first of all, CO_2 from future emissions.

Atmospheric chemistry and biospheric component of the climate system are within the realm of the IGBP, International Geosphere-Biosphere Program. Only concerted and well-coordinated efforts of these and other international and national programs would lead to the goal. In our country in 1991, we started the National Program, "Global Changes of

Environment and Climate," but with all of the perturbations occurring within the country, it remains to be seen what will be left of the program.

In technical terms, our ability to answer the question in the title of this paper is whether, or when, we can extract the climate change signal from the noise, i.e., the natural fluctuations of the system parameters. These fluctuations are of all spatial and time scales. The longest records are for the temperature and globally they are only about a hundred years long. We are having finite short samples for processes which, we know, have variations throughout all Earth's history. The most striking examples are ice ages. We know that the growth of ice sheets occupied several tens of years and decay about 10 thousand years with the global mean temperature difference about 5°C. Comparing this with modern changes of about 0.5°C for this century we see that these changes are an order of magnitude faster than in the past. There were faster changes in the Earth's history than glacial to interglacial changes with smaller amplitude, like Holocene optimum of 6 000–8 000 years BP, or Younger Drias about 10 500 years ago. They were accompanied, as was told above, by changes in the greenhouse-gas concentrations, so we know that the effect works in general, though we are still uncertain about the exact patterns and time course of the changes and about what was causing what. It looks as though the Holocene optimum, which was globally about 1°C warmer than the modern epoch, was caused mainly by the different Earth's orbit parameters then and now, and the Younger Drias, when the temperature, at least in the middle latitudes around Atlantics, was some 2 to 3°C colder than now, was caused by changes in the Atlantics thermohaline circulation due to disintegration of the Laurentide ice sheet supplying a large inflow of fresh water. But even the rate of changes of the global mean temperature seems to have been smaller than now.

The question remains what kind of a spurious trend for a hundred years we may encounter in the observational record. A very simplified and crude climate model by Wigley and Raper (1990), a so-called upwelling-diffusion model (where deep ocean, its upper mixed layer, land and atmosphere are represented by boxes with fluxes of heat and mass among them) could produce trends up to 0.3 K per century in some parts of the 100 000 years simulation record, but this result has only an illustrative, or cautionary value, because such a model is, as scientists say, only a scientific educational toy. It remains to be seen what could be the long-term temporal variability in comprehensive climate models, but the model should wait for much faster computers able to perform integrations for many hundreds or thousands years. There is one unpleasant feature in such models: many of them have so-called "climate drift," a systematic change in the model climate which is cured by some artificial means. So the problem would remain to extract natural variability from such artificial trends.

Another kind of empirical evidence lies in the climate of past warm epochs. Their study is the goal of the IGBP PAGES Project. These studies have already shown that temperature changes are directly associated with the greenhouse gases concentrations in the atmosphere. The ancient pollen distributions can give, with time resolution of a hundred years, regional distributions of temperature and precipitation for warm and cold seasons (see MacGracken et al., 1990). These reconstructions are in general agreement with climate model predictions: more warming is in higher latitudes and in winter than in low latitudes and in summer with more precipitation in warmer epochs. The reconstructions for Holocene optimum (about 1°C warmer) and for previous interglacial (about 125 000 years BP when global temperature was some 1.5–2°C warmer than now) do show consistent regional patterns of changes between the epochs and the present climate changes. Therefore we in our country, devoid of large computers, rely on these reconstructions in assessing future changes regionally. There is a consistency of the patterns and these changes in amplitude and geography. At least these reconstructions can be used to validate climate models (see IPCC I and MacGracken et al., 1990).

The shifts in climate mean characteristics, such as temperature and precipitation, may be of not so large practical importance as changes in the frequency and intensity of extreme events, such as heat waves causing droughts, blocking situations, floods, tropical hurricanes, etc. The assessing of such changes, say, for the 1980s in comparison with the colder 1960s has not yet even started.

One important issue is the soil moisture content, which determines crop yields. Many models now show large dessications in summer for the interior of continents. The data on the soil moisture are virtually absent, but Vinnikov and Yeserkepova (1989) found a considerable amount (about 50 points) of data several decades long for the USSR that do not show such dessication and tend to reveal an increase in soil moisture at several parts of European USSR and our Middle Asia for the last couple of decades. The soil moisture reproduction, therefore, is another important issue to be improved by the climate models.

A striking example of a regional climate change is presented by the Caspian Sea, whose area is larger than that of Germany. From 1977 to 1991 its level has risen by almost two meters, causing multibillion dollar damage to the coastal parts of Russia, Kazakhstan, Turkmenia, Iran, and Aserbaidzhan. The reasons for changes of the sea level for the past hundred years have been analyzed by Golitsyn and Panin (1989). They used data on run-off, precipitation from a dozen meteorological stations on the coast of the sea and its islands, and calculated evaporation from its surface using wind and humidity data. The calculated and observed changes of the sea level were well correlated ($r \approx 0.8$) and a large fall in the 1930s of the level and the modern rise were also reproduced quite well. The reasons for the rise at present are explained by about 40% due to increased river run-off, by another 40% due to increased precipitation on the sea itself, and by decreased evaporation. The cause for the last factor is quite unexpected. The wind data for the last decades revealed a statistically significant decrease of the wind speed by 10 to 30% at most of the Caspian stations. We analyzed wind data on a few dozens of stations in the Volga basin, which supplies about 80% of fresh water into the sea, and found a similar decrease, which also may explain a part of the increased run-off (Golitsyn et al., 1990). Due to large uncertainties in the wind measurements at the meteorological posts, nobody has tried yet to treat the wind data on large time and space scales, but our example shows that here we may expect surprises. Anyway, the Caspian Sea should serve as a good case study for damages expected as the World Ocean levels rise. The last 15–20% of the rise is due to a dam constructed hastily in early 1980s, cutting from the Sea Kara-Bogaz-Gol bay. The bay was lower than the main sea and connected to it with a narrow strait through which several cubic kilometers of water was flowing out of the sea. The idea was to cut this sink from the water budget of the sea, because during 1940–1977 its level was slowly falling (with considerable fluctuations). People responsible for water management at the time had much construction power and resources and were looking for where and how it should be applied. The most obvious large-scale object for them was the Caspian Sea, and they wanted to save it and its ecosystems. So the project was to divert to the Volga some parts of the run-off of the rivers flowing to the North. Another project was the dam. The first project required dozens of billions in roubles and many years of work. But the dam, about 2 km long, was relatively easy and was built in two years. The bay dried up in three years and now salt and dust blow everywhere; Iran was officially complaining about that. In May 1991 a Soviet delegation visited Iran to discuss the problems of the Caspian sea-level rise. Many Iranians believe that the rise is because the Soviet Union made a secret diversion of its northern rivers. This is a good illustration of the economic value of the right climate forecast. Based on the paleoclimate analogy, Budyko et al. (1988) have made a projection that the Caspian level around 2050 could be 5 m higher, which would be a really great catastrophe. Bearing in mind the 2 m rise from 1977 to 1991 and a 40 cm rise only in 1991, these numbers do not seem to be unreasonable.

The answer to the question when we will have better evidence was addressed in Chap. 8.4 of IPCC-I. If the climate sensitivity is high and the natural variability is such that it "explains" 0.5 K warming (with actual greenhouse of 1 K minus 0.5 K of the variability), then we may expect another 0.5 K warming in about 12 years. If the sensitivity is low then we might expect "detection" of the change in some 50 years.

One can always argue that the rise of the temperature may be due to natural long-term variability and has nothing to do with the observed rise of the gg concentration (and aerosol); here we observe transient phenomena, but in the past the changes were much slower and the climate system was closer to equilibrium than now. The observed rise of the ocean level is just the reaction of mountain glaciers and the thermal expansion to the observed climate changes whatever the cause. The other indicators specific to the greenhouse effect, the stratospheric temperature decrease and a decrease of the diurnal temperature amplitude, have not been analyzed in proper time and scale dimension. The first indicator on a quasi-global basis for the period 1958–1989 (see IPCC-I, Fig. 7.17) does show a trend in the right direction though with considerably less statistical significance than for the surface global mean air temperature. The amplitude of the diurnal cycle, or a value close to that (i.e., the difference between maximum day and minimum night temperatures) was looked at only at a few US locations, where it showed a decrease in expectation with theoretical understanding of the greenhouse effect. There is no technical obstacle to looking at the problem on a global basis, and for the whole period since about 1860, it requires only time, resources, and devotion.

Unfortunately, the last two indicators can not be ascribed solely to the anthropogenic greenhouse gases emissions, because water vapor is also a strong gg. If we take a stance that the observed $0.5°C$ increase in the surface temperature is due to natural variability, then the humidity of the air must increase simply because the warmer air can contain more water vapor (Clapeyron-Clausius equation!). This illustrates the difficulties in providing a definitive answer to the question in the title of the paper.

Karl Popper, the Austrian-born philosopher, in his book "The Logic of Scientific Discovery" (many editions in several languages since 1930s) developed the idea that in natural sciences nothing can be proved in any rigorous sense but everything (any hypothesis) can be refuted. As an extreme example of such logic he said that from the fact the Sun has risen today (and about two trillion times since the formation of the Solar System, — G.S.G.) one can not "prove" with a rigor of mathematics that it will rise tomorrow, though it is a plausible hypothesis.

This stance is very useful for the development of science itself, serving as an impetus for the broadening and deepening of our knowledge of nature, but it is of little use from any pragmatic point of view. Anyone of us is sure that the Sun will rise tomorrow.

The greenhouse effect in its detail is much more complicated than the mechanics of the Earth's rotation and motion around the Sun. But the general physics is simple and without gg our mean temperature at the surface would be some 30 K lower than now. Any additions of new amounts of gg must increase the surface temperature, which increases the water vapor in air, enhancing the effect (an example of many feedbacks operating in the climate system). So the question is then transformed into how fast the system would warm, i.e., what is its transient response, how would it be distributed in space, i.e., geographically, and in time. e.g., seasonally. And here all the complexities of the climate system start to matter. When we are trying to quantify the answers we encounter a whole bunch of uncertainties due to imperfect knowledge of the processes and the inherently statistical nature of the climate system and its evolution.

What kind of energy, and in general, development policy could be adopted in the case of large uncertainties? I believe that the options must not be chosen with only climate change in mind. One should always remember that this is only a part of the great problem of global environmental change. All the environmental problems are closely interlinked, and aggravation of one leads to increase in acuteness of many. The general strategic approach is only one: relieve or decrease anthropogenic stress on the environment, use less resources, use them in a more efficient way, share knowledge with others, because this is the global problem of survival for us and future generations. There should be economic incentives to behave in this manner and even now there are very many cases when this path of development could bring quick benefits even at the present fiscal and economic conditions. So if you are in government, business, or education, do not ask, when we will have better evidence? The evidence is enough to start moving in the right direction of conserving natural resources and using them in a more efficient way.

REFERENCES

1. Budyko, M. I., N. A., Efimova, V. V. Lobanov, 1988. Future level of the Caspian Sea. *Meteorology and Hydrology* No. S, 86–94. (English translation of the magazine by AGU is available).
2. Crutzen, P. J., and P. H. Zimmermann, 1991. The changing photochemistry of the troposphere. *Tellus.* **43 AB**, No. 3/4, 147–163.
3. Golitsyn, G. S., and G. N. Panin, 1989. Contemporary changes of the Caspian Sea level. *Meteorology and Hydrology,* No. 1, 57–64.
4. Golitsyn, G. S., A. B. Dzuba, A. G. Osipov, and G. N. Panin, 1990. Regional climate changes and their revealing in the contemporary rise of the Caspian Sea level. *Soviet Physics—Doklady, USSR Ac. Sci.* **313**, No. S, 1224–1228, 1990.
5. IPCC I: Climate Change. The IPCC Scientific Assessment. Cambridge Univ. Press, 1990.
6. Lorius, C., J. Jozel, D. Raynand, J. Hansen, and H. Le Treut, 1990. The ice core record: climate sensitivity and future greenhouse warming. *Nature* **347**, 139–145.
7. MacGracker, M. C., M. I. Budyko, A. D. Hecht, Y. A. Izrael, 1990. Prospects for Future Climate: a special US/USSR report on climate and climate change. UCRL-102506 Preprint, September 1990.
8. MacKenzie, J. J., and M. T. El-Ashri, 1989. Tree and crop injury: a summary of the evidence. In: J. J. MacKenzie and M. T. El-Ashri (Eds.). Air Pollution's Toll on Forests and Crops. Yale Univ. Press, New Haven, Connecticut.
9. Mokhov, I. I., 1991. Global cloudiness: Tendencies of change. In: E. Sindoni and A. Y. Young (Eds.). ISPP-7 "Piero Caldirola." Controlled Active Global Experiments (CAGE). SIF, Bologna, 19–37.
10. Vinnikov, K. Ya., and I. B. Yeserkepova, 1989. Empirical data and the results of modeling soil moisture. *Meteorology and Hydrology,* No. 11, 64–72 .

Observed Global Warming: Are We Sure?

Thomas R. Karl
NOAA, NESDIS, NCDC
Federal Building
Asheville, NC 28801

1. INTRODUCTION

There has been a considerable amount of attention generated over the issue of whether or not the climate record reflects the impact of anthropic emissions of greenhouse gases. Part of this debate has been focused on the quality of our instrumented climate record. Specifically, are we really sure that there has been a global warming over the past 100+ years? I will try to convince you that, indeed, there has been a global warming, and the evidence is beyond reasonable doubt. On the other hand, there remains substantial uncertainty about the details of the warming, whether it is a direct result of anthropic emissions of greenhouse gases, or perhaps has been mitigated or enhanced by other factors including natural climate variability.

2. THE CASE FOR GLOBAL WARMING

The observed climate record is derived from a changing mix of observing systems, numerous and changing methods of observing methods, sparse spatial coverage, and from observing sites which have had significant changes to their local environment. For these reasons many scientists have questioned whether the climate record really reflects climate change, or perhaps better reflects changes in our measurement capabilities. If we only had one source of data to deduce temperature change, then even the most careful and comprehensive analyses would not, in my opinion, lead to evidence of a warming beyond a reasonable doubt. In fact, it is the measurements from disparate data sets covering the past 100+ years that provide us with a consistent picture of global warming.

The observed global warming derived from Fig. 1 amounts to about 0.45°C since the latter half of the nineteenth century to the present time. How does this warming rate compare with independent measurements? Well, Fig. 2 depicts the observed warming rates for several independent measurements of temperature; land surface temperatures, nighttime marine air temperatures, and sea surface temperatures. There are only small differences among the various rates of warming. Other evidence to support the warming is derived from sea-level measurements and alpine glaciers throughout the world (Figs. 3 and 4). Of course, these data sets have many problems of interpretation as well, but they do support the temperature series. Rates of sea level rise are not inconsistent with the rate of global warming, and the global retreat of Alpine glaciers since the nineteenth century also supports the thermometric measurements.

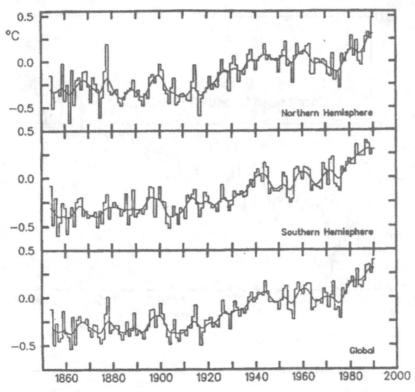

Fig. 1. Global and hemispheric changes of temperature as produced by the Inter-governmental Panel on Climate Change. Smoothed curve is an eleven year binomial filter (from Jones 1991).

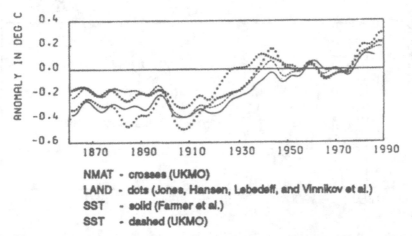

NMAT - crosses (UKMO)
LAND - dots (Jones, Hansen, Lebedeff, and Vinnikov et al.)
SST - solid (Farmer et al.)
SST - dashed (UKMO)

Fig. 2. Changes of nighttime marine air temperatures (NMAT), land air temperatures (LAND), and sea surface temperatures (SSTs), as derived from IPCC 1990.

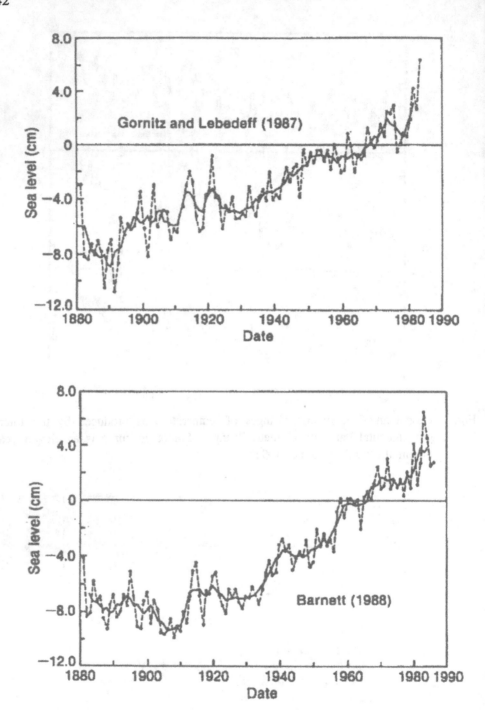

Fig. 3. Changes of sea level (from IPCC, 1990).

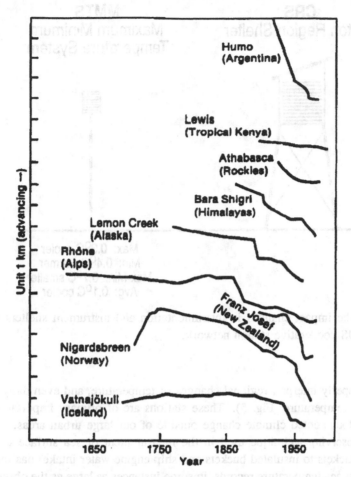

Fig. 4. Worldwide glacier termini fluctuations over the last three centuries (from IPCC, 1990).

3. PROBLEMS OF DATA CONTINUITY

3.1 Temperature

The rate of temperature increase over the past 100+ years has neither been continuous in time or space. This fact takes on added importance when we try to understand why and how the climate has changed including its impact on socioeconomic and biogeophysical systems. Many impediments stand in our way when we try to document these changes.

Figures 5 through 8 depict just a few of these problems. In the mid and late 1980s the United States automated its older observing system at over 3000 stations. This introduced a significant bias in the temperature record that requires certain adjustments to the data

44

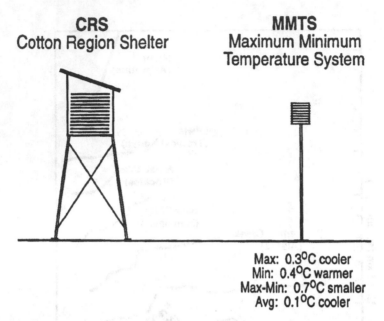

CRS
Cotton Region Shelter

MMTS
Maximum Minimum
Temperature System

Max: 0.3°C cooler
Min: 0.4°C warmer
Max-Min: 0.7°C smaller
Avg: 0.1°C cooler

Fig. 5. The impact changes in instrumentation and instrument shelters in the US cooperative station network.

in order to properly interpret regional changes of temperature, and even daily maximum and minimum temperature (Fig. 5). These stations are of primary importance to those interested in documented climate change outside of our large urban areas. Changes in the type of observing technique used in the measurement of sea surface temperatures (uninsulated buckets to insulated buckets and ship engine water intake) has required scientists to adjust sea temperature records, in some instances as large as the observed signal of global warming (Fig. 6). Even when the thermometer used to measure temperature remains constant, biases can be introduced into the land and marine temperature record because of the varying types of shelters used to house the thermometer. These shelters protect the instrument from direct solar radiation and precipitation. Nonetheless, each of the shelters has differing ventilation rates and protection from direct and indirect solar radiation. Figure 7 depicts an example of some of the commonly used thatched sheds in the tropics during the early part of this century, prior to their replacement by the Stevenson type screen. Table I shows how these changes can effect the temperature record. Notice that these effects are not spatially homogeneous. During the latter part of the nineteenth century much of North America used thermometers mounted on north facing walls, and in other portions of the globe thermometers were housed in shelters that were not properly shielded from solar radiation. As a result, many believe that the summer temperatures are biased high while the winter temperatures are biased low during much of the nineteenth century (Fig. 8). Another example of the biases that have to be addressed

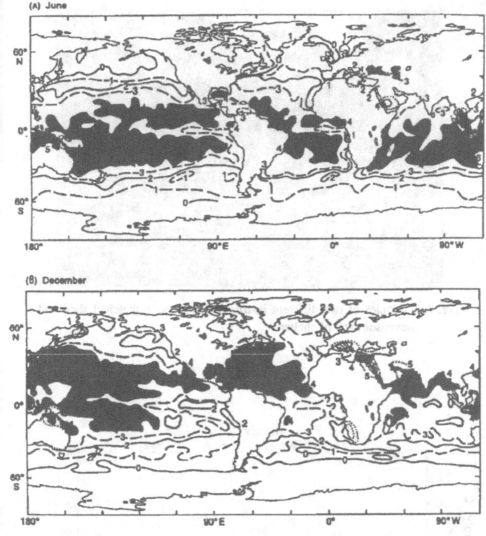

(A) June

(B) December

Fig. 6. Corrections required to uninsulated buckets prior to 1900 (Folland and Parker, 1989).

in our records relates to changes in observing practices. In the USA (Fig. 9) a change in observing times in our cooperative network (consisting of over 5000 stations) from evening to morning has resulted in a bias of between 0.1 and 0.2°C. This must be taken into consideration when compiling regional time series.

One of the other problems of interpretation of the temperature record is related to station relocations and modifications of the local environment around a station. Figure 10 depicts the effects of changes in station location when combined with increases of urbanization. In one instance, for Chicago, Illinois, the urbanization effects are overwhelmed

Fig. 7. Examples of two different instrument shelters: a thatched shed and a Stevenson type of shelter.

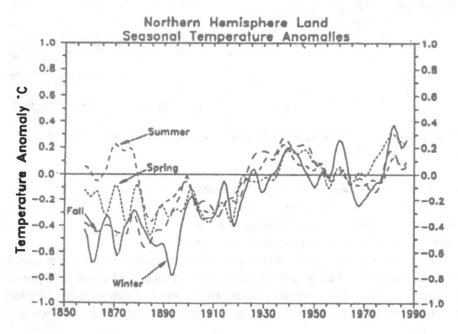

Fig. 8. Changes in northern hemisphere land surface air temperatures.

TABLE I. Temperature differences (°C) of a liquid-in-glass thermometer suspended beneath a felted shed from those in a nearby Stevenson screen (from Bamford (1928) and Chen (1979).

	Sri Lanka	Hong Kong	
Average maximum	−0.5	0.3[a]	0.1[b]
Average minimum	−0.5	0.0[a]	0.0[b]

[a]Data derived from days with >1 hour of sunshine.
[b]Data derived from days with <1 hour of sunshine.

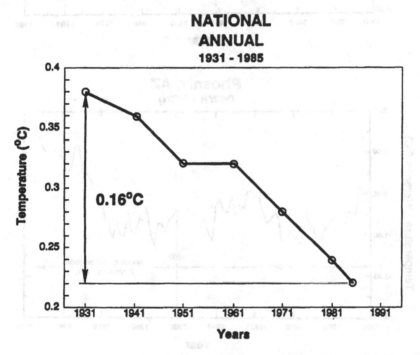

NATIONAL
ANNUAL
1931 - 1985

Fig. 9. The bias associated with historical changes in the observation time in the NOAA cooperative station network.

by station relocations and at another location, Phoenix, Arizona, the effects of increased urbanization are quite apparent.

Perhaps one of the most unsettling characteristics of our climate record is the inadequate spatial coverage of land stations and sea temperature measurements. Figures 11

48

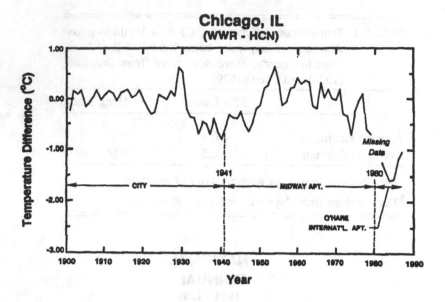

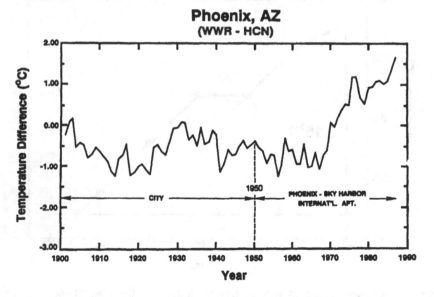

Fig. 10. The effects of changes in station location combined with changes in the local environment around the station.

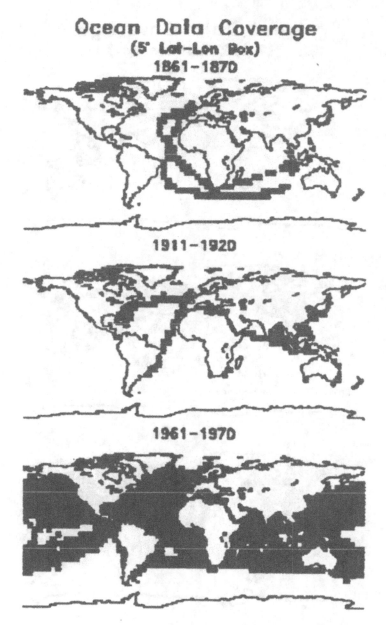

Fig. 11. Spatial coverage of ocean temperature observations for selected decades.

and 12 depict the distribution of stations across the globe for selected decades of the nineteenth and twentieth centuries. Spatial coverage peaked in the 1960s. Our Laboratory has recently completed a series of simulations to determine the error introduced in the time series depicted in Fig. 1 due to the changes in spatial coverage. Results indicate that the errors due to inadequate spatial coverage are still an order of magnitude smaller

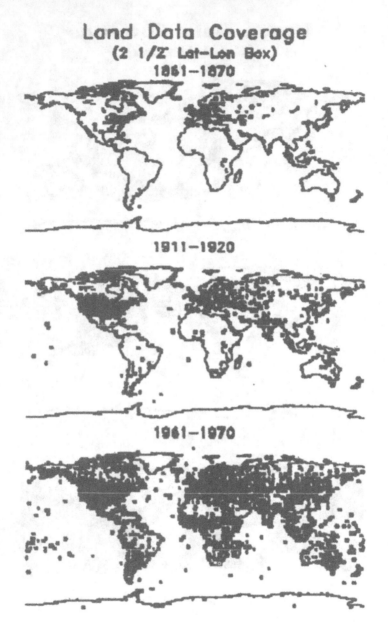

Fig. 12. Spatial coverage of land temperature observations for selected decades.

than the observed global warming (Figs. 13 and 14) when calculating long term trends, but for short term trends (about 30 years) the standard errors of estimate can be quite high (0.2 to 0.3°C).

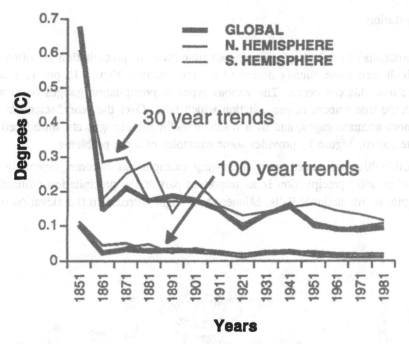

Fig. 13. Standard errors of estimate of global and hemispheric mean temperature anomalies using the Micro-wave Sounding Unit (MSU) data set of
Spencer and Christy (1990) and the observed distribution of observations as depicted in Fig. 11 and 12.

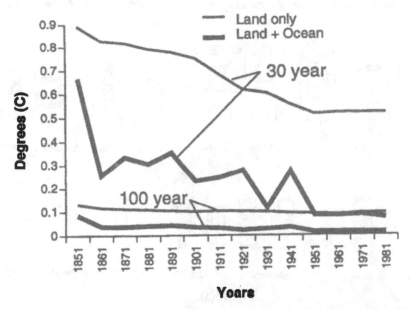

Fig. 14. Same as Fig. 13 only results are based on a 100-year transient Atmo-
spheric/Ocean General Circulation Model perturbed with 1% per year
increases of CO_2 (Manabe et al., 1991).

3.2 Precipitation

The problems of producing homogeneous time series of precipitation are often even more difficult than those already described for temperature. Figure 15 provides an indication of how this can occur. The various types of precipitation gauges consistently undercatch the true amount of precipitation which falls. Over the years, scientists have devised more accurate gages, and as a result non-climatic changes are introduced into the climate record. Figure 16 provides some examples of these problems.

The difficulties associated with measuring precipitation become especially pronounced when solid precipitation is an important part of the precipitation climatology. For example, at International Falls, Minnesota, a slight increase in the elevation of the

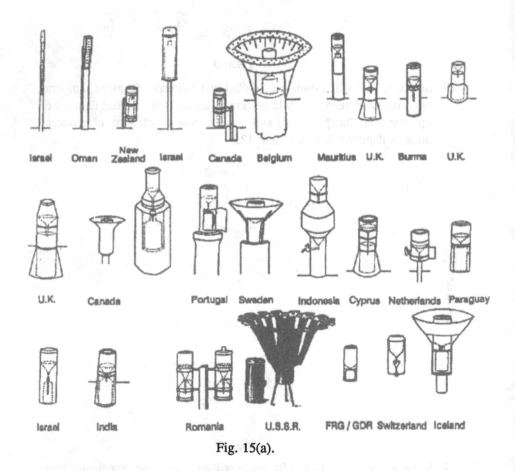

Fig. 15(a).

precipitation gage from 1 to 2 m above the ground reduced the cold season precipitation by 30%. The slight increase in elevation prevented blowing and drifting snow from falling into the gage (Fig. 17).

Much work has been completed for some regions, so that relatively homogeneous time series of precipitation can be produced. Much more work remains. Precipitation changes are likely to be even more important to life on earth than changes of temperature.

One of the ironies of measuring and producing time series of precipitation is that the time series are so heavily influenced by high frequency variability that it is often impossible to detect statistically significant long-term changes of precipitation, prior to encountering practically significant changes.

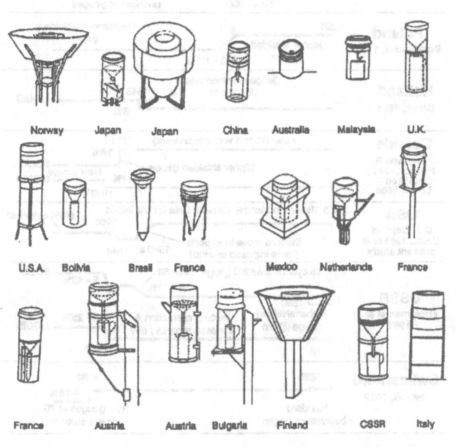

Fig. 15(b).

Fig. 15. Types of manual standard precipitation gauges that have been used throughout the world.

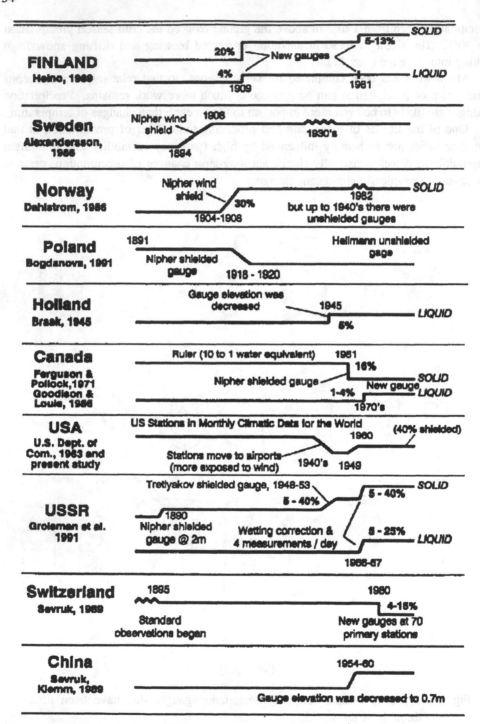

Fig. 16. Examples of the kinds of discontinuities introduced into long-term precipitation time series (Karl et al., 1991).

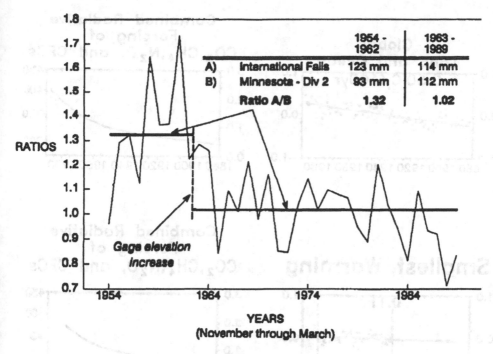

Fig. 17. Ratio of precipitation at International Falls, Minnesota to the precipitation in the North Central Climate Division of Minnesota during the snow season (November through March).

4. CAN WE LINK THE OBSERVED WARMING TO THE GREENHOUSE EFFECT?

Despite the fact that greenhouse gases have unequivocally increased over the past century at the same time global temperatures have increased, it is still not possible to attribute the observed global warming to the anthropic-induced greenhouse effect. Figure 18 depicts part of the dilemma. Much of the observed warming in the record occurred prior to the large increase of greenhouse gases. Furthermore, the rate of warming is significantly smaller than that projected by climate models given the increase of anthropic greenhouse gases.

There are many other inconsistencies in trying to relate directly increases of greenhouse gases to changes of temperature. For example, in Fig. 19 we see that differences between the ocean and land temperatures have not systematically increased, but many of the climate models run with enhanced CO_2 suggest that the land temperatures should increase faster than the ocean temperatures. The pole to equator temperature gradient is also expected to decrease in the Northern Hemisphere, but Fig. 20 suggests that this

56

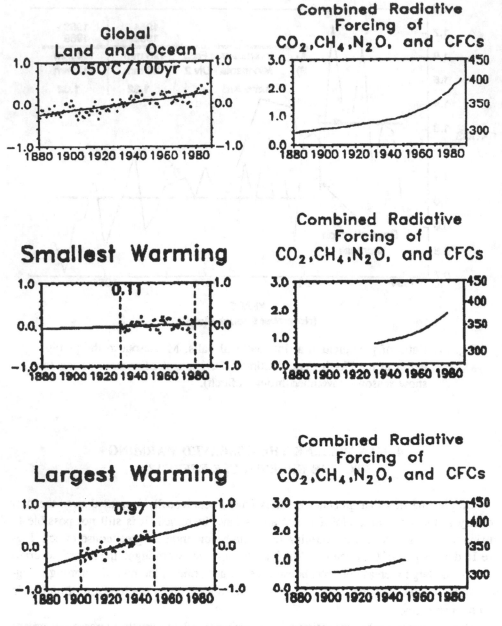

Fig. 18. Changes of temperature and radiative anthropic greenhouse forcing. The 50 years with the smallest and largest warming rates are highlighted.

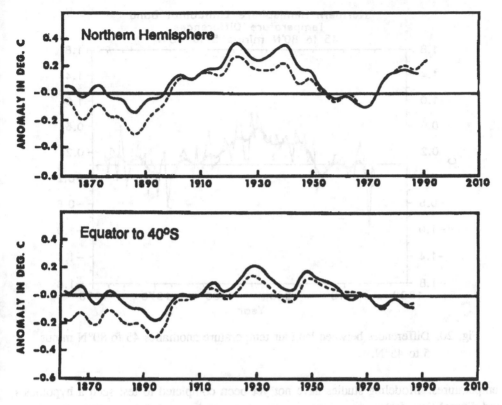

Fig. 19. Differences between the land air and sea surface temperatures anomalies relative to 1951–80 (from IPCC, 1990).

has not occurred. Of course, one of the largest uncertainties about the climate system is the manner in which heat may be stored in the deep waters of the Atlantic and Pacific. Recent observations (Fig. 21) suggest that a significant amount of heat has been stored in the deep waters of the North Atlantic Ocean.

Perhaps some of the most challenging pieces of observational evidence regarding an adequate explanation of the observed warming comes from some analyses of the changes of daily maximum and minimum temperatures. In much of North America and Asia (Canada, Alaska, USSR, PRC) the increase of temperature over the past several decades has been primarily due to an increase of the daily minimum temperature with only a small increase of the daily maximum temperature (Fig. 22). Such a scenario is not projected by climate models with enhanced concentrations of greenhouse gases. If the models are correct, this implies that other forcing factors must be operating. This can include natural variations, but may also include the effects of man-made sulfate aerosols which have been observed in the lower troposphere and often result in hazy skies. These aerosols tend to reflect incoming solar radiation which could lead to reduced daytime

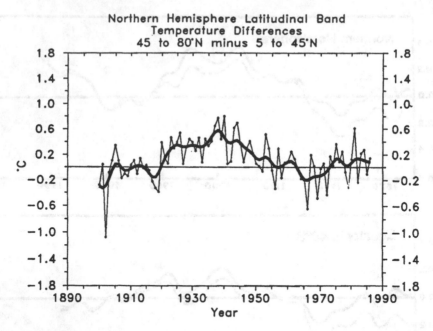

Fig. 20. Differences between land air temperature anomalies 45 to 80°N minus
5 to 45°N.

temperatures. Modeling studies have not yet been completed to test such a hypothesis
and diurnal time-scales.

I should mention that some recent analyses show that there are some similarities between the "greenhouse" projected changes of zonal and vertical temperature and observed
changes. Unfortunately, changes in stratospheric ozone and El-Niño events also produce
similar changes to those observed.

One of the more sobering aspects of our climate system and what we may be able
to predict in the future can be deduced from Fig. 23. Even if we perfectly predicted the
mean winter temperatures in the state of Florida over the past 100 years, we would have
not done very well in predicting killing citrus freezes. Given the present state of climate
models, a perfect seasonal forecast for the state of Florida is not likely to be a reality in
the foreseeable future, let alone the frequency of these extreme events.

5. CONCLUSIONS

The climate has warmed since the turn of the century by several tenths of degrees Celsius. Important details accompanying this warming still require much more
careful documentation and interpretation. Examples of these details include the differential changes of the day and night temperatures, changes in precipitation, and changes in
the extremes of climate as related to changes in the central tendency.

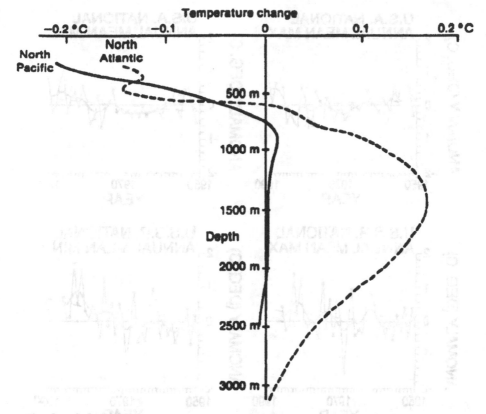

Fig. 21. Sub-surface ocean temperature changes at depth between 1957 and 1981 in the North Atlantic and North Pacific (IPCC, 1990).

Attribution of the observed global warming to an enhanced anthropic greenhouse is not possible. The observed rate of warming is significantly smaller than that expected from atmospheric-ocean climate models. Patterns of observed warming are not altogether consistent with model projections.

The climate is indeed warming, but it is not at all clear that the climate is responding in a manner consistent with projections from atmosphere/ocean climate models perturbed with doubled or slowly increasing concentrations CO_2. Although the state of climate modeling is far from perfect, it would be ill-advised to ignore the possibility of very serious environmental consequences as the concentrations of greenhouse gases continue to increase. On the other hand, it is equally ill-advised to presume global catastrophic environmental consequences will result if greenhouse gases continue to increase at their present rate.

Increased monitoring and modeling efforts are required to narrow our current uncertainties. Given that our climate models are not perfect, it is prudent to rely on the synergy of models and observations to increase our predictive capability.

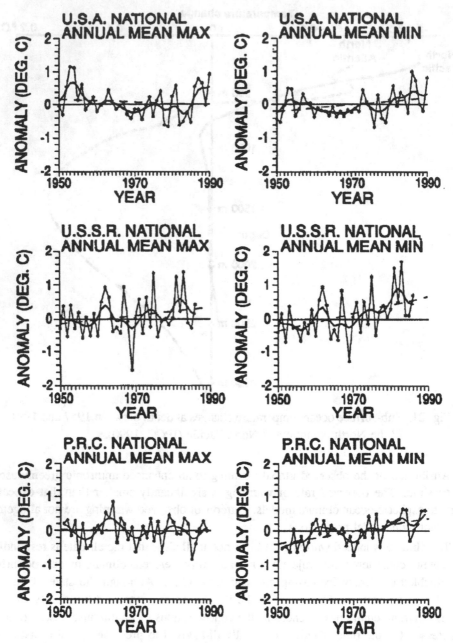

Fig. 22. Variations of temperature for the contiguous United States, the Union of
Soviet Socialists Republics, and the People's Republic of China. Solid
lines are 9 point binomial filters to the annual values and dashed line
is a linear trend. (Karl et al., 1991).

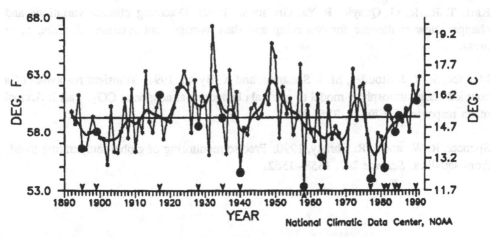

Fig. 23. Mean winter temperatures over the state of Florida. Heavy bold dots represent years which experience extensive citrus damage.

REFERENCES

Bamford, A. J., 1928: On the exposure of thermometers in Ceylon. *Cevlon J. Sci.*, Sect. E, Vol. I, 153–167 plus 3 plates.

Chen, T. Y., 1979: Comparison of air temperatures taken from a thermometer screen, a thatched shed, and a whirling thermometer. Royal Observatory, Hong Kong Technical Note No. 49.

Folland, C. K., and D. E. Parker, 1989: Observed variations of sea surface temperature. NATO Advanced Research Workshop on Climate-Ocean Interaction, Oxford, UK, 26–30 Sept 1988. Kluwer Academic Press, pp. 31–52.

IPCC-Intergovernmental Panel on Climate Change, 1990: Climate Change: The IPCC Scientific Assessment. Eds., J. T. Houghton, G. J. Jenkins, and J. J. Ephraums. Cambridge University Press, 365 pp.

Jones, P. D., and T. M. L. Wigley, 1991: The global temperature record for 1990. USA Dept. of Energy Research Summary, Carbon Dioxide and Information Analysis Center, Oak Ridge, TN, 4 pp.

Karl, T. R., G. Kukla, V. Razuvayev, M. Changery, R. Quayle, R. Heim, Jr., D. Easterling, and Congbin Fu: Global warming: Evidence for asymmetric diurnal temperature change. In Press. *Geophys. Res. Lett.*

62

Karl, T. R., R. G. Quayle, P. Ya. Groisman, 1992: Detecting climate variations and change: New challenges for observing and data management systems. *J. Clim.* **5**, in press.

Manabe, S., R. J. Stouffer, M. J. Spelman, and K. Byran, 1991: Transient responses of a coupled ocean-atmosphere model to gradual changes of atmospheric CO_2. Part I: Annual mean response. *J. Clim.* **4**, 785–818.

Spencer, R. W., and J. R. Christy, 1990: Precise monitoring of global temperature trends from satellites. *Science* **247**, 1558–1562.

Reliability of the Models: Their Match with Observations

Warren M. Washington

National Center for Atmospheric Research*

Boulder, Colorado 80307

The objectives of my talk are to compare climate model simulations with observed climatology and to demonstrate that present state-of-the-art climate models are credible simulators of the true climate system. I will show examples from atmospheric, ocean, and sea-ice models and coupled models. The topic of climate variability is important to policy-makers who have to decide future energy strategies. But strict comparisons of climate models are not possible because observations have errors, the observed climate has natural variability, and the models also have variability. Our main objective is to capture the major features of climate. The models are tuned somewhat and there is some arbitrariness with respect to dissipative mechanisms, clouds, precipitation, and convective processes. We often adjust some of these parameters to move the models closer to observations. Unfortunately, we do not have models in which we specify only gravity, solar constant, and rotation rate and everything comes out right.

To give some idea of how our models handle simulations, Fig. 1 shows the observed and overall zonally averaged temperature structure of the lower part of the atmosphere, the troposphere, and the stratosphere (see Washington and Parkinson, 1986, denoted WP). All present-day climate models are capable of simulating the seasonal change of temperature. As another measure of the climate system, Fig. 2 (WP) shows the geographical surface temperature distributions for winter and summer averages. In a geographical sense, the simulation quality is not as good as the zonal averages. The observed seasonal changes for January and July are quite large and are captured in all general circulation model studies.

Another important measure of the simulation is the zonally averaged east-west component of the wind (Fig. 3WP). We see major westerlies, wind belts in the midlatitudes, and easterly winds in the tropics. We also see stratospheric jets and the easterlies, or trade winds, in the models. Some climate studies with higher-resolution models do an even better job of simulating circulation patterns.

One quantity not well simulated is the precipitation pattern (Fig. 4). The observed precipitation data are not good, especially in the Southern Hemisphere and over the oceans in general. Figure 4 shows several model simulations for long-term averages of June, July, August, and December, January, February from the Canadian Climate Centre (CCC), the Geophysical Fluid Dynamics Laboratory (GFDL), and the United Kingdom

* The National Center for Atmospheric Research is sponsored by the National Science Foundation. Partial funding of the carbon-dioxide-related research is provided by the U.S. Department of Energy's Office of Health and Environmental Research.

64

ZONAL TEMPERATURE

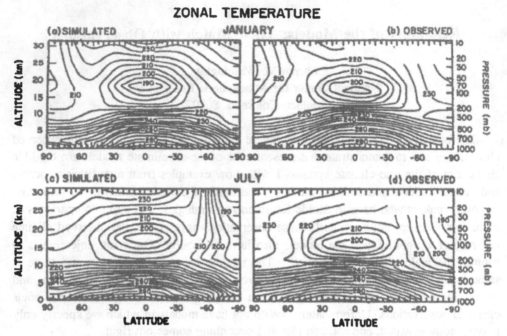

Fig. 1. Zonally averaged temperatures in K computed for perpetual January and July simulations and calculated from observations averaged over December–February and June–August.

SURFACE TEMPERATURE

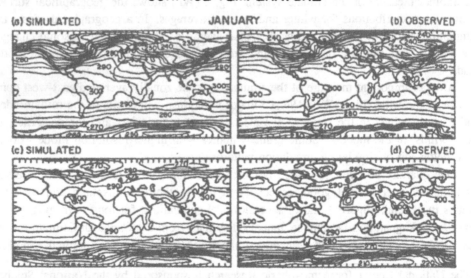

Fig. 2. Global distributions of surface air temperatures in K simulated and observed for January and July.

ZONAL WIND

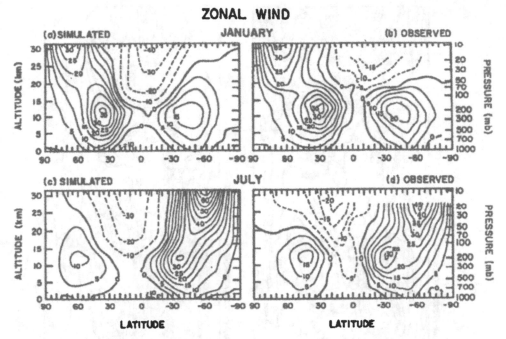

Fig. 3. Zonally averaged u (east-west) component of the wind in m s^{-1} computed for perpetual January and July simulations and calculated from observations averaged over December–February and June–August.

Hadley Research Centre (UKHRC) (IPCC, 1990). The major features are simulated reasonably well in all these models, but there are still factors of two, and maybe even higher, differences between the models in certain geographical regions.

Figure 5 (IPCC, 1990) shows some observed radiation quantities that can be compared with satellite data. The observed outgoing long-wave radiation at the top of the figure compares favorably with several models—the National Center for Atmospheric Research (NCAR), GFDL, UKHRC, Goddard Institute for Space Studies (GISS), GFDL, and CCC. An important satellite quantity is the planetary albedo, the measure of reflected solar radiation. Again, the models compare quite well with the observed.

Another important component of the climate system is the oceans. Figure 6 shows an instantaneous surface velocity field from a recent computation of Semtner and Chervin (1992). The model has observed top forcing and weak forcing in the oceans' lower interior. The resolution of the model is 20 levels in the vertical and half a degree longitude-latitude resolution. This model is one of the computer models to be tested on the massively parallel computers being installed at Los Alamos. We are using a version of this model in our new coupled model at a resolution of one degree. All of these features in the world's oceans are realistic. They are mesoscale eddies transporting momentum, heat, and salinity as part of the oceans' circulation. For example, the Atlantic circulation

Fig. 4(a) – (d).

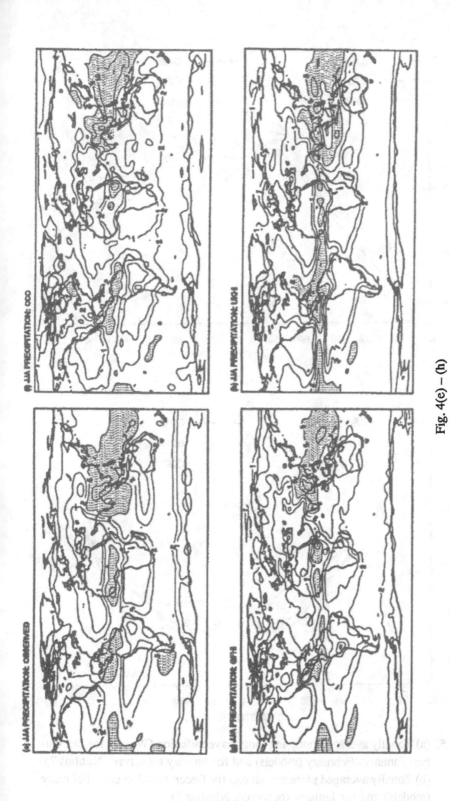

Fig. 4(e) – (h)

Fig. 4. Precipitation (mm day⁻¹) for December – January – February [(a) – (e)] and June – July – August [(e) – (h)]; observed [(a), (e)] and CCC model [(b), (f)], GFD L model [(c), (g)] and UK model [(d), (h)].

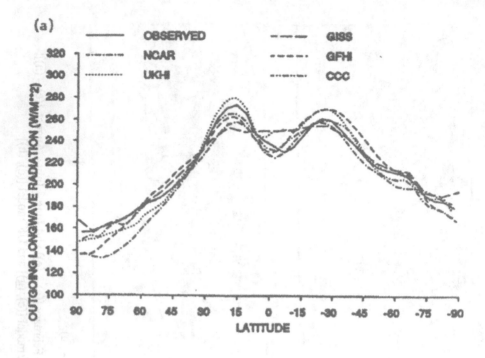

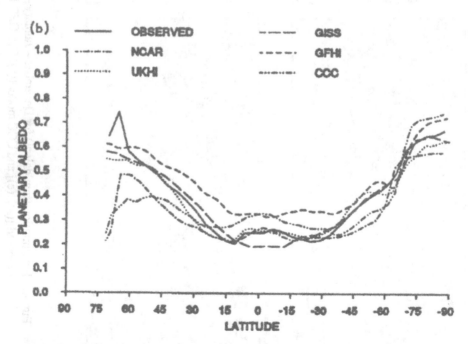

Fig. 5. (a) Zonally averaged outgoing long-wave radiation (W m^{-2} for December–January–February (models) and for January (observed, Nimbus 7). (b) Zonally averaged planetary albedo for December–January–February (models) and for January (observed, Nimbus 7).

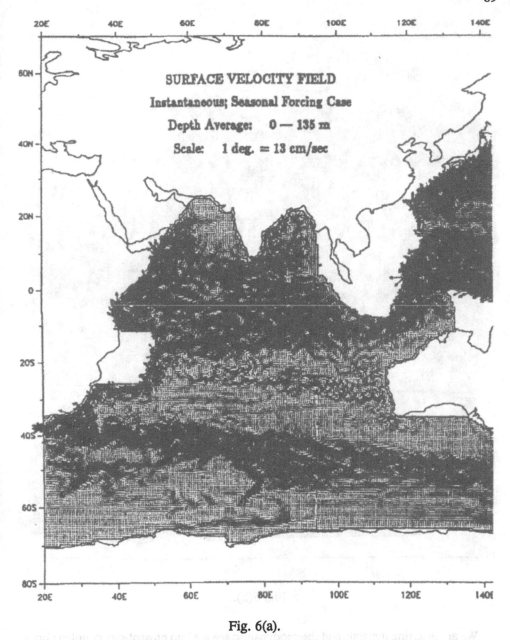

Fig. 6(a).

has a very complex eddy structure of which the Gulf Stream is a part. Note the east Greenland Current and the Labrador Current system. The Southern Hemisphere also contains complex eddy structures. Ocean models have come a long way in simulating the first order of physics of ocean circulation.

Fig. 6(b).

We are inserting dynamic and thermodynamic sea ice into present-day coupled climate models. In earlier studies of climate change at the various modeling centers, sea ice was treated simply—taking into account thermodynamics, by considering only heat flux at the top of the ice and heat flux from the ocean into the ice and by calculating how the ice would grow or decay. In the newer coupled climate models, we account for actual dynamics—wind stress, water stress, and tilt of the oceans' surface, internal ice pressure from the compacting of ice together and the generation of internal pressures.

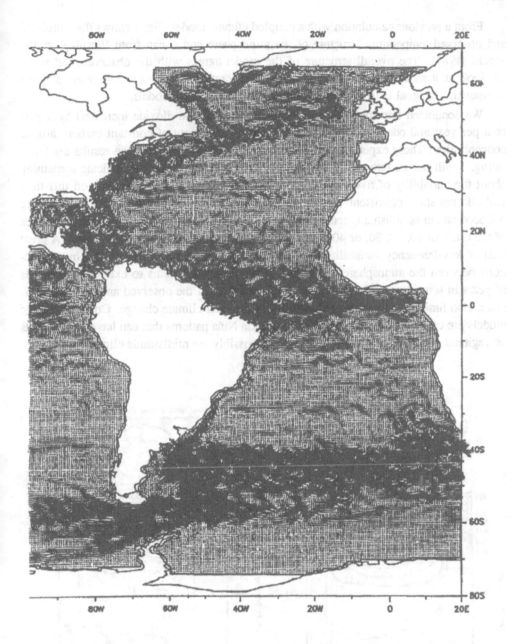

Fig. 6. Instantaneous world ocean velocity field — from 1/2° latitude-longitude
model.

This adds a great deal of complexity to the climate system, but it is necessary because
the determination of ice limits is a key feedback mechanism in the climate system.

From a previous calculation with a coupled climate model, Fig. 7 shows the simulated and observed temperature structure of the atmosphere and ocean from Washington and Meehl (1989). The overall structure of the model agrees with the observed, but when we look at it more closely, we find sizeable discrepancies between observed and the computed. Our goal is to narrow the differences with better models.

We conducted climate-change experiments with carbon dioxide increased by 1 percent per year and compared them with a control experiment of constant carbon dioxide concentration. These experiments were run to nearly 100 years and the results are interesting. I will not go into detail about the experiments, but I want to indicate something about the variability of five-year seasonal averages. The regional patterns on this time scale do not show consistent warming or cooling. It appears that five-year averages are not adequate to establish a greenhouse warming signal. We must take long-term averages of the order of 10, 20, 30, or 40 years in order to arrive at more stable statistics. A great deal of low-frequency variability in these coupled climate models is caused by interactions between the atmosphere and ocean. This variability seems to extend to hundreds of years in some cases. These fluctuations, a real part of the observed and simulated climate, also limit the degree of predictability of regional climate change. Coupled climate models are capable of simulating El Niño and La Niña patterns that can have large effects on regional patterns, such as monsoons, and, possibly, on midlatitude climate anomalies.

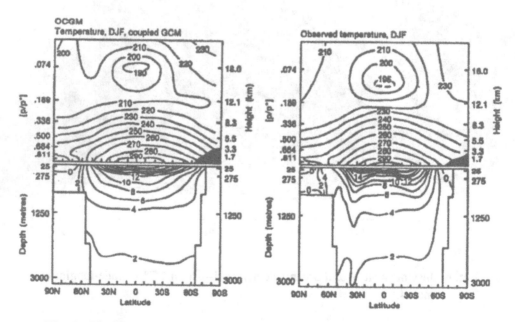

Fig. 7. Zonal mean temperatures for December–January–February for atmosphere and ocean as simulated in a 30-year integration with (left) the NCAR coupled model and (right) from observations.

This makes the task of climate prediction more difficult. And these effects, of course, will be felt in the energy and agricultural industries.

I should say something about the paleoclimate model because of the ways we can use it to verify climate models. By simulating past climates, we have an additional important test of whether or not climate models are working properly. Kutzbach and Guetter (1986) have been conducting experiments to explain the past climates up to the last ice age and in steps of every 3000 years. We find a remarkable agreement in his modeling studies in which he can explain changes in circulation patterns and the wetter and drier regions. Other researchers are trying to use these models for studies of other paleoclimate periods.

Researchers are engaged in a debate about how much resolution is necessary. Ideally, climate modelers want to use the highest resolution possible. In reality, a modeler like me or one of the other climate researchers at this meeting will say, "Gee, I want to do this experiment sometime in my lifetime." Thus, researchers must compromise. Fortunately, we can take advantage of the enormous increase in computer capability and begin to carry out many century-long experiments in weeks instead of years. We can also improve the spatial resolution of the atmosphere on the oceans for a select set of experiments. Physical processes in climate models need to be substantially improved, particularly the cloud-radiation-precipitation mechanisms.

We must be careful not to view climate modeling as an engineering task, and assume that all we need to do is to put the components together and suddenly a better climate model emerges. We need a diversity of approach. It is healthy for the science that researchers are trying different approaches. Some approaches will be successful and some will fail, but I think this is part of the normal scientific methodology.

Are climate models good enough to form the basis for careful energy-policy guidance? If the greenhouse effect is real, even though the amplitude and timing are not known, should we start to provide some policy guidance in our energy strategy? I think the answer is a cautious yes. I believe that there are many good reasons to take steps to lessen the impact of the greenhouse effect. One of these is a sound, sensible energy policy. Climate models can already contribute to the evolution of energy policy.

REFERENCES

Intergovernmental Panel on Climate Change, 1990: *Climate Change: The IPCC Scientific Assessment. Cambridge University Press,* Cambridge, England, 364 pp.

Kutzbach, J .E., and P. J. Guetter, 1986: The influence of changing orbital parameters and surface boundary conditions on climate simulations for the past 18,000 years. *Journal of the Atmospheric Sciences* **43**, 1726–1759.

Semtner, A. R., and R. M. Chervin, 1992: Ocean general circulation from a global eddy-resolving model. *Journal of Geophysical Research,* submitted.

74

Washington, W. M., and G. A. Meehl, 1989: Climate sensitivity due to increased CO_2: experiments with a coupled atmosphere and ocean general circulation model. *Climate Dynamics* **4**, 1–38.

Washington, W. M., and C. L. Parkinson, 1986: *An Introduction to Three-Dimensional Climate Modeling*. University Science Books, Mill Valley, CA, and Oxford University Press, New York, NY, 422 pp.

Implication of Anthropogenic Atmospheric Sulphate for the Sensitivity of the Climate System

Michael E. Schlesinger and Xingjian Jiang*
Department of Atmospheric Sciences
University of Illinois at Urbana-Champaign
105 S. Gregory Avenue
Urbana, Illinois 61801

Robert J. Charlson
Department of Atmospheric Sciences
University of Washington
Seattle, WA 98195

ABSTRACT

The temperature sensitivity of the climate system—characterized by the change in the global-mean, equilibrium surface air temperature induced by a doubling of the pre-industrial CO_2 concentration (ΔT_{2x})—is estimated by comparing the time-dependent temperature changes from 1765 simulated by an energy-balance climate/upwelling-diffusion ocean model with the temperature deviations from 1951–1980-average temperatures observed from 1861 to 1990.

The model is forced by tropopause radiative changes due to greenhouse gases and anthropogenic sulphate aerosols, the former as given by the Intergovernmental Panel on Climate Change and the latter based on estimated emission rates of sulphur in the form of SO_2 (QSO_2-S) from 1861 to 1990. The tropopause radiative forcing corresponding to QSO_2-S in 1978 is estimated for the clear atmosphere based on Charlson et al. (1991) and for the cloudy atmosphere from an estimate of the ratio of the number of anthropogenically produced cloud-condensation nuclei (CCN) advected over the oceans to the natural marine source of CCN, with a resultant value of $\Delta FSO_4^=(1978) = -1.1 \pm 0.9$ Wm^{-2}.

Because of the wide range of uncertainty in $\Delta FSO_4^=(1978)$, ΔT_{2x} is first estimated as a function of $\Delta FSO_4^=(1978)$. This is accomplished by comparing the model's simulated global-mean temperature changes from 1861 to 1990 with the corresponding unfiltered observed global-mean temperature deviations, defined as the average of the unfiltered observed mean temperature deviations for the Northern and Southern Hemispheres. For each prescribed value of $\Delta FSO_4^=(1978)$, a value of ΔT_{2x} is obtained which minimizes the root-mean-square error (RMSE) between the simulated and observed global-mean

* Present affiliation: NASA Goddard Space Flight Center, Institute for Space Studies, 2880 Broadway, New York, NY 10025.

temperature deviations. The resultant relation between $1/\Delta T_{2x}$ and $\Delta FSO_4^=(1978)$ is linear for $0 \geq \Delta FSO_4^=(1978) \geq -1.1$ Wm^2, with values that range from $\Delta T_{2x} = 1.41°C$ for $\Delta FSO_4^=(1978) = 0$ to $\Delta T_{2x} = 5.92°C$ for $\Delta FSO_4^=(1978) = -1.1$ Wm^2. For $\Delta FSO_4^=(1978) < -1.1$ Wm^2 the $1/\Delta T_{2x} - \Delta FSO_4^=(1978)$ relation is nonlinear and yields ΔT_{2x} values that are too large to be physically realistic, for example, $\Delta T_{2x} = 19.4°C$ for -1.4 Wm^{-2} and $\Delta T_{2x} = 42.5°C$ for -1.5 Wm^{-2}.

To estimate unique values for ΔT_{2x} and $\Delta FSO_4^=(1978)$, the differences between the temperature deviations of the Northern and Southern Hemispheres simulated by the model during 1861–1990 are compared with the corresponding observed interhemispheric differences in temperature deviation such that the RMSE is minimized. Estimates of ΔT_{2x} and $\Delta FSO_4^=(1978)$ are obtained for three values of the oceanic interhemispheric heat-exchange coefficient, $\beta_o = 0, 3.5,$ and 7.0 Wm^{-2}/°C. For this range of β_o it is estimated that $1.71 \leq \Delta T_{2x} \leq 2.84°C$ and $-0.25 \geq \Delta FSO_4^=(1978) \geq -0.71$ Wm^{-2}.

To evaluate the robustness of these estimates they were obtained again using the simulated and observed temperature deviations from 1861 to only 1970, 1975, 1980, and 1985. The resultant estimates vary with ending year such that $1.65 \leq \Delta T_{2x} \leq 3.00°C$ and $0 \geq \Delta FSO_4^=(1978) \geq -0.74$ Wm^{-2}. This indicates that future estimates may differ from our 1990 estimates, but will likely not be less than our estimates based on assuming $\Delta FSO_4^=(1978) = 0$. The latter estimates also vary with ending year, from a maximum of $\Delta T_{2x} = 1.87°C$ in 1970 to a minimum of $\Delta T_{2x} = 1.39°C$ in 1985. Accordingly, we broaden our estimates to be $1.39 \leq \Delta T_{2x} \leq 3.00°C$, or $\Delta T_{2x} = 2.2 \pm 0.8°C$, and $0 \geq \Delta FSO_4^=(1978) \geq -0.74$ Wm^{-2}.

1. INTRODUCTION

The most recent assessment of the temperature sensitivity of the climate system — characterized by the change in the global-mean, equilibrium surface air temperature induced by a doubling of the pre-industrial CO_2 concentration, ΔT_{2x} — was performed by the Intergovernmental Panel on Climate Change (IPCC). In this assessment (Houghton, et al., 1990) the IPCC adopted the same range for ΔT_{2x} that was promulgated earlier in a report by the U.S. Academy of Sciences (NAS, 1979), namely, $1.5°C \leq \Delta T_{2x} \leq 4.5°C$. This range of ΔT_{2x} values was, and continues to be, based on simulations by atmospheric general circulation (AGC)/ocean models, the atmospheric component of which is the most-sophisticated member within the climate model hierarchy, while the oceanic component is highly simplified to shorten the simulation time necessary to achieve climatic equilibrium, thereby reducing the computer time required.

During the intervening 11-year period between the 1979 NAS report and the 1990 IPCC report, scientific studies did not reduce the 3.0°C range of uncertainty in the $\Delta T_{2x} = 3.0 \pm 1.5°C$ estimate. In fact, recent AGC/mixed-layer ocean (MLO) model simulations have revealed just how difficult such a model-based reduction in uncertainty is, this because of the inherent model "parameterization problem" — the inclusion of the effects of the unresolved physical processes on the resolved physical scales in terms of the resolved scales alone—that results from the limited horizontal resolution permitted by contemporary "supercomputers" (Schlesinger, 1990). In particular, a series of simulations performed with the United Kingdom Meteorological Office AGC/MLO model obtained $1.9°C \leq \Delta T_{2x} \leq 5.2°C$ simply by changing the parameterization of clouds and their radiative interactions, with $\Delta T_{2x} = 5.2°C$ for the standard model reduced to $\Delta T_{2x} = 1.9°C$ when the model was modified to include the effect of the difference in the fall speeds of ice and liquid water and the effect of variable cloud optical properties—a large negative "cloud optical-depth" feedback (Mitchell et al., 1989). In contrast, a more recent simulation with the Laboratoire de Météorologie Dynamique AGC/MLO model found that cloud optical-depth feedback was essentially zero (Le Treut, personal communication, 1991).

From the simulations of CO_2-induced equilibrium climate change performed during the last 11 years, as well as before, one can conclude that it is not likely that the uncertainty in the estimate of ΔT_{2x} can be substantially reduced in the next decade simply by performing additional simulations with continually revised climate models. Furthermore, even if such future model simulations were to converge to a narrower range of ΔT_{2x}, it does not mean that such a narrower range is correct. Rather, the then-more-uniform results of the models could simply mean that they are all incorrect; in other words, unanimity does not necessarily mean truth. In this sense it has been useful that the models have not produced a narrow range for ΔT_{2x} because its opposite has given us a healthy caution "that not all of these simulations can be correct, and perhaps all could be wrong" (Schlesinger and Mitchell, 1985, 1987).

It has recently been estimated by non-AGC/ocean-model-based means that $\Delta T_{2x} = 0.5°C$ (Lindzen, 1990). In part to incorporate this lower estimate of ΔT_{2x} we (Schlesinger and Jiang, 1991) recently revised the projections of the greenhouse-gas (GHG)-induced global-mean surface temperature change to the year 2100 that we made for the IPCC report (Houghton et al., 1990), thereby effectively using the estimate of $\Delta T_{2x} = 2.5 \pm 2.0°C$. As stated by us (Schlesinger and Jiang, 1991), the revised projections for the IPCC "Business-As-Usual" scenario "clearly show that the magnitude of the potential GHG-induced climate change problem ranges from "catastrophic" (5.9°C warming from 1990 to 2100 for $\Delta T_{2x} = 4.5°C$) "to minor" (0.9°C warming from 1990 to 2100 for $\Delta T_{2x} = 0.5°C$) "depending on the true value of ΔT_{2x} for the climate system," and we concluded that "it is imperative to narrow the range of possible values of ΔT_{2x}."

But how can the range of possible values of ΔT_{2x} be narrowed? One way that this can be done, and perhaps the only credible way, is to use our climate models to simulate past climates for which observations are available, particularly the evolution of temperature during the period of instrumental observations beginning in 1861. Our method for estimating ΔT_{2x} from this past climate is described below.

2. METHOD OF ESTIMATING ΔT_{2x} FROM OBSERVED TEMPERATURE DEVIATIONS

The record from 1861 to 1989 of the deviations of mean global surface temperature (δT_g), mean Northern Hemisphere surface temperature (δT_N), and mean Southern Hemisphere surface temperature (δT_S) from their respective 1951–1980 average surface temperatures has been compiled by the IPCC (Houghton et al., 1990) from land surface air temperatures and sea surface temperatures. These temperature deviations have kindly been provided us and updated by C. Folland, D. Parker, and A. Jackson of the Hadley Centre at the United Kingdom Meteorological Office. Figure 1 presents δT_g from 1861 to 1990, together with a mean global surface temperature deviation defined by $\delta T_G \equiv (\delta T_N + \delta T_S)/2$ and a binomial filtered version of the latter defined by $\delta \overline{T}_G \equiv (\delta \overline{T}_N + \delta \overline{T}_S)/2$. Figure 1 shows that δT_g and δT_G are not identical. In response to our enquiry, the reasons for this difference have been communicated to us by D. Parker as follows:

> "The global values often differ from half the sum of NH and SH because the areal coverages of data in NH and SH differ, more so in some years than in others. Anomalies are computed for 5 Deg lat. × long. boxes for individual months: these are averaged into 3 month seasons, then geographically into NH, SH, Globe using cosine(latitude) weighting for each box with data available, then temporally again into years. Thus, we do not have global and hemisphere reference values." (Parker, 1992, personal communication)

However, δT_g and δT_G differ in absolute value by not more than 0.021°C. The filtered temperature-deviation record $\delta \overline{T}_G$ shows no secular change from 1861 to 1925, a rise of about 0.3°C from 1925 to 1944, a decrease of about 0.1°C from 1944 to 1976, and a rise of about 0.3°C from 1976 to 1990.

Such changes in global-mean surface temperature can be induced by changes in the net (terrestrial plus solar) radiation at the top of the troposphere caused by both natural and anthropogenic changes, the former as exemplified by variations in solar irradiance and volcanic activity, and the latter by changes in atmospheric greenhouse gases and aerosol particles.

If we knew the changes in the net radiation at the tropopause as a function of time, $\Delta F(t)$, we could use a climate model to calculate the changes $\Delta T_G(t)$ of the model's global-mean surface temperature for a given value of the climate sensitivity, ΔT_{2x}. We could then compare $\Delta T_G(t)$ with $\delta T_G(t)$ to determine their agreement quantitatively. By doing this for many values of ΔT_{2x} we could determine the value of ΔT_{2x} which yields the best agreement. This value would then be our "best estimate" of ΔT_{2x}.

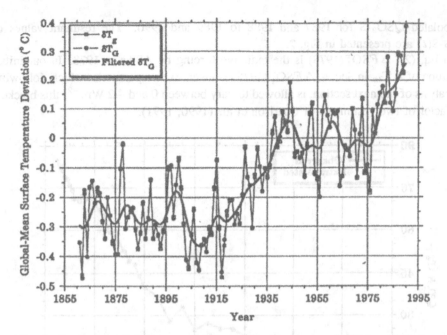

Fig. 1. Evolution of the mean global surface temperature deviations δT_g (open squares), δT_G (filled circles), and $\delta \overline{T}_G$ (heavy line). The data and binomial filter were kindly provided by C. Folland, D. Parker and A. Jackson of the Hadley Centre at the United Kingdom Meteorological Office.

In our initial application of this ΔT_{2x}—estimation strategy, we ignore the radiative forcing due to volcanic activity and any changes in solar irradiance. The total radiative forcing $\Delta F(t)$ is taken to be

$$\Delta F(t) - \Delta F_{GHG}(t) + \Delta FSO_4^=(t) \ , \tag{1}$$

where $\Delta F_{GHG}(t)$ is the radiative forcing by greenhouse gases from 1765 to 1990 as used by Schlesinger and Jiang (1991), and $\Delta FSO_4^=(t)$ is the radiative forcing by anthropogenic sulphate aerosols (ASAs). The latter we take as

$$\Delta FSO_4^=(t) = \Delta FSO_4^=(1978)\frac{QSO_2\text{-}S(t)}{QSO_2\text{-}S(1978)} \ , \tag{2}$$

where $QSO_2\text{-}S(t)$ is the emission rate of sulphur in the form of SO_2 as compiled by Möller (1984) for $t = 1860$ to 1977. We have adjusted Möller's emission rates for 1972–1977 downward to follow the curve of Hameed and Dignon (1988) and used the method of Möller (1984), together with the global annual fossil fuel consumption rates from 1978 to 1988 from Boden et al. (1990), to extend $QSO_2\text{-}S(t)$ through 1988. Because we have temperature observations through 1990 with which to compare, we have linearly extrapolated QSO_2-S for 1987 and 1988 to 1989 and 1990. The resultant values of QSO_2-S(t) are presented in Fig. 2.

In Eq. (2), $\Delta F SO_4^=$(1978) is the radiative forcing by ASAs in 1978. In our initial estimation of ΔT_{2x} in Sec. 4, $\Delta F SO_4^=$(1978) is taken to be a parameter which, following the analysis of the next section, is allowed to vary between 0 and -2 Wm^{-2}; this brackets the "factor of two" estimates of Charlson et al. (1990, 1991).

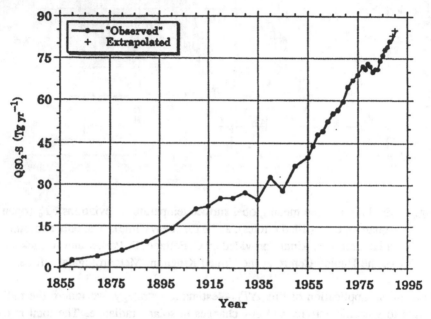

Fig. 2. Evolution of the emission rate of the anthropogenic source of SO_2, QSO_2-S, "observed" from 1860 to 1988 based on Möller (1984), Hameed and Dignon (1988) and Boden et al. (1990), and linearly extrapolated to 1989 and 1990.

3. ESTIMATION OF $\Delta F SO_4^=$(1978)

Anthropogenic increases in the concentrations of sulphate aerosol particles increase the reflection to space of solar radiation, both directly in cloud-free (clear) air and indirectly in cloudy air, the latter as a result of some particles acting as cloud-condensation nuclei (CCN) and thereby increasing the albedo of cloud as per Twomey (1977). Consequently,

$$\Delta F SO_4^=(1978) = \Delta F SO_{4-\text{clear}}^=(1978) + \Delta F SO_{4-\text{cloud}}^=(1978) . \tag{3}$$

Our estimates of the clear-sky and cloudy components of $\Delta F SO_4^=$(1978) are presented below.

3.1. $\Delta F\,SO^=_{4-clear}(1978)$

Charlson et al. (1991) estimated $\Delta F SO^=_{4-clear}$ from

$$\Delta F SO^=_{4-clear} = -K(1 - A_c)(1 - R_s)^2\, b\, \delta SO^=_4 \,, \tag{4}$$

where K is a constant which includes the average incoming solar flux, the average clear-sky transmission and geometric factors; A_c is the average fractional cloud cover; R_s the average surface albedo; b the hemispheric backscatter fraction of the aerosol optical depth; and $\delta SO^=_4$ the aerosol optical depth due to anthropogenic $SO^=_4$. The latter can be expressed in terms of the aerosol column burden, $BSO^=_4$, as

$$\delta SO^=_4 = \alpha\, f(RH)\, BSO^=_4 \,, \tag{5}$$

where α is the empirical scattering efficiency at low relative humidity (RH) per unit mass, and $f(RH)$ is the fractional increase in α due to hygroscopic growth at typical tropospheric RH. Finally, $BSO^=_4$ is estimated from chemical and lifetime considerations to be

$$BSO^=_4 = \frac{3\, QSO_2\text{-S}\, f_{RX}\, \tau SO^=_4}{A} \,, \tag{6}$$

where f_{RX} is the fraction of SO_2 oxidized to $SO^=_4$, $\tau SO^=_4$ is the atmospheric lifetime of anthropogenic sulphate aerosol, A is the area of the Earth (5.1×10^{14} m^2) and the factor of three accounts for the molecular weight of $SO^=_4$ being three times that of sulphur. Combining Eqs. (4)–(6) and substituting the values presented in Table I give

$$\Delta F SO^=_{4-clear}(\text{Wm}^{-2}) = -8.2 \times 10^{-3}\, QSO_2\text{-S}(\text{Tg yr}^{-1})\,. \tag{7}$$

The radiative forcing $\Delta F SO^=_{4-clear}(1978)$ is uncertain due to the uncertainties of the quantities in Eqs. (4)–(6) upon which it is based. Accordingly, in Table I we estimate these uncertainties as standard deviations. The total relative uncertainty of 0.730 gives

$$\Delta F SO^=_{4-clear}(1978) = -0.58 \pm 0.42 \text{ Wm}^{-2}\,, \tag{8}$$

which is about 27% smaller than the uncertainty crudely estimated in Charlson et al. (1991). From Table I it can be seen that the largest contributions to the total uncertainty are due to α, $\tau SO^=_4$, and b. This has implications for the measurements required to reduce the uncertainty in $\Delta F SO^=_{4-clear}(1978)$.

TABLE I. Values and Uncertainties of the Quantities in Eqs. (4)–(6) and their Contributions to the Uncertainty in $\Delta F SO_{4-clear}^{=}$(1978), Ranked in Descending Order of Importance. Uncertainties are estimated as approximate standard deviations.

Symbol	Quantity	Value and Uncertainty	Squared Relative Uncertainty
α	Aerosol mass scattering efficiency (m^2 g^{-1}; White, 1986)	5 ± 2	0.160
$\tau SO_4^{=}$	Lifetime of atmospheric $SO_4^{=}$ (yr; Slinn, 1983)	0.016± 0.006	0.141
b	Aerosol hemispheric backscatter fraction (Vanderpol, 1975)	0.15 ± 0.04	0.071
f_{RX}	Fraction of SO_2 oxidized to $SO_4^{=}$ aerosol (Rodhe & Isaksen, 1980)	0.5 ± 0.1	0.040
K	Proportionality coefficient in Eq. (2.1) (Wm^{-2}, 3-D model of Charlson et al., 1991)	489 ±100	0.042
$f(RH)$	Fractional increase in α due to hygroscopic growth (Charlson et al., 1984)	1.7 ± 0.3	0.031
QSO_2-S	Source strength of SO_2-S in 1978 (Tg yr^{-1}; Charlson et al., 1991)	71 ± 10	0.020
$1 - A_c$	Fraction of earth not covered by cloud (Warren et al., 1988)	0.39 ± 0.05	0.016
$(1 - R_s)^2$	Square of surface co-albedo (1985)	0.72 ± 0.08	0.012

Total Relative Uncertainty = $\sigma[\Delta F SO_{4-clear}^{=}(1978)]/\Delta F SO_{4-clear}^{=}(1978)$ ≅ [Sum Squared Relative Uncertainty]$^{1/2}$ = 0.730.

3.2. $\Delta F SO_{4-cloud}^{=}$(1978)

We now estimate $\Delta F SO_{4-cloud}^{=}$(1978), the change in the net radiation at the tropopause due to the indirect effect of anthropogenic $SO_4^{=}$ particles acting as CCN. Estimating this anthropogenic CCN forcing is fraught with difficulty, largely because there is no fundamental theory that can be used to predict the response of the CCN concentration to either the source strength of SO_2 or the mass concentration of $SO_4^{=}$.

One empirical approach is suggested by the results of Radke and Hobbs (1976) who observed concentrations of CCN active at 0.2% supersaturation emanating from the east coast of North America over the Atlantic Ocean. If we assume that 6×10^{18} CCN per second are advected over the North Atlantic from North America, and a comparable amount from Asia are advected over the North Pacific, the total is about 10^{19} s^{-1}. The total natural marine CCN source over the globe can be estimated crudely by a simple box model as follows. Considering only the oceanic area of the Northern Hemisphere (A_{NH} = 3.6×10^{14} m^2), if the natural sulphate/sea salt CCN concentration is CCN = 70 cm^{-3} (or 7×10^7 m^{-3}), the CCN scale height is $H_{CCN} = 1.2$ km, and the CCN lifetime in the atmosphere is $\tau_{CCN} = 5 \times 10^5$ s, then the natural source strength of CCN, Q_{CCN}, is

$$Q_{CCN} \approx \frac{A_{NH} \, H_{CCN} \, CCN}{\tau_{CCN}} \approx \frac{(3.6 \times 10^{14} \text{ m}^2)(1.2 \times 10^3 \text{ m})(7 \times 10^7 \text{ m}^{-3})}{5 \times 10^5 \text{ s}}$$

$$\approx 6 \times 10^{19} \text{ s}^{-1} . \qquad (9)$$

Thus, the 10^{19} anthropogenic CCN per second that are advected over the oceans where they can nucleate marine stratiform cloud is roughly 20% of the natural source strength. We now employ the estimate by Wigley (1989) that an increase in CCN concentration of 20% would yield a reduction in the net radiation at the tropopause of about 1 Wm^{-2} for the Northern Hemisphere. Taking this to apply in 1978 and assuming the anthropogenic CCN effect is nil in the Southern Hemisphere gives

$$\Delta F \, SO_4^{=}{}_{-cloud}(1978) = -0.5 \pm 0.5 \text{ Wm}^{-2} , \qquad (10)$$

wherein we have assumed a 100% relative uncertainty due to the crudeness of this estimate approach.

3.3. $\Delta F \, SO_4^{=}(1978)$

Substituting Eqs. (8) and (10) into Eq. (3) yields

$$\Delta F \, SO_4^{=}(1978) = -1.1 \pm 0.9 \text{ Wm}^{-2} . \qquad (11)$$

Consequently, our estimate of $\Delta F SO_4^{=}(1978)$ ranges from -0.2 to -2.0 Wm^{-2}.

4. ESTIMATION OF ΔT_{2x} AS A FUNCTION OF USING OBSERVED GLOBAL-MEAN TEMPERATURE DEVIATIONS

To calculate the changes of temperature in response to the combined forcing of GHGs and anthropogenic sulphate aerosol, we have used the simple climate/ocean model of Schlesinger and Jiang (1991). This model determines the changes in the temperatures of the atmosphere and ocean, the latter as a function of depth from the surface to the ocean floor. In the model the ocean is subdivided vertically into 40 layers, with the uppermost being the mixed layer and the deeper layers each being 100 m thick. Also, the ocean is subdivided horizontally into a polar region, where bottom water is formed, and a nonpolar region where there is vertical upwelling. In the nonpolar region heat is transported upwards toward the surface by the water upwelling there and downwards by physical processes whose effects are treated as an equivalent diffusion. Heat is also removed from the mixed layer in the nonpolar region by transport to the polar region and downwelling toward the bottom, this heat being ultimately transported upward from the ocean floor in the nonpolar region.

Five principal quantities must be prescribed in this simple model: (1) the temperature sensitivity of the climate system, ΔT_{2x}; (2) the vertically uniform upwelling velocity for the global ocean, W; (3) the vertically uniform thermal diffusivity, κ, by which all non-advective vertical heat transport in the ocean is parameterized; (4) the depth of the oceanic mixed layer, h; and (5) the warming of the polar ocean relative to the warming of the nonpolar ocean, Π. Here we use the same oceanic values as Schlesinger and Jiang (1991): $W = 4$ m y^{-1}, $\kappa = 0.63$ cm^{-2} s, $h = 70$ m, and $\Pi = 0.4$.

For each prescribed value of $\Delta FSO_4^=(1978)$ the model has been run with the forcing given by Eqs. (1) and (2) for many values of ΔT_{2x}, each run beginning in 1765 with $\Delta T(1765) = 0$ and extending to 1990. For each ΔT_{2x} the root-mean-square "error" (RMSE) of $\Delta\theta_G(t) = \Delta T_G(t) + C_G$ to $\delta\overline{T}_g(t)$ has been calculated over $1861 \leq t \leq 1990$, with ΔT_G taken equal to the oceanic mixed-layer temperature change and C_G chosen such that $\Delta\theta_G(1880)$ equaled the average of $\delta\overline{T}_g(t)$ from 1861 to 1900. The results from this preliminary study have been presented by Schlesinger (1991) and are displayed again here in Fig. 3.

Figure 3 presents a plot of $1/\Delta T_{2x}$ as a function of $\Delta FSO_4^=(1978)$ as determined by our global model (GM), curve labeled "GM (Fil. δT_g, 1861–1900)," the reason for so doing being the evident linear dependence of $1/\Delta T_{2x}$ on for $\Delta FSO_4^=(1978)$ magnitudes less than about 1.1 Wm^{-2}. As described in Appendix A, this is the theoretically expected behavior. On the right-hand ordinate of Fig. 3 we have indicated the values of the feedback f (Schlesinger, 1985, 1988a,b) which satisfies

$$\Delta T_{2x} = \frac{(\Delta T_{2x})_0}{1 - f} , \qquad (12)$$

where $(\Delta T_{2x})_0 \approx 0.3 \; \Delta F_{2x} \approx 1.32°C$ is the sensitivity with zero feedback (only radiative-convective adjustment of temperature without changes in other quantities such as water vapor, clouds and sea ice), and $\Delta F_{2x} = 4.39$ Wm^{-2} is the radiative forcing due to a doubling of CO_2.

In Fig. 3 we also present the corresponding curve labeled "HM (Fil. δT_g, 1861–1900)" determined from our hemispheric model (HM), a detailed description of which

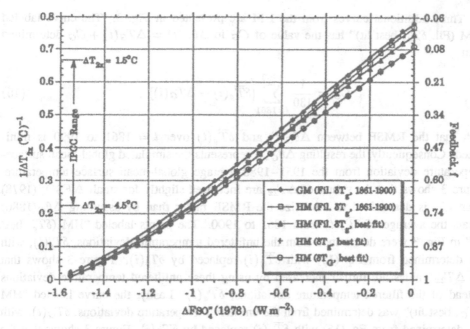

Fig. 3. Inverse of the climate sensitivity ΔT_{2x} plotted versus the change in the net radiative forcing at the top of the atmosphere in 1978 due to anthropogenic sulphate aerosol, $\Delta F SO_4^=(1978)$. The right-hand vertical scale shows the feedback f defined by $\Delta T_{2x}/(\Delta T_{2x})_0 = 1/(1-f)$, where $(\Delta T_{2x})_0 = 1.32°C$ is the temperature sensitivity without feedback.

is given in Appendix B. As described therein, the global-mean temperature changes simulated by the HM differ from those simulated by the GM for two reasons: (1) the ocean fraction in each hemisphere of the HM differs from the total ocean fraction in the GM, and (2) the air-sea heat exchange coefficient $\lambda_{1,0}$ is prescribed in the HM while another coefficient dependent on $\lambda_{a,0}$ is prescribed in the GM. For the HM we define ΔT_G by

$$\Delta T_G = \frac{\Delta T_N + \Delta T_S}{2}, \qquad (13)$$

where

$$\Delta T_i = \sigma_i \, \Delta T_{1,i} + (1-\sigma_i)\Delta T_{a,i}, \quad i = N, S, \qquad (14)$$

with $\Delta T_{1,i}$ and $\Delta T_{a,i}$ being, respectively, the changes in oceanic mixed-layer and atmospheric temperature in hemisphere i, the ocean and land fractions of which are σ_i and $1 - \sigma_i$. Figure 3 shows that the effect of the two differences between the HM and GM is to increase the values of ΔT_{2x} for all $\Delta F SO_4^=(1978)$.

Three additional curves from the HM are presented in Fig. 3. The curve labeled "HM (Fil. δT_g, best fit)" has the value of C_G in $\Delta\theta_G(t) = \Delta T_G(t) + C_G$ determined from

$$C_G = \frac{1}{130} \sum_{i=1861}^{1990} [\delta\overline{T}_g(i) - \Delta T_G(i)] , \qquad (15)$$

such that the RMSE between $\Delta\theta_G(t)$ and $\delta\overline{T}_g(t)$ over t = 1861 to 1990 is minimized. Consequently, the resulting $\Delta\theta_G(t)$ represents the simulated global-mean surface-temperature deviation from the 1951–1980 average global-mean surface temperature. Figure 3 shows that the values of ΔT_{2x} are increased slightly for small $\Delta FSO_4^=(1978)$ when C_G is determined to minimize the RMSE rather than by assuming $\Delta\theta_G(1880)$ equals the average of $\delta\overline{T}_g(t)$ from 1861 to 1900. The points labeled "HM (δT_g, best fit)" in Fig. 3 were determined from the unfiltered temperature deviations, $\delta T_g(t)$, with C_G determined from Eq. (15) with $\delta\overline{T}_g(i)$ replaced by $\delta T_g(i)$. Figure 3 shows that the ΔT_{2x} values are further increased by using these unfiltered temperature deviations instead of the filtered temperature deviations, $\delta\overline{T}_g(t)$. Lastly, the curve labeled "HM (δT_G, best fit)" was determined from the unfiltered temperature deviations, $\delta T_G(t)$, with C_G determined from Eq. (15) with $\delta\overline{T}_g(i)$ replaced by $\delta T_G(i)$. Figure 3 shows that the values of ΔT_{2x} determined from $\delta T_G(t)$ are virtually identical to those determined from $\delta T_g(t)$. Accordingly, henceforth we shall restrict attention to the ΔT_{2x} values determined from $\delta T_G(t)$. A regression equation for the corresponding $1/\Delta T_{2x}$ values in terms of $\Delta FSO_4^=(1978)$ is

$$\frac{1}{\Delta T_{2x}} = 0.7042 + 0.4584x - 0.1005x^2 - 0.068\,47x^3 ,$$

$$x = \Delta FSO_4^=(1978) . \qquad (16)$$

Figure 3 shows that for zero radiative forcing by anthropogenic sulphate aerosol, $\Delta FSO_4^=(1978)$ = 0, the best fit of $\Delta\theta_G(t)$ to $\delta T_G(t)$ over t = 1861 to 1990 is given by $\Delta T_{2x} = 1.41°C$ (f = +0.06). This value is slightly less than the minimum value adopted by the IPCC (Houghton et al., 1990), $\Delta T_{2x} = 1.5°C$ (f = +0.12), and is 2.9 times larger than the value of $\Delta T_{2x} = 0.5°C$ (f = −1.64) proposed by Lindzen (1990).

Admitting our smallest estimate of the anthropogenic sulphate aerosol forcing given by Eq. (11), $\Delta FSO_4^=(1978)$ = −0.2 Wm^{-2}, gives $\Delta T_{2x} = 1.65°C$ (f = +0.20), which is slightly larger than the minimum IPCC value. From this result it appears that the climate system cannot be as insensitive as claimed by Lindzen, but could be almost as insensitive as the minimum IPCC value if the sulphate aerosol forcing has only its minimum estimated value.

On the other hand, if the anthropogenic sulphate aerosol forcing has our estimated nominal value given by Eq. (11), $\Delta FSO_4^=(1978)$ = −1.1 Wm^{-2}, then $\Delta T_{2x} = 5.92°C$

($f = +0.78$). This sensitivity is larger than the maximum value adopted by the IPCC, $\Delta T_{2x} = 4.5°C$ ($f = +0.71$) and is slightly larger than the maximum values simulated by general circulation models, namely, $\Delta T_{2x} = 5.3°C$ ($f = +0.75$) by Li and Le Treut (1991) and $\Delta T_{2x} = 5.4°C$ ($f = +0.76$) by Mitchell and colleagues (Mitchell et al., 1989; Senior and Mitchell, 1992). If $\Delta T_{2x} = 5.4°C$ is considered to be the largest possible value of ΔT_{2x}, then $\Delta FSO_4^=(1978) \geq -1.06$ Wm^{-2}.

In summary we find that a range of $1.65 \leq \Delta T_{2x} \leq 5.92°C$ is consistent with the combined radiative forcing by greenhouse gases and ASAs, the latter, respectively, for $-0.2 \geq \Delta FSO_4^=(1978) \geq -1.1$ Wm^{-2}.

5. ESTIMATION OF ΔT_{2x} AND $\Delta FSO_4^=(1978)$ USING OBSERVED HEMISPHERIC-MEAN TEMPERATURE DEVIATIONS

To estimate unique values of ΔT_{2x} and $\Delta FSO_4^=(1978)$ requires additional observational data and a revision of our model to simulate it. For the data we use the evolution of the mean-surface-temperature deviations for the Northern and Southern Hemispheres, δT_N and δT_S, presented in Fig. 4 together with the corresponding filtered temperature deviations, $\delta \overline{T}_N$ and $\delta \overline{T}_S$, and we generalize our model to simulate the mean hemispheric temperature deviations, ΔT_N and ΔT_S, defined by Eq. (14).

In the generalization of the GM to this HM (Appendix B), two additional parameters must be introduced, namely, the interhemispheric heat-exchange coefficients for the atmosphere and ocean, β_a and β_o. Because the values of these new parameters are not known, we endeavored to determine them empirically by using the HM to simulate the annual cycle of mean northern and southern hemispheric temperatures for the present climate. From these empirical tests, taking $\beta_a = 0$, we found that $\beta_o = 6.5$ Wm^{-2}/°C if we ignored the heat capacity of the atmosphere, as is done for the secular climate change ($C_a = 0$; see Appendix B), and $\beta_o = 1.2$ Wm^{-2}/°C if the atmospheric heat capacity was not ignored ($C_a \neq 0$). Accordingly, we adopted $\beta_a = 0$ and $0 \leq \beta_o \leq 7$ Wm^{-2}/°C.

For each prescribed value of β_o the HM is run for many values of ΔT_{2x}, each run with the forcing for the Northern and Southern Hemispheres given by

$$\Delta F_N(t) = \Delta F_{GHG}(t) + 1.8\,\Delta FSO_4^=(t) \tag{17}$$

and

$$\Delta F_S(t) = \Delta F_{GHG}(t) + 0.2\,\Delta FSO_4^=(t) \,, \tag{18}$$

with $\Delta F_{GHG}(t)$ and $\Delta FSO_4^=(t)$ as in the GM, the latter as given by Eq. (2) with $\Delta FSO_4^=(1978)$ given by Eq. (16) for each value of ΔT_{2x}. The hemispheric forcing defined by Eqs. (17) and (18) gives the same global-mean forcing as in the GM [Eq. (1)], with 90% of the global-mean sulphate forcing located in the Northern Hemisphere and

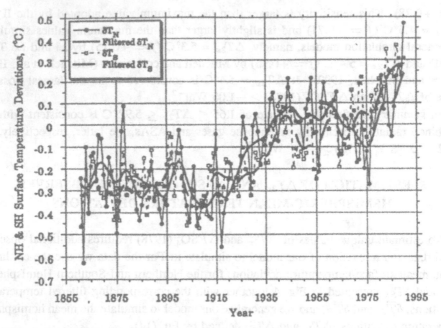

Fig. 4. Evolution of the mean northern-hemisphere surface-temperature deviations, δT_N (filled circles) and $\delta \overline{T}_N$ (heavy solid line), and mean southern-hemisphere surface-temperature deviations, δT_S (open squares) and $\delta \overline{T}_S$ (heavy dashed line). The data and binomial filter were kindly provided by C. Folland, D. Parker and A. Jackson of the Hadley Centre at the United Kingdom Meteorological Office.

10% in the Southern Hemisphere (Charlson et al., 1991). Each HM run begins in 1765, with the temperature deviations for each hemisphere equal to zero and extends to 1990.

For each ΔT_{2x} the RMSE of $\Delta \theta_{N-S}(t) = \Delta T_{N-S}(t) + C_{N-S}$ to $\delta T_{N-S}(t)$ is calculated over $1861 \leq t \leq 1990$, where

$$\Delta T_{N-S} = \Delta T_N - \Delta T_S , \tag{19}$$

with ΔT_N and ΔT_S as defined in Eq. (14),

$$\delta T_{N-S} = \delta T_N - \delta T_S , \tag{20}$$

as shown in Fig. 5, and with C_{N-S} determined from

$$C_{N-S} = \frac{1}{130} \sum_{i=1861}^{1990} [\delta T_{N-S}(i) - \Delta T_{N-S}(i)] , \tag{21}$$

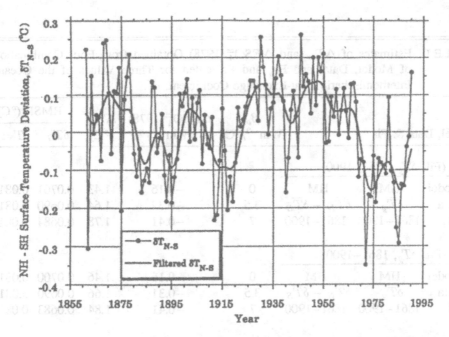

Fig. 5. Evolution of the mean Northern-Hemisphere minus Southern-Hemisphere surface-temperature deviation, δT_{N-S} (filled circles) and $\delta \overline{T}_{N-S}$ (heavy solid line). The data and binomial filter were kindly provided by C. Folland, D. Parker, and A. Jackson of the Hadley Centre at the United Kingdom Meteorological Office.

such that the RMSE between $\Delta\theta_{N-S}(t)$ and $\delta T_{N-S}(t)$ over $t = 1861$ to 1990 is minimized. Consequently, the resulting $\Delta\theta_{N-S}(t)$ represents the simulated surface-temperature deviation $\delta T_{N-S}(t)$. (In the preliminary calculation reported by Schlesinger (1991), C_{N-S} was determined such that $\Delta\theta_{N-S}(1880)$ equaled the average of $\delta\overline{T}_{N-S}(t)$ from 1861 to 1900.)

The results of this estimation of ΔT_{2x} and $\Delta FSO_4^=(1978)$ are presented in Table II under the entry "HM (δT_G, best fit)," together with estimates from three of the other combinations of model, data, and fit presented in Fig. 3. The results for "HM (δT_G, best fit)" are $\Delta T_{2x} = 2.28 \pm 0.56°C$, with the larger value corresponding to $\beta_o = 7$ and $\Delta FSO_4^=(1978) = -0.71$ Wm^{-2}, and the smaller value to $\beta_o = 0$ and $\Delta FSO_4^=(1978) = -0.25$ Wm^{-2}. Physically this occurs because as the interhemispheric heat exchange coefficient β_o increases, the model-simulated interhemispheric difference in temperature deviation decreases, so a larger sulphate radiative forcing $\Delta FSO_4^=(1978)$ is required to reproduce the observed interhemispheric difference in temperature deviation; in turn, this larger $\Delta FSO_4^=(1978)$ requires a larger ΔT_{2x} for the model to reproduce the observed global-mean temperature deviation.

TABLE II. Estimates of ΔT_{2x} and $\Delta FSO_4^=(1978)$ Obtained from Four Combinations of Model, Data, and Fit, and Presented for Three Values of the Oceanic Interhemispheric Heat-Exchange Coefficient, β_o.

Model, Data & Fit			β_o (Wm^{-2}/°C)	$\Delta FSO_4^=(1978)$ (Wm^{-2})	ΔT_{2x} (°C)	RMSE (°C)	
						δT_G	δT_{N-S}
GM (Fil. δT_g, 1861–1900)							
Model	GM	HM	0	−0.16	1.43	0.0701	0.0816
Data	$\delta \overline{T}_g$	$\delta \overline{T}_N - \delta \overline{T}_S$	3.5	−0.32	1.64	0.0690	0.0812
Fit	1861–1900	1861–1900	7	−0.41	1.78	0.0684	0.0811
HM (Fil. δT_g, 1861–1900)							
Model	HM	HM	0	−0.16	1.46	0.0700	0.0816
Data	$\delta \overline{T}_g$	$\delta \overline{T}_N - \delta \overline{T}_S$	3.5	−0.31	1.66	0.0690	0.0812
Fit	1861–1900	1861–1900	7	−0.41	1.84	0.0683	0.0811
HM (Fil. δT_g, best fit)							
Model	HM	HM	0	−0.26	1.63	0.0692	0.0781
Data	$\delta \overline{T}_g$	$\delta \overline{T}_N - \delta \overline{T}_S$	3.5	−0.57	2.23	0.0669	0.0775
Fit	Best Fit	Best Fit	7	−0.74	2.78	0.0655	0.0775
HM (δT_G, best fit)							
Model	HM	HM	0	−0.25	1.71	0.1134	0.1282
Data	δT_G	$\delta T_N - \delta T_S$	3.5	−0.54	2.28	0.1120	0.1279
Fit	Best Fit	Best Fit	7	−0.71	2.84	0.1110	0.1280

The results in Table II from the other combinations of model, data, and fit indicate that our estimates of ΔT_{2x} and $\Delta FSO_4^=(1978)$: (1) do not depend on the choice of global-mean temperature deviation, δT_G or δT_g, or on the choice of unfiltered or filtered temperature deviation, δT_G or $\delta \overline{T}_G$ [compare HM (δT_G, best fit) and HM (Fil. δT_g, 1861–1900)], although the RMSE of ΔT_G and ΔT_{N-S} are of course larger for the unfiltered data than for the filtered data; (2) do not depend on the choice of model, GM or HM [compare GM (Fil. δT_g, 1861–1900) and HM (Fil. δT_g, 1861–1900)]; and (3) do depend on the choice of fit, 1861–1900 or Best Fit [compare HM (Fil. δT_g, 1861–1900) and HM (Fil. δT_g, Best Fit)]. It is this dependence which is responsible for the difference between the estimate of $\Delta T_{2x} = 2.28 \pm 0.56°C$ from HM (δT_G, best fit) and the estimate of $\Delta T_{2x} \approx 1.64 \pm 0.18°C$ from GM (Fil. δT_g, 1861–1900), the preliminary presentation of which by Schlesinger (1991) was $\Delta T_{2x} = 1.7 \pm 0.2°C$. The cause of this narrower range in the preliminary estimate of ΔT_{2x} was the choice of C_G such that $\Delta\theta_G(1880)$ equaled the average of $\delta\overline{T}_g(t)$ from 1861 to 1900, and the choice of C_{N-S} such that $\Delta\theta_{N-S}(1880)$ equaled the average of $\delta\overline{T}_{N-S}(t)$ from 1861 to 1900. Accordingly, the preliminary estimate of ΔT_{2x} presented by Schlesinger (1991) is superseded by our self-consistent estimate of $\Delta T_{2x} = 2.28 \pm 0.56°C$.

6. COMPARISON OF SIMULATED AND OBSERVED TEMPERATURE DEVIATIONS

The temperature deviations $\Delta\theta_G(t) = \Delta T_G(t) + C_G$ and $\Delta\theta_{N-S}(t) = \Delta T_{N-S}(t) + C_{N-S}$ simulated by HM (δT_G, best fit) with $\beta_o = 3.5$ Wm^{-2}/°C ($C_G = -0.440°C$ and $C_{N-S} = 0.0464°C$) are presented in Figs. 6(a) and 6(c) in comparison with the observed temperature deviations, $\delta T_G(t)$ and $\delta T_{N-S}(t)$, and the corresponding residual temperature deviations, $\Delta\theta_G(t) - \delta T_G(t)$ and $\Delta\theta_{N-S}(t) - \delta T_{N-S}(t)$, are presented in Figs. 6(b) and 6(d). Figure 6 also presents the temperature deviations simulated by the HM for the case where the radiative forcing by ASAs is neglected, $\Delta FSO_4^=(1978) = 0$ ($C_G = -0.397°C$ and $C_{N-S} = 0.0028°C$), together with the corresponding residual temperature deviations.

Figures 6(a) and (b) show that the global-mean temperature deviations simulated by the HM with and without the ASAs are nearly identical, albeit the former having $\Delta T_{2x} = 2.28°C$ and the latter $\Delta T_{2x} = 1.41°C$. Figure 6(a) shows that the model-simulated global-mean temperature deviations forced by the GHGs with and without ASAs do reproduce the overall warming trend of the observed temperature deviations, but do not reproduce the relative extrema. Figure 6(b) shows that the residual global-mean temperature deviations, $\Delta\theta_G(t) - \delta T_G(t)$, oscillate around zero (their means with [without] ASAs are $3 \times 10^{-7}°C$ ($-4 \times 10^{-6}°C$]), with extremes of $-0.264°C$ [$-0.274°C$] and $0.290°C$ [$0.297°C$]). These residual global-mean temperature deviations represent a combination of the response of the climate system to external forcing not included here, such as that due to volcanoes and solar irradiance variations, plus the internal variability of the climate system. Further work is required to determine the contributions of these external and internal factors to the residual global-mean temperature deviations shown in Fig. 6(b).

Figure 6(c) shows that the interhemispheric-temperature-difference deviations simulated by the HM with and without ASAs are quite different. The deviations simulated without ASAs are always positive, hence the simulated temperature deviations in the Northern Hemisphere always exceed those in the Southern Hemisphere and gradually increase from 1861 to 1990. In contrast, the deviations simulated with ASAs gradually decrease from 1861 to 1990 such that they are positive only until about 1950; thereafter they are negative as a consequence of the larger observed deviations in the Southern Hemisphere than the Northern Hemisphere during the 1970s and 1980s.

92

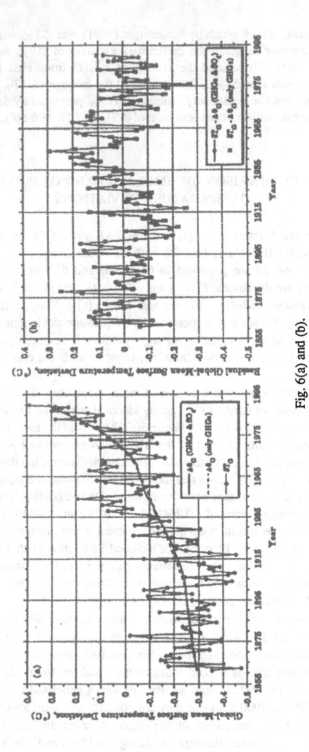

Fig. 6(a) and (b).

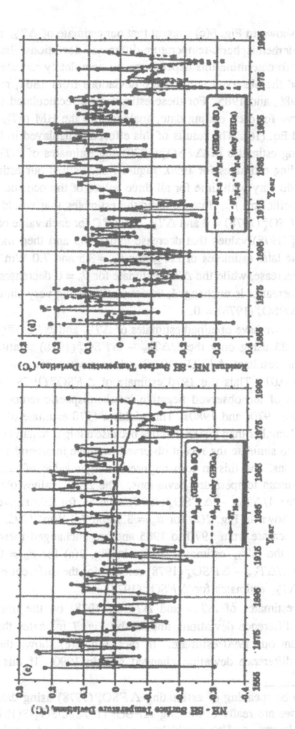

Fig. 6(c) and (d).

Fig. 6. Temperature deviations $\Delta\theta_G(t) = \Delta T_G(t) + C_G$ [panel (a)] and $\Delta\theta_{N-S}(t) = \Delta T_{N-S}(t) + C_{N-S}$ [panel (c)] simulated by the HM for $\beta_o = 3.5$ Wm^{-2} /°C with sulphate radiative forcing [HM (δT_G, best fit], $C_G = -0.440$°C and $C_{N-S} = -0.0464$°C; heavy solid line] and without [$C_G = -0.397$°C and $C_{N-S} = -0.0028$°C; dashed line], in comparison with the observed temperature deviations, $\delta T_G(t)$ and $\delta T_{N-S}(t)$. The corresponding residual temperature deviations, $\Delta\theta_G(t) - \delta T_G(t)$ and $\Delta\theta_{N-S}(t) - \delta T_{N-S}(t)$ are shown in panels (b) and (d).

The simulated deviations shown in Fig. 6(c) suggest that our estimate of ΔT_{2x} might depend significantly on the interhemispheric-temperature-difference deviations observed during the past twenty years. To determine this dependence we completely repeated our estimations of ΔT_{2x}, but used the observed temperature deviations from 1861, respectively, only to 1970, 1975, 1980, and 1985. For these estimates we recalculated a new $1/\Delta T_{2x} - \Delta F\mathrm{SO}_4^=(1978)$ curve for each ending date, analogous to the HM (δT_G, best fit) curve shown in Fig. 3 and Eq. (16). The results of this effort are displayed in Fig. 7 together with the corresponding estimates of $\Delta F\mathrm{SO}_4^=(1978)$ and estimates of ΔT_{2x} for $\Delta F\mathrm{SO}_4^=(1978) = 0$, all including estimates for 1990. Figure 7 shows that our estimates of ΔT_{2x} and $\Delta F\mathrm{SO}_4^=(1978)$ do vary with time for all three values of the oceanic inter-hemispheric heat exchange coefficient, β_o. Interestingly, the estimates that would have been obtained in 1970 give $\Delta F\mathrm{SO}_4^=(1978) = 0$ and $\Delta T_{2x} = 1.87°\mathrm{C}$ for each value of β_o.* Later estimates yield $\Delta F\mathrm{SO}_4^=(1978)$ values that decrease until 1985 and then increase slightly in 1990. Similarly, the later estimates of ΔT_{2x} for $\beta_o = 3.5$ and 7.0 $\mathrm{Wm}^{-2}/°\mathrm{C}$ increase until 1985 and then decrease, while the ΔT_{2x} estimate for $\beta_o = 0$ decreases until 1980, increases to 1985 and thereafter is unchanged, this evolution being very similar to that of the ΔT_{2x} values for $\Delta F\mathrm{SO}_4^=(1978) = 0$.

To understand this behavior we have obtained estimates of ΔT_{2x} and $\Delta F\mathrm{SO}_4^=(1978)$ for 1970, 1975, 1980, and 1985 using only the $1/\Delta T_{2x} - \Delta F\mathrm{SO}_4^=(1978)$ relation for 1990 given by Eq. (16). The resultant $\Delta F\mathrm{SO}_4^=(1978)$ values are the same as those shown in Fig. 7(b) to within 0.01. Thus the 1970 estimate of $\Delta F\mathrm{SO}_4^=(1978) = 0$ is a consequence of the absence of the observed negative interhemispheric-temperature-difference deviations during the 1970s and 1980s. This yields a 1970 estimate of ΔT_{2x} for $\beta_o = 3.5$ and 7.0 $\mathrm{Wm}^{-2}/°\mathrm{C}$ smaller than later estimates because a larger negative value of $\Delta F\mathrm{SO}_4^=(1978)$ is required to simulate the recent observed negative interhemispheric-temperature-difference deviations, and this in turn requires a larger value of ΔT_{2x} to simulate the observed global-mean temperature deviations. The ΔT_{2x} values obtained for each ending date using the $1/\Delta T_{2x} - \Delta F\mathrm{SO}_4^=(1978)$ relation for 1990 given by Eq. (16) are similar to those shown in Fig. 7(a) for $\beta_o = 3.5$ and 7.0 $\mathrm{Wm}^{-2}/°\mathrm{C}$, while the ΔT_{2x} values for $\beta_o = 0$ increase from 1970 to 1985 and are unchanged thereafter. Consequently, the decrease in the ΔT_{2x} estimates shown in Fig. 7(a) for $\beta_o = 0$ is a result of the differences in the $1/\Delta T_{2x} - \Delta F\mathrm{SO}_4^=(1978)$ relation for the different ending dates, as is the evolution of ΔT_{2x} estimates for $\Delta F\mathrm{SO}_4^=(1978) = 0$.

The dependence of our estimates of ΔT_{2x} and $\Delta F\mathrm{SO}_4^=(1978)$ on the observed interhemispheric-temperature-difference deviations implied by Fig. 7 indicates that future estimates may differ from our 1990 estimate. In fact, Fig. 6(c) shows that the interhemispheric-temperature-difference deviation changed sign in 1990. If this posi-

---

* It may seem paradoxical to be speaking of estimating $\Delta F\mathrm{SO}_4^=(1978)$ using data that end before 1978. However, we are really estimating $\Delta F\mathrm{SO}_4^=(1978)/Q\mathrm{SO}_2\text{-}S(1978)$ in Eq. (2), with $Q\mathrm{SO}_2\text{-}S(1978)$ known, and so would have obtained the same results had we used a reference year earlier than 1978.

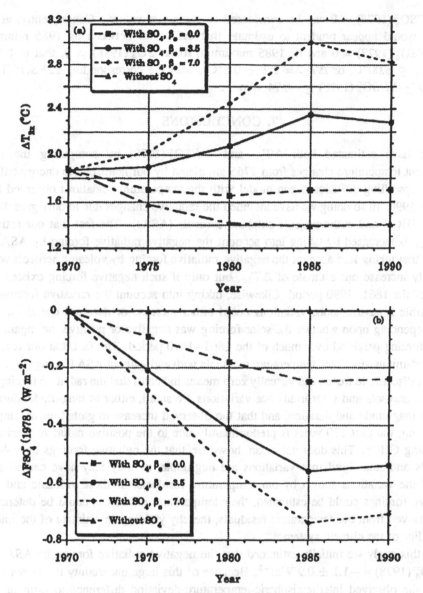

Fig. 7. Estimate of ΔT_{2x} (a) and $\Delta FSO_4^=(1978)$ (b) versus the year in which
the estimate could have been made using observational data from 1861
to that year. β_o is the oceanic interhemispheric heat exchange coefficient
in $Wm^{-2}/°C$.

tive deviation were to continue, it would tend to reduce a future estimate of ΔT_{2x}
below our 1990 value. However, because such future estimates cannot yield a positive
value for $\Delta FSO_4^=(1978)$ and still be considered physically realistic, our ΔT_{2x} estimates

for $\Delta F\mathrm{SO_4^=}(1978) = 0$ can be considered as minimum values. Consequently, at this time it would appear prudent to estimate that ΔT_{2x} lies between its 1985 minimum for $\Delta F\mathrm{SO_4^=}(1978) = 0$ and its 1985 maximum for $\Delta F\mathrm{SO_4^=}(1978) \neq 0$, that is 1.39°C $\leq \Delta T_{2x} \leq 3.00$°C, or $\Delta T_{2x} = 2.2 \pm 0.8$°C), with the corresponding $\Delta F\mathrm{SO_4^=}(1978)$ being $0 \geq \Delta F\mathrm{SO_4^=}(1978) \geq -0.74$ Wm^{-2}.

7. CONCLUSIONS

We have estimated both ΔT_{2x} and $\Delta F\mathrm{SO_4^=}(1978)$ by comparing the time-dependent temperature changes from 1765 simulated by our hemispheric energy-balance climate/upwelling-diffusion ocean model with the temperature deviations observed from 1861 to 1990. In so doing we have included the radiative changes due to only greenhouse gases (GHGs) and anthropogenic sulphate aerosols (ASAs). The fact that our estimate of ΔT_{2x} is increased by taking into account the negative radiative forcing by ASAs indicates that taking into account the negative radiative forcing by volcanic aerosols would similarly increase our estimate of ΔT_{2x}, but only if such negative forcing existed over much of the 1861–1990 period. Likewise, taking into account the radiative forcing due to possible solar-irradiance variations could either increase or decrease our ΔT_{2x} estimate depending upon whether the solar forcing was negative or positive, but again only if this forcing persisted over much of the 1861–1990 period. The fact that our resultant observed-minus-simulated temperature residuals with and without ASA forcing [Fig. 6(b)] have no discernible trends and virtually zero means indicates that the radiative forcings by volcanic aerosols and solar irradiance variations are small, either in magnitude, duration or both magnitude and duration, and that the observed increase in global-mean temperature during the past 130 years is predominantly due to the positive radiative forcing by increasing GHGs. This does not mean, however, that the radiative forcings by volcanic aerosols and solar irradiance variations are negligible, as they may have caused some part of the variations shown by our temperature residuals. If these volcanic and solar radiative forcings could be estimated, their temperature influences could be determined and removed from our temperature residuals, thereby yielding an estimate of the internal variability of the climate system.

In this study we initially estimated that the negative radiative forcing by ASAs was $\Delta F\mathrm{SO_4^=}(1978) = -1.1 \pm 0.9$ Wm^{-2}. Because of this large uncertainty it was necessary to use the observed interhemispheric-temperature-deviation difference to estimate both ΔT_{2x} and $\Delta F\mathrm{SO_4^=}(1978)$, this in addition to using the global-mean temperature deviation to determine ΔT_{2x} as a function of $\Delta F\mathrm{SO_4^=}(1978)$. Our results show a significant dependence on the observed interhemispheric-temperature-deviation difference. This suggests that future estimates of ΔT_{2x} may vary from our estimate based on data for the 1861–1990 period, unless an independent estimate of $\Delta F\mathrm{SO_4^=}(1978)$ is obtained. Such an independent estimate would allow ΔT_{2x} to be estimated directly from the global-mean temperature deviations alone. Such an independent estimate would have to include both the ASA clear-sky radiative forcing, $\Delta F\mathrm{SO_{4-clear}^=}(1978)$, and the ASA cloud radiative forcing, $\Delta F \text{SO}_{4-\text{cloud}}^=(1978)$. The results presented in Table I indicate that the uncertainty for $\Delta F \text{SO}_{4-\text{clear}}^=(1978)$ could be significantly reduced by reducing the uncertainties in the aerosol mass scattering efficiency, the lifetime of atmospheric $\text{SO}_4^=$, and the aerosol hemispheric backscatter fraction. However, reduction in the uncertainty of $\Delta F \text{SO}_{4-\text{cloud}}^=(1978)$ will likely be considerably more difficult. In this endeavor our estimate of the maximum $\Delta F \text{SO}_4^=(1978) = -0.74 \text{ Wm}^{-2}$ may be of use. In particular, if the estimate in Table I is realistic for $\Delta F \text{SO}_{4-\text{clear}}^=(1978)$, then $\Delta F \text{SO}_{4-\text{cloud}}^=(1978)$ must be small.

Because our estimates of both ΔT_{2x} and $\Delta F \text{SO}_4^=(1978)$ display a dependence on the observed interhemispheric-temperature-deviation differences, we increased the uncertainty of our ΔT_{2x} estimate by decreasing the minimum value to that given by assuming $\Delta F \text{SO}_4^=(1978) = 0$. Nevertheless, the resultant range of our $\Delta T_{2x} = 2.2 \pm 0.8°\text{C}$ estimate is about half the 1.5°C range of IPCC estimate, while our nominal estimate of $\Delta T_{2x} = 2.2°\text{C}$ is quite close to the IPCC "best estimate" of $\Delta T_{2x} = 2.5°\text{C}$. Our resultant minimum estimate of $\Delta T_{2x} = 1.4°\text{C}$ is virtually three times larger than the $\Delta T_{2x} = 0.5°\text{C}$ estimate of Lindzen (1990), while our maximum estimate of $\Delta T_{2x} = 3.0°\text{C}$ is six times larger.

ACKNOWLEDGMENTS

The authors thank C. Folland, D. Parker and A. Jackson (Hadley Centre at the United Kingdom Meteorological Office) for providing the IPCC surface temperature data and binomial filter used to construct Figs. 1 and 4, and A. Ghanem (University of Illinois) for his assistance in running the simple climate/ocean model. This research was supported by the US National Science Foundation and the US Department of Energy, Carbon Dioxide Research Program, Office of Health and Environmental Research, under Grant ATM-9001310.

APPENDIX A

RELATIONSHIP BETWEEN $1/\Delta T_{2x}$ AND $\Delta F \text{SO}_4^=(1978)$

The model-simulated global temperature change at time t can be represented by the convolution integral

$$\Delta T_G(t) = \frac{\Delta T_{2x}}{\Delta F_{2x}} \int_{t_0}^{t} \frac{d \Delta F(\tau)}{d\tau} R_c(t - \tau) \, d\tau , \qquad (A.1)$$

where ΔF_{2x} is the forcing due to a doubling of the CO_2 concentration, R_c is a climate response function which characterizes the temporal response of the climate system to an instantaneous CO_2 doubling (Schlesinger et al., 1985), and t_0 is the initial time (1765) when ΔT_G and ΔF are both zero. If the climate system had no thermal inertia, then R_c would be unity and the system would always be in equilibrium with its forcing, that is,

$$\Delta T_G(t) = \Delta T_{G,eq}(t) \equiv \frac{\Delta T_{2x}}{\Delta F_{2x}} \Delta F(t) \ . \qquad (A.2)$$

However, the climate system has thermal inertia because of the ocean, hence the response lags the forcing and $0 < R_c < 1$. Accordingly, Eq. (A.1) can be rewritten as

$$\Delta T_G(t) = \Delta T_{G,eq}(t) \, \overline{R}_c(t) \ , \qquad (A.3)$$

where

$$\overline{R}_c(t) = \frac{1}{\Delta F(t)} \int_{t_0}^{t} \frac{d\,\Delta F(\tau)}{d\tau} \, R_c(t - \tau) \, d\tau \ . \qquad (A.4)$$

In the text we fit $\Delta T_G(t) + C_G$ to the observed temperature deviation $\delta T_G(t)$ by determining C_G and ΔT_{2x} to minimize the RMSE, thereby approximately yielding $\Delta T_G(t) + C_G \approx \delta T_G(t)$. Using this relation with $t = 1978$ together with Eqs. (A.2) and (A.3) gives

$$\frac{1}{\Delta T_{2x}} = C \left[\Delta F_{GHG}(1978) + \Delta F\mathrm{SO}_4^=(1978) \right] \ , \qquad (A.5)$$

where

$$C = \frac{\overline{R}_c(1978)}{\Delta F_{2x}[\delta T_G(1978) - C_G]} \ . \qquad (A.6)$$

From Eq. (A.5) it is evident that $1/\Delta T_{2x}$ linearly decreases as the magnitude of the negative sulphate forcing $\Delta F\mathrm{SO}_4^=(1978)$ increases. However, the climate response function $\overline{R}_c(1978)$ decreases as ΔT_{2x} increases (Schlesinger, 1989). Consequently, as ΔT_{2x} increases A eventually begins to decrease significantly and thereafter $1/\Delta T_{2x}$ decreases less rapidly with $\Delta F\mathrm{SO}_4^=(1978)$ than linearly.

The value of $\overline{R}_c(1978)$ can be estimated from Eq. (A.6). Using $C = 0.4584$ (°C Wm^{-2})$^{-1}$ obtained Eq. (16), $\Delta F_{2x} = 4.39$ Wm^{-2}, $\delta T_G(1978) = 0.019$°C, and $C_G = -0.440$, we obtain $\overline{R}_c(1978) = 0.924$. Because this estimate of $\overline{R}_c(1978)$ is close to unity, it suggests that the response of the climate system is close to its equilibrium response. Other studies such as that of Schlesinger and Jiang (1990) indicate the contrary. Clearly these contradictory findings deserve further investigation.

APPENDIX B

HEMISPHERIC ENERGY-BALANCE
CLIMATE/UPWELLING-DIFFUSION OCEAN MODEL

The hemispheric energy-balance climate/upwelling-diffusion ocean model shown schematically in Fig. B.1 is similar to its global counterpart (Schlesinger and Jiang, 1991). In the model the ocean in each hemisphere is subdivided vertically into L layers, with the uppermost being the mixed layer. Also, in each hemisphere the ocean is subdivided horizontally into a polar region where bottom water is formed, and a nonpolar region where there is vertical upwelling. In the nonpolar region heat is transported upwards toward the surface by the water upwelling there and downwards by physical processes whose effects are treated as an equivalent diffusion. Heat is also removed from the mixed layer in the nonpolar region in each hemisphere by a transport to the polar region and downwelling toward the bottom, this heat being ultimately transported upward from the ocean floor in the nonpolar region.

The governing equations for the change in the mean atmospheric temperatures for the northern and southern hemispheres induced by hemispheric external radiative forcing ΔF_N and ΔF_S are

$$
\begin{aligned}
C_a \frac{\partial \Delta T_{a,N}}{\partial t} = \Delta F_N - \lambda \Delta T_{a,N} - \sigma_N \, \lambda_{a,o}(\Delta T_{a,N} - \Delta T_{1,N}) \\
- \beta_a(\Delta T_{a,N} - \Delta T_{a,S}) \, ,
\end{aligned}
\tag{B.1}
$$

and

$$
\begin{aligned}
C_a \frac{\partial \Delta T_{a,S}}{\partial T} = \Delta F_S - \lambda \Delta T_{a,S} - \sigma_S \, \lambda_{a,o}(\Delta T_{a,S} - \Delta T_{1,S}) \\
+ \beta_a(\Delta T_{a,N} - \Delta T_{a,S}) \, ,
\end{aligned}
\tag{B.2}
$$

where C_a is the atmospheric heat capacity, σ_i the fractional area of the ocean in hemisphere i, $\lambda = \Delta F_{2x}/\Delta T_{2x}$ is a climate sensitivity parameter with ΔT_{2x} the change in equilibrium surface temperature induced by a CO_2 doubling which creates a radiative forcing ΔF_{2x}, $\lambda_{a,o}$ is the air-sea heat-exchange coefficient, $\Delta T_{1,i}$ is the change in the oceanic mixed-layer temperature in hemisphere i, and β_a is the interhemispheric atmospheric heat exchange coefficient. The second term on the right-hand sides of Eqs. (B.1) and (B.2) represents the outgoing longwave radiation and climate feedback, the third term the heat exchange between the atmosphere and the ocean, and the last term the interhemispheric heat exchange.

As in our global model (Schlesinger and Jiang, 1991), we assume that the atmosphere is in equilibrium with the ocean because of its much smaller heat capacity (Wigley and

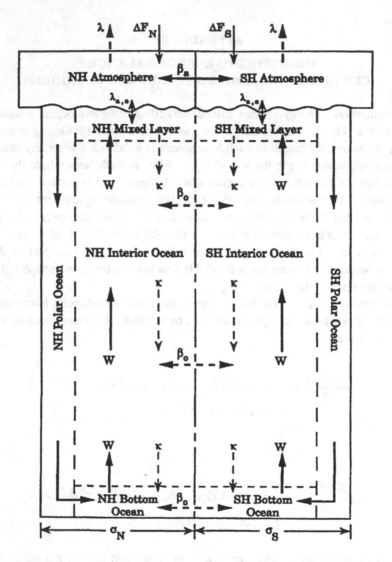

Fig. B.1. Schematic diagram of the hemispheric model (HM). The symbols by the arrows indicate the following physical processes: ΔF_N and ΔF_S: tropopause radiative forcing of the Northern and Southern Hemispheres, respectively; λ: radiative-plus-feedback temperature response of the climate system; β_a: atmospheric interhemispheric heat exchange; $\lambda_{a,o}$: air-sea heat exchange; W: vertical heat transport by upwelling; κ: vertical heat transport by all processes other than upwelling, treated as a diffusive process; β_o: oceanic interhemispheric heat exchange. The quantities σ_N and σ_S denote the fractions of Northern and Southern Hemispheres covered by ocean, respectively.

Schlesinger, 1985). Consequently, we set the left-hand sides of Eqs. (B.1) and (B.2) equal to zero and obtain

$$\Delta T_{a,N} = \frac{\Delta F_N + \sigma_N \, \lambda_{a,o} \, \Delta T_{1,N} + \frac{\beta_a}{f_S}(\Delta F_S + \sigma_S \, \lambda_{a,o} \, \Delta T_{1,S})}{f_N - \frac{\beta_a^2}{f_S}} , \qquad (B.3)$$

and

$$\Delta T_{a,S} = \frac{\Delta F_S + \sigma_S \, \lambda_{a,o} \, \Delta T_{1,S} + \frac{\beta_a}{f_N}(\Delta F_N + \sigma_N \, \lambda_{a,o} \, \Delta T_{1,N})}{f_S - \frac{\beta_a^2}{f_N}} , \qquad (B.4)$$

where

$$f_i = \lambda + \sigma_i \, \lambda_{a,o} + \beta_a \, , \quad i = N, S \, . \qquad (B.5)$$

The governing equations for the hemispheric-mean oceanic mixed-layer temperature changes are

$$\rho c \Delta z_1 \, \frac{\partial \Delta T_{1,N}}{\partial t} = \lambda_{a,o}(\Delta T_{a,N} - \Delta T_{1,N}) + \rho c \kappa \, \frac{\Delta T_{2,N} - \Delta T_{1,N}}{0.5 \Delta z_2}$$

$$+ \rho c W (\Delta T_{1,N} - \Delta T_{P,N}) - \beta_o(\Delta T_{1,N} - \Delta T_{1,S}) \, , \qquad (B.6)$$

and

$$\rho c \Delta z_1 \frac{\partial \Delta T_{1,S}}{\partial t} = \lambda_{a,o}(\Delta T_{a,S} - \Delta T_{1,S}) + \rho c \kappa \, \frac{\Delta T_{2,S} - \Delta T_{1,S}}{0.5 \Delta z_2}$$

$$+ \rho c W (\Delta T_{1,S} - \Delta T_{P,S}) + r_{NS} \, \beta_o(\Delta T_{1,N} - \Delta T_{1,S}) \, , \, (B.7)$$

where Δz_k is the thickness of layer k, with $k = 1$ being the mixed layer; κ is the vertically uniform vertical heat diffusivity, W the vertically uniform upwelling velocity, β_o the vertically uniform oceanic interhemispheric heat exchange coefficient, r_{NS} the ratio of the ocean areas in the Northern and Southern Hemispheres, and ΔT_P the temperature change in the polar region defined by

$$\Delta T_{P,i} = \Pi \Delta T_{1,i} \, , \quad i = N, S \, , \qquad (B.8)$$

with Π being a parameter. The first term on the right-hand sides of Eqs. (B.6) and (B.7) represents the air-sea heat flux, the second term the vertical heat transport as parameterized by diffusion, the third term the vertical advective heat transport and the heat exchange between the oceanic mixed layer and the polar ocean, and the last term the interhemispheric heat exchange.

Substituting Eq. (B.3) into Eq. (B.6) and Eq. (B.4) into Eq. (B.7) gives

$$\gamma_N \, \rho c \Delta z_1 \frac{\partial \Delta T_{1,N}}{\partial t} = \Delta F_N - \lambda \Delta T_{1,N} + \gamma_N \, \rho c \kappa \frac{\Delta T_{2,N} - \Delta T_{1,N}}{0.5 \Delta z_2}$$

$$+ \gamma_N \, \rho c W (\Delta T_{1,N} - \Delta T_{P,N}) - \gamma_N \, \beta_o (\Delta T_{1,N} - \Delta T_{1,S})$$

$$+ \frac{\beta_o}{f_S} [\Delta F_S + \sigma_S \, \lambda_{a,o} \, \Delta T_{1,S} - (\lambda + \sigma_S \, \lambda_{a,o}) \Delta T_{1,N}] \,, \quad (B.9)$$

and

$$\gamma_S \, \rho c \Delta z_1 \frac{\partial \Delta T_{1,S}}{\partial t} = \Delta F_S - \lambda \Delta T_{1,S} + \gamma_S \, \rho c \kappa \frac{\Delta T_{2,S} - \Delta T_{1,S}}{0.5 \Delta z_2}$$

$$+ \gamma_S \, \rho c W (\Delta T_{1,S} - \Delta T_{P,S}) + r_{NS} \, \gamma_S \, \beta_o (\Delta T_{1,N} - \Delta T_{1,S})$$

$$+ \frac{\beta_a}{f_N} [\Delta F_N + \sigma_N \, \lambda_{a,o} \, \Delta T_{1,N} - (\lambda + \sigma_N \, \lambda_{a,o}) \Delta T_{1,S}] \,, \quad (B.10)$$

where

$$\gamma_N = \sigma_N + \frac{\lambda}{\lambda_{a,o}} + \frac{\beta_a}{\lambda_{a,o}} \left(1 - \frac{\beta_a}{f_S} \right) \,, \qquad (B.11)$$

and

$$\gamma_S = \sigma_S + \frac{\lambda}{\lambda_{a,o}} + \frac{\beta_a}{\lambda_{a,o}} \left(1 - \frac{\beta_a}{f_N} \right) \,. \qquad (B.12)$$

Equation (B.9) with Eq. (B.11) and Eq. (B.10) with Eq. (B.12) differ from the governing equation for the oceanic mixed-layer temperature changes in our global model in that they each contain interhemispheric heat transports for the atmosphere and ocean. However, even if these interhemispheric heat transports are made zero by defining $\beta_a - \beta_o = 0$ and the hemispheric forcings ΔF_N and ΔF_S are taken to be equal, the resulting equations will differ from that in the global model unless the ocean fractions in each hemisphere, σ_N and σ_S, are taken equal to the value used in the global model, $\sigma_G = 0.71$. Consequently, because we take $\sigma_N = 0.61$ and $\sigma_S = 0.81$, the global-mean temperature changes from our hemispheric model given by $(\Delta T_N + \Delta T_S)/2$ will in general not equal the mean temperature changes from our global model, even when there is no interhemispheric heat transport and the forcing in each hemisphere is identical.

There is yet another reason why results from the hemispheric and global models differ. In the global model $\gamma_G = \sigma_G + \lambda/\lambda_{a,o}$ and σ_G were fixed, hence $\lambda_{a,o}$ varied with $\lambda = \Delta F_{2x}/\Delta T_{2x}$ as the prescribed value of ΔT_{2x} was varied. In the hemispheric model we instead fix $\lambda_{a,o}$, σ_N, σ_S, and β_a so that γ_N and γ_S vary with λ as the prescribed value of ΔT_{2x} is varied.

The governing equations for the mean interior ocean temperature changes for the Northern and Southern Hemispheres are

$$
\rho c \Delta z_k \, \frac{\partial \Delta T_{k,N}}{\partial t} = \rho c \left[\kappa \frac{\Delta T_{k+1,N} - \Delta T_{k,N}}{0.5(\Delta z_k + \Delta z_{k+1})} + W \frac{\Delta T_{k,N} + \Delta T_{k+1,N}}{2} \right]
$$
$$
- \rho c \left[\kappa \frac{\Delta T_{k,N} - \Delta T_{k-1,N}}{0.5(\delta_k \Delta z_{k-1} + \Delta z_k)} + W \frac{\Delta T_{k-1,N} + \Delta T_{k,N}}{2} \right]
$$
$$
- \beta_o (\Delta T_{k,N} - \Delta T_{k,S}) \,, \quad k = 2, \ldots, \quad L - 1 \,, \tag{B.13}
$$

and

$$
\rho c \Delta z_k \, \frac{\partial \Delta T_{k,S}}{\partial t} = \rho c \left[\kappa \frac{\Delta T_{k+1,S} - \Delta T_{k,S}}{0.5(\Delta z_k + \Delta z_{k+1})} + W \frac{\Delta T_{k,S} + \Delta T_{k+1,S}}{2} \right]
$$
$$
- \rho c \left[\kappa \frac{\Delta T_{k,S} - \Delta T_{k-1,S}}{0.5(\delta_k \Delta z_{k-1} + \Delta z_k)} + W \frac{\Delta T_{k-1,S} + \Delta T_{k,S}}{2} \right]
$$
$$
+ r_{NS} \beta_o (\Delta T_{k,N} - \Delta T_{k,S}) \,, \quad k = 2, \ldots, \quad L - 1 \,, \tag{B.14}
$$

where $\delta_k = 0$ for $k = 2$ and $\delta_k = 1$ for $k > 2$. Heat is transported in the interior ocean by the vertical diffusion, vertical advection, and interhemispheric exchange.

The governing equations for the mean temperature changes of the bottom ocean for the Northern and Southern Hemispheres are

$$
\rho c \Delta z_L \, \frac{\partial \Delta T_{L,N}}{\partial t} = -\rho c \left[\kappa \frac{\Delta T_{L,N} - \Delta T_{L-1,N}}{0.5(\Delta z_{L-1} + \Delta z_L)} + W \frac{\Delta T_{L-1,N} + \Delta T_{L,N}}{2} \right]
$$
$$
+ \rho c W \Delta T_{P,N} - \beta_o (\Delta T_{L,N} - \Delta T_{L,S}) \,, \tag{B.15}
$$

and

$$
\rho c \Delta z_L \, \frac{\partial \Delta T_{L,S}}{\partial t} = -\rho c \left[\kappa \frac{\Delta T_{L,S} - \Delta T_{L-1,S}}{0.5(\Delta z_{L-1} + \Delta z_L)} + W \frac{\Delta T_{L-1,S} + \Delta T_{L,S}}{2} \right]
$$
$$
+ \rho c W \Delta T_{P,S} + r_{NS} \beta_o (\Delta T_{L,N} - \Delta T_{L,S}) \,. \tag{B.16}
$$

The governing Eqs. (B.9), (B.10) and (B.13)–(B.16) are time-integrated using backward-implicit differencing for the vertical diffusion and advection terms, and forward differencing for the forcing and interhemispheric heat-exchange terms. The values of the parameters in these equations are presented in Table B.I.

TABLE B.I. Parameters Used in the Energy-balance Climate/Upwelling-diffusion Ocean Model.

Quantity	Symbol	Value
Specific heat capacity of water, cal $g^{-1}°C^{-1}$	c	0.955[a]
Number of ocean layers	L	40[a]
Vertical velocity, m yr-1	W	4[a]
Ratio of the ocean areas in the Northern and Southern Hemispheres	r_{NS}	0.7531
Total depth of ocean, m	z_L	4000[a]
Atmospheric interhemispheric heat-exchange coefficient, $Wm^{-2}/°C$	β_a	0[b]
Oceanic interhemispheric heat-exchange coefficient, $Wm^{-2}/°C$	β_o	0 to 7[b]
Radiative forcing due to doubled CO_2, Wm^{-2}	ΔF_{2x}	4.39[a]
Change in equilibrium surface temperature induced by ΔF_{2x}	ΔT_{2x}	adjustable
Time step, month	Δt	1[a]
Oceanic mixed-layer depth, meter	Δz_1	70[a]
Depth of interior oceanic layer, meter	Δz_k	100[a]
Vertical heat diffusivity, $cm^2 s^{-1}$	κ	0.63[a]
Air-sea heat-exchange coefficient, $Wm^{-2}°C^{-1}$	λ_{ao}	16[a]
Polar region temperature change relative to nonpolar region temperature change	Π	0.4[a]
Density of water, g cm^{-3}	ρ	1.030[a]
Fraction of Northern Hemisphere covered by ocean	σ_N	0.61
Fraction of Southern Hemisphere covered by ocean	σ_S	0.81

[a] As used in the global model of Schlesinger and Jiang (1991).
[b] Choice discussed in Sec. 5.

REFERENCES

Boden, T. A., P. Kanciruk, P., and M. P. Farrell, 1990: in *Trends '90* 3 (Carbon Dioxide Information Analysis Center, Oak Ridge, TN, 1990).

Charlson, R. J., D. S. Covert, and T. V. Larson, T. 1984: In *Hygroscopic Aerosols*, Ruhnke, L. H., and A. Deepak (eds.), A. Deepak Publishing, Hampton, VA, 35–44.

Charlson, R. J., J. Langner, and H. Rodhe, 1990: Sulphate aerosol and climate. *Nature* **348**, 22.

Charlson, R. J., J. Langner, H. Rodhe, H., C. B. Leovy, and S. G. Warren, 1991: Perturbation of the northern hemisphere radiative balance by backscattering from anthropogenic sulfate aerosols. *Tellus* **43**, 152–163.

Hameed, S., and J. Dignon, 1988: Changes in the geographical distributions of global emissions of NO_x and SO_x from fossil-fuel combustion between 1966 and 1980. *Atmos. Environ.* **22**, 441–449.

Houghton, J. T., G. J. Jenkins, and J. J. Ephraums (eds.), 1990: *Climate Change: The IPCC Scientific Assessment*, Cambridge University Press, Cambridge, 364 pp.

Li, Z.-X., and H. Le Treut, 1991: Cloud-radiation feedbacks in a general circulation model and their dependence on cloud modelling assumptions. *Climate Dynamics* (submitted).

Lindzen, R. S., 1990: Some coolness about global warming. *Bull. Amer. Meteorol. Soc.* **71**, 288–299.

Mitchell, J. F. B., C. A. Senior, and W. J. Ingram, 1989: CO_2 and climate: a missing feedback? *Nature* **341**, 132–134.

NAS, 1979: Carbon dioxide and climate: A scientific assessment. Report of an ad hoc study group on carbon dioxide and climate. U.S. National Academy of Sciences, Washington D.C., 22 pp.

Radke, L. F., and P. V. Hobbs, 1976: Cloud condensation nuclei on the Atlantic seaboard of the United States. *Science* **193**, 999–1002.

Rodhe, H., and I. Isaksen, 1980: Global distribution of surface components in the troposphere estimated in a height/latitude transport model. *J. Geophys. Res.* **85**, 7408–7409.

Schlesinger, M. E., 1985: Feedback analysis of results from energy balance and radiative-convective models. In *The Potential Climatic Effects of Increasing Carbon Dioxide*,

M. C. MacCracken and F. M. Luther (eds.), U.S. Department of Energy, DOE/ER-0237, pp. 280–319.

Schlesinger, M. E., 1988a: Quantitative analysis of feedbacks in climate model simulations of CO_2-induced warming. In *Physically-Based Modelling and Simulation of Climate and Climatic Change*, M. E. Schlesinger (ed.), NATO Advanced Study Institute Series, Kluwer, Dordrecht, pp. 653–736.

Schlesinger, M. E., 1988b: How to make models for behaviour of clouds. *Nature* 336, 315–316.

Schlesinger, M. E., 1989: Model Projections of the Climatic Changes Induced by Increased Atmospheric CO_2. In *Climate and the Geo-Sciences: A Challenge for Science and Society in the 21st Century*, A Berger, S. Schneider and J. Cl. Duplessy, Eds., Kluwer Academic Publishers, Dordrecht, 375–415.

Schlesinger, M. E., 1990: Theoretical Estimates of Greenhouse-Gas-Induced Climate Change. In *Prospects for Future Climate: A Special US/USSR Report on Climate and Climate Change*, M. C. MacCracken, M. I. Budyko, A. D. Hecht and Y. A. Izrael (eds.), Lewis Publishers, Chelsea, Michigan, pp. 113–156.

Schlesinger, M. E., 1991: Model simulation-observation intercomparison strategies for estimating the sensitivity of the climate system to increases in greenhouse gases. In Proceedings of the First DEMETRA Meeting on Climate Variability and Global Change, Chianciano Terme, Italy, 28 October–1 November 1991 (in press).

Schlesinger, M. E., and X. Jiang, 1990: Simple model representation of atmosphere-ocean GCMs and estimation of the timescale of CO_2-induced climate change. *J. Climate* 3, 1297–1315.

Schlesinger, M. E., and X. Jiang, 1991: Revised projection of greenhouse warming. *Nature* 350, 219–221.

Schlesinger, M. E., and J. F. B. Mitchell, 1985: Model projections of the equilibrium climatic response to increased CO_2. In *The Potential Climatic Effects of Increasing Carbon Dioxide*, M. C. MacCracken and F. M. Luther (eds.), 81–147, U.S. Department of Energy, DOE/ER-0237.

Schlesinger, M. E., and J. F. B. Mitchell, 1987: Model projections of the equilibrium climatic response to increased CO_2. *Rev. Geophys.* 25, 760–798.

Schlesinger, M. E., W. L. Gates and Y.-J. Han, 1985: The role of the ocean in CO_2-induced climatic warming: Preliminary results from the OSU coupled atmosphere-ocean GCM. In *Coupled Ocean-Atmosphere Models*, J. C. J. Nihoul (ed.), Elsevier, Amsterdam, 447–478 pp.

Senior, C. A., and J. F. B. Mitchell, 1992: CO_2 and climate: The impact of cloud parameterization. *J. Climate*, submitted.

Slinn, W. G. N., 1983: In *Air-to-Sea Exchange of Gases and Particles*, P. S. Liss and W. G. N. Slinn (eds.), NATO ASI Series C108, Reidel, Dordrecht, 299–405.

Twomey, S., 1977: *Atmospheric Aerosols*. Elsevier, Amsterdam, 302 pp

Vanderpol, A. H., 1975: A Systematic Approach for Computer Analysis of Air Chemistry Data. Ph.D. dissertation, Univ. of Washington, Seattle, WA, 110 pp.

Warren, S. G., C. J. Hahn, J. London, R. M. Chervin and R. L. Jenne, 1988: Global Distribution of Total Cloud Cover and Cloud Type Amount over the Ocean. NCAR Technical Note TN-317+STR, NCAR, Boulder, CO, 42 pages + 170 maps.

White, W. H., 1986: *Atmos. Environ.* **20**, 1659–1672.

Wigley, T. M. L., 1989: Possible climate change due to SO_2-derived cloud condensation nuclei. *Nature* **339**, 365–367.

Wigley, T. M. L., and M. E. Schlesinger, 1985: Analytical solution for the effect of increasing CO_2 on global mean temperature. *Nature* **315**, 649–652.

Toward Regional Climate-Change Scenarios: How Far Can We Go?

P. H. Whetton and A. B. Pittock

Climate Impact Group

CSIRO Division of Atmospheric Research

P.B. No. 1, Mordialloc 3195, Victoria, Australia

ABSTRACT

This paper outlines the approach being used by the Climate Impact Group at CSIRO in providing regional climate change scenarios for use in impact studies in Australia. The paper provides an Australian region example of how far it is possible to go at present in assessing possible regional climate change due to the enhanced greenhouse effect. The approach is based on critical examination of the regional output of global climate models, but information from other sources, such as palaeoclimatic studies and limited area modelling, are also taken into account. In this discussion we concentrate on regional rainfall change. Our assessment, at this stage, is that rainfall could increase in those regions of Australia where summer rainfall dominates, but that decreases may occur in the winter rainfall region. In addition there may be a general increase in the intensity of extreme rainfall events. However, because of the large uncertainty associated with this assessment, we are recommending that a range of scenarios are used in impact studies in Australia.

INTRODUCTION

The main tool atmospheric scientists have for producing more detailed estimates of how climate may change, taking into account the complexities of the climate system, is the general circulation model (GCM). At various research centers around the world, GCM experiments have been performed, where the model is run to equilibrium for present levels of greenhouse gases ($1 \times CO_2$), and then again for greenhouse gas levels equivalent to a doubling of carbon dioxide ($2 \times CO_2$). Generally models agree reasonably well in their simulated climate change on a global scale.[1] However, for smaller regions, model results often disagree considerably, and can show changes of differing sign in important parameters such as rainfall. This is partly due to deficiencies in GCMs, particularly relating to how they represent oceans, clouds and surface hydrology, but is also due to their coarse horizontal resolution and their consequent poor representation of local scale topographic features.

Clearly these large differences make the task of estimating regional climate change very difficult. One approach to this problem is to examine the whole range of available model results and consider as significant only those changes which are common to all or most models. For example, this consensus approach has been used by Kellogg and Zhao.[2] The problem with consensus is that equal weight is given to a model regardless of whether its regional control climate simulation is very poor, or very good, when compared to observations. It can be argued that inclusion of a model with very poor control performance may needlessly increase the range of $2 \times CO_2$ climates to be considered.

In approaching our task of estimating climate change for the Australian region, we have taken the view that a minimum, if not sufficient, condition for taking GCM results seriously is that the model does an acceptable job in simulating the present climate in the region of interest. For some seven GCM experiments for which we had data, we compared the control ($1 \times CO_2$) simulation of mean sea level (MSL) pressure, surface temperature and rainfall with present Australian region climate, and then excluded from consideration those experiments found to have a control climate that was unacceptably poor. It was our assessment[3] that the CSIRO 4-level model (CSIRO[4]) produced the most acceptable simulation of regional climate, and that of the UKMO low resolution simulation[4] produced a poorer but also acceptable simulation. The CSIRO[4] experiment was conducted by H. B. Gordon and B. G. Hunt of this Division (see Ref. 5). We considered that the other simulations we looked at were unacceptable for the Australian region (although these models may perform well in other regions). For the two models retained, we briefly examined their global control performance, and found this to be comparable to other low-resolution models.[3] An intercomparison of model results in the broader Australian-New Zealand region by Mullan and Renwick[6] also found the CSIRO[4] model simulation to be the best, with the one by the Geophysical Fluid Dynamics Laboratory model[7] the next best.

Given the result of this intercomparison, and given our greater access to the results of the local model, the results of the CSIRO[4] model have been the main basis of our assessments of climate change in the region, although where the results of the UKMO model have differed significantly from those of CSIRO,[4] this has been taken into account. We have also used information from other sources that we consider relevant, such as patterns of climate change in the regional palaeoclimatic and historical record. Also, for assessing possible changes in atmospheric circulation features of regional importance (e.g., tropical cyclones, east coast lows) which are not simulated, or poorly simulated, by the GCMs, we have been drawing on results from relevant studies with higher resolution-limited area models.

Results we have obtained using this approach are the basis for regional climate change assessment reports CSIRO is providing annually for various Australian State Governments.[8-10] In this paper, we outline some of these results and our assessments. This provides an Australian region example of how far we feel it is possible to go at present in estimating regional climate change.

To limit the discussion, we focus on rainfall, as this is a critical parameter in the region, and one for which the sign of the local response to global warming is not clear (unlike temperature, which models show will increase across the globe in the long term average). We also include an examination of changes in daily rainfall intensity simulated by the CSIRO[4] model for enhanced greenhouse conditions. These results are presented globally, as we feel they are directly relevant to climate change assessment in other regions as well as in Australia.

ASSESSMENT OF POSSIBLE CHANGES
IN AVERAGE RAINFALL OVER AUSTRALIA

The CSIRO[4] model control simulation gives the correct seasonal cycle of rainfall in the northeast, northwest and southwest of the continent, but gives too much summer rain in the southeast. Figure 1 shows the winter- and summer-rainfall dominated regions of Australia as "defined" by the model. In view of its poor horizontal resolution and crude representation of surface topography, we have averaged the rainfall changes due to an effective doubling of CO_2 over all grid points in each of the two regions defined in Fig. 1.

Percentage changes in rainfall between the $1 \times CO_2$ and $2 \times CO_2$ runs in the two areas for various averaging periods are shown in Table I. The results are based on ten years of control and ten years of $2 \times CO_2$ simulations. Increases are found in the summer rainfall region, and those in the summer half year are statistically significant at the 95% level. Decreases are apparent in the winter rainfall region, especially in winter, although these are only marginally significant statistically (90% confidence level).

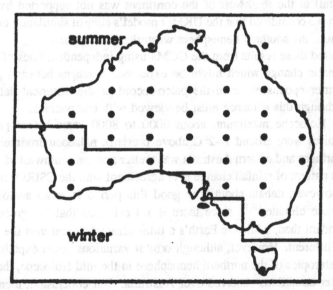

Fig. 1. Winter and summer rainfall regions over Australia as defined in the control simulation of the CSIRO (Ref. 4) model. A grid point is located in the summer rainfall region if more than 50% of total annual rainfall at that grid point occurs during November–April.

TABLE I. Percentage change in rainfall given by the CSIRO (Ref. 4) model ($2 \times CO_2$ simulation relative to the $1 \times CO_2$ simulation) for winter- and summer-rainfall regions and various averaging periods. The two regions are shown in Fig. 1.

	November–April	May–October	Year
Summer rainfall region	+7.6[a]	+4.3	+6.6[a]
Winter rainfall region	−2.1	−7.6[b]	−5.3[b]

[a]Significant at the 95% level.
[b]Significant at the 90% level.

These results are generally in agreement with changes in MSL pressure simulated by CSIRO.[4] The model indicates a significant reduction in MSL pressure over the Australian continent in both summer and winter and some weakening of the midlatitude westerlies over southern Australia. Observational studies of regional associations between rainfall and MSL pressure anomalies would suggest that these pressure changes would increase rainfall over northern, eastern and inland Australia, but reduce it in southern coastal areas, particularly in the southwest.[11]

The UKMO model agreed with the CSIRO[4] model in showing increases in rainfall over the north and east of the continent most markedly in the summer months. Declining winter rainfall in the southwest of the continent was not supported by the UKMO model, although we considered that the UKMO model's control simulation of the relevant circulation feature, the southern hemisphere westerlies, was poorer.

We have tested these results from the GCMs using independent lines of evidence for patterns of climatic change which might be expected. Changes between past globally cooler and warmer epochs either in the paleo-record or the historical data period are possibilities, although this evidence must be viewed with reservations.

During the Holocene maximum, about 6000 to 8000 years before present (when global temperatures were around $1-2°C$ above present), palaeoenvironmental evidence suggests that northern and eastern Australia was wetter and the southwest of the continent drier.[12–14] This pattern of rainfall change is in agreement with the CSIRO[4] model results.

There is, however, debate about how good this period is as an analogy of an enhanced greenhouse climate,[15,16] since there is no evidence that the greenhouse gases were more abundant then, while the Earth's orbital parameters and thus the solar forcing were certainly different. However, although orbital variations could explain wetter conditions in the subtropics of the northern hemisphere in the mid Holocene, they would not predict the wetter conditions observed over Australia. Our conclusion, therefore, is that the Holocene Maximum may offer a partial analogy to enhanced greenhouse conditions, since the direct forcing mechanism seems to have been mediated through some globally symmetric process, rendering the exact cause of the warming irrelevant to the resulting spatial pattern.

The last 100 years have also seen a small globally averaged warming.[1] This means that any systematic change in rainfall patterns over this time may be in part a response to global warming (whatever the reason for the warming). Analysis of rainfall trends over Australia since 1913[10,17] shows a marked decrease in rainfall over southwest Western Australia. The more summer dominated regions of northern and eastern Australia have shown a slight increase. Of course, observed trends in rainfall may be unrelated to the enhanced greenhouse effect. However, the agreement between the observed trend and the CSIRO[4] simulated change for southwest Western Australia, inevitably forces one to consider more seriously the prospect of greenhouse-induced drying for that region.

As a result of looking at these other lines of evidence some further assessment could be made of the CSIRO[4] rainfall results. The region of increasing rainfall in northern and eastern Australia in the CSIRO[4] model was statistically significant (95% level), supported by the UKMO results, palaeoclimatic evidence, and partly by observed rainfall trends. The region of decreasing rainfall, most marked in southwest Western Australia, was supported by palaeoclimatic evidence and observed rainfall trends, but was only marginally statistically significant (90% level) and was not supported by the UKMO results. Thus we feel more confident in the simulated rainfall increases than the decreases.

As there is large uncertainty associated with the lines of evidence used, this assessment of rainfall change in the Australian region is very tentative. We are recommending to researchers wishing to use this information in impact studies to test sensitivity of impacts to a range of scenarios. However, we believe this range should be centered on our current tentative estimates of change as described above. For example, for rainfall averaged over the summer rainfall region, we are currently recommending that increases of 0-20% be considered, with an expanded range for local changes within the region.

To give further insight into the range of possible change in rainfall on local scale, we have been using statistical techniques to relate changes in broadscale flow given by the GCM to local rainfall changes (a technique similar to that of Ref. 18). We have done this for regions in southwestern and southeastern Australia (which have climatically important local topography not represented in GCMs) using multiple linear regression to relate observed station rainfall and grid point MSL pressure data sets. The work has shown the high sensitivity of rainfall in these regions to relatively small changes in the broadscale flow. This has further demonstrated to us the need for climate change scenarios to contain a broad range of possible rainfall change at a local scale, until the broadscale patterns of atmospheric circulation change given by the GCMs become more reliable.

ASSESSMENT OF CHANGES IN RAINFALL INTENSITY

The possibility of changes in the frequency or severity of extreme rainfall events in response to an enhanced greenhouse effect has received little attention in the literature on climate change. However, one only needs to consider the present impact of floods (and droughts) in the Australian environment, or in many other regions, to realize that any change in the frequency of extreme rainfall events could be one of the most important regional impacts of global warming.

Theoretically, an increase in heavy rainfall events can be expected in response to global warming. Satellite observations and theory indicate that the amount of precipitable water in a vertical column over the oceans increases, non-linearly, with increasing sea surface temperature.[19] Thus the amount of water vapor available for precipitation is greater for higher base temperatures typical of the tropics or enhanced greenhouse conditions. Where rainfall over land is due to moisture transported from over the oceans, this relationship would be expected to also generally apply. Furthermore, in the case of convective rainfall systems, the increase in rainfall associated with the increase in precipitable water should be amplified. This is because the condensation of the available water vapour in the system releases latent heat which further increases the strength of the convective activity. These factors are reflected in the observed general tendency for a greater frequency of heavy rainfall and associated convective activity in the tropics compared to the midlatitudes, and in summer compared to winter.

A tendency for increased convective activity can be seen in the results obtained from enhanced greenhouse experiments with general circulation models (as was noted by Houghton et al., Ref. 1, p. 153). For example, Noda and Tokioka,[20] who examined changes in various rainfall types in an enhanced greenhouse experiment, found a general increase in the frequency of intense convective rainfall and a decrease in less intense non-convective rainfall. Hansen et al.[21] noted that increases in precipitation simulated by the Goddard Institute of Space Studies model under enhanced greenhouse conditions were due to an increase in moist penetrating convection. Wetherald and Manabe[22] found in their GCM that convection penetrated to higher levels in response to increased CO_2.

We have analyzed daily rainfall results, both in the Australian region and globally, from the CSIRO[4] model with regard to changes in rainfall type[23,24] and, like Noda and Tokioka,[20] we found general increases in convective rainfall, and decreases in non-convective rainfall.

We present here results which demonstrate the implications of these changes in rainfall type for daily rainfall intensity within the CSIRO[4] model. To best demonstrate their significance, these results are presented globally. In Fig. 2 we compare maps of the change in annual total rainfall, the annual number of rain days (days ≤ 0.2 mm), and average annual rain per rain day. Ten years of control and doubled CO_2 data are used, and changes significant at the 95% level (using a simple t-test) are shaded.

In the pattern of change in annual rainfall [Fig. 2(a)] note particularly the regions of both increase and decrease in midlatitudes with few regions of significant change. This complicated response is found also in the results of other models, but without agreement in the pattern detail (see Ref. 1). It is, as we noted earlier, a major reason why regional assessments of rainfall are very difficult at present.

The change in the number of rain days [Fig. 2(b)] presents a different picture. The number of rain days increases in high latitudes (where the rain is primarily stratiform)

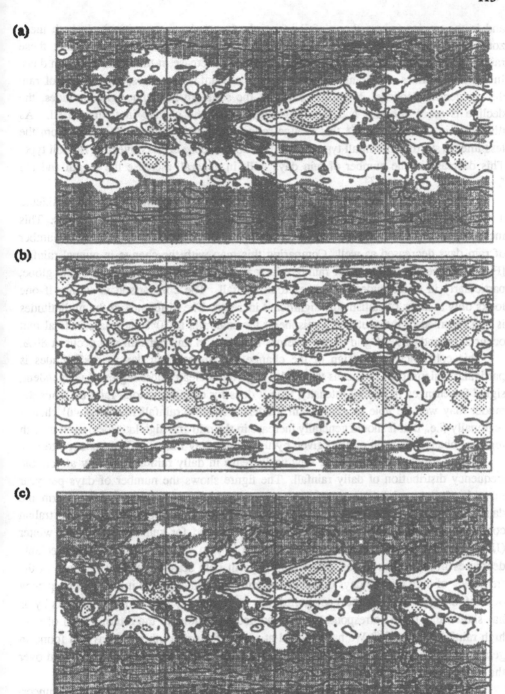

Fig. 2. Annual change ($2 \times CO_2$ - control) in (a) rainfall (contour interval 0.5 mm/day), (b) number of rain days (contour interval 20 days), and (c) rain per rain day (contour interval 0.5 mm/day). Areas of significant increase (95% level) are shown with heavy stippling, and of decrease with light stippling.

and in the tropics (where the rainfall is mainly due to deep convection). In both these zones the pattern of increase is similar to that in total rainfall, indicating that these rainfall increases are at least partly explained by an increase in the number of rain days. In the midlatitudes, including regions where total rainfall increases, the number of rain days generally declines. Most noticeably in the southern hemisphere midlatitudes, the decline is more coherent and more significant than the change in total rainfall. As discussed in Gordon et al.,[24] this change reflects a shift in midlatitudes away from the less intense stratiform rainfall type toward the more intense deep convective rainfall type. This decrease in the number of rain days will, in general, increase the length and the frequency of dry spells.[23]

Figure 2(c) shows changes in the annual rain per rain day, which can be considered a measure of daily rainfall intensity. Rain per rain day increases in most regions. This includes many midlatitude regions where total rainfall decreased, but where the number of rain days decreased as well. Comparing this map with the change in annual rainfall [Fig. 2(a)], we see that rain per rain day increases over a much larger area of the globe, particularly in midlatitudes, than does total rainfall. The comparison is similar if one looks only at areas of statistically significant increase. The difference in the midlatitudes is most marked in the southern hemisphere; for example, at $30°S$ increases in total rain occur at around 50% of grid points whereas increases in rain per rain day occur at 80%.

This coherence in the sign of the change in rain per rain day in midlatitudes is particularly significant. It means the model is giving a clearer, less regionally dependent, signal for changes in rainfall intensity, than it does for changes in total rainfall. Thus the model may well be able to give useful guidance on how rainfall intensity will change regionally (i.e., that it should generally increase in the mid and high latitudes) even though regional changes in total rainfall given by the model may not be reliable enough to use.

Figure 3 gives an example of how an increase in daily rainfall intensity affects the frequency distribution of daily rainfall. The figure shows the number of days per year in each of eight rainfall intervals (which are on a logarithmic scale to transform the data to near normal distributions) combined for all model grid points over the Australian continent, for control and doubled CO_2 conditions, and for summer (DJF) and winter (JJA). The change in the average rainfall rate is also shown. There are appreciable decreases in the frequency of low rainfall days in both seasons, and an increase in the frequency of high rainfall days especially in summer. Although the increase in frequency of rainfall events in the highest categories may not look very large, further analysis by us has shown that these correspond to quite marked reductions in the return period of very high rainfall events.[24,25] We have looked at results for some other regions and found in general the same pattern of response.[24] For example, Fig. 4 shows results averaged over the midwestern United States.

These potentially very important results need to be considered in light of the uncertainty associated with them. Firstly, they are the results of just one GCM, the only GCM for which we had daily data. The result needs to be tested against the corresponding results of other models, particularly ones that use different convective parameterization

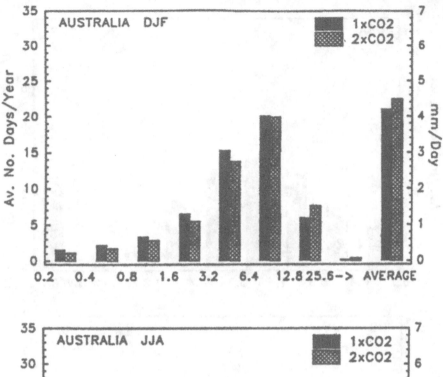

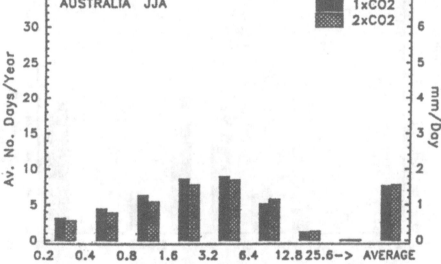

Fig. 3. Histograms of daily rainfall frequency in the control (solid) and doubled
CO$_2$ (stippled) runs of the CSIRO (Ref. 4) model for June–August (JJA)
and December to February (DJF) averaged over all grid points in the
Australian region (112.5–152°E, 11–43°S). Only rain days (days of at
least 0.2 mm precipitation) are graphed. The column on the extreme
right of each panel represents average rainfall rate for the region.

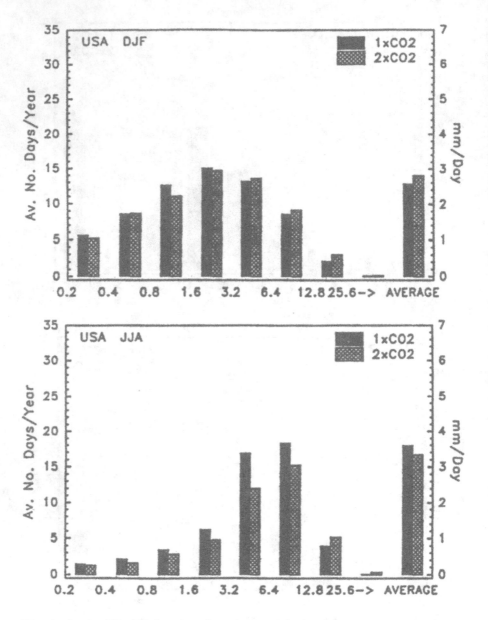

Fig. 4. As for Fig. 3, but for the midwestern United States (33.6–40.0°N, 90.0–101.25°W).

schemes (CSIRO[4] uses a modified Arakawa moist-adjustment scheme which generates a mass flux[5]). We cited some results earlier on changes in rainfall type in other models that are consistent with our results, but global analyses of daily rainfall frequencies are presently not available in the literature for other models. Furthermore, we have not widely validated the CSIRO[4] model's performance at simulating the observed characteristics of daily rainfall. Apart from the difficulties of assembling the appropriate observed global data sets, validation is greatly complicated by the mismatch of scales of observed and simulated rainfall systems. Convective storms in the model can have an areal extent no smaller than a model grid square (around 500 km square), whereas in the real atmosphere their extent is typically orders of magnitude less than this. This means that one can not expect model output to reproduce rainfall frequency distributions (and particularly the most intense rainfall totals) seen at individual stations in the real world. Nevertheless, for some selected Australian grid points, we have compared the CSIRO[4] grid point rainfall frequency distribution with that produced by averaging observed data over a number of stations within the grid square, and found the model performance good in some cases, but poor in others. This scaling problem means that changes in model generated daily rainfall frequency distributions, such as those shown in Fig. 3 can only, at best, be viewed as providing qualitative and not quantitative guidance on how observed daily rainfall frequency distributions may change with global warming.

Work by Climate Impact Group in conjunction with the Australian Bureau of Meteorology using a high resolution limited area model is also providing evidence of possible marked increases in rainfall intensity in association with global warming. McInnes et al.[26] used the Bureau of Meteorology Research Centre (BMRC) limited area model to simulate four recent examples of cut-off lows observed on the eastern Australian coast. These systems, known as 'east coast lows,' are a major source of flood rainfalls and gale force winds on the southeast coast of Australia. The experiment was run with sea-surface temperature (SST) as it was observed, and then again with SST increased by around 3°C. With the enhanced SST, peak 24-hour rainfall values were found to increase by 45–80%. In addition, the central pressure of these systems deepened, and the areal extent of gale force winds increased by 50–70%. Similar experiments are underway on the sensitivity of tropical cyclones in the Australian region to enhanced greenhouse conditions, using the CSIRO limited area model. It must be noted, however, that such experiments test the impact of only one element of the changed climate; changes in other aspects of the broadscale atmospheric circulation may lead to different results.

Despite the caveats associated with the GCM and limited area modelling results, we feel that these results, combined with the physical arguments discussed earlier, point clearly toward an increase in the frequency of heavy rainfall events. Assessing these changes in quantitative terms will require significant improvements in GCM simulations. We believe a particularly useful way forward would be to nest a limited area model within a GCM run in an enhanced greenhouse experiment, so that high resolution daily rainfall statistics may be collected over multi-year simulation runs for a region of interest. This approach is being developed at CSIRO. In the meantime we believe it is appropriate to include, in the regional climate change assessment we are providing for the Australian region, an indication that extreme rainfall events may increase in frequency.

CONCLUDING COMMENTS

This paper has given some outline of the approach that the Climate Impact Group in CSIRO is pursuing in fulfilling its ongoing aim of providing the best available climate change advice for the Australian region. It does not cover, by any means, the full scope of the group's activities or the full scope of climate change advice which it is presently providing to Australian State Governments. We focused here on rainfall because of its critical regional importance, and the difficulty of assessing how it may change. Our assessment outlined in this paper gives an indication of how far we feel we can go at present in assessing rainfall change in the Australian region. Use of a similar approach for other regions of the globe may lead to a more certain, or less certain, conclusion; it would depend on the climate of the region and the nature of the evidence available. However, we do feel that the evidence for possible increases in the frequency of heavy rainfall events is of general applicability, and potentially very important in impact studies. So far it has received little attention in regional climate change assessment (e.g., Ref. 1).

REFERENCES

1. J. T. Houghton, G. J. Jenkins, and J. J. Ephraums, *Climate Change—The IPCC scientific assessment* (Cambridge University Press, 1969), 365 pp.
2. W. W. Kellogg and Z.-c. Zhao, *J. Climate* **1**, 348 (1988).
3. P. H. Whetton and A. B. Pittock, CSIRO Division of Atmospheric Research Technical Paper No. 21, 73 pp. (1991).
4. C. A. Wilson and J. F. B. Mitchell, *J. Geophys. Res.* **92**, 13315 (1987).
5. H. B. Gordon and B. G. Hunt, *Int. J. Climatol.* **4**, 347 (1991).
6. A. B. Mullan and J. A. Renwick, Climate change in the New Zealand region inferred from general circulation models (New Zealand Meteorological Service Report prepared for the Ministry for the Environment, 1991). 142 pp.
7. S. Manabe and R. T. Wetherald, *J. Atmos. Sci.* **44**, 1211 (1987).
8. A. B. Pittock and R. J. Allan, (eds.), The greenhouse effect: regional implications for Western Australia (EPA, Western Australian Government, Perth, 1990), 65 pp.
9. A. B. Pittock and J. L. Evans (eds.), Regional impact of the enhanced greenhouse effect on the Northern Territory (Conservation Commission of the Northern Territory, Darwin, 1990), 68 pp.
10. A. B. Pittock and P. H. Whetton, (eds.), Regional impact of the enhanced greenhouse effect on Victoria (Office of the Environment, Victorian Government, Melbourne, 1990), 70 pp.
11. A. B. Pittock, *Search* **6**, 498 (1975).
12. R. J. Wasson, *Geol. Soc. Aust. Symp. Proc.* **1**, 83 (1990).
13. G. Singh and J. Luly, *Palaeogeogr., Palaeoclimatol., Palaeoecol.* **84**, 75 (1991).
14. R. J. Wasson and T. H. Donnelly, CSIRO Division of Water Resources, Technical Paper 91/3, 48 pp., (1991).

15. J. F. B. Mitchell, *J. Climate* **3**, 1177 (1990).
16. T. J. Crowley, *J. Climate* **3**, 1282 (1990).
17. A. B. Pittock, *Climatic Change* **5**, 320 (1983).
18. T. M. L. Wigley, P. D. Jones, K. R. Briffa and G. Smith, *J. Geophys. Res.* **95**, 1943 (1990).
19. G. L. Stephens, *J. Climate* **3**, 634 (1990).
20. A. Noda and T. Tokioka, *J. Meteor. Soc. Japan* **67**, 1057 (1989).
21. J. Hansen, D. Rind, A. DelGenio, A. Lacis, S. Lebedeff, M. Prather, R. Ruedy, Coping with Climate Change, Proc. Second N. Amer. Conf. on Preparing for Clim. Change, Dec. 6–8, 1988 (Climate Institute, Wash. DC, 1989), p. 130.
22. R. T. Wetherald and S. Manabe, *J. Atmos. Sci.* **45**, 1397 (1988).
23. P. H. Whetton, A. M. Fowler, M. R. Haylock and A. B. Pittock, Submitted to *Austg Met. Mag.* (1991).
24. H. B. Gordon, P. H. Whetton, A. B. Pittock, A. M. Fowler and M. R. Haylock, to be submitted to *Climate Dynamics*.
25. A. B. Pittock, A. M. Fowler and P. H. Whetton, International Hydrology and Water Resources Symposium 1991: Challenges for Sustainable Development, Preprints of Papers, Volume 1, Perth 2–4 October (Institute of Engineers, Australia, 1991).
26. K. L McInnes, L. M. Leslie and J. L. McBride, in preparation, (CSIRO Div. of Atmos. Res.).

Regional-Scale Simulations of the Western U.S. Climate

J. E. Bossert, C.-Y. J. Kao, J. Winterkamp
Los Alamos National Laboratory
Los Alamos, NM 87545

J. O. Roads, S. C. Chen, K. Ueyoshi
Scripps Institution of Oceanography
University of California at San Diego
La Jolla, CA 92093

INTRODUCTION

Over the past two decades the meteorological community has witnessed the evolution of general circulation models (GCMs) from studies attempting to simulate realistic large-scale dynamical regimes and energy transports to present investigations examining future climate-change scenarios. This evolution is certain to continue over the next decade, as demands are placed upon the GCM community to generate realistic scenarios for future climate change on regional as well as global scales. However, fine-scale regional climate and climate variability cannot be properly simulated with present coarse resolution GCMs. Even observations of regional climatology are poorly represented with the existing widely scattered rawinsonde network and relatively coarse-scale global analyses. For example, the National Meteorological Center's (NMCs) global analysis is tabulated on a 2.5° by 2.5° grid after conversion from a spherical harmonics data set truncated at wavenumber 30. In the mountainous western U.S., where climate scales are strongly coupled to the underlying topography, this 2.5° by 2.5° grid is clearly inadequate.

One approach to dealing with this regional climatology question has been to use existing coarse-scale GCM output to provide the large-scale background climate for specific regions (Rind 1988; Gutowski et al. 1991). Another alternative is to locate a regional model with a finer resolution mesh over the region of interest within the GCM. For example, a limited area model could be forced on the boundaries by the large-scale GCM output over an extended period. The feasibility of this approach has recently been demonstrated by Dickinson et al. (1989), Anthes et al. (1989), Giorgi (1989), and Giorgi and Bates (1989), using the Penn State/NCAR mesoscale model. The most encouraging aspect of these studies is that we now know that mesoscale models can provide a sufficiently detailed regional climatology.

From these pioneering studies, we were inspired to begin to develop regional climatologies with the Colorado State University Regional Atmospheric Modeling System (CSU-RAMS). Our major goal is to develop a better understanding of the hydrologic cycle in the mountainous arid west. An advantage of using the RAMS code is that we can generate detailed descriptions of precipitation processes, which will hopefully translate into realistic surface yields of both rain and snow. In the ensuing sections, we first describe the model and its microphysics parameterizations, then continue with our methodology for incorporating large-scale data into the model grid. Preliminary results demonstrating the mesoscale variation of precipitation over the mountainous western U.S. are then presented.

THE MESOSCALE MODEL

The RAMS mesoscale model is a highly flexible modeling system, capable of simulating a wide variety of mesoscale phenomena. The basic model structure is described in Tripoli and Cotton (1982). More recent model developments are described in Tremback et al. (1986) and Cotton et al. (1988). The model framework for the present study incorporates a three-dimensional, terrain-following non-hydrostatic version of the code. The simulation includes topography derived from a 5-minute global data set with a silhouette averaging scheme that preserves realistic topography heights. This height data is then interpolated to the model grid which has 0.5° horizontal resolution at the tangent point of the polar stereographic grid at 40.0°N and 112.5°W. In these experiments, we cover the geographical domain from 127.5°W to 97.5°W and 27.5°N to 52.5°N. In the vertical we use 21 levels, corresponding to a resolution of 300 meters near the surface and 1000 meters at the top of the model. Non-hydrostatic equations are used so that nested grids capable of resolving cloud-scale phenomena can eventually be implemented.

At the surface, temperature and moisture fluxes are determined from the surface energy balance, which includes both short- and longwave fluxes (Chen and Cotton 1983), latent and sensible fluxes, and sub-surface heat conduction from a soil temperature model (Tremback and Kessler 1985). A modified Kuo-type cumulus parameterization is incorporated in the model, although in the present January simulation cumulus precipitation is much less important than stratiform precipitation, which is treated with a bulk microphysical parameterization. The microphysics parameterization, outlined in detail in Flatau et al. (1989), describes the physical processes leading to the formation and growth of precipitation particles within a cloud. The cloud particles can be liquid or ice, or some combination, and may have a regular or irregular shape. The scheme categorizes these particles as cloud droplets, rain drops, ice crystals, snow crystals, aggregates of ice crystals, and graupel or hail. Each species can grow independently from vapor and self-collection, or interact with other species through collision and coalescence processes. In the configuration used for this study, the mixing ratio of each species is predicted and the total concentration is diagnosed, using a specified size distribution. It is intended that the rain and snow fields will eventually be coupled to a detailed surface hydrology model.

124

DEVELOPMENT OF A REGIONAL CLIMATOLOGY

For these preliminary experiments, we are concentrating on the period 1-31 Jan 1988. We use only NMCs 2.5° by 2.5° twice daily global analyses to drive the regional model. The analyses consist of winds, temperatures, and geopotential heights at 11 standard pressure levels and relative humidity for 6 levels at and below 30 kPa. Virtual temperature and surface pressure are also available. From this global analysis, a region of the data is acquired which is large enough to totally encompass the eventual model domain by 12.5° on all sides. An example of the NMC data obtained for our simulations is given in Fig. 1 for 0000Z 5 Jan 1988. The wind vectors [Fig. 1(a)] give some idea of the relative coarseness of the data set, which is still of much higher resolution than most GCMs. The 50-kPa height field [Fig. 1(b)] reveals the presence of two deep lows at this time, one centered in the North Pacific and the other just south of Hudson Bay. Between these vortices lies a narrow high-amplitude ridge over western North America.

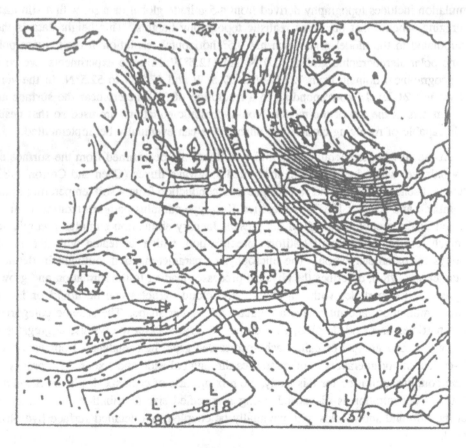

Fig. 1(a).

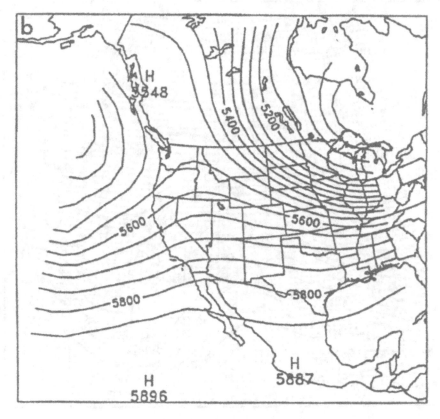

Fig. 1(b).

Fig. 1. NMC 2.5° data at 50 kPa over a portion of northwestern hemisphere at 0000Z 5 Jan 1988 for (a) wind speed every 3.0 ms^{-1}, and (b) geopotential heights every 50 m.

The next stage of the data assimilation process involves the preparation of a gridded isentropic vertical coordinate data set, obtained by vertically interpolating the NMC vertical pressure coordinate data set to specified isentropic levels, and then horizontally interpolating this data onto a higher resolution 0.6° grid. The use of an isentropic vertical coordinate has some advantages, namely, that synoptic flow is adiabatic to a first-order approximation, and hence, this analysis should give a realistic estimate of the flow between the NMC data points. The disadvantages of isentropic coordinates, such as having isentropes which intersect the ground, is not a factor in this analysis. The isentropic analysis is performed over a smaller area than the previous pressure stage, to avoid potential boundary problems, but is still 5° larger on all sides than the model grid. In the vertical, 30 isentropes are specified to increase the data resolution, particularly near the surface. A plan view of the isentropic stage pressure data for the 320°K isentrope is

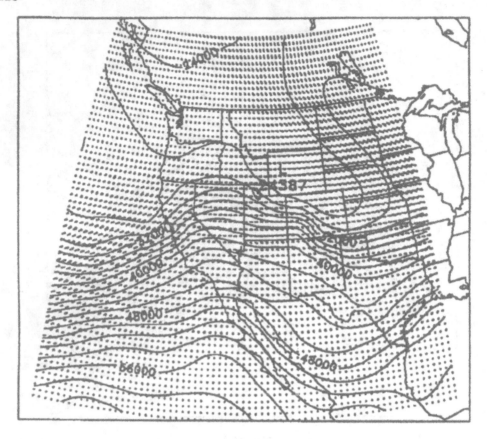

Fig. 2. Wind vectors and pressure (Pa) on the 320 K surface at 0000Z 5 Jan 1988.

given in Fig. 2. Steeply sloping isentropes near 40°N signify the location of the polar front. These sloping isentropes, as well as the mixing ratio, are depicted in a north-south cross-section in Fig. 3. This figure shows that very dry conditions exist near the surface north of 34°N, which is over the Great Basin, while moister conditions are present south of that latitude, and in a broad tongue between 70 and 90 kPa.

With the isentropic data set, we can now initialize the RAMS model. This is done by interpolating the isentropic data onto the model grid (0.5° resolution) to obtain a full set of prognostic fields for model integration. The model domain, along with the topography used, is shown in Fig. 4. The topography is smoothed to contain only wavelengths greater than or equal to 4 times the horizontal resolution, so as not to introduce unresolvable modes into the simulation. Lateral boundary conditions of the domain are updated each time-step by linearly interpolating between successive 12-hour large-scale analyses.

We compare two diverse methods for generating the regional climatology applicable to January 1988. In the first method the model is initialized from the NMC analysis and

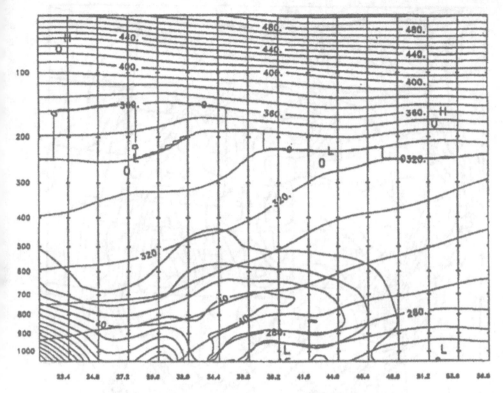

Fig. 3. South-to-north cross section at 117.0°W of potential temperature θ (in K) and mixing ratio r_v (in g/kg $\times$ 10^1) for 0000Z 5 Jan 1988.

then integrated for 12 hours. These simulations are re-initialized every 12 hours, but this could be extended to longer periods if further analysis indicates that this technique does not allow the model to fully develop a realistic mesoscale circulation within this time frame. We will present comparisons for the average of sixty-two 12-hour simulations at 3, 6, 9, 12 hours to address this question. The second method initializes only the first 12-hour simulation and then updates the lateral boundary conditions toward each successive 12-hour analysis, throughout the rest of the month. This methodology is similar to that adopted by Giorgi and Bates (1989). We are currently investigating methods of slightly nudging interior model grid points as well, to keep the mesoscale simulation more closely adjusted to the large-scale conditions for the month-long run.

Preliminary analyses have shown that after an initial spin-up time, the re-initialized climatology begins to resemble the month-long climatology run in many aspects; the advantage of the first method is that many individual 12-hour runs can be done simultaneously (i.e., in parallel), while the second method requires sequential job submission, greatly increasing the real-time cost of the experiment.

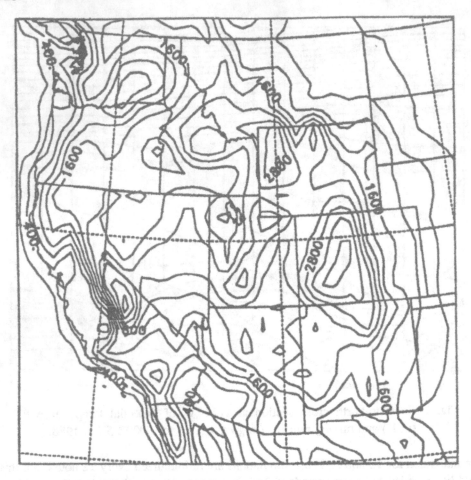

Fig. 4. Model domain over the western U.S. and topography heights in meters. Contour interval 300 m.

PRELIMINARY RESULTS

In this section, we show several fields from two 12-hour simulations to demonstrate how the model is capable of capturing essential mesoscale weather features over the western U.S., the ensemble of which constitutes the regional climate. In Fig. 5 the accumulated precipitation fields over the model domain for the 12 hours ending at 0000Z 5 Jan 1988 are shown. The large-scale flow at this time was shown in Figs. 1–3. The figure shows that up to 5 mm of rain has accumulated over parts of the Sacramento Valley in California [Fig. 5(a)], with lesser amounts over the lower elevations of the Great Basin. Snowfall is widespread over a large portion of the intermountain west [Fig. 5(b)], with a maximum of 5.7 mm (water equivalent) over northeastern Nevada. Note the well-defined

Fig. 5(a).

rain/snow line over the northern Sierra Nevada in California at about 1600 msl. Snow amounts decrease significantly in the lee of the Sierra Nevada and other major ranges, due to mountain wave effects which produce strong subsidence in these areas.

Over the following 24 hours, the system responsible for the precipitation in Fig. 5 progressed onshore and southward. This is reflected in the accumulated rain and snow from the 12-hour simulation ending at 0000Z 6 Jan 1988 (Fig. 6). Rainfall during this period [Fig. 6(a)] is concentrated in the Central Valley and Sierra Nevada foothills in California, with lesser amounts over northern Arizona. Accumulated snowfall totals of up to 9 mm occur over the central Sierra Nevada [Fig. 6(b)], with a belt of snow stretching eastward into the Colorado Rockies. The areal coverage of the precipitation in both of these simulations compares favorably with published observations. However, these are only preliminary comparisons, and more rigorous comparisons are necessary and forthcoming.

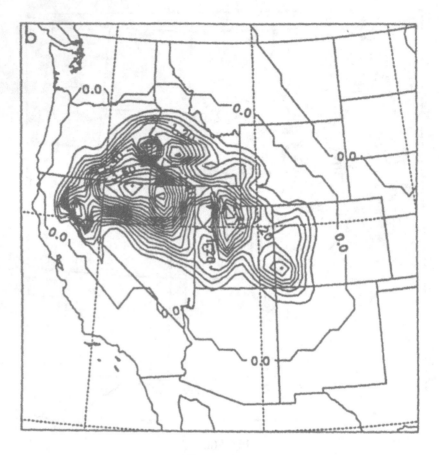

Fig. 5(b).

Fig. 5. Accumulated (a) rain, and (b) snow in mm at 0000Z 5 Jan 1988. Contour interval 0.3 mm.

SUMMARY

Preliminary results from our regional climatology study for January 1988 are encouraging. The dependence of precipitation on topography over the complex terrain of the model domain is clearly demonstrated, as is the ability of the microphysics parameterization to produce realistic liquid and ice phase precipitation fields and accumulations. However, further judgments on the capabilities of the model await additional analysis for the entire month-long January climatology. Averages of the precipitation and other relevant fields will be presented at the conference. Ultimately, we plan to extend our study to include the grid nesting capabilities of the RAMS model and generate even higher

Fig. 6(a).

resolution climatologies for various western U.S. river-basins, while still maintaining our 0.5° grid for assimilating large-scale data. Work is currently in progress to assimilate data from the Los Alamos GCM (Kao et al. 1990) to provide the boundary forcing for the RAMS model. We are also planning a regional climatology study of summertime wind and precipitation patterns with RAMS. This season appears to be intrinsically more difficult to simulate, due to the very localized nature of convective storm development.

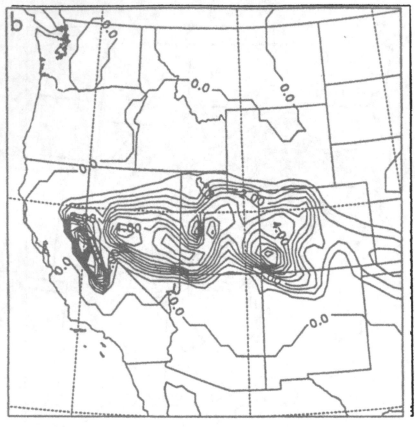

Fig. 6(b).

Fig. 6(b). As in Fig. 4, but for 0000Z 6 Jan 1988, and contour interval of 0.5 mm in (b).

REFERENCES

Anthes, R. A., Y.-H. Kuo, E.-U. Hsie, S. Low-Nam, and T. W. Bettge, 1989: Estimation of skill and uncertainty in regional numerical models. *Quart. J. Roy. Meteor. Soc.* **115**, 763–806.

Chen C., and W. R. Cotton, 1983: A one-dimensional simulation of the stratocumulus-capped mixed layer. *Bound.-Layer Meteor.* **25**, 289–321.

Cotton, W. R., C. J. Tremback, and R. L. Walko, 1988: CSU RAMS—A cloud model goes regional. *Proc. NCAR Workshop on Limited-Area Modeling Intercomparison*, Nov. 15–18, NCAR, Boulder, CO, 202–211.

Dickinson, R. E., R. M. Errico, F. Giorgi, and G. T. Bates, 1989: A regional climate model for the western U.S. *Clim. Change* **15**, 383–422.

Flatau, P. J., G. J. Tripoli, J. Verlinde and W. R. Cotton, 1989: The CSU-RAMS Cloud Microphysics Module: General theory and code documentation. Atmospheric Science Paper No. 451, Colorado State University, Dept. of Atmospheric Science, Fort Collins, CO 80523, 88 pp.

Giorgi, F., 1989: On the simulation of regional climate using a limited area model nested in a general circulation model. *J. Climate* **3**, 941–963.

Giorgi, F. and G. T. Bates, 1989: The climatological skill of a regional model over complex terrain. *Mon. Wea. Rev.* **117**, 2325–2347.

Gutowski, W. J., D. S. Gutzler, and W.-C. Wang, 1991: Surface energy balances of three general circulation models: Implications for simulating regional climate change. *J. Climate* **4**, 121–134.

Kao, C.-Y. J., G. A. Glatzmaier, and R. C. Malone, 1990: Global three-dimensional simulations of ozone depletion under postwar conditions. *J. Geophys. Res.* **95**, 22495–22512.

Rind, D., 1988: Dependence of warm and cold climate depiction on climate model resolution. *J. Climate* **1**, 965–997.

Tremback, C. J. and R. Kessler, 1985: A surface temperature and moisture parameterization for use in mesoscale numerical models. *Preprints, Seventh Conf. on Numerical Weather Prediction*, Montreal, Amer. Meteor. Soc.

Tremback, C. J., G. J. Tripoli, R. Arritt, W. R. Cotton, and R. A. Pielke, 1986: The Regional Atmospheric Modelling System. *Proc. Internat. Conf. Development and Application of Computer Techniques to Environmental Studies,* November, Los Angeles, California, P. Zannetti, Ed., Computational Mechanics Publications, Boston, 601–607.

Tripoli, G. J., and W. R. Cotton, 1982: The Colorado State University three-dimensional cloud/mesoscale model-1982. Part I: General theoretical framework and sensitivity experiments. *J. Rech. Atmos.* **16**, 185–220.

Simulations of Greenhouse Trace Gases
Using the Los Alamos Chemical Tracer Model

Chih-Yue Jim Kao and Eugene Mroz
Los Alamos National Laboratory
Los Alamos, NM 87544

Xuexi Tie
Scripps Institution of Oceanography, University of California
San Diego, CA

ABSTRACT

Through three-dimensional global model studies on atmospheric composition and transport, we are improving our quantitative understanding of the origins and behavior of trace gases that affect Earth's radiative energy balance and climate. We will focus, in this paper, on the simulations of three individual trace gases including CFC-11, methyl chloroform, and methane. We first used our chemical tracer model to study the global distribution and trend of chemically inert CFC-11 observed by the Atmospheric Lifetime Experiment. The results show that the model has the ability to reproduce the time-series of the observations. The purpose of this CFC-11 simulation was to test the transport of the model. We then used the model to introduce methyl chloroform into the atmosphere according to the known emission patterns and iteratively varied OH fields so that the observed concentrations of methyl chloroform from the observations could be simulated well. The rationale behind this approach is that the reaction with OH is the dominant sink for methyl chloroform and the transport of the model has been tested in the previous CFC-11 study. Finally, using the inferred OH distribution from the methyl chloroform study and the best estimated methane source strength and distribution, we conducted a steady-state simulation to reproduce the current methane distribution. The general agreement between the modeled and observed methane surface concentrations has laid a foundation for the simulation of the transient increase of methane.

INTRODUCTION

The prospect of climate change arises from observations that the chemical composition of the atmosphere is being altered by emissions resulting from human activities. Approximately thirty trace gases in the atmosphere are possibly contributing to climate through their greenhouse warming potential or stratospheric ozone depletion (Ramanathan et al., 1985). Some of these gases are of strictly human origin while others have both human and natural sources. The distribution of these gases in the earth atmosphere is determined by complex processes associated with atmospheric chemistry and dynamics. This logically invites the use of three-dimensional (3-D) general circulation models (GCMs) as a major tool to study these trace gases. In fact, it is our ultimate goal that atmospheric dynamics and chemistry are combined together to formulate an interactive system for the long-term predictive purposes of climate change. However, prior to entering that stage, there are a few formidable tasks associated with atmospheric chemistry alone. For example, trace gases such as CH_4, O_3, NO_x, CO, H_2O, and other hydrocarbons are all chemically interactive with each other and are all involved with sources and sinks and stratosphere/troposphere exchange processes. More importantly, they all react with the concentrations of OH, which is considered, in some respects, the most reactive species in the atmospheric chemistry. Thus, it is generally recognized that the near-term goal is to develop a multiple-family chemical tracer model to simulate simultaneously the above mentioned species in a chemically interactive fashion. This naturally leads us to an immediate task; namely, a series of tests of the model on the behavior of each individual trace gas or a subset of those tracers. The tool for these near-term tasks is the 3-D chemical tracer models (CTMs) (see a review in Prather et al., 1987). Here, we define the CTM as a model which uses winds and temperature simulated from a GCM to drive the species continuity equations with chemistry sink and source terms.

In this paper, we first examine the role of atmospheric circulations in the dispersal of industrial CFC-11 to provide a baseline test for our 3-D global CTM. CFC-11 is an attractive candidate for this test, not only because it is a greenhouse gas, but also because its emission rates and locations can be estimated from industrial sales figures. Furthermore, CFC-11 is essentially inert in the troposphere and has a well known photolysis rate in the stratosphere. Measurements of the trends in CFC-11 surface concentrations have been carried out for several years at five selected locations by the Atmospheric Lifetime Experiment (ALE) network.

The second aspect of this paper is the simulation of methyl chloroform (CH_3CCl_3). Methyl chloroform, a minor greenhouse gas, is an industrial solvent whose geographical distribution of emission sources is essentially identical to CFC-11. The ALE network also measures the surface concentrations of methyl chloroform. However, unlike CFC-11, methyl chloroform is removed from the troposphere by reaction with OH radicals. Therefore, we are able to use our modeled global distribution, seasonal cycle and trend of methyl chloroform to infer the global distribution, seasonal cycle and trend of atmospheric OH field.

The third aspect of this paper is the simulation of atmosphric methane with the emphasis on its geographical distribution. Like methyl chloroform, methane is mainly destroyed by reaction with OH. Thus, the OH field inferred from the previous methyl chloroform simulation is used as in input in the methane simulation. Individual methane source strength and its distribution are estimated from various studies (e.g., Cicerone and Oremland, 1988; Aselmann and Crutzen, 1989). The model results are compared with the measurements conducted from 1983 to 1985 (Steele et al., 1987; Boden et al., 1990).

Section 2 of this paper is devoted to a description of the Los Alamos GCM and CTM. Section 3 discusses the model results and their comparison with the observations. Finally, a summary is given in Sec. 4.

2. MODEL DESCRIPTION

2.1 The Basic GCM

The Los Alamos National Laboratory started its climate modeling program with the "Nuclear Winter" project (Malone et al., 1986). The basic GCM with which we began that work is the Community Climate Model (CCM) (Version 0-A) of the National Center for Atmospheric Research (NCAR). The model variables are represented by a sum of spherical harmonics with a rhomboidal function at wave number 15. This gives a latitudinal resolution of 4.5° and longitudinal resolution of 7.5°. The vertical computation domain is from the surface up to about 33 km, using a σ-coordinate. The climate simulation characteristics of the basic model have been documented by Pitcher et al. (1983), Ramanathan et al. (1983), and Malone et al. (1984). During the course of nuclear winter simulations, a number of modifications to the NCAR CCM0-A were made (Malone et al., 1986; Kao et al., 1990), which include: (1) annual and diurnal cycles of insolation, (2) subsurface storage and diffusion of heat and moisture, (3) stability-dependent vertical diffusion in the atmosphere using the scheme of Louis (1979), (4) a finer vertical resolution with 12 layers in the troposphere and 8 layers in the stratosphere, (5) stratospheric chemistry with a seven-family scheme (Kao et al., 1990), and (6) a more realistic distribution of sea surface temperature and sea ice with their annual cycle. With these modifications, more recent climate research includes: (1) the incorporation of the Arakawa-Schubert cumulus parameterization (Kao, 1990), (2) the simulation and diagnosis of large-scale atmospheric moisture and surface hydrology over North America (Roads et al., 1991), (3) the simulation of anomalous light extinction from large area forest fires (Porch and Kao, 1990), and (4) the simulation of Antarctic circulation with numerical tracer experiments (Mroz et al., 1989). The basic model climate is obtained from a ten-year simulation with a fixed but seasonally-varying sea surface temperature.

2.2 Transport in CTM

With the model described above as a parent GCM, its simulated 3-D winds, temperature, and sub-grid diffusion on the Fourier-Legendre grid system (i.e., 48 equally spaced longitude points and 40 Gaussian quadrature latitude points) are input to the CTM to solve the continuity equation for CFC-11. We selected a time-split method that computes the advection of the tracer in the longitudinal, latitudinal, and vertical directions sequentially, alternating the order every time step. The horizontal transport is accomplished via a finite element ("Chapeau function") method (Pepper and Baker, 1979). A time-centered implicit integration scheme results in a tridiagonal system of equations. This scheme is second-order accurate in time and fourth-order accurate in space for a uniformly spaced mesh, as is the case for the longitudinal transport. The method has low phase error and is relatively nondispersive. An advantage of this scheme is that once the transport of a single species has been determined for a model time step, other species can be transported at small additional cost. This was a strong motivation for choosing this scheme, since we plan to treat multiple species in the future.

For the transport in the vertical direction, we need an algorithm that inherently produces nonnegative concentrations and accurately defines a tracer front in the vertical direction since the sources of most trace gases are at the surface. The method we selected is an improved flux-corrected transport (FCT) algorithm formulated by Zalesak (1979). Its five-point interpolation of species column mass produces fourth-order accurate fluxes on the unequally spaced vertical levels used in the CTM. We added a correction to the flux limiter that prevents very small negative concentrations (Chock and Dunker, 1983) that otherwise would have been generated by numerical truncation. It should be noted that the advantages of the FCT algorithm are better realized if a higher resolution is selected, as is the case in the current CTM.

3. MODEL RESULTS

3.1 CFC-11

The only identified source for CFC-11 is anthropogenic. Figure 1 shows an example of surface source strength for 1980. The annual CFC-11 emission rates estimated by

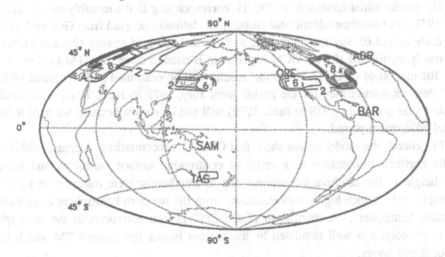

Fig. 1. CFC-11 Source strength for 1980 with units of gram/year/km^2. The contour interval is 2.0×10^7 gram/year/km^2. Also shown are the five ALE sites.

Chemical Manufacturers Association (CMA) is shown in Table I. The daily emissions in a certain year are determined using a smoothed function of the annual emissions during the previous year and the following year, as designed by Golombek and Prinn (1986). The mass of the daily emissions is then added into the layer from the surface to about 1.5 km, assuming the mass mixing ratio is constant with height in the layer. The chemical sink of CFC-11 is mainly through photolysis in the stratosphere. The potential sink due to ocean is neglected in this study (Prather et al., 1987).

TABLE I. CFC-11 Annual Release Rate (Kt).	
Year	Release
1978	294.6
1979	276.1
1980	264.8
1981	264.3
1982	258.8
1983	274.8
1984	296.1
1985	309.3

The model initial condition of CFC-11, corresponding to the monthly-mean values of July 1978, is a two-dimensional field (altitude vs. latitude) adapted from Golombk (1982). The daily output of winds, temperature, and vertical sub-grid mixing simulated from the previously mentioned 10-year GCM experiment is used to drive the CTM to simulate the 3-D distribution of CFC-11. The model results will be examined and compared with the ALE data-set documented for the period from July, 1979 to June, 1985. The result of the first year (i.e., July, 1978 to June, 1979) will not be shown because we treat it as the model-adjustment period.

In general, the model results show that CFC-11 concentrations decrease with height in the northern troposphere as a result of continuous surface emissions and increase with height in the southern troposphere due to interhemispheric transport in the upper troposphere, by which higher concentrations from the northern hemisphere enter into the southern hemisphere. A sharp drop of the CFC-11 concentrations in the stratosphere due to photolysis is well simulated by the current higher resolution CTM which has 8 stratospheric layers.

Figure 2 shows the comparisons between the simulated (solid lines) and measured (dotted lines) surface concentrations for the five ALE sites from July 1979 to June 1985. At Adrigole, Ireland (ADR), it is found [Fig. 2(a)] that severe fluctuations appear in the model results throughout the entire six years (actually, they were also found at the

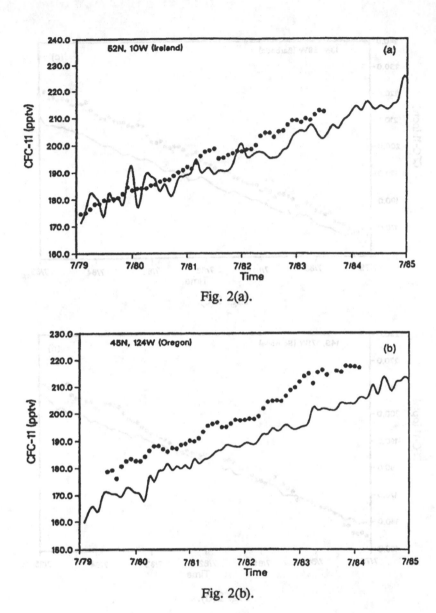

Fig. 2(a).

Fig. 2(b).

Oregon site, as will be shown later). This is mainly ascribed to the circulation of the parent GCM which tends to overestimate the high-frequency variations in mid-latitudes. At Cape Mear, Oregon (ORE), it is seen [Fig. 2(b)] that the model underestimates the observations by about 10 ppt in a consistent manner. One obvious reason is that the emissions from North American sources cannot effectively be transported to ORE due to the prevailing mid-latitude westerlies. The relationship between ORE and the North American sources is essentially a regional pollutant transport problem. As suggested in Prather et al. (1987), a finer horizontal resolution (window calculation) is necessary to ob-

140

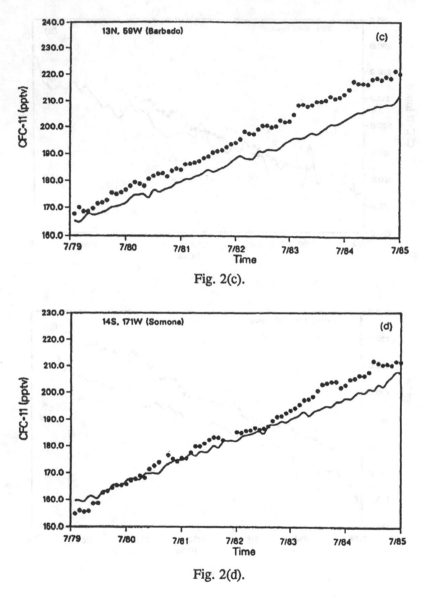

Fig. 2(c).

Fig. 2(d).

tain a successful simulation at ORE due to regional pollution. Nonetheless, the seasonal cycle in terms of wintertime maxima and summertime minima is well simulated by the model. As for the site at Barbados (BAR), model values show a distinct growing under-estimation of the measured values, which is not the case for other ALE sites. The reason for this particular discrepancy at BAR is not immediately clear. However, considering that BAR is consistently under the influence of the Atlantic subtropic high, the anthro-pogenic gases produced in the Northern America can enter into the Barbados region only through an indirect circulation, that is, the divergent flow leaving the southern part of the

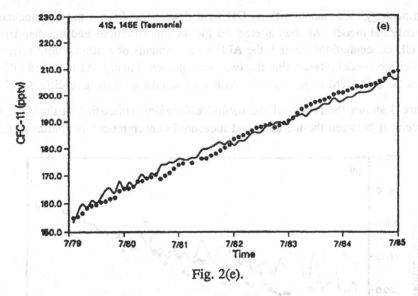

Fig. 2(e).

Fig. 2. Comparisons between the simulated (solid lines) and measured (dotted lines) surface concentrations for the five ALE sites: (a) Adrigole, Ireland (ADR), (b) Cape Mears, Oregon (ORE), (c) Ragged point, Barbados (BAR), (d) American Samoa (SAM), and (e) Tasmania (TAS).

mid-latitudes westerlies and then merging into the northeasterlies of the subtropic high. Due to this circuitous route, the coarse-resolution GCM may not accurately describe such a complicated flow pattern. It is encouraging to see in Figs. 2(d) and 2(e) that the agreement between the simulated and observed concentrations in the southern hemisphere ALE sites; namely, Samoa (SAM) and Tasmania (TAS), is generally good, reflecting a reasonable interhemispheric transport.

3.2 Methyl Chloroform

The surface source pattern of methyl chloroform is similar to that of CFC-11. The annual emission rates are also obtained from the CMA. In addition, the transport mechanism used here is identical to that in the CFC-11 simulation. The major atmospheric sink of methyl chloroform is through the reaction with OH radicals. However, the global distribution of OH is not well known. Routine measurements of OH concentrations that are representative of large portions of the troposphere are not yet technically feasible. A methodology in terms of using simulated methyl chloroform to infer the OH field has been developed, taking advantage of: (1) flux of methyl chloroform to the atmosphere can be closely estimated, (2) the destruction of methyl chloroform through OH can be expressed as simple bimolecular reaction rate, and (3) the measurements of methyl chloroform conducted by the ALE. Obviously, it implies that an iterative process is involved with

142

this methodology. We started with an OH field determined from a pure photochemical two-dimensional model. We then altered the global concentrations and interhemispheric ratio of OH concentrations through the ALE measurements of methyl chloroform and a simple two-box model representing the two hemispheres. Finally, the modified OH field is used in the 3-D CTM to simulate the ALE concentrations at the five different sites.

Figure 3 shows the results of the methyl chloroform simulation at the ALE sites. The agreement between the modeled and measured concentrations is similar to that of

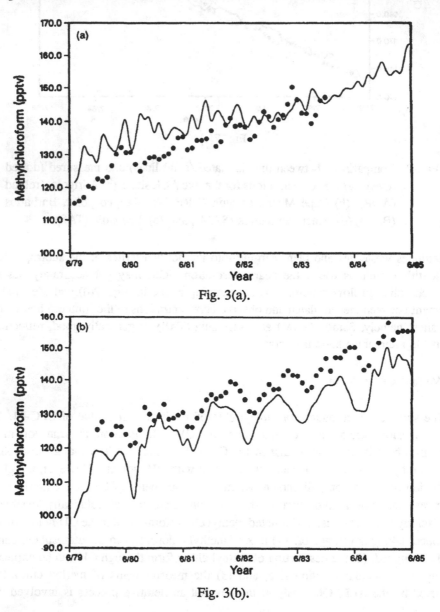

Fig. 3(a).

Fig. 3(b).

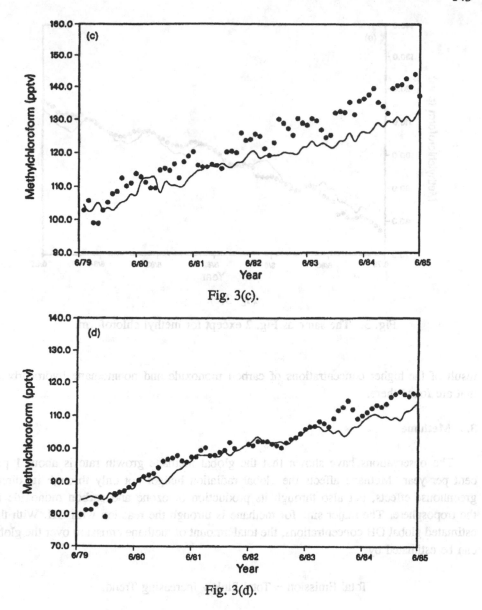

Fig. 3(c).

Fig. 3(d).

the CFC-11 simulation, indicating the importance of global transport mechanisms on both species. The effects of OH on methyl chloroform are clearly identified through the distinct seasonal cycles shown in Fig. 3(e) (for TAS) in comparison with its counterpart shown in Fig. 2(e). The global OH field used to produce the above simulation has an interhemispheric ratio ($[OH]_N/[OH]_S$) of 0.46. This hemispheric asymmetry results from the additional destruction of OH in the northern hemisphere, that can be expected as a

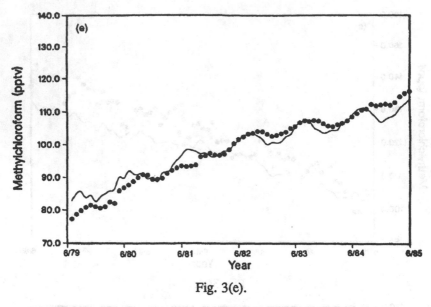

Fig. 3(e).

Fig. 3. The same as Fig. 2 except for methyl chloroform.

result of the higher concentrations of carbon monoxide and nonmethane hydrocarbons that are found there.

3.3 Methane

The observations have shown that the global methane growth rate is about 1 per cent per year. Methane affects the global radiation budget not only through its direct greenhouse effects, but also through its production of ozone and carbon monoxide in the troposphere. The major sink for methane is through the reaction with OH. With the estimated global OH concentrations, the total amount of methane emission over the globe can be estimated by

Total Emission = Total Sink + Increasing Trend,

as proposed by Cicerone and Oremland (1988). Since the causes of the trend are uncertain, we wll concentrate, in this study, only on the global methane distribution by assuming that the total emission equals the total sink for a four-year simulation. With the total atmospheric burden of methane estimated in 1984 and the previously estimated total OH amount, the total sink is about 385 Tg/yr and so is the total emission. We then redistribute this total emission into nine different types of methane sources with different geographical distributions, according to various studies (e.g., Khalil and Rasmussen,

1983; Seiler, 1984; Crutzen et al., 1986; and Cicerone and Oremland, 1988). The nine different types of methane sources are animal, rice paddy, wetland, biomass burning, natural gas, coal mining, landfill, termite, and ocean. The OH distribution inferred from the methyl chloroform simulation is used as input to the methane simulation. The initial condition is constructed from 1984 data. The model was run for four years under the above mentioned steady-state assumption, and the results of the later three years will be compared with the measurements averaged from 1983 to 1985 (Steele et al., 1987; Boden et al., 1990).

Figure 4 shows the comparison between the modeled and measured surface concentrations at the locations where measurements are available. Since the longitudinal gradient is generally small, Fig. 4 is plotted as concentrations versus latitudes. We note that the model results agree with the observations fairly well, especially for the southern hemisphere. For the northern hemisphere, the influence due to local sources cannot be well simulated by the current resolution. This acounts for much of the underestimation of methane by the model in the northern hemisphere. Overall, the model underestimates the observations by approximately 3 per cent.

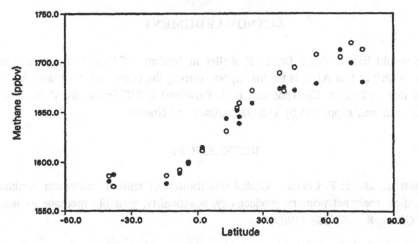

Fig. 4. Comparison between the modeled (closed circles) and measured (open circles) methane concentrations at the locations where measurements were available.

4. SUMMARY

We have developed a 3-D chemical tracer model (CTM), which uses winds, temperature, and sub-grid-scale mixing simulated from the LANL 20-layer GCM to drive the species continuity equations with surface sources and chemical sinks. We first used our CTM to simulate the global distribution and trend of chemically inert CFC-11 observed by the Atmospheric Lifetime Experiment (ALE) from 1978 to 1985. The results show that the CTM not only closely simulates the interhemispheric gradient of CFC-11, but also demonstrates the ability to reproduce the time-series of the 7-year ALE observations. The overall purpose of this CFC-11 study was to test the transport of the CTM.

We then completed another important step toward the understanding of OH radicals, a species central to the atmospheric chemistry. We used the CTM to introduce methyl chloroform into the atmosphere according to known emission patterns and iteratively varied OH fields so that the observed concentrations of methylchloroform from the ALE could be simulated well. The rationale behind this approach is that the reaction with OH is the dominant sink for methyl chloroform and the transport of the CTM has been successfully tested in the previous CFC-11 simulation.

Using the inferred OH distribution from the methyl chloroform simulation and the best estimated methane source strength and distribution, we conducted a steady-state simulation for methane distribution. The general agreement between the modeled and measured methane concentrations has laid a foundation for the simulation of the transient increase of methane.

ACKNOWLEDGMENT

We would like to thank Dr. C. F. Keller at Institute of Geophysics and Planetary Physics (IGPP) at Los Alamos for his support during the course of this research. Thanks are also due to Drs. R. Cicerone and F. S. Rowland at UC Irvine for their assistance. This research was supported by U.S. Department of Energy.

REFERENCES

Aselmann, I., and P. F. Crutzen, Global distribution of natural freshwater wetland and rice paddies, their net primary productivity, seasonality, possible methane emission, *J. Atmos. Chem.* **8,** 307–358 (1989).

Chock, D. P., and A. M. Dunker, A comparison of numerical methods for solving the advection equation, *Atmos. Environ.,* **17,** 11–24 (1983).

Cicerone, R. and R. Oremland, Biogeochemical aspects of atmospheric methane, *Global Biogeochem. Cycles* **2,** 299–327 (1988).

Crutzen, P. J., I. Aselmann, and W. Seiler, Methane production by domestic animals, wild ruminants, other herbivorous fauna, and human, *Tellus* **38B,** 271,284 (1986).

Golombek, A., A global 3-D model of the circulation and chemistry of long-lived atmospheric species, Ph.D. dissertation, Mass. Inst. of Technol., Cambridge, 1982.

Golombeck, A., and R. G. Prinn, A global 3-D model of the circulation and chemistry of $CFCl_3$, CF_2Cl_2, CH_3CCl_3, CCl_4, and N_2O, *J. Geophys. Res.* **91**, 3985–4001 (1986).

Khalil, M. A. K., and R. A. Rasmussen, Sources, sinks, and seasonal cycles of atmospheric methane, *J. Geophys. Res.* **88**, 5131–5144 (1983).

Kao, C.-Y. J., G. A. Glatzmaier, R. C. Malone, and R. P. Turco, Global three-dimensional simulations of ozone depletion under postwar conditions, *J. Geophys. Res.* **95**, 22495–22512 (1990).

Kao, C.-Y. J., A sensitivity study of the Arakawa-Schubert Cumulus parameterization in the Los Alamos GCM. Cloud Physics Conference, American Meteorological Society, 505–509, San Francisco, California, July 23–27, 1990.

Levy, H. II., J. D. Mahlman, W. J. Moxim, Tropospheric N_2O variability, *J. Geophys. Res.* **87**, 3061–3080 (1982).

Louis, J.-F., A parametric model of vertical eddy fluxes in the atmosphere, *Boundary Layer Meteo.* **17**, 187–202 (1979).

Malone, R. C., L. H. Auer, G. A. Glatzmaier, M. C. Wood, and O. B. Toon, Nuclear winter: Three-dimensional simulations including interactive transport, scavenging, and solar heating of smoke, *J. Geophys. Res.* **91**, 1039–1053 (1986).

Mroz, E. J., M. Alei, J. H. Cappis, P. R. Guthals, A. S. Mason, and D. J. Rokop, Antarctic atmospheric tracer experiment, *J. Geophys. Res.* **94**, 8577–8583 (1989).

Pepper, D. W. and A. J. Baker, A simple one-dimensional finite-element algorithm with multidimensional capabilities, *Numer. Heat Transf.* **2**, 81–95 (1979).

Picher, E. J., R. C. Malone, V. Ramanathan, M. L. Blackmon, K. Puri, and W. Bourke, January and July simulations with a spectral general circulation model, *J. Atmos. Sci.* **40**, 580–604 (1983).

Porch, W. M., and C.-Y. J. Kao, Anomalous light extinction from large area forest fires. Champman Conference on Global Biomass Burning: Atmospheric, Climatic, and Biopheric Implications, Williamsburg, Virginia, March 19–23, 1990).

Prather, M., M. B. McElroy, S. Wofsy, G. Russell, and D. Rind, Chemistry of the global troposphere: Fluorocarbons as tracers of air motion, *J. Geophys. Res.* **92**, 6579–6613 (1987).

Ramanathan, V., E. J. Pither, R. C. Malone, and M. L. Blackmon, The response of a spectral general circulation model to improvements in radiative processes, *J. Atmos. Sci.* **40**, 605–630 (1983).

Ramanathan, V., R. J. Cicerone, H. B. Sighn, and J. T. Kiehl, Trace gas trends and their potential role in climate change, *J. Geophys. Res.* **90**, 5547–5566 (1985).

Roads, J. O., S.-C. Chen, C.-Y. J. Kao, D. Langley, and G. Glatzmaier, The Los Alamos general circulation model hydrological cycle, *J. Geophys. Res.*, submitted (1991).

Seiler, W., Contribution of biological processes to the global budget of methane in the atmosphere, Current Perspectives in Microbial Ecology, Edited by M. J. Klug and C. A. Reddy, American Society of Microbiology, 1984.

Tie, X., A three-dimensional global dynamical and chemical model of methane, Ph.D. Dissertation, School of Geophysical Sciences, Georgia Institute of Technology, 1990.

Zalesak, S. T., Fully multidimensional flux-corrected transport algorithms for fluids, *J. Comput. Phys.* **31**, 335–362 (1979).

Snow Hydrology

Susan Marshall, Gary Glatzmaier
Los Alamos National Laboratory, Los Alamos, NM 87545

John O. Roads
Scripps Institute of Oceanography, La Jolla, CA 92093

ABSTRACT

Snow covers from 4 to 25% of the Northern Hemisphere seasonally and contributes up to a third of the world's irrigation water as well as runoff for hydroelectric power generation and urban water supplies. Snow also is a critical component of the global climate and enhances climate sensitivity in mid to high latitudes.

Even with simple parameterizations of snowfall and snow-mass budgets, global climate models capture well the geographic distribution of snow cover and snowfall. However, this does not inspire confidence in model results from climate change scenarios. Snow-mass budget models and snow thermal and physical properties need to be better modeled, prognostic and physically-based and not tuned to the current climate. In order to accomplish this, we need to inquire further into our understanding of snow processes.

A realistic snow hydrology has been implemented into the Los Alamos General Circulation Model (GCM). The snow hydrology includes parameterizations for snowfall and a budget equation for snow mass. Snow ablation is parameterized through an energy-balance equation. Snow cover influences the ground temperatures and surface water storage as well as the surface albedo, which is dependent upon the snow depth and surface roughness height.

INTRODUCTION

Snow is the most pervasive land surface type in the mid to high latitudes. In the northern hemisphere, snow cover varies greatly between its minimum and maximum extents. In winter, snow and ice cover approximately 25% of the Northern Hemisphere while the summer snow covers only 4% of the hemisphere.[1] The water equivalent of seasonal snow cover is a major factor in the surface hydrology. Steppuhn (1981) estimates that a third of the global irrigation water comes from snowmelt.[2] Spring snowmelt also provides runoff for irrigation, hydroelectric power generation and drinking water for much of the mid latitudes.

Several characteristics of snow cover are important for any discussion of climate. The high surface albedo of snow is probably the most significant component of the energy budget over snow. A clean, fresh snow cover can reflect more than 80% of the solar radiation incident upon its surface. The surface climate is sensitive to the value of the surface albedo. The 'ice/snow albedo feedback' refers to the positive feedback to the climate system by a melting of high albedo surfaces. As the surface albedo decreases with melt, the surface is able to absorb more radiation and warms. This mechanism is responsible for the enhanced response at higher latitudes to a global temperature warming. Although dependent on the age and properties of the snow cover and spectrum of incoming radiation, the surface albedo of snow is generally parameterized in climate models in terms of temperature and solar zenith angle.

The low thermal conductivity of a snow cover reduces the heat exchange between the soil and the atmosphere (much like sea ice decouples the ocean-atmosphere heat exchanges). The thermal conductivity of snow also depends on the age and properties of the snow but is often kept constant in climate models.

For the hydrologic cycle, we are most concerned with the processes which contribute to the extent and mass of the snow cover, including snowfall, melt and sublimation. In climate models, the parameterizations of such processes are often crude. Distinction between snowfall and rainfall is based on a simple temperature criterion of 0° Centigrade. Sublimation is often not considered and snowmelt is computed from the surface energy balance assuming a constant snow temperature of 0° Centigrade.

MODELING EFFORTS

We have recently included an interactive snow hydrology in the Los Alamos GCM (Fig. 1). Snow melt is modeled by solving the surface energy budget. The snow temperature is a prognostic variable in the model, determined both from the thermal diffusion and the energy budget at the surface. Any excess energies at the surface layer are first used to raise the temperature of the snow. Once the snow layer is at its melting temperature (0° Centigrade), excess energy can then be used to melt the snow.

Snow meltwater is used to recharge the top layer soil moisture. The snow mass is determined by solving the mass budget. Snow cover grows by the addition of snowfall, and is reduced by losses from melt and sublimation (generally only a small percentage of the total ablation).

The geographic distribution of modeled and observed snow cover is shown in Fig. 2 for the Northern Hemisphere winter. These maps compare model results for 5 'winter' (December, January and February) averages and observed probabilities of snow cover extent for the end of December (from Dickson and Posey[3]). The modeled and observed distributions of snow cover correspond well. The model's most apparent inconsistency is in the coarse resolution of the results. At this resolution (approximately 4° latitude by 7° longitude), the model is unable to represent correctly the smaller mountain ranges of the Alps and Caucasus.

Figure 3 shows the distribution of model precipitation, broken into the components of snowfall and rain. The maps show the annual averages of these fields. These results correspond to observed patterns of snow and rain. Much of the rain occurs along the

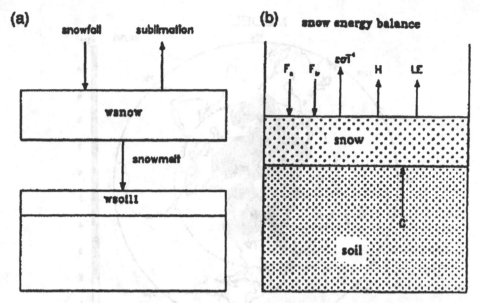

Fig. 1. Schematics illustrating the (a) snow-mass budget and (b) snow energy balance as modeled in the Los Alamos GCM. F_s and F_{ir} are the downward solar and infrared radiative fluxes at the surface; $\varepsilon \alpha T^4$ is the upward emitted longwave flux by the snow surface, H and LE are the fluxes of sensible and latent heat at the surface and G is the ground heat flux. All units are in W m^{-2}.

Inter-Tropical Convergence Zone with the maximum snowfall occurring in mountainous regions and along the storm tracks of the Western Atlantic and Pacific Oceans.

CONCLUSIONS

General Circulation Models (GCMs) with an interactive snow and surface hydrology are useful in investigating theories of climate change. They are especially important to experiments of glaciation and ice-age climates where the seasonal and inter-annual variations in the snow cover are crucial to forcing the onset of glaciation.

Inclusion of an interactive snow hydrology is also necessary in experiments of a climate warming due to increased greenhouse gases, in two respects. First, initial model results show an enhanced warming effect in the higher latitudes as a result of the ice/snow albedo feedback. Studies by Harvey[4] and Shine and Henderson-Sellers[5] indicate the magnitude of this feedback is dependent upon the parameterization of snow and ice used in the model. Second, the water-vapor feedback mechanism indicates an enhancement of the hydrologic cycle with higher temperatures. This has been postulated to result in increased snowfall at higher latitudes.[6] This coupled with the seasonality of the warming

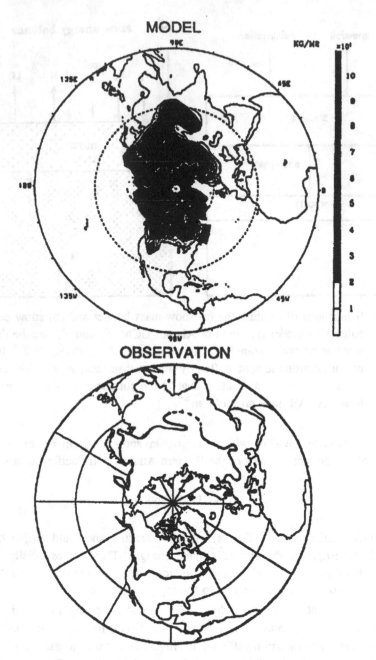

Fig. 2. Modeled and observed Northern Hemisphere winter snow cover. Model results are an average of 5 December, January and February snow depths over land and sea ice in units of kg m^{-2} ($\times 10^1$). Observed data show the 50% probability of snow cover for the end of December (taken from Dickson and Posey). Observed data do not include snow over sea ice.

ANNUAL AVERAGE SNOWFALL (kg m⁻²d⁻¹ x 10⁻¹)

ANNUAL AVERAGE RAINFALL (kg m⁻²d⁻¹)

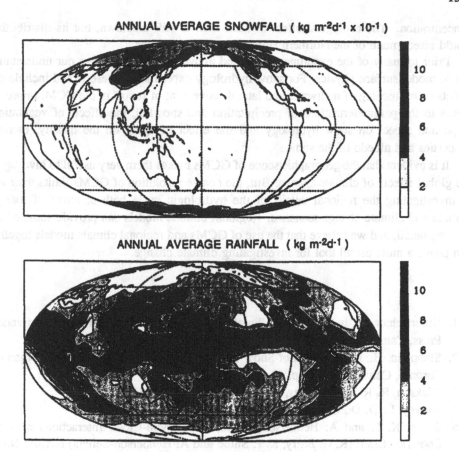

Fig. 3. Global distribution of model snowfall and rainfall. The results indicate
an annual average value.

(greater in winter and less in summer) is a possible precursor to the formation of mountain glaciers.

Experiments of aerosol and particulate pollution in the atmosphere have concentrated on the effects on the radiation budget. Pollution of this sort falls into two categories, light-absorbing and non-light-absorbing impurities. Light-absorbing impurities act to reduce the snow albedo, resulting in an increase in absorbed radiation at the surface. This leads to increased melt and should result in a shorter snow cover duration, allowing for enhanced absorbed energy over the melt-free period. Brown[7] estimates an approximate two week increase in the melt-free period with the inclusion of light-absorbing impurities in the Alaskan snow cover.

Non-light-absorbing impurities effect the quality of the melt water. Acidic precipitation in the form of snowfall is stored in the snow cover until melt occurs. Since the acid is soluble, it is very often flushed out in the first or second melt event, resulting in the phenomenon known as 'acid pulse.' These first few melts release melt water of high acid

154

concentration. The ecological effect of such melt is still unknown, but its distribution would affect much of the Northern Hemisphere mid latitudes.

Prior to many of the experiments outlined above, we must improve our understanding of model surface physics. For snow hydrology experiments, these would include the effects of re-freezing of meltwater and lateral movement of runoff in the GCM, improvements to the parameterizations of precipitation and snowfall, the effects of vegetation, slope and aspect on snow hydrology and the effect of aging on the density, thermal properties and albedo of the snow.

It is evident that the geographic scope of GCMs makes them very useful to investigate the global effects of climate change. But, the coarse resolution of GCMs limits their use in investigating the regional details of the hydrologic cycle both in terms of today's climate and climate change scenarios. Regional climate models can provide some of the missing detail, and we believe that the use of GCMs and regional climate models together can prove a most useful tool for investigating climate change.

REFERENCES

1. Untersteiner, N., in The Global Climate (J. T. Houghton, ed., Cambridge University Press, Cambridge, 1984), pp. 121–140.
2. Steppuhn, H., in Handbook of Snow (D. M. Gray and D. H. Male, eds., Pergamon, Toronto, Canada, 1981), pp. 60–125.
3. Dickson, R. R. and J. Posey, *Mon. Wea. Rev.* **95**, 347 (1967).
4. Harvey, L. D. D., *Climatic Change* **13**, 191 (1989).
5. Shine, K. P. and A. Henderson-Sellers, Cryosphere-Cloud Interactions near the Snow/Ice Limit (R. G. Barry, K. P. Shine and A. Henderson-Sellers) NASA: NAG-5-142, 78 (1984).
6. Warren, S. G. and S. Frankenstein, *Annals of Glaciology* **14**, 361 (1990).
7. Brown, B. E., *J. Geophys. Res.* **87**, 1347 (1982).

Global Simulations of Smoke from Kuwaiti Oil Fires and Possible Effects on Climate

Gary A. Glatzmaier, Robert C. Malone, C.-Y. Jim Kao

Los Alamos National Laboratory

Los Alamos, NM 87545

ABSTRACT

The Los Alamos Global Climate Model has been used to simulate the global evolution of the Kuwaiti oil fire smoke and its potential effects on the climate. The initial simulations were done shortly before the fires were lit in January 1991. They indicated that such an event would not result in a "Mini Nuclear Winter" as some people were suggesting. Further simulations during the year suggested that the smoke could be responsible for subtle regional climate changes in the spring such as a 5°C decrease in the surface temperature in Kuwait, a 10% decrease in precipitation in Saudi Arabia and a 10% increase in precipitation in the Tibetan Plateau region. These results are in qualitative agreement with the observations this year.

1. INTRODUCTION

Early in January 1991 the media began to give considerable attention to the possibility that Iraqi forces might intentionally ignite a large number of Kuwaiti oil wells. Some speculated that the resulting smoke could have significant effects on the global climate akin to "Nuclear Winter" (Turco et al. 1983; Malone et al. 1986). They expected that at least there would be regional effects such as a reduction in the monsoon rainfall that is so critical to agriculture in the Indian subcontinent. At that time, before the Kuwaiti oil fires were lit, Los Alamos conducted several computer simulations of a Kuwaiti oil fire scenario using the Los Alamos Global Climate Model (GCM).

This three-dimensional time-dependent atmospheric model (Malone et al. 1985, 1986; Gifford et al. 1988; Kao et al. 1990) is a modified version of the National Center for Atmospheric Research (NCAR) Community Climate Model, version 0 (Pitcher et al. 1983; Ramanathan et al. 1983; Malone et al. 1984). It is a spectral code, using a spherical-harmonic spectral-transform method, that solves the atmospheric flow equations in a 3D spherical shell with continents, oceans, and sea ice. It self-consistently computes clouds, precipitation, and solar and infrared radiation fluxes. The modifications made at Los Alamos to the original NCAR model are the following. The vertical diffusion coefficients for heat, moisture and momentum are calculated at each grid point and timestep based on the local stability of the atmosphere. Heat and moisture storage are computed below the soil surface. Both diurnal and annual cycles of solar radiation are included.

The model has the capability of tracking the 3D evolution of gases and aerosols in the atmosphere, their absorption of solar and infrared radiation, their scavenging by precipitation, and their photo-chemical interactions with several families of chemical species. This model has latitude and longitude resolutions of 4.5 and 7.5 degrees, respectively, and has 20 vertical levels in the atmosphere up to 33 km and 6 levels in the soil down to 5 m. It runs with a 15-minute timestep. Thirty years have been simulated with this model and the results are in good agreement with observations.

The Los Alamos GCM was initially developed in the 1980s to study the "Nuclear Winter" hypothesis, which suggests that surface temperatures would drop several degrees for several weeks following a nuclear war due to the large quantities of black smoke that would be injected into the atmosphere from burning cities. Our detailed model simulations (Malone et al. 1985, 1986) showed that, for large injections of smoke, temperatures would not drop as much as earlier one-dimensional models predicted (Turco et al. 1983) and that roughly half of the smoke would be washed out of the atmosphere within the first week. However, the model results indicated that temperatures in the summer in the United States, for example, could drop below freezing at times and therefore cause crop damage. In addition, our results showed that the smoke that was not washed out would be buoyantly lofted (by solar heating) into the stratosphere where it would remain for years. The model indicated that the resulting higher stratospheric temperatures (due to the solar heated smoke) and the greatly enhanced quantities of nitrogen oxides (produced in the nuclear explosions) would significantly affect the photochemical reactions that maintain the stratospheric ozone which protects life on Earth from ultraviolet radiation. The results predict that as much as 15% of the ozone would be depleted in the northern hemisphere in the first 20 days (Kao et al. 1990).

2. THE SIMULATIONS

The global three-dimensional transport of smoke is computed in the GCM with a Lagrangian tracer particle scheme that is well suited for the treatment of small-area sources (Gifford et al. 1988). The simulated soot is assumed to be injected at a constant rate of 6.1×1010 grams per day and assumed to have a visible absorption coefficient of 10 meters^2/gram and an infrared extinction coefficient of 1 m^2/gram (D. Engi & B. Zak 1991, Sandia National Laboratory, Private Communication). It is injected up to 3 km above the surface before the model-predicted winds transport it and the model-predicted precipitation removes it.

We ran this scenario several times, each for 30 simulated days. After 30 days, the simulated smoke is represented by over 14 000 numerical tracer particles. Weather simulated by the model for January conditions, which is very representative of observed weather patterns, was used because the actual weather conditions that might occur could not be known in advance. To increase the confidence in the model results, several 30-day simulations were done starting with different, albeit typical, initial weather conditions.

The model results showed that after 30 days about 15% of the total smoke that has been injected still remains in the atmosphere. The amount of smoke in the atmosphere at any given time depends on the local weather conditions (simulated in the model). Transient excursions of the smoke plume over India and the central Soviet Union are observed in the simulations (Fig. 1), but the effects of the smoke are confined primarily to within 1000–2000 kilometers of Kuwait. On the average, a maximum visible sunlight attenuation of 50% occurs directly over Kuwait, but only about 4% reduction occurs in the vicinity of India. Although some heating of smoke by sunlight does occur, the amount of smoke lofted into the stratosphere is small compared to "Nuclear Winter" scenarios because the Kuwaiti smoke is injected much more slowly and in much smaller amounts. After about a week, the steady injection of smoke is approximately balanced by its removal by rain, so that about 0.3 Tg of smoke remains in the atmosphere, mainly over the Middle East. The average lifetime of the smoke predicted by the model (based on rainfall predicted by the model itself) is about 5 days, which is in good agreement with observed lifetimes of aerosols in the lower atmosphere (Malone et al. 1985). Although the regional climate around Kuwait is affected by the smoke, the model results indicate that the smoke has no significant effect on the global climate.

Subtle regional climate changes in the spring were assessed by conducting several 30-day computer simulations for June conditions. Ten different June cases were modelled, all with the same amount of smoke injected, and were compared to the same ten Junes without smoke. The results indicated that the absorption of sunlight by the smoke causes the surface within about 1000 km of Kuwait to be cooled by an average of about 5 degrees centigrade. The model results also show about a 10% decrease in precipitation in Saudi Arabia and about a 10% increase over the Tibetan Plateau.

3. SUMMARY

The Los Alamos Global Climate Model, which was originally developed to study the "Nuclear Winter" hypothesis during the 1980s, was used to simulate the Kuwaiti oil fire smoke and assess its potential impact on the global climate. The measurements made this summer by the National Oceanic and Atmospheric Administration (NOAA) have shown that the injection rate estimated by Sandia and used in all these computer simulations may have been high by about 50%. However, the Los Alamos GCM results that indicated a reduction in surface temperature and precipitation near Kuwait and an increase in precipitation in southwest China are in qualitative agreement with this summer's observations. Certainly the model prediction in early January that there would be no significant global climate effect has been confirmed.

This work was supported by the Los Alamos National Laboratory which is operated by the University of California for the U.S. Department of Energy.

158

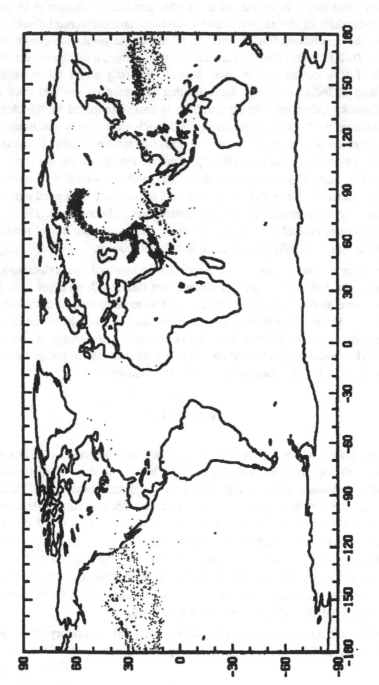

Fig. 1. A snapshot of the very time-dependent smoke distribution simulated with the Los Alamos Global Climate Model.

REFERENCES

Gifford, F. A., Barr, S., Malone, R. C. & Mroz, E. J. (1988) "Tropospheric relative diffusion to hemispheric scales" *Atmos. Environment* **22**, 1871.

Kao, C.-Y. J., Glatzmaier, G. A., Malone, R. C. & Turco, R. P. (1990) "Global three-dimensional simulations of ozone depletion under post-war conditions" *J. Geophys. Res.* **95**, 22495.

Malone, R. C., Auer, L. H., Glatzmaier, G. A., Wood, M. C. & Toon, O. B. (1986) "Nuclear winter: Three dimensional simulations including interactive transport, scavenging, and solar heating of smoke" *J. Geophys. Res.* **91**, 1039.

Malone, R. C., Auer, L. H., Glatzmaier, G. A., Wood, M. C. & Toon, O. B. (1985) "The influence of solar heating and precipitation scavenging on the simulated lifetime of post-nuclear war smoke" *Science* **230**, 317.

Malone, R. C., Pitcher, E. J., Blackman, M. L., Puri, K. & Bourke, W. (1984) "The simulation of stationary and transient geopotential-height eddies in January and July with a spectral general circulation model" *J.'Atmos. Sci.* **41**, 1394.

Pitcher, E. J., Malone, R. C., Ramanathan, V., Blackmon, M. L., Puri, K. & Bourke, W. (1983) "January and July simulations with a spectral general circulation model" *J. Atmos. Sci.* **40**, 580.

Ramanathan, V., Pitcher, E. J., Malone, R. C. & Blackmon, M. L. (1983) "The response of a spectral general circulation model to improvements in radiative processes" *J. Atmos. Sci.* **40**, 605.

Turco, R. P., Toon, O. B., Ackerman, T. P., Pollack, J. B. & Sagan, C. (1983) "Nuclear winter: Global consequences of multiple nuclear explosions" *Science* **222**, 1283.

Possible Effects of Smoke from Kuwaiti Oil Fires on Climate and Society

A. Ginzburg, V. Haritonenko, O. Kurliandskaya
Institute of Atmospheric Physics
3, Pyzheusky
Moscow 109017 Russia

As a result of rapid and global changes in the world, a different understanding of security emerges in the international arena. Earlier the understanding was based on a military approach. Today the bipolar world has gone. Instead, a lot of new centers of power emerge before our eyes. And those of the so-called Third World are of the greatest concern.

A chain of ecological catastrophes has shaken the world. It was learned that the ecological situation does not happen by itself but depends on concrete actions of definite groups of people. The War in the Gulf is the last and probably the "best" example.

It is the unique geographic location of the arena of the Gulf War that made absolutely senseless any attempt to preserve a local character of the conflict under the conditions of the multinational armed force involvement. And it was the first time in contemporary history that regional conflict emerged up to the international level.

In particular, it was obvious from the very beginning that the ecological consequences of the war would, in no case, be contained inside the Iraqi and Kuwaiti borders.

Long ago Saddam promised to separate Iraq and its 19th province, Kuwait, from the entire world with a 'burning wall' in case of a war. And, several days before January 15, Iraqi oil tankers were placed close to the Kuwaiti shore.

Those definite preparations for an ecological war did not seem to attract enough attention from the side of the so-called civilized states. In general, about 500 million tons of oil is mined in the Middle East region per year. In particular, about 250 Mtons in Saudi Arabia, 90 Mtons in Iraq, 70 Mtons in Kuwait. The Mid-Eastern oil stocks already discovered represent more than half of the total oil reserves of the world.

It was not easy before and during the Gulf War to make a precise assessment concerning the amount of oil that possibly can be burnt in case of either accidental or planned actions. Having a goal like that (carrying out an assessment), it is worth working with maximum numbers, as it is much more dangerous to diminish the danger of an ecological disaster than to exaggerate it.

The first estimates show that, in the case of many fires, the annual oil output can be eliminated in a month. The amount of smoke fromf the burning Kuwaiti oil can reach 3 million tons, and about 25 million tons of smoke can be provided in case of oil burning in the whole region. The latter number is close to that of the total amount of soot loaded into the atmosphere of the Earth in case of a global nuclear conflict.

If the smoke from the region smoothly covers the interval from 20 to 40 degrees of Northern Latitude, then all the territory covered will become absolutely isolated from solar radiation, and as a result the temperature of the soil will become 20 degrees lower (A. Ginzburg, 10 Jan 1991).

In Winter the atmospheric circulation in the Northern hemisphere is more intensive than in Summer, so that the smoke will be moved from the subtropics to middle and tropical latitudes. This possibly enables the minimizing of climate effects in the region, but vastly broadens it throughout the Earth.

In several weeks or months the smoke will fall down. But even such climate influence will inevitably impose a long-standing disorder both in agriculture and in the economy of the region as well as that of the entire community of the world.

Just before, during, and after the Gulf War there were a lot of estimates, according with numbers of burning wells, of the amount of oil and smoke going into the environment. Some of them are shown below.

One week before beginning the Gulf War, the British Meteorological Office convened a conference in London that suggested that fires from Kuwait's oil wells could cause an environmental catastrophe (Financial Times of Jan 91, Aldous, 1991).

Before the Iraqi invasion, Kuwait produced more than two million barrels of oil a day from 365 active wells; most of them would continue to burn if they were ignited. J. Cox, a British chemical engineer, said 3 million barrels per day (bpd) could burn. Abdulluh Toukan, science adviser to King Hussein of Jordan, said that as much as 10 million bpd could burn if a proportion of Kuwait's many out-of-commission wells were also set alight.

K. Browning, director of research of the British Meteorological Office, showed first results of computer calculations based on the burning of 2 million bpd.

P. Crutzen, taking Toukan's figure, has produced some rough calculations that predict a cloud of soot covering half of the Northern Hemisphere within 100 days. Land temperature cloud could be reduced by 5–10°C in this short time.

At the end of January Sir F. Warner estimated that at the time of invasion there were produced 1.5 million bpd from 800 wells in Kuwait. From 2 to 10 million bpd could burn, producing from 30 000 to 150 000 tons of smoke per day.

Smoke goes up into the atmosphere and reaches the heights determined by its initial temperature. Nuclear winter theory shows that smoke can penetrate the tropopause and rise into the stratosphere. Smoke would mix and be transported around the Earth, absorb sunlight, and cool the land surface.

In February, the land temperature in the oil fields was 10–15°C cooler than in nearby Kuwait City, because sunlight was not reaching the desert (see Arkin et al., 1991).

Many teams — at least three from the US, two from Germany, and one each from Great Britain, Canada, and the Soviet Union — used computer models of General Atmospheric Circulation for modeling climatic effects of the Kuwaiti oil fires.

In Ginzburg, Trosnikov, Haritonenko (1990) there were considered the effects of large aerosol pollutions on the large-scale meteorological characteristics of the Earth's atmosphere for winter conditions to evaluate the possible impact on the climate as a result of a nuclear war. It was detected that in winter the net cooling of the land surface is less than in a similar summer case.

The main effect in the winter case occurs in the tropical latitudes of the Northern hemisphere, where the decreasing of the surface temperature due to the presence of the solar-heated aerosol in the atmosphere is stronger than in summer. In addition, aerosol is intensively transported over the hemisphere increasing the thermal contrasts in the atmosphere, thus increasing the zonal circulation. Thus the winter case has some specific features.

Of course, the assumptions about value and space distribution of initial aerosol pollutions were rather sophisticated.

Practically simultaneously, different numerical models were used to predict and evaluate possible atmospheric effects from burning oil wells in Kuwait. So a group from Germany completed a series of experiments with the Hamburg global-coupled atmosphere-ocean climatic model (19 levels, 5,6 resolution) with an interactive soot-transport model and extended radiation scheme (S. Bakan et al. 1991).

P. Carl used a similar modification of two-level coarse resolution ($12° \times 15°$) general circulation model (P. Carl, 1991).

In K. A. Browning et al. (1991) a set of models, including a simple model of plume rise, a mesoscale weather-prediction model for reproduction of the local effects, and long-range dispersion and general circulation models for the longer-term response, were used to investigate environmental consequences from the fires of the oil wells.

I. Trosnikov and Ye. Egorova (Hydrometeocenter) with A. Ginzburg and V. Harito-nenko (Institute of Atmospheric Physics) also completed numerical experiments with the general circulation model with a permanent source of aerosol pollutions, restricted to one horizontal cell of the model grid (12×15) and located in the Gulf region.

It was marked and supposed by the first satellite observations that the smoke plumes are not rising higher than 2–4 kilometers in the troposphere. The transport of aerosol into the stratosphere is negligible. Thus most of the aerosol particles are washed out by rain within a few days and their residence time in the atmosphere is small and was assumed in the models to be from 10 to 30 days.

I. Trosnikov and Ye. Egorova restricted experiments to the first three winter weeks. Nevertheless, due to the fast horizontal spreading of the smoke in winter, as was shown in the previous nuclear-winter experiments (Ginzburg et al., 1990), in a few days a dense smoke cloud forms over the Gulf region, and a thin smoke veil spreads over low latitudes of the Northern Hemisphere as far north as China.

It was determined that the smoke distribution becomes quasi stationary one or two months after the beginning of the oil burning (February, 1991). The expected reduction in surface temperatures caused by the smoke's sun-dimming effects occurs in a region of radius several hundred kilometers around Kuwait (up to $10°C$) with only $1–2°C$ drops within some thousand kilometers zone. Although the surface temperature response isn't distinguishable at longer distances from the source with respect to the control experiment, the distribution of the aerosol concentration in the veil isn't uniform.

It was also determined that the separate spots of smoke being transported to the East can produce local deviations in the weather regimes. So in [2], anomalous heating was found over the Tibetan Plateau in May, which can be explained by the proximity of the absorbing soot layers in the atmosphere to the surface at high altitudes.

In all models mentioned above, a constant burning rate of the oil was assumed, which in most experiments corresponded to the pre-war production of oil in Kuwait. But as was remarked in P. Carl (1991) the current data suggest the fears that a more catastrophic scenario is realized. The firefighters are not managing to put out the fires on wells as quickly as was thought possible. Kuwait is plagued by oil lakes, some several kilometers across and more than a meter deep, created by oil spilling from unlit wells.

The post-war official Kuwaiti upper estimate of the rate of oil burning exceeds the pre-war one by about four times, as in one of the scenarios, studied in P. Carl, 1991, where a significant disturbance in the tropical and subtropical hydrology was obtained. Also, the model sensitivity studies, carried out by a group from the Max Planck Institute in Hamburg, with the rate of oil burning raised to 15 million barrels a day resulted in highly significant global effects including extreme temperature and precipitation anomalies and unusually frequent storms.

It can't be excluded that Kuwaiti fires can generate not only regional-scale environmental effects, but they can be a source of global atmospheric anomalies.

An approach to international relations now is under revision. For decades, the Gulf has served as an arena for the Soviet-American contest. The recent situation, for the first time in modern history, gives an opportunity for a democratic approach based on negotiations and conflict resolution.

More definitely the Gulf War has proven the absence of a satisfactory international mechanism for conflict prevention and regulation through non-military ways.

The Gulf War has also proven the global interdependence of the modern world. For instance, the war, characterized as a limited one in military terms, practically leads to some hardly predictable and obviously unplanned social and ecological long-term consequences.

Thus it is a challenge to the security not only of neighboring states, but of superpowers as well. While the notorious superpower contest is not on the agenda any more, there is a unique chance for Soviet-American cooperation in this region and in the World.

The Gulf War put a lot of questions on the world agenda. A few of them: why did the politicians and media start to discuss the ecological catastrophe only after it had already happened? Is it a result of an unwillingness of the anti-Iraqi coalition participants to feel any responsibility for some "by-products" of military actions, or merely an understandable human quality to hope for the best?

Days ago some witty guys called the nuclear winter theory "political physics." But, in our opinion, the neglect of researches of this type can seriously influence the readiness of the international community to estimate and understand the threatening regional and global catastrophes.

REFERENCES

P. Aldous. Oil-well Climate Catastrophe? *Nature* **349,** 96 10 January 1991.

W. Arkin, D. Durrant, M. Cherni. Modern Warfare and the Environment. A Case Study of the Gulf War. Greenpeace study, May 1991, 171 pp.

S. Bakan et al., Climate Response to Smoke from the Burning Oil Wells in Kuwait. *Nature* **351,** 367–371 (1991).

K. A. Browning et al., Environmental Effects from Burning Oil Wells in Kuwait. *Nature* **351,** 363–367 (1991).

P. Carl, Notes on the Climate Response in the Aftermath of Gulf War II. Submitted to *Zeitschrift fur Meteorogie*, 1991.

A. Ginzburg, Climate and Atmospheric Consequences of Nuclear War. *Ambio.* **XVIII** (7) 384–390 (1989).

A. Ginzburg, Climate Shock from Possible Oil Fires in the Gulf Region, Draft paper for British Petroleum Moscow office, 10 January 1991).

A. Ginzburg and O. Kurliandskaya, Gulf War Climatic Shock, Draft paper presented at organizing meeting of International Congress, "Challenges," Berlin, 23 February 1991.

A. Ginzburg, I. Trosnikov, V. Haritonenko, Large-scale Atmospheric Effects of the Large Aerosol Pollutions in the General Circulation Models, Preprint No. 12. IAP AS USSR, Moscow 1990, 33 pp.

J. Hogran. Burning Questions. *Scientific American* (July 1991). pp. 25–27.

F. Warner, Oil Fires and Nuclear Winter, Press release, 28 January 1991.

Part Two:

Climate Change, Energy Technologies,

and

Economics

National Energy Strategies and the Greenhouse Problem

Irving M. Mintzer
Stockholm Environment Institute
Silver Spring, Maryland 20901

ABSTRACT

Energy use is a critical input to many economically important activities. National governments formulate strategies and policies for assuring the availability and reliability of adequate supplies of energy for these activities. But the extraction, mobilization, and conversion of energy resources inevitably generates risks to human societies and to the environment. Among these risks are those associated with the emission of radiatively-active trace substances, commonly called "greenhouse gases." If current policies and strategies continue unchanged, and if the resulting trends in the emissions of these gases continues, there is a significant risk of rapid and disruptive climate change in the decades ahead. To reduce the risks of rapid climate change while preserving the prospects for economic development, national energy strategies must be modified to increase the efficiency of energy use and to develop cleaner, safer, and less-carbon intensive supplies of energy.*

INTRODUCTION

Energy policy plays a vital and dual role in modern societies. The combustion of fuels and the conversion of energy from one form to another fires the engines of economic activity in modern societies. National, state, and local governments choose strategies and implement policies to assure continuous, reliable, and adequate supplies of energy. But the activities that assure supplies of commercial fuels and the processes by which potential energy in the fuel is converted to useful work on the job also produce negative impacts. The effluents and residuals from these activities damage the environment and reduce human well-being. Among the most important risks of energy use are those associated with the release into the atmosphere of invisible, radiatively-active trace gases. These gases are popularly referred to as greenhouse gases. They are transparent to incoming sunlight but absorb and re-emit the earth's outbound infrared radiation, trapping heat close to the planet's surface.[1]

The most important greenhouse gases include water vapor, carbon dioxide (CO_2), nitrous oxide (NO_2), methane (CH_4), the chlorofluorocarbons (CFCs), and tropospheric

---

* This paper is based on a study prepared by the author for the *Journal of the Marine Technology Society* of Washington, DC.

ozone. Human activities are contributing to the buildup of all these gases.[2] The increasing concentrations of these gases are changing the composition and behavior of the atmosphere. Following a major international assessment involving 300 scientists from more than thirty countries, a consensus now exists in the atmospheric science community that if current emissions trends continue the atmospheric buildup of these gases may lead to rapid and dangerous changes in global and regional climates.[3]

Some gases contribute more than others to the warming effect, however. These differences are significant on cumulative, current, and per-molecule bases of comparison. The warming effects of the various gases are sometimes expressed as an index, relative to the effect of an instantaneous emission of 1 kg of CO_2.[4] Assuming a 100-year time horizon for the analysis, and assigning the value of 1.0 on such an index to the global warming potential (GWP) of 1 kg of CO_2, then the GWP of methane is approximately 20, the GWP of NO_2 about 300, and the GWP of the CFCs between 1500 and 3000. (See Table I.) More recent analyses suggest that the IPCC Assessment may have overstated the global warming potential of methane because it ignored certain aspects of chemical interactions in the atmosphere.

TABLE I. Relative Cumulative Climate Effect of 1990 Man-Made Emissions.

Greenhouse Gas	GWP (100-yr basis)	1980 Emissions (teragrams)
CO_2	1	26 000
CH_4	21	300
N_2O	290	6
CFCs	750–3000	0.9
Others	Various	Various

It is not only the relative strengths of the gases but the quantity of emissions that determines their total warming effect. The emissions of CO_2 during the last century have had a cumulative effect that is approximately two-thirds of the total contribution from the buildup of all greenhouse gases. On a cumulative basis, methane buildup has contributed about 15% of the total, CFCs about 6%, NO_2 about 3%, and other gases the remaining 10%.[5] At present rates of emission, CO_2 contributes about 50% of the annual commitment to future global warming, CH_4 contributes 18%, CFCs about 14%, NO_2 approximately 6%, and other gases 13%. (See Fig. 1.)

ENERGY USE AND GREENHOUSE GAS EMISSIONS

Many greenhouse gases are released as a result of humankind's consumption and use of energy.[6] In 1990, the aggregate global use of commercial energy occurred at the rate

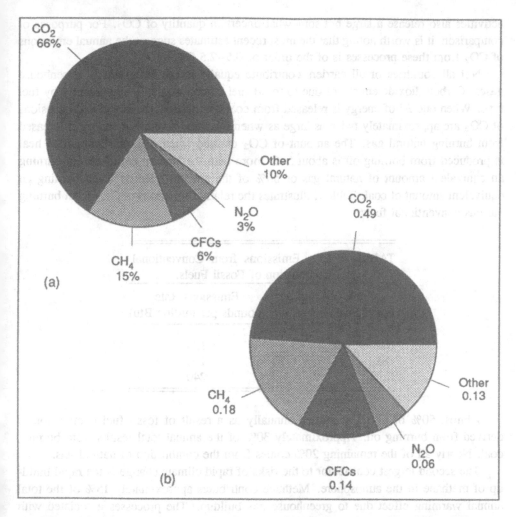

Fig. 1. Greenhouse Gas Contributions to Global Warming: (a) 1880–1980,
(b) 1980s.

of approximately 300 exajoules (EJ) per year or a continuous rate of about 10 terawatts of average power use. (One EJ equals 10^{18} joules or approximately one quadrillion BTU.) About 80% of commercial energy is derived from fossil-fuel resources. These fuels embody solar energy captured and stored through photosynthesis millions of years ago. Of the remainder, about 10% is from traditional biomass fuels, 5% from hydropower, and 5% from nuclear fission.

The use of fossil fuels represents a one-time conversion of complex hydrocarbons into CO_2 and water vapor, most often by processes of rapid oxidation. These processes release approximately six gigatons (GT) of carbon as CO_2 into the atmosphere annually. (One GT equals one billion tons.) Biological processes, deforestation, and land-use conversion activities also release a large but somewhat uncertain quantity of CO_2. For purposes of comparison, it is worth noting that the most recent estimates suggest the annual emissions of CO_2 from these processes is of the order of 0.5–2.5 GT.[7]

Not all countries or all carriers contribute equally to the emissions of greenhouse gases. Carbon dioxide emissions due to fossil-fuel combustion vary significantly by fuel type. When one EJ of energy is released from coal combustion, the associated emissions of CO_2 are approximately twice as large as when the same amount of energy is released from burning natural gas. The amount of CO_2 emitted when a similar amount of heat is produced from burning oil is about 50% more than the amount emitted from burning an equivalent amount of natural gas or 66% of the amount released when burning an equivalent amount of coal. Table II illustrates the relative emissions of CO_2 from burning various conventional fuels.

TABLE II. CO_2 Emissions from Conventional Combustion of Fossil Fuels.

Fossil Fuel	Emissions Rate (pounds per million Btu)
Natural gas	120
Oil	170
Coal	240

Almost 50% of all CO_2 emitted annually as a result of fossil fuel combustion is derived from burning oil. Approximately 30% of the annual total results from burning coal. Nearly all of the remaining 20% comes from the combustion of natural gas.

The second largest contributor to the risks of rapid climate change is the rapid build-up of methane in the atmosphere. Methane contributes approximately 15% of the total annual warming effect due to greenhouse gas buildup. The processes associated with extraction, transportation, and use of fuels are a major source of methane emissions. Approximately 15% of methane emissions are associated with fossil fuel production and use. Underground coal mines are a major source of methane emissions. From 1975–1985, emissions of methane from the ten largest U.S. underground mines increased by 75%, from 69 to 110 million cubic feet per day. Furthermore, the 25 mines that emitted the most methane produced 60% of the total emissions but only 11% of the underground coal mined in the U.S.

Another 15% of methane emissions worldwide results from the burning of biomass in less-than-stoichiometric conditions. Most of these emissions are associated with de-forestation and land-clearing activities. A large fraction of the remainder results from fuelwood use for cooking and water heating.

Energy use is connected to atmospheric pollution in another important way as well. When gasoline is burned in an automobile or wood is burned in a stove, carbon monoxide

(CO) and other Products of Incomplete Combustion (PICs) are emitted. Carbon monoxide is not itself a greenhouse gas, but it reacts strongly with a naturally present species in the atmosphere, the hydroxyl radical (OH). The hydroxyl radical normally reacts with and is the principal sink for methane in the lower atmosphere. If the stock of hydroxyl radicals is depleted by reaction with CO or other pollutants, then there are less of these natural cleansing agents available to react with molecules of methane. As a consequence, the residence time of methane molecules in the atmosphere lengthens and the concentration of methane rises even faster than the estimated increase in methane emissions.

DISTRIBUTION OF GREENHOUSE GAS EMISSIONS BY REGION

Taken together, the total global contribution to future warming also varies by region. Approximately two-thirds of total global primary energy consumption results from activities undertaken by the 20% of the world's population that lives in the advanced industrial economies. These countries also burn the largest fractions of the most valuable or premium fuels—including more than 70% of the petroleum and 85% of the natural gas.

Figure 2 illustrates the regional contribution to the global warming potential during the 1980s. The United States, which represents approximately 5% of the global population, contributed about 21% of the annual commitment to future greenhouse warming in the 1980s. The USSR, with a slightly larger population than the U.S. but a less advanced economy, added about 14% of the total. The European Economic Community contributed

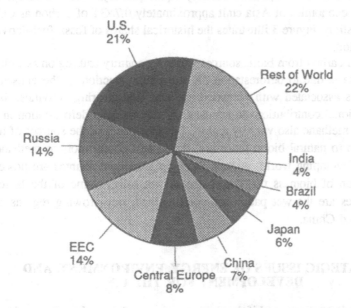

Fig. 2. International Contributions to the Greenhouse Effect, 1980s.

14%, Japan about 6%. Central Europe added about 8%. The developing world, with approximately 80% of the global population contributed the remainder. China alone produced emissions equivalent to about 7% of the total; the remainder of the developing world generated roughly 30% of the total global commitment to future warming.

Although the developed world still leads in emissions of greenhouse gases, the rate of emissions growth is more rapid in the developing world. These societies are involved in a fundamental and historic transition from principally agricultural societies, whose main fuel is an annually cycled supply of biomass, to industrial economies running on commercial energy carriers. In the past the energy economy of these countries was fueled by a flow of carbon that began with the production of biological materials by photosynthesis followed by a subsequent release of carbon through oxidation of those materials back to CO_2 (by either combustion or digestion). The CO_2 was returned to biomass in the following year through a repetition of the photosynthetic process in another plant. But more recently these societies have moved into a period of rapid industrialization in which they have become increasingly dependent on fossil fuel resources. Now, as the per capita use of fossil fuels continues to grow, and the populations of developing countries continue to increase, the regional emissions of CO_2 and other energy-related greenhouse gases is bound to increase.

Nonetheless, it is important from a political standpoint to note that the emissions of carbon dioxide vary not only by source but also by geographic region. The United States which uses about 30% of the world's primary energy, releases almost 25% of the CO_2 derived from fossil fuels each year. The formerly centrally planned economies of the Soviet Union and Central Europe emit a nearly equal quantity. Western Europe and the centrally-planned economies of Asia emit approximately 0.7 GT of carbon as CO_2 from fossil fuel combustion. Figure 3 illustrates the historical shares of fossil fuel-derived CO_2 emissions by region.

The release of carbon from biotic sources varies by country and region as well. Much of these emissions is due to deforestation. A substantial fraction of the emissions due to deforestation is associated with fuelwood use and land-clearing activities. Figure 4 illustrates the national contributions to annual CO_2 release from deforestation in 1980.

Emissions of methane also vary by region. Because many of the sources of methane emissions are due to natural biotic processes, the largest contributors to methane emissions are in the developing world. Among the most important sources are processes of anaerobic digestion of biomass that take place in wet soils. Some of the largest sites for these processes are the wet paddy soils of low-land, rice-growing regions of Asia, including India and China.

STRATEGIC ISSUES OF ENERGY, ENVIRONMENT, AND DEVELOPMENT FOR THE U.S.

If current trends continue, and if developing countries follow the pattern of economic development pioneered by the Western industrialized nations, the atmospheric buildup of

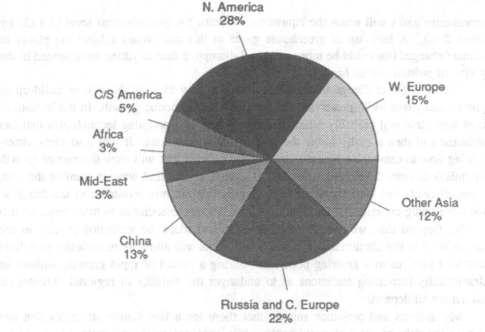

Fig. 3. Regional Distribution of CO$_2$ Emissions from Fossil Fuel Combustion, 1986.

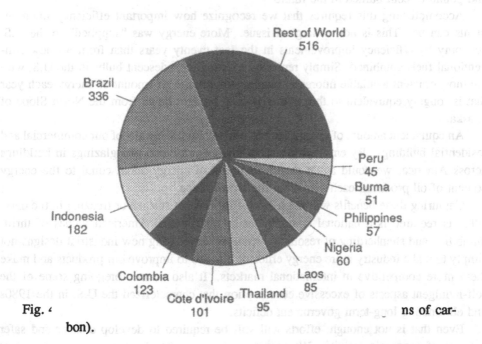

Fig. 4 ... ns of carbon).

greenhouse gases will reach the equivalent of twice the pre-industrial level of CO_2 by about 2030.[8] A build-up of greenhouse gases at this rate would subject the planet to climate changes that could be more rapid and disruptive than anything experienced in the period of written human history.[9]

The strategic challenge that we face today is how to slow the rate of build-up of greenhouse gases while preserving the prospects for economic growth. In the industrialized West, this will probably mean developing new, less-polluting technologies that can stabilize and then actually lower the level of CO_2 emissions. It will also mean determining how to establish trade, aid, and energy policies that will slow the rate of growth in emissions from developing countries. If indeed we can find ways to transfer the best, most efficient, and least polluting technologies that are now available in the North to the developing countries of the South, we can help these societies to meet basic human needs. Beyond that, we can, together, develop and adapt the technologies now in use in the West to the circumstances of the South. This will allow these societies to deliver essential services to a growing population during a period of rapid growth, without so dramatically increasing emissions as to endanger the stability of regional climates on which we all depend.

My analysis and prejudice suggest that there are a few simple strategies that can reduce the risks of rapid climate change while encouraging the prospects for economic growth. First, and most important, is to increase the efficiency of energy supply and use, deriving more physical work and economic value from each barrel of oil, tank of gas, and pound of coal burned in the future.

Accomplishing this requires that we recognize how important efficiency improvements can be. This is not a marginal issue. More energy was "supplied" to the U.S. economy by efficiency improvements in the last twenty years than from all new conventional fuels combined. Simply replacing existing incandescent bulbs in the U.S. with the most efficient available fluorescent lamps would save an amount of energy each year that is roughly equivalent to the oil that is now brought down from the North Slope of Alaska.

An equivalent amount of energy can be "mined" from the walls of our commercial and residential buildings. By employing state-of-the-art insulation and glazings in buildings across America, we could again save an amount of energy about equal to the energy content of oil produced each year from the North Slope.

Capturing these benefits will not require either rocket science or freezing in the dark. What is required is a national commitment to fundamental American values of thrift, durability, and shepherding of resources. It means developing new industrial designs not simply to make industry more energy efficient but also to improve our products and make them more competitive in international markets. It also means foregoing some of the self-indulgent aspects of excessive consumption that characterized the U.S. in the 1980s and eliminating long-term government deficits.

Even that is not enough; efforts will still be required to develop cleaner and safer patterns of economic activity. We cannot save our way to heaven. Instead, we must increase the capacity of biological systems to absorb carbon dioxide. Humankind must preserve the fertility of the Earth's soil and reforest significant areas of the U.S. In parallel, we must discourage deforestation in developing countries and work together toward a situation where there will be no net loss of forests globally by the early part of the next century.

Although all of these actions are necessary in the U.S. to reduce the global risk of rapid climate change, it will not be sufficient. Humankind must also shift the mix of fuels used from carbon-intensive fuels like coal to hydrogen-intensive fuels like natural gas and ultimately to renewable fuels — to hydroelectric systems, to solar systems, and to wind systems. If operated in a safe manner with adequate provision for waste disposal, nuclear-electric systems can continue to be part of the energy mix for decades to come. If this approach is followed, improving the hardware of energy technologies while changing the cultural values and patterns of behavior that determine the levels of energy demand, then humankind may be able to preserve the prospects for economic growth while minimizing the long-term environmental damages from energy supply and use.

REFERENCES

1. World Meteorological Organization, *Changing Climate: The IPCC Scientific Assessment,* Report of the Scientific Working Group of the Intergonermental Panel on Climate Change, Cambridge University Press, Cambridge, UK, 1991.
2. World Meteorological Organization (WMO), *ibid.*
3. WMO, *ibid.*
4. WMO, 1990, *ibid.*
5. Lashof, D., and D. Tirpak, ed., *Policies for Stabilizing Global Climate Change,* U.S. Environmental Protection Agency, Washington, D.C., 1990.
6. Mintzer, I. M., 1990. Energy, Greenhouse Gases, and Climate Change, in *Annual Review of Energy 1990* 15, 513–50, Palo Alto, California; Annual Reviews Inc.
7. Lashof, D., and D. Tirpak, ed., *ibid.*
8. WMO, *op cit.*
9. Lashof, D., and D. Tirpak, *ibid.*

An Overview of Energy Technology Options for Carbon Dioxide Mitigation

D. F. Spencer
Electric Power Research Institute
Palo Alto, CA

I. ABSTRACT

The purpose of this paper* is to perform a preliminary assessment of the options and costs of various carbon dioxide mitigation approaches and to place potential requirements to control global carbon dioxide emissions in perspective. In order to limit carbon dioxide levels in the earth's atmosphere to no more than twice pre-anthropogenic levels, it will be necessary to limit CO_2 emissions to approximately 10 gigatons per year by 2050. The implications of such a constraint to the a) developed countries, b) developing countries, and c) international community are assessed. It is clear that international priorities must be established and specific approaches developed in the first quarter of the 21st century to define the necessary, minimum-cost mitigation strategies.

II. INTRODUCTION

Over the last 100 years, anthropogenic sources of carbon dioxide from fossil fuel use have added approximately 150 gigatons of carbon to the earth's atmosphere. This represents approximately a 25% increase over the pre-anthropogenic level. The present global fossil fuel flux is approximately 6 gigatons of carbon annually and the "airborne fraction" is approximately 0.6. Thus, we are adding 3.6 gigatons of carbon to the atmosphere annually from fossil fuel use. Recent data indicates that deforestation may be a source term of 0.5 to 5.0 gigatons of carbon annually, also in the form of carbon dioxide. With this large range of uncertainty in the source term, it is unclear what the sink for this additional CO_2 flux may be, as the fossil fuel "airborne fraction" can account for the full net increase in CO_2 within the atmosphere.

In order to place the potential fossil fuel CO_2 source term in perspective, it should be realized that stored fossil resources in the earth are equivalent to approximately 5 to 10 times the present atmospheric carbon loading.[1] In addition, the ocean contains 50 to 60 times the atmospheric carbon loading, primarily in the form of dissolved inorganic carbon. Thus, potential carbon releases to the atmosphere are many times the present level, and the earth's overall carbon cycle must be carefully assessed and controlled.

---

* This paper is an excerpt, with minor modifications, from Spencer, D. F., "A Preliminary Assessment of Carbon Dioxide Mitigation Options," *Annual Reviews of Energy and the Environment*, 1991. **16**:259–73.

The recent assessment by the Intergovernmental Panel on Climate Changes[2] discusses two emissions scenarios and three control policies scenarios. The so-called "Business as Usual" or "2030 High Emissions Scenario" estimates CO_2 emissions of approximately 16.5 gigatons of carbon by 2050 and approximately 25 gigatons of carbon by 2100. This leads to CO_2 doubling by 2050 and essentially tripling by 2100 (without consideration for other emission species).

The IPCC study is based on an average annual growth rate in CO_2 emissions over the period 1985–2025 of 0.8 percent in Western Europe, 1.4 percent in North America and Pacific OECD countries, and 3.6 percent in developing countries. In this analysis, I have utilized slightly higher growth rates, based on recent evidence from the last five years, that energy growth is stronger than expected. Therefore, I have adopted a 1.5 percent growth rate in the U.S., OECD, USSR and Eastern European countries, and 4.0 percent for the developing countries. With the recent political changes in Central and Eastern Europe, and their new focus on a market-driven economy, the projections used in this paper may be more representative. Finally, I expect that with the exception of the European and Japanese economies, coal will be the predominant added energy source. This, of course, will maximize CO_2 emissions per unit of energy output.

As a result, unabated CO_2 emissions result in approximately 28 gigatons of carbon emitted by 2050, with CO_2 doubling from fossil fuel usage expected by 2040.

In either case, CO_2 doubling is likely to occur sometime in the period from 2040 to 2060, regardless of one's assumptions, at least in a "Business as Usual" scenario. Therefore, the essence of the arguments put forth in this paper should be valid irrespective of the exact time frame for implementation. Should the recent trends in CO_2 increases from deforestation be as high as recently reported, and the unknown sink for this CO_2 become saturated in its absorption capability for CO_2, doubling could occur earlier than 2040. The combined addition of carbon from both fossil fuel use and deforestation is growing at an increasingly alarming rate, and requires our close monitoring and appropriate mitigation planning.

The consequences of CO_2 doubling are generally the focus for climatic modeling impact studies; however, climatic changes may already be occurring due to increases in the CO_2 concentration, as well as other greenhouse gas emissions. There is presently significant controversy over the issue of carbon dioxide induced temperature increases in the earth's atmospheric temperature, or whether these temperature increases result from natural processes. It is likely that a decade or more of detailed measurements coupled to more specific models will be required to determine clear effects of CO_2 increases. Until we obtain better coupled atmosphere-ocean models, better spatial resolution, a much better understanding of CO_2 sources and sinks, and a better understanding of biogeochemical, cloud, and other feedbacks, the uncertainties in the effects of increasing CO_2 in the atmosphere will remain.

Further, the costs associated with attempting to control CO_2 emissions or sequestering atmospheric CO_2 are much more easily estimated than the benefits. Modest increases in atmospheric temperature ($1-3°C$), sea level rises of less than 1 meter, increased precipitation, etc., may be acceptable. Careful economic analysis of acceptable, consequential, and potential cataclysmic effects of carbon dioxide increases must parallel the global climate change predictions. A near-term focus on developing bounding impacts would be highly beneficial in assessing "acceptable" CO_2 atmospheric concentrations.

With these considerations in mind and based on the projected growth in world population and energy demands, it is expected that CO_2 levels will reach approximately 500 ppm by the mid 21st century, even with very substantial pressure on reducing fossil fuel use and conservation. Although we do not know that this concentration of CO_2 in the atmosphere is acceptable—i.e., we are not certain of the concomitant global change implications—I have chosen a target control level of 500 ppm of CO_2 by 2050.

Assuming only a net increase in CO_2 from fossil fuel use, this level is consistent with emissions of approximately 10 gigatons of carbon in 2050. This increase over today's world levels has been projected to account only for the planned increase of coal use within the Republic of China; i.e., all other countries would have to maintain CO_2 emissions at today's levels. Obviously, some equitably shared carbon emissions budgets must be developed.

The author is very aware of the arbitrary establishment of a 10 gigatons of carbon emission target by 2050. This was done to force consideration of the real implications of such a constraint. In addition, the author is aware of the uncertainties surrounding each estimate of potential carbon displacement or sequestering and associated costs. However, again the intent of this paper is to force specific consideration of each of these mitigation strategies, their limitations, and potential costs. All too often, each strategy is discussed out of context, not on a comparative basis, where it can better be assessed relative to competing approaches.

Beyond 2050, we must seek to eliminate net CO_2 emissions by developing a global strategy to produce and use equal amounts of carbon annually, i.e., no further fossil fuel use. Implementation of such a strategy would most likely require another 50 years to fully implement. This would result in the "maximum" CO_2 concentration of the atmosphere reaching approximately 600–650 ppm by the end of the 21st century.

Although this is a prodigious scientific and technical challenge, I believe we must be developing the framework for such a strategy now. The first steps in such a framework are to assess the various CO_2 mitigation options available, their potential for reducing or eliminating CO_2 emissions, and the approximate cost per ton of carbon displaced or sequestered. The remainder of this paper focuses on these considerations.

In addition, with the significant uncertainties which presently exist relative to the effects of increasing concentrations of CO_2 as well as other greenhouse gases such as nitrous oxides, methane, and chloro-fluorocarbons, we must continuously reassess whether such "target CO_2 control" levels are acceptable. Further strategies which further curtail CO_2 concentrations and their concomitant costs should be assessed. However, as will be seen in this paper, achieving a 500-ppm CO_2 concentration in the atmosphere by 2050 will be extremely difficult to attain.

III. CO$_2$ MITIGATION POTENTIAL

A. Perspective

Recent international attention has been focused on a potential carbon tax to limit CO$_2$ emissions. One such analysis, Ref. 3, indicates that long-run equilibrium values of such a tax would be approximately 250 dollars per ton of carbon, with greater costs to the U.S. and Soviet Union/Eastern European countries in the early 21st century. The analysis in Ref. 3 projects that such a tax would lead to an estimated carbon emission level in 2100 of approximately 8 gigatons/year. Since the airborne fraction is only 0.6, the "effective cost" of reducing "airborne carbon" is 400 dollars per ton of C.

To place these costs in perspective, a 100 dollar per ton control cost would increase coal-based electricity costs by approximately 50%. If we consider this "250 dollar per ton of carbon tax disincentive" as a basis for defining acceptable mitigation costs, up to 100 dollars per ton of carbon would be acceptable and still be significantly less than the proposed carbon tax. I believe this is a maximum acceptable upper bound for increases in cost due to CO$_2$ mitigation requirements. Further, a ton of actual carbon removed from the atmosphere is worth approximately 160 dollars per ton compared to a reduction in the fossil fuel source term. Thus, the "acceptable cost" of various mitigation strategies depend on the approach being used.

Therefore, I will adopt a target cost control range of 100 dollars per ton of reduced fossil fuel emissions and 160 dollars per ton for carbon sequestering approaches. Of course, the following analysis of mitigation options, and associated costs, will reflect our present best understanding of the approximate costs of each option; however, continuing enhanced understanding of each option is necessary. Obviously the lowest cost mitigation options are preferred and are only limited by their estimated CO$_2$ mitigation potential.

B. Mitigation Options

There are numerous approaches which have been proposed for reducing carbon dioxide emissions. In this paper we will classify the various control strategies as applicable to a) developed countries, b) developing countries, and c) the entire international community. It is clear that certain mitigation approaches are more appropriate to the state of economic development of a particular country/region than others.

1. Developed Countries

The U.S., OECD, U.S.S.R. and Eastern European countries essentially represent the economically developed world and presently produce approximately two thirds of worldwide CO$_2$ emissions. With present trends in energy growth within developed countries, unabated energy growth of approximately 1.5 percent per annum is expected. Thus, energy use in the developed countries would double by 2050. This would result in fossil fuel emissions of approximately 8 gigatons per year. It is expected, lacking CO_2 emission constraints, that these countries will represent approximately thirty percent of CO_2 emissions by 2050. If we adopt a target global constraint of 10 gigatons of emissions by 2050, the developed world would be "allocated" 3 gigatons of carbon in 2050—a reduction of approximately 63 percent from uncontrolled levels or 33 percent below today's annual carbon emissions.

The primary approaches for reducing carbon emissions available to developed countries include:

    a) Major enhancement in end-use efficiencies of electricity, oil, and gas;

    b) Enhancement in electricity generation efficiency; and

    c) Powerplant controls of CO_2 emissions including (1) substitution for carbonaceous fuels, i.e., nuclear, solar, other, (2) scrubbing, compression, and sequestering of CO_2 from stack gases, and (3) partial carbon recycling.

Enhancements in End-Use Efficiencies. Since the oil embargo of 1973, energy intensity—the amount of energy required to produce a dollar of U.S. gross national product—has fallen by 28 percent.[4] Further, estimates for reductions in consumption resulting from improvements in end-use efficiency range from 20–75 percent within the U.S.[4,5] In addition, electricity is becoming an increasingly important component of our energy system, now representing 40 percent of our annual energy resource use and projected to grow to well over 50 percent in the future. Increased electrification is particularly desirable where substitution leads to overall efficiency gains, e.g., microwave ovens and freeze concentration. Although not all developed countries will have the same upward efficiency improvement potential, increases in world energy prices, such as those caused by the recent Iraqi invasion of Kuwait, will provide major incentives to conserve.

In the U.S., the growing gap between oil production and demand presents a major energy security threat and opportunity for significant energy savings. Between 1974 and 1988, transportation's share of U.S. petroleum consumption rose from 51 percent to more than 63 percent, and, as we know, transportation continues to be almost totally dependent on petroleum-based fuels. Many believe that the technical and economic potential for doubling fuel economy of cars and light trucks exists. Cars twice as efficient as current generation cars are projected to cost only 200–800 dollars more.[6] And though this first cost would be more than recovered through fuel savings over the life of the cars, manufacturers probably will not install these features in the short term. However, these efficiency improvements should be realized over the next 20–25 years. This would produce a potential savings of 30–35 percent in energy use with current growth rates of annual vehicle miles.

Although other OECD countries do not have this same efficiency improvement potential, there are opportunities to increase fuel efficiency somewhat. In addition, some economies may result from improved processing of crude oil into petrochemicals and further improvements in use of oil and gas for industrial process heat and combined electricity generation. However, it is not expected that improvements of more than five to ten percent will be achieved.

The biggest savings in electricity can be attained in a few areas: lights, motor systems and the refrigeration of food and rooms. In the U.S., lighting consumes about a quarter of electricity, about 20 percent directly, plus another five percent in cooling equipment. In a typical existing commercial building, lighting uses about two-fifths of all electricity directly, or more than half including cooling load. EPRI believes that as much as 55 percent could be saved through cost-effective means.

Exploiting the full menu of efficiency opportunities can double the quantity and more than halve the cost of savings. To capture major electricity savings cheaply, one must not only install new technologies but also rethink the engineering of whole systems, paying meticulous attention to detail.

The U.S. natural gas industry and gas consumers are at the threshold of a great period of challenge and opportunity. Futher R&D can lead to the development of technologies that enhance gas usage. Two areas of particular interest in this regard are gas-fueled cooling technologies (e.g., gas heat pumps) and gas-fueled cogeneration systems (e.g., small gas turbines or fuel cells).

Significant cost savings and energy efficiency improvements are also possible for industrial gas customers by 1) using advanced gas-fueled processes for manufacturing, 2) using improved burner systems, controls, and components, 3) incorporating advanced high-temperature materials, and 4) applying advanced technologies and materials to recover and reuse waste energy.

During the 21st century, natural gas usage throughout the world will be greatly expanded; therefore these end-use technology improvements are crucial, if 25–50 percent overall energy efficiency improvement targets are to be achieved.

Again without carrying out a specific nation-by-nation analysis, we will adopt an expected realizable energy efficiency improvement target of 25–50% by 2050 in developed countries. Thus, end-use energy efficiency improvements could provide a potential carbon emission reduction of two to four gigatons of carbon annually (Table I).

The costs for various end-use efficiency improvements will obviously vary with the specific situation; however, most inexpensive approaches have already been adopted. Studies conducted for EPRI indicate an approximate initial cost of 400–600 dollars per kwe of installed electric end-use technology. Although this may not pertain to other energy systems, we will use it to obtain an approximate estimate of the cost of reducing carbon emissions. These capital costs are equivalent to approximately 40–60 dollars per ton of carbon displaced (Table I). In addition, we have eliminated the need for this electricity equivalent generation/transmission system, i.e., the cost of using carbon.

In this analysis, we differentiate between the estimated cost of avoiding the production of carbon, through increased efficiency, etc. and the "net cost to society." The net cost is the differential cost between the cost of CO_2 reduction and the costs which would have been incurred to meet the energy or electricity demand. For purposes of analysis, I have utilized the costs associated with a coal-fired powerplant as the costs which would have

TABLE I. Potential Carbon Emission Reduction for Developed Countries.

Mitigation Approach	Potential Carbon Displacement/ by 2050	Approximate Cost (Equivalent) ($/ton of C)	Approximate Net Cost (Equivalent) ($/ton of C)
End-use efficiency improvement	2.0 −4.0	40−60	−140 to +20
Generation efficiency improvement	0.2 −1.2	20−60	0− 40
Substitution for carbonaceous electricity production			
• Nuclear	0.2 −2.0	N/A	35− 70
• Solar and other	0.05−0.1	N/A	25−200
CO_2 scrubbing/compression (sequestering costs not included)	1−3	100−200	100−200

been incurred, if the efficiency improvement, or fuel substitution approach, had not been adopted.

If we credit the entire equivalent cost of the generation and transmission system and the coal cost, we have a credit of 180 dollars per ton. If we only credit the conservation system with the coal cost, the credit is approximately 40 dollars per ton of carbon. Thus, the net cost for enhanced end-use efficiency is minus 140 to plus 20 dollars per ton of carbon (Table I). Thus, there is a tremendous economic incentive to define additional opportunities to increase end-use efficiency in the developed countries.

Enhancement of Generation Efficiency. Again, we will focus our attention on electricity generation efficiency improvements. If we assume that approximately 50% of the developed countries' energy use will be in the form of electricity over the next 60 years, the electricity system will be responsible for approximately 50 percent of the carbon emissions. If the conservation measures discussed above are achieved, the total carbon emissions are 4 to 6 gigatons. Over this period, we can anticipate power generation efficiency improvements from such advanced systems as coal gasification — combined cycle or fuel cells, magneto hydrodynamic power generation, or other advanced coal systems in the range of 20–40 percent. If one half of the coal plants in 2050 use advanced coal technology, we have the potential to reduce carbon emissions by 0.2 to 1.2 gigatons (Table I).

Even with these higher efficiencies, it is expected that these advanced coal technologies will have 10 to 30 percent higher capital costs than conventional coal. This translates into a 20 to 60 dollar per ton additional cost for carbon control (Table I). The improved efficiency results in net costs of 0 to 50 dollars per ton of carbon.

Substitution for Carbonaceous Materials. The electricity sector also provides great fuel flexibility from the generation standpoint. Nuclear, solar, biomass, hydro, etc., are all potential substitution approaches for the use of carbonaceous materials. It is difficult to estimate the substitution potential for nuclear power. It could provide the preponderance of electricity generation within the developed world. The technology is available; however, costs, public concerns, and waste disposal are limiting factors. We could certainly envisage anywhere from 100 000 to nearly 1 000 000 MWe being deployed over the next 60 years. This range results in a carbon substitution potential of 0.2 to 2.0 gigatons per year. The costs of nuclear powerplants are expected to be nearly competitive with advanced coal; however, presently we estimate their costs could be substantially greater—perhaps $500-$1,000/kwe; however, fuel costs are approximately one-half that of coal. These translate into an equivalent net carbon cost of 35 to 70 dollars per ton.

Direct solar and other non-carbonaceous energy systems are expected to have only a limited role within the next 50–60 years in the developed countries. (Potential Biomass Substitution is discussed in Sec. 3.) This fundamental limitation is the average low solar availability in these countries and the resulting poor economics of solar systems compared with other energy technologies. Further, most of the solar systems, including wind, are limited to low average annual capacity factors. I can project no more than 50 000 to 100 000 MWe of capacity with an average capacity factor of 0.3. This translates into a potential carbon substitution of 0.05 to 0.1 gigatons. The capital costs of these systems are likely to be $2,000 to $3,000/kwe greater than comparable fossil fired units; however, these fossil units consume natural gas or distillate at prices of 2 to 5 dollars per million Btu. The approximate net cost range is 35 to 200 dollars per ton of C.

CO_2 Scrubbing/Compression/Sequestering. EPRI has conducted a number of studies to estimate the net power loss and costs associated with scrubbing carbon dioxide from coal fired plants. These studies indicate a net power loss of approximately 35% and plant incremental capital costs of $1,000-$2,000/kwe.[7] This translates into a net cost of carbon control of $100-$200/ton. Coal consumption for power production in the developed countries is approximately 1.5 billion tons per year and will likely double over the next 60 years. The degree to which carbon dioxide scrubbing, compression, and sequestering is applicable to these plants is very uncertain; however, we estimate that 1 to 3 gigatons of carbon could be controlled in this manner. Further, it is unclear just how the CO_2 would be sequestered; although recent research work indicates the potential for sequestering CO_2 in a clathrate form at ocean depths greater than 3000 feet. This scrubbing approach is clearly uneconomical, with limited applicability, and certainly would be a "last resort."

There are other technological approaches being considered such as partial carbon recycle through photo-catalytic processes, forest waste substitution for coal, etc. It is extremely difficult to estimate any significant contribution from these sources.

Assessment. It is clear from Table I that the least cost approach to CO_2 mitigation is enhanced end-use efficiency and improvements in generation efficiency to the extent achievable. If a major commitment were made by the developed countries to conservation and these potential carbon displacements could be realized, there would be great global carbon benefit.

Returning to our allocation of 3 gigatons to the developed world by 2050, versus an unconstrained level of 8 gigatons, this could conceptually be achieved within the least cost options in Table I, at net costs no greater than \$70/ton and perhaps less. This certainly provides us a basis for focusing on these options; however, a major commitment to a policy emphasizing these options is required. The maximum annual cost to the developed world would be \$350 billion, well within its economic capability.

2. Developing Countries

The global carbon control situation is much different from the perspective of the developing countries (DC). Total unconstrained growth is projected to produce nearly 20 gigatons of carbon by 2050. If we again are attempting to limit carbon emissions to 10 gigatons by 2050, the "allocation" to the developing world would be 7 gigatons or a limitation in growth by 65%. In addition, the number of options available to the developing world are greatly reduced from that of the developed world.

Enhancement in End-Use Efficiencies. Although end-use efficiency improvements have long-range potential, in many cases, not even today's state-of-the-art commercial products are available nor is the average individual per capita income available to purchase high efficiency, high-cost appliances or systems. Our ability to have these markets served by state-of-the-art appliances and systems is essential in order to limit energy growth, with minimal impact on the economic development of DC's.

Although a careful analysis is necessary to verify these estimates, it is projected that a 10–15% reduction in energy growth can be made with little effect on the economies of these countries, particularly if the primary energy use is in the form of electricity. Beyond this level of conservation, the gross developed products (GDP) of individual DC's is likely to be impacted.

In order to curtail CO_2 emissions, we establish an overall target of 25% reduction in energy growth in the developing countries, from four percent per annum to three percent per annum from now to 2050. This would translate into a savings of five gigatons of carbon annually by 2050. For purposes of estimating the cost of this "control strategy," we assume the gross developed product is decreased by an additional 12.5 percent

(Energy/GDP of 1.0) beyond the level which would have no significant impact on the economies of these countries.

If this approach could be achieved, it would cost developing countries nearly 2.5 to 3.0 trillion dollars of GDP by 2050, or a cost of \$800 to 1000/ton of carbon. This clearly is an unacceptable cost. Thus, the development of an energy conservation strategy which meets global CO_2 control needs, without inordinately taxing the lesser developed countries, must be carefully designed. At this point, energy conservation in developing countries cannot be expected to be much greater than 12.5% or a reduction from 4% per annum growth to 3.5% Thus, energy conservation will only decrease carbon emissions from the uncontrolled projection by 2.5 gigatons annually.

Enhancement of Generation/Transmission Efficiency. Here we focus on efficiency improvements which could be made in electricity generation and transmission systems. If we assume that 50% of the developing countries' energy use will be in the form of electricity by 2050 (Ref. 5), the electricity system will be responsible for approximately 50 percent of carbon emissions. If the conservation measures which have little impact on the economy are achieved, the total carbon emissions in the LDC's would be 17.5 gigatons. In addition to the advanced coal generation efficiency improvements discussed previously, providing potential efficiency improvements of 20–40 percent, there are opportunities to improve transmission efficiencies by an additional 10–20 percent. If one half of the coal plants in 2050 use advanced coal technology, we have the potential to reduce carbon emissions by 1.75 to 3.5 gigatons.

Assuming the costs of these plants are comparable to those within the developed countries, the costs of carbon control are 20 to 60 dollars per ton, with net costs of 0 to 50 dollars per ton.

Substitution for Carbonaceous Materials. Nuclear, solar, biomass, hydro, etc., are all potential substitution approaches for producing electricity rather than using carbonaceous fuels. It is difficult to estimate the nuclear potential in developed countries, but even more so in developing countries. Most of the developing countries which will contribute to major increased energy use — China, India, Thailand, Indonesia, etc., have large coal or other carbonaceous fuel resources. The use of these resources is clearly their "low cost option."

Again, one can envisage nuclear power substitution of 100 000 to 1 000 000 MWe in the DC's during the next 60 years. It may behoove the developed world to subsidize such an approach, since major nuclear substitution plays a significant role in reducing carbon emissions. On the other hand, the potential for nuclear proliferation is greatly increased, and may be a key limiting factor. Further, in the DC's, nuclear plant cost premiums may be greater than in the developed world, so we estimate additional capital costs of \$1000–1500/kwe. Fuel costs may be comparable to coal in these countries; therefore, net carbon costs will be 80 to 120 dollars per ton.

In general, solar and other non-carbonaceous energy systems have a greater potential use in the developing countries due to their geographical location. A major exception to this statement is China, with its large coal resources. However, unless significant cost reductions are made in solar systems, their high specific capital costs per unit energy output will remain a major limitation. Therefore, I believe that solar and other non-carbonaceous energy systems will have only a limited role within the next 50–60 years. (Biomass substitution is discussed in Sec. 3.) The solar contribution is expected to be comparable to that in the developed world and should have comparable cost premiums, or perhaps higher. Thus, carbon substitution of 0.05 to 0.1 gigatons is estimated, at net costs of 50 to 200 dollars per ton of C.

CO_2 Scrubbing/Compression/Sequestering. This option has a great carbon control potential within the DC's; however, costs are prodigiously high (Table II) and this approach is even less likely to be accepted by developing countries.

Assessment. It is clear from Table II that achieving 13 gigatons of carbon reduction from unabated energy development in the DC's is impossible at costs less than 100 dollars

TABLE II. Potential Carbon Emission Reductions for Developing Countries.

Mitigation Approach	Potential Carbon Displacement/ Sequestering by 2050 (gigatons of Carbon)	Approximate Cost (Equivalent) ($/ton of C)	Approximate Net Cost (Equivalent) ($/ton of C)
Conservation/ end-use efficiency improvement	2.5 2.5	Essentialy no cost 800–1000	Essentially no cost 800–1000
Generation transmission system efficiency improvement	1.75–3.5	20–60	0– 50
Substitution for carbonaceous electricity production			
• Nuclear	0.2 –2.0	N/A	80–120
• Solar and other	0.05–0.1	N/A	50–200
CO_2 scrubbing/compression (sequestering costs not included)	5.0 –7.5	100–200	100–200

per ton (maximum potential approximately 8 gigatons). Even the 8 gigaton estimate assumes strong nuclear substitution and adoption of advanced coal technologies. A more likely estimate is 6.0 gigatons. If our carbon control budget is to be achieved, it appears that a global mitigation strategy must be considered which could sequester 5.0–7.0 gigatons of carbon annually, if costs of less than 100 dollars per ton of carbon can be achieved.

3. International/Global Mitigation Potential

The primary mitigation approaches available on a global basis are ecological macro fixation of carbon. These may be terrestrial phytomass species or oceanic phytomass. The enhancement of net primary productivity and ultimate sequestering or substitution of biomass for fossil fuels provides a significant carbon mitigation potential. However, even enhanced biomass production results in required areas of approximately 1 million square miles per gigaton of carbon fixed. Of course, if biomass is utilized for carbon fixation, it must substitute for fossil fuel use in order to reduce net CO_2 fluxes to the atmosphere. Therefore, harvestering, transportation and utilization must be considered in estimating the realizable CO_2 reduction potential.

Terrestrial Phytomass. EPRI has recently supported two assessments of the potential to store carbon in terrestrial systems, increases in forest growth and area and production of halophytes in marginal lands such as salt marshes and deserts. Kulp[8] has estimated that increased forest growth could store 1.1 gigatons of carbon by 2050 and plantations to increase forest area could store an additional 2.5 gigatons (Table III). The analysis indicates the world could meet increased demand for wood products and energy by that time, with half the wood going into long-term storage and half for energy. Delivered costs of such products are highly dependent on harvesting and transportation costs, but are likely to be in the range of 30 to over 100 dollars per ton.

Halophytes can be grown as managed systems in marshes, coastal deserts, and inland salt deserts.[9] It is estimated that perhaps 0.5 million square miles could be managed to store and sequester 0.5 gigatons of carbon. These halophytes could be sequestered initially by turning them into the soil; however, as carbon levels in the soil increase, they would have to be harvested and used for fuel. Halophyte biomass for power generation would cost 100–160 dollars per dry ton of carbon, not including transportation costs. Total costs are likely to be 150–250 dollars per dry ton delivered.

Oceanic Phytomass. The open ocean area of the earth covers over 100 million square miles. Although marine primary productivity in the ocean is estimated to be 25–50 gigatons of carbon annually,[1] the net carbon flux to the bottom of the ocean is only 2–3 gigatons annually. By artificially stimulating production and harvesting or sequestering the oceanic phytomass, very significant amounts of airborne carbon could be stored. (Although it is expected that most of the carbon fixed in the oceanic phytomass

TABLE III. Ecological Macro Fixation.

	Potential Carbon Storage or Sequestering by 2050	Approximate Cost to Combustion/Conversion Site ($/ton of C)
Terrestrial phytomass[a]		
• Increased forest growth	1.1	30 to >150
• Plantations	2.5	
Increased forest area		
• Halophytes	0.5	150–250
Oceanic phytomass[a]		
• Macro-algal species		
Continental Shelf		
Harvested	0.5	300[b]
• Open ocean		(30– 70)[c]
Natural system simulation	1–2	50–100
Artificial systems	1–5	30–210
• Phytoplankton[d]	1	50–100

[a] Area requirements approximately 10^6 square miles per gigaton of carbon stored.
[b] Present state-of-the-art being practiced in the Far East.
[c] Projected long term based on open ocean cost estimates.
[d] Area requirements approximately 18×10^6 square miles.

will be atmospheric carbon, there is some uncertainty as to the fraction which may be dissolved inorganic or organic carbon contained within the sea water).

A wide variety of macro-algal species such as Macrocystis, Laminaria, Gracilaria, Euchema etc., are produced naturally in the ocean. Most of the large kelp beds are found on the continental shelves. It is estimated that 5 percent of the continental shelves could be planted with various macro-algal species,[9] storing approximately 0.5 gigatons of carbon annully. These most likely would be harvested and converted to specialty chemicals, synthetic natural gas, etc. Based on today's yield and harvesting approaches, production costs are estimated at 300 dollars per ton of carbon. These estimates appear very high and should have great potential for reduction as our understanding of artificial macro-algal farms develops.

With the production limitations on the continental shelves, open ocean stimulated macro-algal growth is clearly required for substantial carbon fixation to occur. A recent workshop sponsored by EPRI[10] reviewed the open ocean production and sequestering potential of macro-algal systems. Initially, the focus for carbon fixation would be to stimulate growth of natural macroalgae in major currents and gyres such as the Sargasso Sea. Plant disposal may be by sinking or conversion to fossil fuel substitutes; e.g., methane, and possibly sequestering carbon as CO_2 hydrates (clathrates). The sequestering approaches would result in real carbon extraction of 1–2 gigatons, if major productivity increases can be achieved. The costs associated with this stimulation would mainly be "seed" farms and nutrient addition if the carbon is sequestered by sinking. Minimum nutrient supply from upwelling would cost approximately 30 dollars per ton of carbon, and sequestered costs would be 50–100 dollars per ton.

H. Wilcox has estimated a range of costs for an open ocean orbiting tensioned grid system[10] with costs for the system ranging from 40 to 220 dollars per ton, not including upwelling or nutrient costs. In this case, valuable chemicals, synthetic natural gas, etc., are produced. If we include the costs of upwelling and estimated value of the SNG produced, costs are estimated to be 30 to 210 dollars per ton of carbon substituted. These estimates are derived from an early conceptual design and, of course, require much more rigid analysis to establish them with any confidence.

Finally, studies to stimulate phytoplankton production have been conducted.[11,12] If iron addition can be made uniformly and function as projected, up to 1 gigaton of carbon could be sequestered annually with this approach. However, the entire southern Atlantic and Pacific Oceans (some 18 million square miles) would have to be "fertilized" with iron, and once this experiment is initiated, it must be continued to keep the carbon sequestered; i.e., hundreds of years of fertilization. In addition, the frequency and effectiveness of the iron addition has yet to be established; thus, no accurate overall costs have been estimated. However, costs of 50–100 dollars per ton of C are consistent with costs estimated for a 400-km^2 experiment. If this experiment establishes the feasibility of the approach, these estimates should be approximately valid.

IV. CONCLUSIONS

Controlling net CO_2 emissions in 2050 to 10 gigatons per year will be a prodigious task, if the world population and energy demands grow as projected. The developed world must focus on major end-use and generation system energy improvements. A major nuclear powerplant substitution policy may be necessary. If clear priorities and directions are established, the developed countries could reduce CO_2 emissions to their fair pro rata share at reasonable annual costs (350 billion dollars per year or less).

The developing world cannot both meet its economic growth objectives and comply with global CO_2 emission limitations at reasonable cost. The cost of energy conservation to the economies of these countries becomes totally unacceptable by the mid 21st century. Even with major generation efficiency improvements and nuclear power substitution, CO_2 emissions will be nearly twice the desired level.

It appears that a major international focus on increasing both terrestrial and marine phytomass is necessary to achieve global stability in CO_2 emissions. This focus is necessary not only to meet the CO_2 emission objectives targeted here, but also to provide a firm basis for substitution of annual produced carbon for fossil carbon by the end of the 21st century. This is the only approach for achieving long-term stability in the atmospheric carbon burden.

Further, the imposition of a carbon tax, absent a specific plan for CO_2 mitigation, is not likely to achieve a satisfactory carbon limitation in the earth's atmosphere. The problem is much more complex than a simple tax can solve. Careful attention should be paid to the proper incentives, world CO_2 cost sharing, etc., necessary to meaningfully limit atmospheric carbon levels.

Finally, this assessment is based on current information relating to potentially acceptable atmospheric CO_2 burdens. As additional information becomes available, this analysis and attendant mitigation strategies should be reviewed and modified. However, the framework and analysis results provide both a valuable policy analysis approach and a methodology to establish meaningful research, development, and commercial mitigation strategies.

REFERENCES

1. Sundquist, E. T. and Broecker, W. S. (editors), The Carbon Cycle and Atmospheric CO_2: Natural Variations Archean to Present, *Geophysical Monograph* 32, American Geophysical Union, Washington, D.C., 1985.
2. Climate Change: The IPCC Response Strategies, World Meteorological Organization/United Nations Environment Program, Island Press, Washington, D.C., 1991.
3. Manne, A. S. (Stanford) and Richels, R. G. (Electric Power Research Institute), Global CO_2 Emission Reductions — The Impact of Rising Energy Costs, June 1990 (forthcoming publication in The Energy Journal).
4. Fickett, A. P. and Gellings, C. N. (Electric Power Research Institute), "Efficient Use of Electricity," *Scientific American*, September 1990.
5. Starr, Chauncey and Searl, M. F., Global Energy and Electricity Futures, Chapter VI Energy Policy Book, The Atlantic Council, 29 June 1990 (revised October 2, 1990).
6. Chandler, W. U., Geller, H. S., and Ledbetter, M. R., Energy Efficiency: A New Agenda, American Council for an Energy-Efficient Economy, July 1988.
7. Engineering and Economic Evaluation of CO_2 Removal From Fossil Fuel Powerplants, Vol. I. Pulverized Coal Fired Powerplants, EPRI Report IE 7365 Vols. I and II.
8. Kulp, J. L., The Phytosystem as a Sink for Carbon Dioxide, EPRI EN-6786 Special Report, May 1990.
9. Seaweeds and Halophytes to Remove Carbon From the Atmosphere. EPRI N-7177, February 1991.
10. A Summary Description of the Second Workshop on the Role of Macroalgal Oceanic Farming in Global Change, Newport Beach, California, 23–24 July 1990 (internal EPRI report).

11. Martin, J. H., Glacial-Interglacial CO_2 Change: The Iron Hypothesis, *Paleoceanography* 5 (1), 1–13.
12. Sarmiento, J. L., Modeling Biological and Chemical Controls of Carbon Dioxide, Symposium on Global Change Systems, American Meteorological Society, February 1990.

The Technology and Economics of Energy Alternatives

John Weyant
Director of Energy Modeling Forum
Stanford University
Stanford, CA 94305

I'd like to start by describing what the Energy Modeling Forum (EMF) has been doing for the last fifteen years. We've institutionalized the comparison of energy models. This idea was stimulated by the observation that important energy issues were being analyzed by a number of different groups who were trying to apply different methodologies using different input assumptions to different scenarios of interest.

Generally, we pick a new topic every two years and try to recruit everyone doing analysis in that issue area and put them together with people who use (or could use) such analyses in business and government. We get the whole group together and let them decide what they are interested in studying. They decide how much the models can do and whether or not there are other means of analysis or data that they can use to study the issues of interest. They essentially design their own study by specifying a reference case and a number of different scenarios, and then they spend a lot of time interpreting results and writing a final report.

Our study on the impacts of greenhouse gas control strategies is about halfway finished and a new study will be done on energy conservation. We also observed in our recently completed international oil study that most of the differences in the models were due to differences in the demand-side behavior, and upon further decomposition and analysis, the big differences were in energy conservation behavior. So in both the oil and climate studies, we highlighted the differences in assumptions about energy conservation and illustrated the importance of these differences for policy.

In our climate-changes study we chose to focus on three countries individually: the U.S., U.S.S.R, and China. We lumped all the other OECD countries together.

We also had to think of some reference case assumptions to benchmark the models. We borrowed population and economic growth assumptions from the Intergovernmental Panel on Climate Change (IPCC) but modified significantly the economic story for China and the U.S.S.R. Those of you who have looked at the IPCC numbers probably remember that world population goes up by a factor of 2 between now and 2100, and economic output by a factor of 10, so GDP per capita goes up by a factor of 5.

One fact I remember about these assumptions is that the average Chinese consumer in 2100 is as well off as the average U.S. consumer is today. Of course, we've already heard earlier in this conference a lot about the possibility of using population growth as a policy variable. We could talk for two more days on that issue alone. In fact, we have a raging debate on this issue within our study on U.S. economic growth forecasts. So with huge differences on such a well-studied economy, if you move to other regions and really took seriously the uncertainties involved the range of reasonable projections would be quite big.

We also have adopted assumptions about worldwide fossil fuel availability. For models that use the backstop technology concept we specified three alternatives: a coal-based liquid synfuel priced at fifty dollars a barrel, a non-carbon producing liquid fuel — which might be biomass — based at a hundred dollars a barrel, and a non-carbon producing electric technology that could be either nuclear- or solar-based at 75 mills per kilowatt hour. We also considered an accelerated technology case where the price of the non-carbon liquid fuel was reduced to 50 dollars a barrel and that of the non-carbon electric source down to 50 mills per kilowatt hour.

So we made three classes of assumptions: demographic, economic activity assumptions, natural resource availability assumptions, and technology assumptions. We really didn't do more to standardize inputs than that. Some people had quite a bit more technology detail than we specified and some didn't.

If we had enough time we could probably do something relatively inexpensive to reduce carbon emissions if you are convinced that it is a big enough problem to worry about. It's the getting from here to there that's really going to be the problem. To illustrate that I would like to refer to a chart from the National Academy of Sciences' Climate-Change Mitigation Panel Report that was put together in June or July of this year. The Academy argued that if everything is a hundred percent implemented immediately and technology costs are at the low end of the range of estimates, you get a great deal of negative or very low cost energy conservation. If only twenty five percent of the available options get implemented and costs are at the high end of the range, energy conservation would be much more expensive. Most of the models come in between these assumptions. So what the energy models try to factor in are the institutional barriers that are left out of the full hundred percent low-cost analysis.

I think there are a number of things that are significant. Two of them are explicitly highlighted in the Academy report.

First, no feedbacks to the energy markets or the economy are considered. One negative feedback that can occur when you are trying to reduce carbon is that if you reduce your energy consumption in the developed countries, that is going to push down the price of energy so you might see the rest of the world comsuming a lot more. We had a group talk about carbon taxes in Vienna this June and were accosted by a dozen OPEC representatives who wanted to know why we were proposing to harm the oil-exporting countries by forcing down energy prices via a carbon tax. So that is one feedback.

There are other feedbacks as well. In fact, if we are successful in making these conservation investments, we ought to stimulate economic growth, and that will mean more economic activity, which means more energy demand. I think that is a rather small negative feedback compared to the fuel-markets feedback.

Another constraint on the implementation of conservation technologies are institutional barriers and implementation costs. Let me give you two examples of seemingly easy-to-fix institutional barriers. One is unemployment in the United States. Why do you think we have unemployed people in the United States? It seems nonsensical that people who want to work can't work, yet that seems to happen. Well, that is one thing the macro economists have studied a lot. What causes it primarily is that sharp changes in relative prices make labor less valuable this year than it was last year. Therefore, wage rates should drop when labor demand declines.

Think about that. What actually happens is that the cost-of-living increases and wage contracts and price agreements keep prices and wages from falling. Economists call this "sticky wages and prices," and this actually explains why we have unemployment in the U.S. and why we have more unemployment when we get energy shocks or other kinds of macro shocks.

A second example of pervasive institutional constraint is the way water is allocated in California. Over 90% of the water is controlled by the farmers, less than 10% by all residential and commercial consumers. The average farmer pays one-thousandth as much for each unit of water as a residential/commercial consumer of water. If they could pay one-hundredth as much as we do, we would never have any water shortages like we have had consistently over the past four years. That should be an easy problem to fix, and you would even think that we could go bribe the farmers to give us more water, but for some reason this tends to be one of those things that is very hard to do, in part because there is an incredibly strong farm lobby in the state.

In addition, I learned from last October's LBL conference on experience with energy conservation in the developing countries that it's not correct to assume that these kinds of institutional barriers only pertain in the U.S. There are lots of barriers to be knocked down in all regions of the world. One fellow from India listed ten constraints that you would think of as being totally ridiculous.

Now, the economists will, in their own way, break these institutional constraints down into things that are driven by informational decision-making costs and those that are just truly screwed up, just politically driven. There is obviously some political benefit to someone from those more irrational factors, but I think once the irrationalities are pointed out to more people, they will be eliminated.

I think the revolution in electric utility rate-making started by Pacific Gas and Electric Company in California is instructive. After fifteen years of debate they finally got the idea that if you reward the conservation investments the same way you do the supply options, both rate-payers and investors will be a lot better off. It took a long time to accomplish this revolution and probably it's going to take some time for this idea to catch on. Maybe it will spread to New England Electric and Southern California Edison soon. I'm sure it will even be adopted in the Midwest very soon, but perhaps not outside the U.S. Remember that in many developing countries energy is highly subsidized, so it is priced way below market levels. I can't even guess how long it is going to be until we get all of the institutional barriers out of the way.

Let's look at the impact on the U.S. from a target of a 20 percent reduction in carbon emissions relative to 1990, achieved solely through the energy sector (i.e., not by planting trees or collecting coalbed methane or anything like that). The carbon tax required peaks at anywhere from 60 dollars per ton of carbon up to about 600 dollars a ton according to our study. Unfortunately, the lower end of that spectrum is people who believe economic growth is going to be very low anyway—much lower than the 2.3 percent annual GNP growth rate assumed for the U.S. through 2020. They actually project roughly between 4 percent and about 2 percent economic growth over that period. That's on the margin. You do see a lot of very high prices that decline by 2030, and this is mostly driven by the constraints on the rate of implementation of the new technology options that I mentioned before.

What is interesting is that even though you have high marginal costs in the early years, you are closer to the baseline to begin with, so the total costs are less than in the longer term. Remember, the target is a 20 percent reduction in carbon emissions relative to 1990 levels, not relative to the baseline projection in each year.

In the short run you've got to adopt higher-cost options, but you've got to do less of them because you are closer to the constraint to begin with. In the long run, you have time to implement all the new technologies with no constraints. You sort of freeze technology in 2015 or something like that. So what happens in the energy models is that the supply guys pick it up and you start to level off at about the $200 per ton of carbon range that Dwain Spencer mentioned earlier.

You have to go through a large number of additional calculations to determine which technologies will be on the margin. If you've got an aggregate model, you know it's either going to be the electric or non-electric backstop that will be on the margin. It turns out it is generally the non-electric backstop because the non-carbon electric backstop actually overtakes the carbon-based one in the base case before you get to 2020. There are tons and tons of graphs and descriptive results emerging from our study.

I would like to conclude by giving my views on the strengths and weaknesses of the end-use engineering approach relative to the more macro-economic approach. Now, obviously each approach has its own strengths and weaknesses, and this means that integrating the two approaches together in a sensible way would be the best of both worlds.

I know some people who have tried to do that. Irving Mintzer, for example, took an aggregate Edmond Reilly-type structure and added detailed demand-side technologies; then he got the macro feedback and the energy-market feedbacks. However, I think the real wild card is going to be the implementation constraints, the institutional barriers.

Let me end with my favorite policy recommendations. I think there is one strategy on which both the energy efficiency advocates and the economists would agree: instituting a small carbon tax (say $10–$15 per ton) that might escalate over time, possibly up to the $200 per ton level, say, 25–30 years from now. You may sacrifice some emissions in the short run with the strategy, but if you get more energy efficiency than the economists think, you may actually be tracking under the 20% reduction from 1990 emissions levels.

If you believe institutional constraints are impeding further implementation of energy efficiency options, you would then tilt the equation from whatever starting point, $15 per ton or whatever, towards those options. In addition, at the same time you could be working on reducing the institutional barriers directly, but I think that's really a no-regrets policy. Moreover, I think on the impact side—since the climate change problem is basically a carbon stock problem, not a carbon flow problem—you are not going to cause much higher stock of carbon in 2030 or 2040 by pursuing this kind of policy action as opposed to a fixed target by 2010.

Global Trends in Energy Use: Indications for Research

Robert N. Schock

Energy Program Leader

Lawrence Livermore National Laboratory

In order to understand energy, we must also understand its relation to those other aspects of society with which it is intimately related. Few studies of world energy futures consider the entire dynamic system of energy, economies, population, and environmental impacts. However, these aspects are crucial if we hope to understand future needs and take actions to meet those needs.

In 1990, the world used more than 340 quads of energy (quad = 10^{15} Btu @ 172 Mbbl oil). How much more is open to some debate because of an unrecorded amount of biomass in forms such as wood, plant material, algae, and animal byproducts. In general, biomass is thought to comprise 10 to 15 percent of the world's total energy use. Almost 90% of the remaining energy used was fossil energy as oil, gas, and coal (Fig. 1). The 340 quads represent an increase of almost 20% in the past 10 years (British Petroleum Statistical Review of World Energy, London, 1991). At this same period, energy use increased 53%

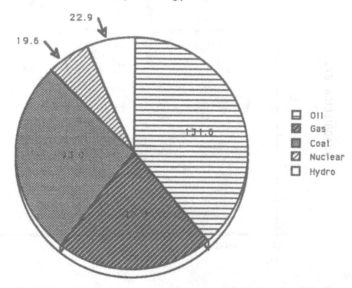

1990 Global Primary Energy (341 Quads)

22.9

19.6

Oil
Gas
Coal
Nuclear
Hydro

Fig. 1. Global energy used in 1990 from a variety of sources. Biomass is not included. (*Statistical Review of Energy*, 1991.)

in Asia, 43% in Africa, and 34% in Latin America. Over the same time, the U.S. and Western Europe increased their energy use by 8% and 6%, respectively. These trends of rapid increases in energy use are indicated in the data in Fig. 2.

What is clear is that while developing countries with almost 80% of the world's population now use 30% of the world's energy this percentage of energy use is increasing rapidly. A result will be increased competition for the limited world energy supplies of the future.

Where will the world's energy come from? At present rates of consumption, world oil reserves will be consumed in about 40 years (C. Masters et al., Science 253, 1991, p. 146). If world use increases at 5% per year (the present energy growth rate in the developing world), this figure drops to 25 years. If use should increase by 10% per year, the world would have less than a 20-year supply.

We will find new oil and begin to produce known oil that is currently too expensive to produce. Yet these sources are limited and are expected to become exceedingly scarce in the early part of the next century. Natural gas is somewhat more abundant than oil, but its use rate is increasing faster and the reserves (C. Masters et al., ibid.) indicate that gas will become scarce even before the middle of the next century.

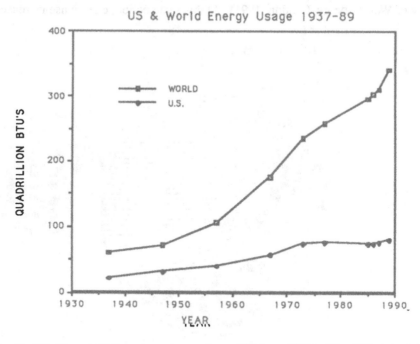

Fig. 2. World and U.S. energy usage from 1937 to 1989. The U.S. trend is typical of the industrialized world. Most of the rapid increases in the past 30 years are due to consumption in the developing world. (Oak Ridge National Laboratory Report, ORNL-6541/V1, 1989.)

Two other sources of energy offer proven options for the world's energy needs. One is the world's large reserves of coal, located primarily in the U.S., the former Soviet Union (Siberia and the Ukraine), and the People's Republic of China. These supplies would support the entire world's energy economy for hundreds of years and if used only to generate electricity for much longer. However, when burned, the higher carbon content in coal results in much more carbon in the atmosphere than is produced from the burning of either gas or oil. This potential for global warming raises questions about its future desirability. Coal use is technically and economically feasible. If breakthroughs can be made that allow coal to be used without as much environmental harm, then its likely use can occur with mitigated environmental damage. More research on technologies to do this must begin now.

The other proven option is nuclear energy from fission. The U.S. notwithstanding, present trends indicate that along with coal, nuclear energy will become a mainstay of the world in the next century. Innovative forms of energy such as fusion, even if they were proven to work today, will take time and resources before they can penetrate the market. It has taken 40 years for nuclear energy to acquire its 6% of the world energy market, an historically rapid rate of penetration.

Several analyses of world energy futures serve as useful guidelines for planning energy research to cope with energy demands. These analyses need not be accurate for our purposes. They merely illustrate the breadth, complexity, and trends of the problem.

Ten years ago, IIASA published one of the earliest studies of all aspects of the energy system (W. Häfele et al., *Energy in a Finite World,* International Institute for Applied Systems Analysis, Ballinger Publishing Co., Cambridge, Mass., 1981). In addition to an examination of the multidimensional aspects of the problem, this study now provides us with a ten-year perspective of projection versus actual use.

Figure 3 shows the lower of two IIASA projections for the year 2030. There were a number of assumptions that went into this projection. Most notable were a world GDP growth rate that decreased from 5% to less than 2% by 2030, an energy intensity ratio that was below recent historical lows, and penetration rates of new technologies (such as solar and biomass) that were somewhat greater than historical rates for new energy technologies.

In the IIASA analysis, oil and gas were used at maximum rates, and coal and nuclear were used to make up the difference. Hydropower, solar, and biomass (almost all of the category "other") are maximized by what were considered to be feasible growth rates. In the intervening ten years, actual world energy consumption has increased at a rate almost 7% beyond this analysis's ten-year projection.

Another scenario produced recently by Häfele considered the effect of possible global warming on energy use (W. Häfele, *Energy Systems Under Stress,* World Energy Conference, Montreal, 1989). For this analysis, Häfele proposed to reduce CO_2 production by one-third before 2030. To achieve this reduction, Häfele assumed that world GDP growth rates decreased to 2.5% now and that world population peaks at 8 billion in 2030. Energy conservation and efficiency improvements are assumed to take place at rates that

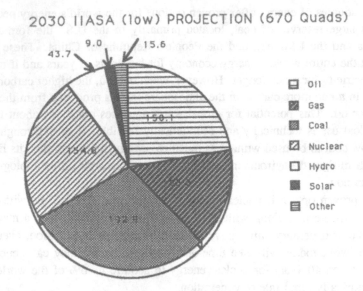

Fig. 3. A projection for world energy use in the year 2030. (W. Häfele et al., 1981.)

are present in the most optimistic environmentalist scenarios. The resulting energy use is shown in Fig. 4.

It is apparent that the goal of CO_2 reduction can be achieved through a significant reduction in the total amount of energy used per capita through the extensive use of new reserves of natural gas, decreased use of coal (a major CO_2 producer), use of solar and biomass at rates that demand very rapid penetration rates into the marketplace, and extensive use of nuclear fission energy as well as slower economic and population growth. Other requirements are too numerous to mention here. However, if coal use is to be limited for environmental reasons, then, in this analysis, the nuclear energy needed 30 or 40 years from now will require construction of hundreds of new nuclear power stations to meet electrical demands that, despite the world's energy conservation, are growing at the same rates as fifteen years ago. Construction of nuclear power stations on this scale would take unprecedented resources and would need to be done on short time scales, a combination that would amount to an enormous worldwide human effort.

Finally, Fig. 5 shows an analysis by J. Goldemberg et al. (*Energy for a Sustainable World*, World Resources Institute, Washington, 1987) for 2020. This scenario uses less energy than today and requires a drastic turnaround in our energy habits. Vastly increased conservation, energy efficiency improvements, very little nuclear fission energy, and extensive use of biomass characterize this scenario. All three scenarios are compared with recent historical trends in Fig. 6.

These analyses indicate that over the next half century the world will have to rely on all energy sources in order to provide the developing world with an opportunity to

2020 SUSTAINABLE WORLD (336 Quads)

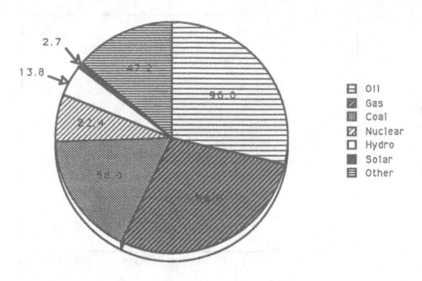

Fig. 4. Projected world energy use in 2030 to achieve CO_2 reduction. (W. Häfele, 1989.)

2030 CO_2 REDUCTION (478 QUADS)

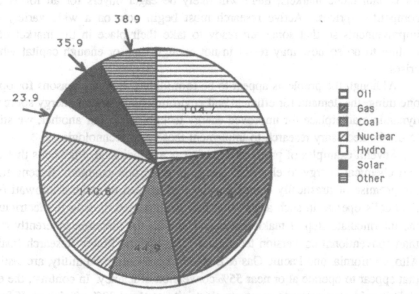

Fig. 5. Projected world energy use in 2020. (J. Goldemberg et al., 1987.)

WORLD ENERGY

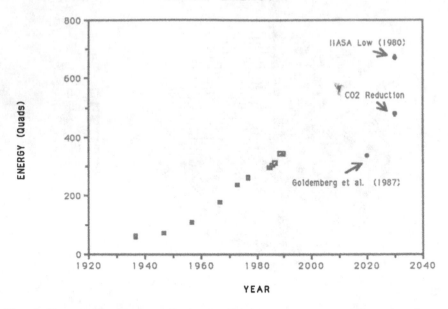

Fig. 6. Recent historical trends in world energy use compared to the three projections from Figs. 3–5.

achieve a reasonable standard of living. Although various forms of energy may compete for certain niche markets, there will likely be eager buyers for all forms of energy at competitive prices. Active research must begin now on a wide variety of technical improvements so that some are ready to take their place in the market of the future. Failure to do so now may result in not enough time or enough capital when the need arises.

Although the problems appear to be formidable, there are reasons for optimism. For one thing, the demand for efficient and environmentally sound energy sources foretells a dynamic marketplace for improved energy technologies. For another, we still have time to do the necessary research to implement these new technologies.

Several examples of prime research areas are indicated. Fuel cells that directly convert chemical energy to electrical energy provide one example. Recent advances hold the promise of drastically decreasing the fuel needed to make a kilowatt of electricity. Fuel cells operate in such a way that fuel is converted directly to electricity, bypassing the intermediate step of making steam. The cells are therefore inherently more efficient than conventional conversion techniques. The Electric Power Research Institute in Palo Alto, California, and Pacific Gas & Electric Co., a California utility, are testing fuel cells that appear to operate at or near 55% conversion efficiency. In contrast, the existing U.S. electric grid now converts energy to electricity at about 32% efficiency. If fuel cells were available now at a competitive cost, we could save over $20 billion each year in fuel costs alone, or 10% of total U.S. energy costs. The U.S. government investment is about

$40 million, or two-tenths of one percent of the annual potential savings. Other promising energy technologies—such as alternative liquid fuels, batteries, advanced electric vehicles, solar-energy conversion, increasing the efficiency of biomass conversion—all receive minimal funding. Even conservative world-energy scenarios show the need for more energy from clean sources beyond what coal and nuclear are likely to provide. We must invest financial resources now in a broad range of technologies to ensure that these sources are available.

Not to act now is to abandon our technical and competitive leadership and to risk a world where increasing numbers of disadvantaged peoples abandon their faith in economic means to raise their standard of living.

The Contribution of Energy Efficiency to Sustainable Development in Developing Countries

Mark D. Levine and Steven Meyers
Lawrence Berkeley Laboratory
Berkeley, California 94720 USA

ABSTRACT

The demand for energy services is growing rapidly in the developing countries. Low levels of energy efficiency in electric power supply and in energy end use mean that the energy sector threatens to absorb an intolerably high share of available financial resources. Energy inefficiency also contributes to local and global environmental problems. A strategy of vigorously improving energy efficiency is thus a key element of a sustainable development path. Results reported in this article show that there is considerable room for efficiency improvement in both existing and new capital stocks, but a much larger effort than presently underway is essential if the potential is to be realized. Assistance from the industrialized countries can play a major role in such an effort.

INTRODUCTION

Economic and social development will not occur in the world's lower-income countries unless they are able to get substantially more energy services than they receive at present. Energy services enable improvements in labor productivity, added mobility, more comfort and convenience—contributing to the development process itself and to an enjoyment of the fruits of this process. Meeting the needs of the developing countries for these services over the next several decades, however, is likely to test the limits of the world's economic, political, and environmental systems; and sustained development is in jeopardy as a result.

In many ways, this poses one of the great technological and policy challenges of the next several generations: increasing energy services for development in Asia, Africa, Latin American and the Caribbean,the Middle East, and Eastern Europe—in many cases by multiples of four or more—during a time when conventional energy paths are no longer adequate, because they are too expensive and too damaging to the global environment. In essence, as a global society we need to find ways to deliver more services while at the same time shift from the delivery systems of the past to the systems of the future.[1]

The features of these systems of the future are still fuzzy at this point, as we learn more about the potentials of renewable energy technologies and other options, but one feature is clear. In industrialized and developing countries alike, the energy production and delivery systems must be far more efficient than they are now. Energy-efficiency improvements will be a key, because in a great many cases they will deliver additional services more cheaply than supply additions, both economically and environmentally.[2]

Moreover, we know that many of the potentials for efficiency improvement are technologically feasible and economically attractive right now, and further potentials can be made attractive with a reasonable amount of additional policy emphasis.[3] While the rest of the vision of our energy future comes into focus, this is one place where we can proceed with confidence: reducing capital requirements for expanding services, reducing environmental impacts associated with those services, and encouraging more attention to the efficient use of resources as an integral part of any strategy for sustainable development.

This paper briefly explores efficiency improvement as an energy strategy for developing countries. First, it describes the evolving energy patterns in these countries and outlines the role that energy efficiency can play in the near and mid-range future. Next, it summarizes the main impacts of these patterns and the issues that must be confronted in improving energy efficiency in many countries, together with the growing record of success in this regard. Finally, it suggests what needs to be done to accelerate energy efficiency improvement in the developing world, as a fundamental contribution to economic and social development.

PATTERNS OF ENERGY CONSUMPTION IN DEVELOPING COUNTRIES

(a) Growth in Energy Demand

Most of the developing world currently faces a growing need for energy services in order to support economic and social development, but inadequate resources to meet that need. On a per capita basis, developing countries in 1987 consumed an average of 18 million Btu in commercial fuels and an additional 8 million Btu in biofuels, compared with more than 130 million Btu in Western Europe and 305 million Btu in the United States. Total primary commercial energy consumption in 1987 was 66 quads in the developing countries compared to 77 quads in the U.S. and 75 quads in the rest of the OECD countries.*

---

* The source of energy data in this section is the International Energy Agency, *World Energy Statistics and Balances 1971–1987*, Paris, 1989. Primary energy consumption includes all non-energy uses; final energy consumption does not include petroleum products used for non-energy purposes, but does include feedstocks for the petrochemical industry. Consumption does not include biomass, use of which is not well-documented. Within the developing world, China accounted for 36 percent of total consumption, Latin America for 24 percent, Asia (excluding China) for 20 percent, Africa for 12 percent, and the Middle East for 9 percent. We have not included as developing countries a number of countries with relatively high per capita GNP that have often been classified as part of the developing world. These countries are Hong Kong, Singapore, Taiwan, South Korea, Israel, Kuwait, Oman, Qatar, Saudi Arabia, and United Arab Emirates. OECD is the Organization for Economic Cooperation and Development and includes Canada, the United States, Western Europe, Japan, Australia, and New Zealand.

Although per capita and total energy consumption in the developing world is less than half that of the developed world, the *rate of growth* in energy consumption has been far higher in the developing countries for the last two decades (Fig. 1). As a result, the share of the LDCs in world commercial energy consumption grew from 14 percent in 1973 to 22 percent in 1987 (Fig. 2). Growth in the LDCs averaged 5.3 percent per year, compared to 0.7 percent per year in the OECD countries, and 3.2 percent per year in the USSR and Eastern Europe together. Reasons for the rapid increase in primary energy consumption in the developing economies include: (1) faster economic growth than in the OECD countries, particularly in the 1970s; (2) migration from rural areas, where energy needs are met primarily with biomass, to urban areas where commercial fuels predominate; (3) penetration of energy-intensive technologies (e.g., increasing use of fertilizers,personal vehicles, and electric appliances); (4) limited capability and resources to improve energy efficiency; and (5) development of energy-intensive industries.

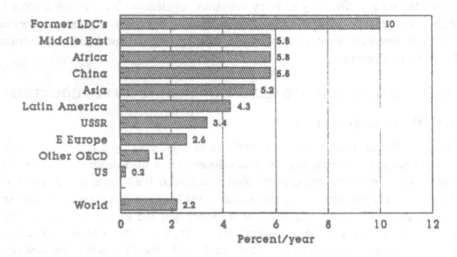

Fig. 1. Average growth rates, 1973–87, primary energy consumption.

(b) Energy Consumption and GDP

Overall, LDC commercial energy consumption* in the developing world grew around 20 percent more than gross domestic product (GDP) between 1973 and 1987. The relationship between energy consumption and GDP has varied between regions and over time (Fig. 3). In Asia (excluding China), energy consumption has grown slightly faster than GDP, especially since 1984. In China, which experienced even more rapid GDP

---

* Through the text we will use the expressions "commercial energy consumption" and energy consumption interchangeably. The expressions exclude traditional (biomass) fuels. If biomass is included, this will be noted explicitly.

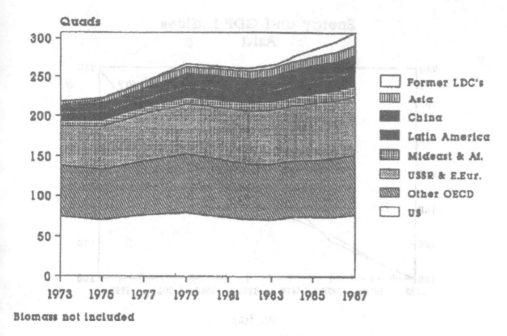

Fig. 2. World primary energy consumption.

growth from the mid-1970s through the late-1980s, increases in energy consumption slowed markedly after 1979, mostly due to improvements in the industrial sector brought about by policies to encourage energy efficiency and major programs to allocate capital to energy efficiency investments.[4] In Latin America, energy consumption has grown somewhat faster than GDP, especially during the recession of the early 1980s. In Africa, energy consumption has risen much faster than GDP, which has increased only slightly since 1980. Under-utilization of industrial capacity may have lowered efficiency in Latin America, while growth in the use of commercial fuels for residential purposes, such as cooking, has contributed to the observed trend in Africa.

(c) Sectoral Energy Consumption

For the developing countries as a group, industry accounted for 50 percent of final energy consumption in 1986, transportation for 22 percent, and buildings and agriculture for 29 percent (Table I). The share of industry ranged from a high of 59 percent in China to a low of 30 percent in the Middle East. Transportation is nearly as important as industry in Latin America. Its relative importance is low only in China, where private transportation is much less used than elsewhere. Energy consumption in buildings and agriculture ranged in share from 36 percent in China, which has much more need for space heating, to 18 percent in the rest of Asia.

Energy and GDP Indices
Asia

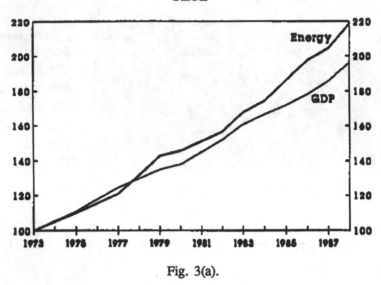

Fig. 3(a).

Energy and GDP Indices
Latin America

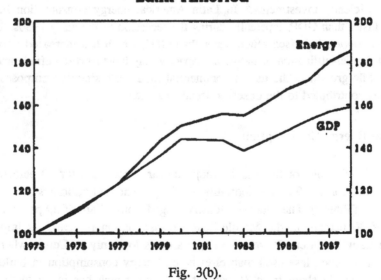

Fig. 3(b).

Energy and GDP Indices
China

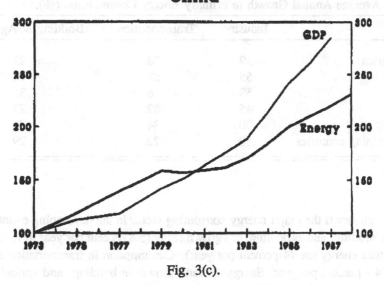

Fig. 3(c).

Energy and GDP Indices
Africa

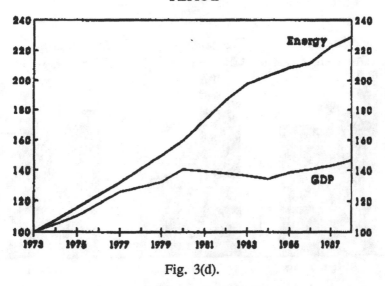

Fig. 3(d).

Fig. 3. Energy and GDP indices.

TABLE I. Average Annual Growth in Primary Energy Consumption (%).

Region	Industry	Transportation	Building & Agriculture
Latin America	39	38	23
Asia	55	27	18
China	59	6	36
Africa	45	32	23
Middle East	30	34	36
Total developing countries	50	22	29

While industry is the major energy-consuming sector in the developing countries, energy use in buildings and agriculture has grown faster (5.8 percent per year in 1973–1986) than industrial energy use (4 percent per year). Consumption in transportation averaged growth of 4.1 percent per year. Energy demand growth in buildings and agriculture was fastest in China and Africa, while it was about the same as growth in industry in Latin America and Asia (Fig. 4). Transport was the fastest growing sector only in the Middle East. Growth in energy use in buildings has been primarily due to increase in appliance ownership, switching from biomass to oil-based fuels for cooking, and construction of modern commercial buildings.

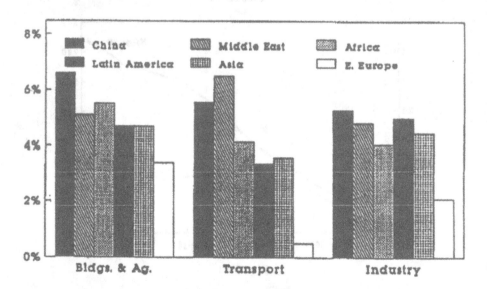

Fig. 4. Sectoral energy use; annual growth rates 1973–86.

(d) Inefficiency of Energy Use

Overall, developing countries require substantially greater quantities of energy resources to deliver one unit of useful energy than do industrialized countries. One analyst estimates that the energy intensity (measured as commercial and non-commercial energy per unit GDP) of developing and Eastern European countries is approximately 60 percent greater than that of industrialized countries.[5]

It is, however, dangerous to infer very much about energy efficiency in developing countries from a comparison of their energy intensity with that of developed countries. Because exchange rates for developing country currencies are often misleading indications of the value of the currency—and other means of valuing local currencies are also uncertain—energy use per unit GDP is an especially uncertain indicator. Also, the structure of economies varies both from one country to another and between developing and developed countries. Thus, even if energy/GDP ratios were accurate and meaningful, they would not necessarily be a good measure of energy efficiency for developing countries.

Nonetheless, it is certain that energy is used very inefficiently in developing countries.[6] The following examples for electricity production and use make this clear:

- The efficiency of generating electricity is often low in developing countries. Typical existing baseload power plants in OECD nations have heat rates (Btus of fuel per kilowatt hour) ranging from 9000 to 11 000. New plants in developing countries are also often in this range. However, as power plants in developing countries age, deterioration in performance is accelerated by a lack of proper maintenance and spare parts, inadequately trained personnel, and by the use of low-quality fuel. The result is that typical existing power plants often operate at 13 000 Btu per kilowatt hour, thereby increasing fuel requirements by 18 to 44 percent.

- Transmission and distribution (T&D) losses are very high in developing countries. A recent survey of 100 developing countries by the World Bank estimated average T&D losses at 17 percent.[7] In OECD countries, typical T&D losses are in the range of 6 to 8 percent. Even correcting losses in developing countries for theft and inadequate billing procedures, it is likely that T&D losses in developing countries are at least 50 percent greater than those in OECD countries.

- The efficiency of use of electricity is often low. Typically, industrial processes — the largest users of electricity in most developing countries — are in the range of 30 percent less efficient than their counterparts in industrialized countries.[6,8,9] Studies of industrial energy and electricity savings opportunities in such countries as Thailand,[10,11] Indonesia,[12] and Egypt[13] indicate reductions in process energy use of 15 to 30 percent with paybacks of one to three years. Comparable gains can be achieved in efficiency of electrical appliances. For example, a recent study in Indonesia suggests that new appliances could have 20 to 30 percent higher efficiencies than current levels with short paybacks, with much greater efficiency gains possible over time.[14]

These losses in the electrical system are compounded. If electricity generation and use are each 30 percent less efficient in developing than in developed countries, and transmission losses (not counting theft) are 16 percent instead of 8 percent, then twice as much energy resources are required to provide end-use service in a developing as in a developed country. Even though the price of electricity to a consumer is highly subsidized in most developing countries, the high economic cost of this power must be borne by the country. It is in many ways a great tragedy that the nations that can least afford to waste economic resources are often unable to avoid such losses, because of inadequate technology, infrastructure, human resources, and essential facilities. Energy — because it is so essential to industrial and economic development, because it demands so much in the way of capital resources — is a prime example of this situation in developing countries.

(e) Electricity

Electricity use merits special attention because growth rates and attendant capital requirements have been especially dramatic in the developing world. Despite the growth in electricity supply, developing countries still use only 500 kWh of electricity per capita per year, compared with 5000 kWh for Western Europe and 10 000 kWh for the United States. Growth in electricity consumption in the developing countries has been higher (7.8 percent per year between 1973–1987) than the increase in overall final energy consumption, which averaged 5.1 percent. Increases in electricity consumption have been well above the economic growth rate. As incomes rise, the demand for amenities associated with electricity services (e.g., comfort, convenience, and increased labor productivity) grows even faster.

REQUIREMENTS FOR AND IMPACTS OF ENERGY EFFICIENCY IMPROVEMENTS

(a) Requirements

Energy efficiency programs, policies, and investments can have a major impact on the growth of energy demand; air pollution emissions felt at the local level (e.g., SO_x, NO_x, and particulates), regional level (e.g., acid rain), and globally (CO_2); and economic development in developing countries.

Figure 5 shows several recent scenarios of energy demand in developing countries through 2025. The Intergovernmental Panel on Climate Change (IPCC) scenario shows primary commercial energy demand in developing countries increasing by a factor of almost four over the forth-year period 1985 to 2025. The World Energy Conference moderate scenario (labelled WEC-M in Fig. 5) and the Global Energy Efficiency Initiative

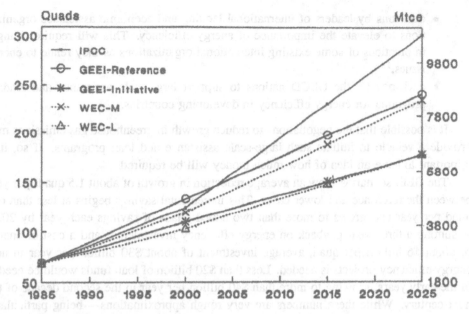

WEC 2020 projections are extrapolated to 2025. All WEC projections are adjusted.

Fig. 5. Primary energy use in developing countries; projections from various
scenarios (WEC 2020 projections are extrapolated to 2025. All WEC
projections are adjusted.)

(GEEI) reference case both show energy use in developing countries increasing about
threefold over forty years.*

The scenarios labelled WEC-L and GEEI-Initiative both represent cases in which
substantial gains in energy efficiency take place over the next decades. Both of these
cases show energy demand growth reduced by about 40 percent from the reference cases.
This is accomplished through efficiency gains alone in the GEEI-Initiative scenario, which
makes explicit many of the requirements associated with this magnitude of reduction of
energy growth. The savings, amounting to some 60 quads per year (75 percent of total
U.S. energy consumption today or about $400 billion per year at current energy prices)
requires:

- decisions by leaders of developing countries to place very high priority on energy
  efficiency which must be considered of equal or greater importance than increased
  energy supply; this also means that energy efficiency must be elevated on the
  economic and social development agendas.

* The descriptions of the scenarios in Fig. 5, including the assumptions underlying them,
are found in World Energy Conference (1990), IPCC (1990) and Levine, et al. (1990). For
purposes of comparison, we have extrapolated the results of the World Energy Conference
Scenarios to 2025.

- decisions by leaders of international lending and economic assistance organizations to elevate the importance of energy efficiency. This will require changes in practices of some existing international organizations as they relate to energy issues.[15]
- decisions by the OECD nations to support large-scale assistance and lending programs for energy efficiency in developing countries.

It is possible that the negotiations to reduce growth in greenhouse gas emissions may provide a vehicle to initiate such large-scale assistance and loan programs. If so, it is important to have an idea of how much money will be required.

The GEEI scenarios show an average reduction in growth of about 1.5 quads per year between the reference and lower cases. This incremental savings begins at less than one quad per year and grows to more than two added quads of savings each year by 2025. Assuming a three-year payback on energy efficiency investments* and a cost of energy of about $6 billion per quad, average investment of about $30 billion per year in new energy efficiency projects is needed. Less than $20 billion of loan funds would be needed in the early years, growing to more than $40 billion per year in the second decade of the next century. While these numbers are very rough approximations — being particularly sensitive to the ability of numerous institutions in the developing and developed world to execute efficiency projects on a very broad scale — they provide a rough measure of the magnitude of the problem.

Of this $30 billion per year investment required, some fraction would come from (1) local sources in developing countries, (2) the private sector of industrialized countries, (3) existing international lending agencies, and (4) OECD sources, whether on a bilateral or collective basis. It is at this time not clear how much of this share each of these four sources is able to likely to bear.

In addition to the loan requirements, assistance funds are needed so that the investment in energy efficiency is productively employed. How much is needed to make such an energy efficiency strategy possible? We do not know the answer. However, two different approaches to thinking about the problem yield somewhat comparable results. The first approach is by analogy with the costs of spurring energy efficiency in advanced industrial countries, such as the United States, through utility demand-side programs, federal and state information programs (e.g., appliance labeling and state energy office activities, home weatherization funds, and federal energy conservation R&D). The second approach is by building up and costing a set of activities — such as large-scale training

---

* This may be a somewhat pessimistic assumption. Experience in the United States has shown a substantial potential for energy savings with paybacks of two years and less. However, considering the virtual non-existence of industries to support energy-efficiency projects in developing countries and the costs of establishing such activities through an economy, we believe this is a reasonable assumption. Note that the assumption of a two-year payback would reduce the capital requirements to achieve the same level of energy savings as shown in the scenario.

and institution building—needed to support developing countries to create markets for energy efficiency. Our best guesses, following both of these approaches, is that *on the order $1 to $2 billion per year of energy efficiency assistance is required for developing countries to create and administer energy efficiency policies and programs* of the magnitude envisioned in this paper. This is on the order of 5 percent of the total required investments.

(b) Impacts

A comparison of the two cases (for either the GEEI or WEC scenario) shown in Fig. 5 shows dramatic differences between them. The efficiency case reduces emissions of CO_2 from developing countries by about 25 percent.[6] Because these cases do not consider either fuel switching or increased growth of renewable energy systems, they do show continued growth in CO_2 emissions from energy production and use at about 2.5 percent per year. However, a long-term effort that reduced growth in coal use by substituting natural gas, renewable energy, and/or nuclear power could reduce CO_2 emissions even further.

From the point of view of developing countries, the major benefit to such an efficiency program is the reduction in capital requirements. It is likely that average capital requirements for energy supply will be cut by about $50 to $60 billion—$25 to $30 billion per year in the near term and $70 to $80 billion per year after 40 years. Thus, the *net* annual savings in investment, after the costs of efficiency investments are subtracted, is $20 to $40 billion per year. Making this magnitude of capital resources available for other development needs in the developing world would have a profound impact on the economic and social conditions of countries throughout the third and fourth worlds. It is difficult to overstate the magnitude of impact that the freeing up of such funds could have. It is as if the entire resources of the World Bank were to more than double in the near term, and then see accelerated growth over many decades. the challenge of making certain that such increased capital goes to productive and valuable projects in the developing world would be substantial.

CONCLUSIONS

Making major progress in these respects will call for wrenching changes of direction in many cases. As noted, "in developing countries, it will need an openness to policy reform and external assistance. . . . In industrial countries, it will need an unprecedented commitment to support energy efficiency improvement through technical, financial, and policy initiatives."[3]

(a) The Agenda for the Industrialized Countries

Global progress depends largely on actions by the high-income countries to set the pace.[2] These actions include making tough policy decisions in order to realize potentials for efficiency improvement. Where global issues such as climate change are concerned, relatively small-percentage improvements in energy efficiencies in affluent countries can make more difference than relatively large-percentage improvements in poor countries. Moreover, the industrialized countries have more resources to invest in innovation and risk-taking.

In a similar vein, industrialized countries must provide substantial collaboration in a spirit of partnership. In many cases, substantial progress with efficiency improvement in lower-income countries will depend on access to resources that in turn depend largely on the policies and actions of higher-income countries. The main opportunities for initiatives on the part of industrialized countries are related to training, to upgrade technical and managerial capabilities in developing countries; access to capital, to make financing available for cost-effective energy efficiency-improving investments through a combination of public and private-sector sources; and support for institutional development, both national and international, related to a commitment to energy efficiency improvement.

(b) The Agenda for the Developing Countries

The starting point for governing agencies in developing countries is for national leaders and energy sector decision makers to give efficiency improvement careful attention as a development investment option and to pursue it. Moreover, developing countries must improve institutional performance by building a management structure that is goal-oriented, adaptable, and resilient. Efficiency improvements require effective indigenous institutions to implement far-reaching plans for improvement. In spite of the inherent difficulties of institutional change,[16] a strategy for gap-filling and strengthening is urgently needed.

Improving incentives for energy efficiency through policy reforms in such areas as energy efficiency and private-sector roles will require facing challenges for policy reform. Those political economies that need reform are often the most resistant to it, because the required degree of painful adjustment is greater.[3] Regardless, we know from the experience of many countries that incentives such as market signals are a key to efficiency improvement.[17]

Most developing countries understand very well that traditional energy paths — the paths followed by the industrialized countries in support of their economic growth — are unlikely to be feasible in the future. But the social and economic development of these countries still depends on providing more and more energy services. The challenge to the global community is to identify, explore, and implement alternative paths that are at the same time better for the environment, better for developing economies, and better-suited to institutional realities.

Entering the 1990s, with every energy and economic forecast pervaded by uncertainties, the nearest thing to a certainty about these alternative paths is that, in nearly every country, at nearly every level of development, in nearly every context, every efficiency improvement will be a central component of an energy strategy for sustained development. But realizing the potential of efficiency improvement will call for forceful action by North and South alike, working together; and the longer we wait to act, the more serious will be the adverse impacts on developing economies and the global environment.

REFERENCES CITED

1. Wilbanks, T., 1988. "Impacts of Energy Development and Use, 1888–2088," in *Earth '88*, Proceedings of the National Geographic Society Centennial Symposium, Washington: 96–114.
2. Helsinki Symposium, 1990, "Energy and Electricity Supply and Demand: Implications for the Global Environment," issue paper, Senior Expert Symposium on Electricity and the Environment (Helsinki, May 1991), Vienna, October 1990.
3. Katzman, H., L. Hill, M. Levine, and T. Wilbanks, 1990. "The Prospects for Energy Efficiency Improvements in Developing Countries," *The Energy Journal*, forthcoming.
4. Levine, M. D., and X. Lui., 1990. "Energy Conservation Programs in the People's Republic of China," Lawrence Berkeley Laboratory, Berkeley, CA.
5. Goldemberg, J., 1990. "Policy Responses to Global Warming," in J. Leggett, ed., *Global Warming: The Greenpeace Report,"* Oxford: Oxford University Press.
6. Levine, M. D., A. Gadgil, S. Meyers, J. Sathaye, J. Stafurik, and T. Wilbanks, 1990. "Energy Efficiency, Developing Nations, and Eastern Europe," A Report to the U.S. Working Group on Global Energy Efficiency. International Institute for Energy Conservation (IIEC), Washington, D.C.
7. Escay, J. R., *Summary Data Sheets of 1987 Power and Commercial Energy Statistics for 100 Developing Countries*, World Bank, 1990.
8. Gamba, J. R., et al., 1986. "Industrial Energy Rationalization in Developing Countries," World Bank, Johns Hopkins University Press.
9. Goldemberg, J., T. B. Johansson, A. K. N. Reddy, and R. H. Williams, 1988. Energy for a Sustainable World, New Delhi: Wiley Eastern Ltd.
10. Tectakeaw, P., 1988. "Energy Conservation in Thailand" in *Energy Conservation: Proceedings of the International Energy Conservation Symposium*, October 1988, Islamabad, Pakistan.
11. U.S. Agency for International Development, 1988. *Power Shortages in Developing Countries: Magnitude, Impacts, Solutions, and the Role of the Private Sector*, Washington, D.C., March.
12. Chatab, I.N., 1988. "Current Energy Conservation Status in Indonesia," in *Energy Conservation: Proceedings of the International Energy Conservation Symposium*, October 1988, Islamabad, Pakistan.

218

13. Gelil, I.A., 1988 "Egypt's National Energy Conservation Strategy" in *Energy Conservation: Proceedings of the International Energy Conservation Symposium,* October 1988, Islamabad, Pakistan.

14. Schipper, L., and S. Meyers, 1990. "Improving Appliance Efficiency in Indonesia." *Energy Policy* 19 (6), 578–588 (July/Aug 1991).

15. Philips, M., 1990. "Alternative Roles for the Energy Sector Management Assistance Program in End-Use Energy Efficiency" (draft), International Institute for Energy Conservation (IIEC), October 1990.

16. Wilbanks, T., 1990. "Institutional Issues in Energy R&D Strategies for Developing Countries," prepared for Workshop on Energy Research and Development for Developing Countries, Oak Ridge, TN, November 1990.

17. Hirst and Greene et al., 1981. *Energy Use from 1973 to 1980: The Role of Improved Energy Efficiency,* Oak Ridge, Tennessee, Oak Ridge National Laboratory ORNL-CON-79.

Brazil—Energy Policies and Economic Development[*]

José Roberto Moreira and Alan Douglas Poole
Secretariate of Science and Technology
Federal Government of Brazil, Brasilia, Brazil

ABSTRACT

This paper summarizes the evolution of Brazil's energy system since 1970 and how policy was influenced by the shocks of 1973/74, 1979/80, and 1986 in the world oil market. The overriding objective, reduction of oil imports, was largely achieved. However, it is argued that key elements of the strategy—emphasis on substitution of petroleum rather than overall energy efficiency and the objective of self-sufficiency—were exaggerated and will need to be modified. This change would bring both economic and environmental benefits. The importance of improved efficiency for environmental objectives is highlighted. New official planing suggests wider recognition of this, but supply-side planning remains dominated by self-sufficiency objectives and is generally timid in relation to changing increasingly obsolete priorities. In terms of greenhouse-gas emissions, Brazil's energy system is relatively favorable, but could be improved. Some lines of action to reconcile better energy development and environmental objectives are summarized: deepen the incorporation of social/environmental concerns throughout the cycle of energy planning, include externalities in energy prices, seek to improve energy efficiency, establish and enforce environmental norms adapted to Brazilian conditions and priorities, and seek international co-operation for the transition in policy—investments in the reduction of greenhouse-gas emission (which may be cheaper in Brazil) are a good example.

I. PRESENT SITUATION

The evolution of the energy system and the historical priorities in Brazil since 1970 were a consequence of three international events: the two "oil shocks" in 1973/74 and 79/80, where strong and instantaneous oil price increases occurred, and the collapse of oil prices in 1986. These events coincided approximately with changes in the Federal Government, increasing even further the political importance of these dates. We can define, for our analysis, four time spans around these events (see Fig. 1).

- the "miracle" of 1970/73, characterized by a very high rate of economic development (average of 12.4% /year and low oil prices);
- the era of "growth with debt" in 1973/79 still with high rate of economical growth (6.7%/year) and high oil prices, especially in the first years;

* "Brazil—Energy Policies, Economic and Constraints Development," prepared for a Comprehensive Approach to Climate Change Policy Workshop, Organized by CICERO— Center for International Climate and Energy Research, Oslo, Norway, July 1–3, 1991.

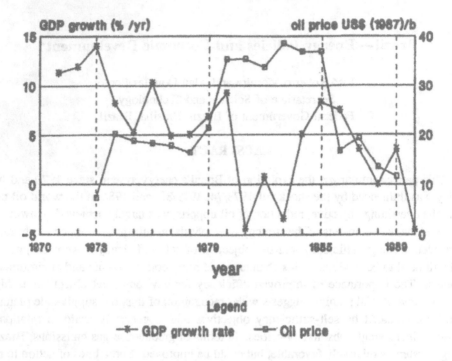

Fig. 1. Economic Growth and Oil Prices.

- the era of "external adjustment and energy substitution" from 1979/85, with low economical growth (2.4%/year) and high oil prices;
- the era of "lost opportunities" in 1985/89 when Brazil was unable to take advantage of the oil price collapse. The average economic growth was low (3.7%/year). While better than the preceding period, it was achieved at the expense of postponing problems. Hyper-inflation showed up at the end of 1989 and the energy system was driven to severe economic difficulties, as shown most visibly in the crisis of the "Proalcool Program"* in 1990.

The year 1990 was characterized by another important international event, the Gulf War, and a new President of the country. The Gulf War consolidated the oil price collapse which occurred in 1986.

Significant investments have been made in the energy system. During the years 1978/87 a total amount of US$ 98.3 billion dollars (in 1987)—that is 15.6% of the "Gross Capital Formation" or 3% of the GNP—was invested to develop the electric, oil, alcohol, and nuclear sectors (see Fig. 2). Figure 3 shows how the money was distributed inside the electric and oil sectors and the evolution of the electric generation

---

* The Proalcool Program is responsible for the large use of straight ethanol (in 5 million cars) and gasoline blended with 22% ethanol (in 6 million cars) in Brazil. The Program started in 1975 and presently ethanol use is larger than use of gasoline.

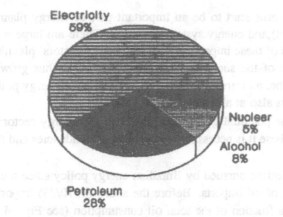

Fig. 2. Energy Sector Investments (1978–1987). 15.6% of gross fixed capital formation, 3% of GDP—total: U.S.$ 98.3 billion.

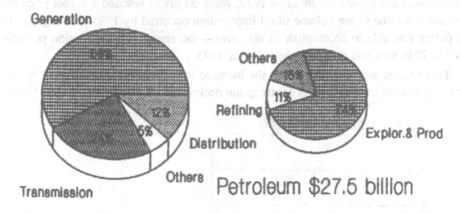

Petroleum $27.5 billion

Electricity $57.8 billion

Fig. 3. Investments and Results (1978–1987). Electricity capacity, 27 → 50 GW; petroleum production, 8.0 → 28.5 mm TPE.

capacity (from 27 to 50 GW) and the evolution of national oil production (from 160 000 to 570 000 barrels/day). Oil importation was another significant source of expenditures and added to US$ 68 billion in the same period. So a total* of US$ 167 billion (1987 dollar)[1] has been spent in the energy sector—an amount one and one-half times higher than our present external debt.

For most of the historical period under consideration, social and environmental impacts were not taken into account, or were treated with low priority. Only recently, after

---

* We are excluding from this figure hard currency expenditures for coal used in the iron and steel industry and some other small earnings from the exportation of gasoline and alcohol.

1985/86, did such concerns start to be an important part of energy planning (from the beginning of the project),and energy system operation. There are large variations in the degree of incorporation of these impacts in different energy sectors' planning. Since then the political dimension of the social and environmental issues has grown. Today this dimension is, and will be, an important boundary condition for energy policy not only at the level of projects but also at a strategic level.

What are the major potential guideline changes in the energy sector with this new approach? To discuss these it is necessary to consider past tendencies and their associated policies.

The overriding objective pursued by Brazilian energy policy since the first oil shock has been the reduction of oil imports. Before the shock (1970/73) imports rose fast, in absolute value and as a fraction of the total oil consumption (see Figs. 4 and 5). In the time span 1973/79 imports continued to grow, but roughly stabilized in relation to total consumption and to the economy's activity (TPE/unit of GDP) (Fig. 6). The situation changed after the second oil shock in 1979, when oil prices reached a higher price level. A massive decline in the volume of oil importation occurred by 1985 (see Fig. 5). Thus the policy was able to accomplish its objective—the amount of importation per unit of GNP in 1989 was less than half the value of 1973.

This decline was due, in part, to the increase in internal oil production (Fig. 7) after 1979. Simultaneously, the oil consumption declined in absolute amount until 1986 (see

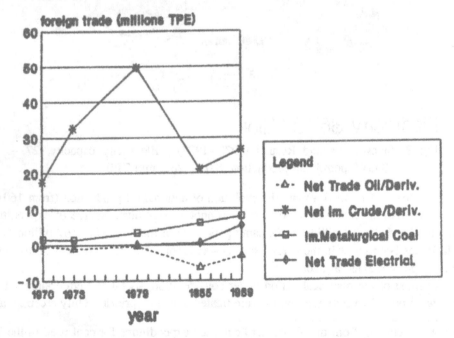

Fig. 4. Foreign Trade in Energy. TPE = ton (metric), petroleum equivalent; (+) imports, (−) exports.

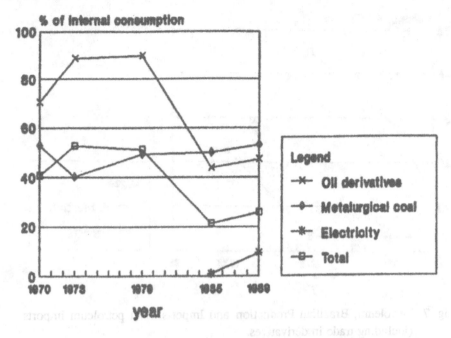

Fig. 5. Net Imports in Relation to Internal Consumption.

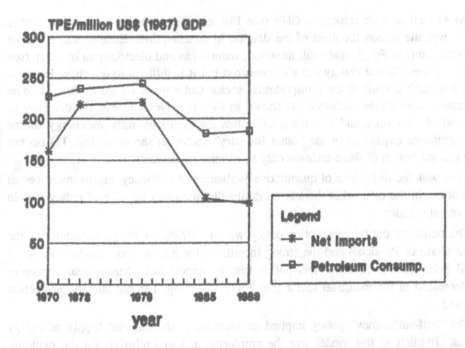

Fig. 6. Petroleum Imports and Consumption in Relation to GDP.

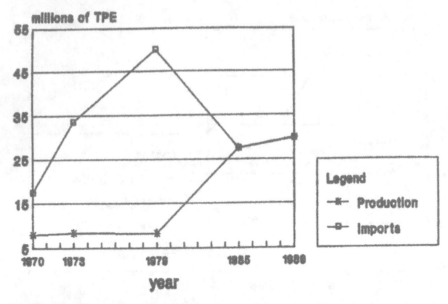

Fig. 7. Petroleum, Brazilian Production and Imports. Net petroleum imports (including trade in derivatives.

Fig. 8) as well as with respect to GNP (see Fig. 6). Oil replacement by other energy sources was the reason for most of the decline in consumption. Ethanol was used for transportation (see Fig. 9) and coal, firewood, natural gas and electricity in industry (see Table I). More efficient energy use also occurred but it is difficult to quantify. Fuel use showed a small decline in the transportation sector and a more visible decline in some industrial sectors ("other industry," as shown in Fig. 10). Nevertheless, the diversity of fuels and of their uses, and also the possibilities for electricity shift (an energy source with significant expansion in the "other industry" sector as shown in Fig. 11) do not allow the attribution of these changes only to efficiency increase.

Even with the difficulties of quantitative evaluation of efficiency improvement versus substitution, in the oil market there is no doubt that the main impulse of policy was in favor of substitution.

The height of the oil substitution policy was in 1979/85, with the launching of the second phase of Proalcool and the strong incentives for national coal production. With the oil price decline in 1986, this policy lost its importance. Steam coal consumption decreased in the industrial sector (see Table I and Fig. 12) and ethanol production stagnated.

The "self-sufficiency" policy implied an increase in the national supply of energy sources. Implicit in this model was the emphasis, almost exclusive, on the exploitation of Brazilian energy resources—neglecting the exchange of energy with neighboring

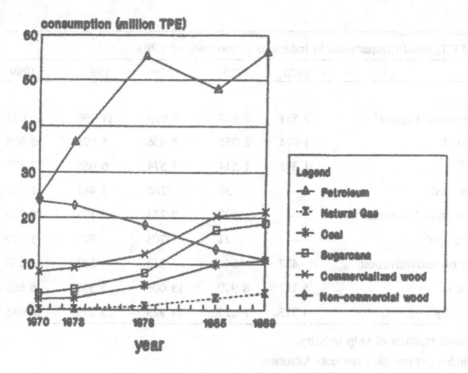

Fig. 8.   Consumption of Primary Fuels.

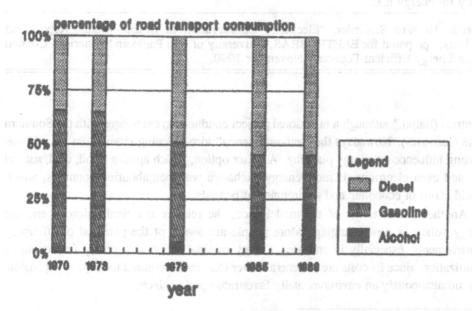

Fig. 9.   Profile of Energy Supply for Road Transport.

TABLE I. Final Consumption in Industries (Thousands of TPE).

	1970	1973	1979	1985	1989
Wood and sugarcane bagasse[a]	7 551	7 868	8 035	11 990	11 071
Charcoal	1 074	2 056	2 656	5 095	6 505
Coke[b]	1 335	1 514	3 574	6 059	7 393
Steam coal	—	59	286	1 481	1 130
Petrochemical materials[c]	—	927	2 224	5 125	6 005
Natural gas[d]	—	21	203	870	1 719
Other oil derivatives/oil	437	640	1 333	882	1 050
Fuel oil	5 117	8 970	13 003	5 351	6 589
Electricity	4 713	7 122	14 923	23 259	28 035

[a]Includes residues of pulp industry.

[b]Includes gas and tar from coke factories.

[c]Includes natural gas and nafta for non-energy use.

[d]Only for energy use.

Source: Howard S. Geller, "Electricity Conservation in Brazil—Status Report and Analysis, prepared for ELETROBRÁS, University of São Paulo and American Council for an Energy-Efficient Economy, November 1990.

countries (Itaipu,* although a bi-national project conducts no exchanges with the Southern Cone Countries). Nowadays the "self-sufficiency" idea is losing momentum but still has a strong influence on energy planning. A better option, which applies to oil, coal, natural gas, and even electricity, is more energy exchange with neighbouring countries, which should result in economic and environmental benefits.

Another characteristic of the model used, the relative marginalization of end-use energy policy, is now changing. More people are aware of the potential of efficiency improvements, especially in end-use. Energy rationalization is a policy for economic optimization, since its costs are in general lower than those required for supply expansion. It is simultaneously an environmentally favourable policy since:

* Itaipu is the largest hydropower plant in the world with an annual production of 70 TWh (1/3 of the Brazilian demand) and 12 600 MW of power.

227

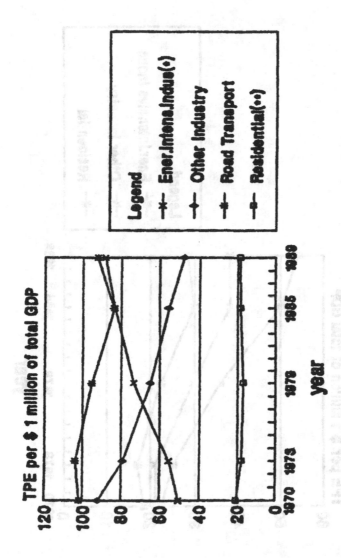

Fig. 10. Intensity of Sectorial Consumption of Commercial Fuels: (*) includes petrochemical feedstock, (**) excludes fuel wood.

228

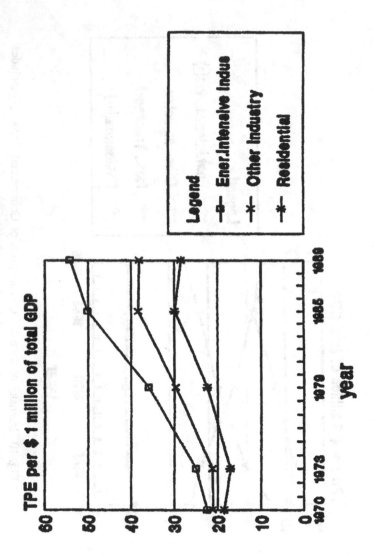

Fig. 11. Intensity of Sectorial Consumption of Electricity.

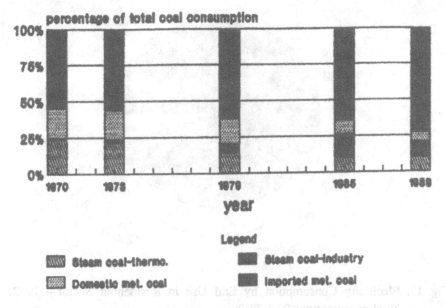

Fig. 12.   Market Profile of Coal Sector.

- energy supply always produces environmental and social impacts; though obviously the degree and types of impact change with the energy source;
- with a low rate of energy demand growth, cleaner energy resources (natural gas, hydroelectric plants with small reservoirs) can have a larger participation. Also, the more environmentally aggressive energy sources (coal, part of the hydroelectric potential of Amazonia), will have more time for the development of improved strategies of exploitation. As such, the effect of an energy demand reduction can be more than proportional in reducing environmental impact;
- measures to increase efficiency are in general synergistic with environmental goals — smoke from the exhaust of a truck is also an energy waste. More generally, major technological advances often deal with several goals, such as higher productivity and less use of natural resources.

A rational policy for the use of energy must address the supply of specific energy services. Figure 13 illustrates, for the residential sector, the shares of these required energy services. The policy should also include the appropriate profile of the economic development model, since it can have a substantial impact on the economy and energy consumption.

There are industrial activities that are energy intensive, that is they consume a high amount of energy in comparison to their aggregate economic value-added or to the employment generated (see Fig. 14). Commonly they are basic materials-processing industries (iron to steel, bauxite to aluminum, etc.). Brazil has already crossed a phase of high growth of these sectors, with an important impact on energy demand (see Figs. 10

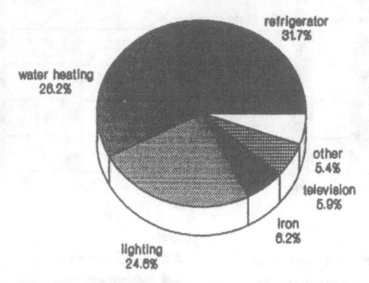

Fig. 13. Electricity Consumption by End Use in Residential Sector—1982.
Total consumption: 27.1 TWh.

and 11). The expectation is that the growth rate of energy-intensive industries will slow down.

The transportation sector is also energy intensive. Generating 4.3% of the GNP, it consumes 20% of the total commercial energy and 52% of the liquid fuels. There exist large differences in the energy intensity of modes. When taking into consideration the possibility of changing the use of modes, it is important for analysis to modify the traditional categories—road, rail, and water transport—especially the first, responsible for 86% of the transportation sector's energy consumption (cargo and passenger). The difference in energy consumption between the private automobile and buses is more significant than the difference between bus and train. Big trucks have characteristics and infrastructure requirements different from medium size trucks. Significant alteration in modal demand trends for transportation will not occur only for energy reasons. However, these can be important in the list of arguments to improve, say, urban collective transport. Historical trends in the modal demand profile and transport infrastructure have been negative as far as energy consumption is concerned, but they have been counterweighted by the favourable increase in vehicles' efficiency, reflecting, with some lag, the world's evolution.

The spectrum of energy policies taken as a whole promoted the increased energy intensity of our economy, as shown in Fig. 15. The most important immediate factor was the large penetration of electricity in the energy matrix. The demand for commercial fuels

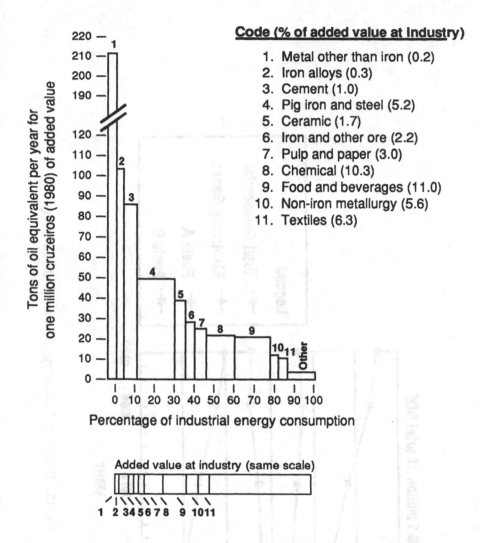

Fig. 14. Energy Consumption and Added Value—Brazil Industrial Sector, 1980.

(excluding firewood for residential and agricultural uses)* per unit of GNP has remained constant or decreased (excluding energy sector self-consumption and its use for sectors other than energy as in "Fuels B" in Fig. 15). The big challenge for the next few years

---

* Residential and agricultural use of firewood is taken as a rough proxy for non-commercial fuel use. Data availability is poor and imprecise. The use of these fuels is still relatively large due to the very low efficiency, and the strong decline can distort interpretation of fuel data.

232

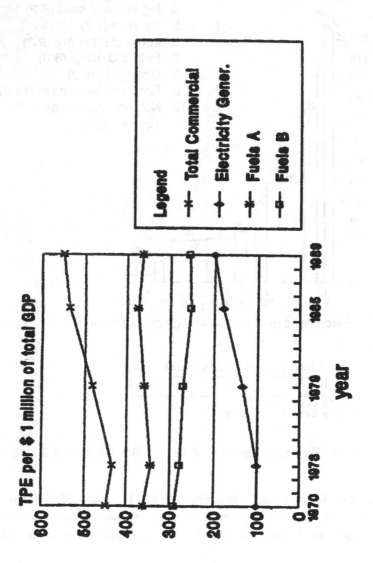

Fig. 15. Evolution of the Energy Intensity of the Economy.

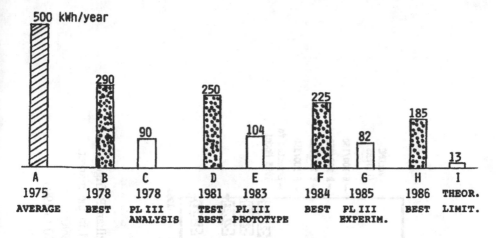

Fig. 16(a). Evolution of Annual Electricity Consumption for a 200-liter Refrigerator with Automatic Defrost without Freezer. PL III is the prototype
developed in the Physics Laboratory of the Technical University of Denmark. (Source: Jorgen S. Norgard, Energy Scenarios in Scandinavia,
Especially an Alternative Energy Plan for Denmark, Physics Laboratory 3, Technical University of Denmark, ABRIL/1983).

is to invert the overall tendency, especially in the use of electricity. The realization of
such a task will bring various advantages for the economy and the environment.

THE FUTURE

Energy growth is a function of the economic growth of the country, of the model
of development and the efficiency of energy use—which is strongly tied to technological advances. There are large opportunities to improve energy efficiency, especially in a country where there is a technological lag of one decade or more
with respect to developed countries in most of the equipment offered to the market. Figure 16 shows the evolution of energy efficient one-door refrigerators in
Brazil as compared with one particular model of refrigerator in Denmark. Table II
lists opportunities for energy conservation in commercial and residential buildings in
Brazil. The major task is the realization of this enormous potential of rationalization. Policies and programmes with this aim showed very modest results in the past.*

* PROCEL—Program for the Conservation of Electricity claims that during the 1986/90
period, a total of 1.5% of total electricity consumption was saved (Ref. 3).

234

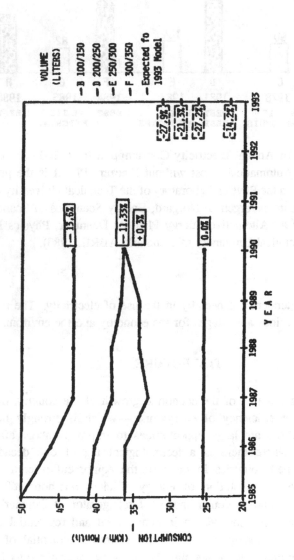

Fig. 16(b). Evolution of Electricity Consumption, One-Door Brazilian Refrigerators. OBS: Figures are average value of products sold in the market.
(——) percentage of reduction between 1985 and 1990;
(– – –) percentage of reduction to be achieved, 1990 to 1993.

TABLE II. Electricity Conservation Supply Curve in 2010 (Improved Technology Scenario).

Sector	Efficiency Measure	Cost of Saved Energy[a] ($/KWh)	Savings Potential (TWh/yr)	Cumulative Savings[b] (TWh/yr)
COM	More efficient refrigeration	0.004	3.0	3.0
IND	More efficient furnaces/boilers	0.011	7.5	10.5
COM	More efficient air conditioning	0.012	5.6	16.1
RES	Heat pump for water heating	0.013	3.4	19.5
IND	More efficient motors	0.014	5.8	25.3
IND	Low-cost measures	0.015	15.6	40.9
IND	More efficient electrochemical procedures	0.016	2.8	43.7
PI	Replacing incandescent lamps	0.016	0.3	44.0
COM	More efficient fluorescent fixtures	0.019	9.3	53.3
IND	More efficient lighting	0.024	2.7	56.0
PI	Replacing self-ballast lamps	0.025	2.6	58.6
RES	Energy-saving incandescent lamps	0.027	1.1	59.7
RES	More efficient air conditioners	0.027	2.4	62.1
RES	More efficient refrigerators	0.029	13.1	75.2
RES	More efficient freezers	0.029	2.8	78.0
PI	Replacing mercury vapor lamps	0.030	3.4	81.4
RES	Power control for el. shower	0.031	4.0	85.4
COM	Conversion to fluorescent lamps[c]	0.031	9.3	94.7
IND	Motor speed controls	0.042	9.9	104.6
RES	Conversion to fluorescent lamps[c]	0.044	6.5	111.1

[a]The cost of saved energy is calculated using a 10% real discount rate (see Appendix for further details).

[b]For reference, 204 TWh of electricity were consumed in Brazil in 1988 and 469 TWh of electricity consumption is projected in 2010 in the base case.

[c]Cost of saved energy based on half of savings from standard and half from compact fluorescent lamps.

Source: Howard S. Geller, "Electricity Conservation in Brazil—Status Report and Analysis, prepared for ELETROBRÁS, University of São Paulo and American Council for an Energy-Efficient Economy, November 1990.

Several basic conditions for the realization of this task are parallel to the necessary conditions for an economic take off—a large reduction in the inflation level,* an increase in investments, reduction in the power of economic cartels, opening up to foreign technology and more demanding consumers. Energy rationalization has restricted prospects during periods of economic stagnation and confusion. In principle, the more dynamic the economy, the better are the opportunities for energy rationalization to occur. However, a set of policies is necessary for the realization of these opportunities. Among these are: non-subsidized energy prices, appropriate tax policy, financing mechanisms, normalization and regulation. Many of these programs are being activated or redesigned by GERE—Executive Group for the Rational Production and Use of Energy.**

A very recent study was performed under the auspices of the Federal Government by the "Commission to Re-examine The National Energy Matrix."[4] This work tried to identify trends and possible strategies to develop the energy sector as well as to provide recommendations to guide the national energy policy and all sectorial plans in the energy area. The conclusions of the study are fundamental to define the country's position in energy—a related question which will be debated during the UNCED-92 Conference in Rio de Janeiro. To better explain the scenarios succinctly we will describe the major aspects of them for the year 2010.

The scenarios are differentiated at two levels. First there are two hypotheses of economic growth:

Year	Low Growth (%)	High Growth (%)
1990 (real)	−4.5	−4.5
1991	0	1
1992	1	3
1993	3	5
1994	5	5.5
1995–2010	5	6

The economy in the year 2010 in the high-growth scenario is 23% larger than in the low-growth scenario.

---

* Inflation in Brazil reached a historic record of 1000% per year in 1990. This was mostly due to the first trimester when the level reached 80% per month. Since the new government began (mid-March, 1990) inflation has grown 400% in more than one year (up to May, 91).

** GERE—is an interministerial group of 3 high-level Secretaries of the Federal Government and 2 representatives of private enterprise. The group has the mandate to recommend improvements and new procedures in the supply and demand side of all energy activities.

Second, for each economic hypothesis there is a "business as usual" and an "alternative" scenario for energy. The "alternative" scenario assumes efficient use of energy resources, greater participation of private capital, larger space for alternative sources of energy—mainly renewables and decentralized production—and larger participation of natural gas. It implies also a smaller than present percentage participation of energy-intensive industries. As an example, in the high-growth economic scenario nafta consumption (by petrochemical industry), coke and charcoal (by the iron and steel industry) would be 20% less than in the equivalent "business as usual" scenario.

Concerning the rational use of energy, the alternative scenario (the preferred one of the Commission) sets very ambitious targets, especially when compared with old forecasts. As an example, electric generation would grow significantly less than GNP when compared with the electric sector's last official long-term plan, the Plan 2010[5] (elasticity 0.70 against 0.90). This reduction is the main factor responsible for very large economies in the volume of investment on the supply side: US$ 26 billion up to the year 2000 and US$ 59 billion from 2001 to 2010. With an economy 23% larger, the alternative scenario for high economic growth consumes 5% less energy and 10% less electricity than the "business as usual" low-growth scenario. Obviously, these results reduce pressure on the environment and natural resources and would be a significant step to adjust energy and environmental goals. The alternative scenario does not exhaust all the potential for increasing the efficiency of the use of energy, in particular in some sectors like transportation, but it is a significant advance in the identification of new goals. The important task is how to translate goals into results and how to implement the necessary policies to mobilize investments in the end-use sector. Such investments are not negligible, but nevertheless are much lower than those required on the supply side and will come mainly from the private sector.*

On the supply side the alternative scenario is less innovative. Natural gas importation is not considered** and the increase in consumption, with respect to the business as usual scenario very modest. The same posture of seeking self-sufficiency is kept for electricity and no international exchange is assumed other than the Itaipu "importation."† A reduction of 27% in oil supply requirements is mainly transferred from the importation market with little change in internal production activities. Overall import dependency falls to 10–12% of today's not very dangerous level of 26%, and is more concentrated in metallurgical coal. The alternative scenario keeps the same target as the

---

* Brazil is suffering an economic recession—the perspectives for 1991, up to now (May) confirm zero or even negative growth.

** Officially known natural gas reserves in Brazil are quite modest and natural gas use is even more modest. Associated and non-associated reserves are 116 billion m^3 and its consumption share on the energy matrix is presently 2%. There is great interest in Brazil to import natural gas from Bolivia, Argentina and Africa.

† Itaipu is a bi-national hydro plant; 50% of its electricity belongs to Paraguay and 90% of this fraction is sold to Brazil. Electricity grid interconnection between Brazil and Argentina, and Brazil and Venezuela is being discussed in an unofficial way.

238

business-as-usual one for the production of national coal (which goes up almost fivefold). This target conflicts with environment goals and economic viability. Another problematic target is the construction of more than 3.7 GW of nuclear energy up to 2010.*

The alternative scenario has some new proposals with regard to the profile of electricity generation. Hydroelectricity decreases its share from 90% to 84%. For the first time sugarcane bagasse receives a significant share as a fuel for electricity generation and generation/cogeneration using natural gas as fuel is considered. Nevertheless thermoelectricity continues to be mainly produced with coal and nuclear. For the generation of electricity from sugarcane, the possibility of the new technology based on the use of gasifiers and gas turbines[6]** is not taken into account, even for sensitivity analysis . The small decrease in hydroelectricity participation together with the major reduction in electricity demand, reduce significantly the pressure to export the hydroelectric potential of Amazonia to other regions of the country, as compared with the last official document (Plano 2010) or the business as usual scenario. But we believe there is still space for further reduction.

The preferred scenario emphasizes two renewable energy forms: ethanol and charcoal. The former keeps its present participation in the transportation sector, doubling production in absolute value to the year 2010. The latter, despite a modest growth (2%/year), increases its share in the iron and steel industry. This target is unlikely to be well accepted by environmentalists. It depends on a total transition from deforestation[†] to man-made forest and appropriate forest management, as well as modernization of the charcoal and iron industries.[‡] Care is justified. Legislation of forest use has existed for

---

* Presently only one commercial nuclear reactor of 600 MW is in operation. With respect to the agreement with West Germany made in 1976, only one 1200-MW reactor is more than half completed and a second one has not yet received permission to restart construction.

** BIG/STIG (Biomass Gasified Steam Injected Gas Turbines) is a new technology, not yet fully proved, which allows the generation of almost 500 kWh/t per tonne of sugarcane processed in sugar and ethanol mills. Using sugarcane bagasse plus sugarcane residues it may be possible to generate around 900 kWh/t. At this upper limit, the sugarcane cultivated today in the country could generate almost as much electricity as the existing hydro system. The cost of investment is expected to be low (US$ 1000/kW out of which gas turbine and electric alternator will cost US$ 260). Up to now there is no record of commercial operation of a pressurized fixed or fluidized-bed gasifier, for biomass, producing a suitable gas to be used in gas turbines. However, experimental results are encouraging.

† Presently it is calculated that approximately 2/3 of the total wood used as energy sources in industrial activities (including charcoal production) comes from deforestation and forest residues.

‡ The efficiency of charcoal production and the amount of charcoal used by the iron industry can be significantly improved. Low-cost investments could decrease primary wood use by 30% per ton of iron produced.

more than a decade and, even so, deforestation remains the main source of wood for charcoal production.

From the perspective of greenhouse-gas emissions, Brazil's energy system is relatively well-placed and the prognosis is good.[7] Brazil's greenhouse-gas problem involves deforestation which is only partially involved with energy. Fossil fuels occupy a relatively small proportion of total commercial energy supply. The alternative scenario would consolidate this situation, especially if the link between charcoal production and deforestation can be broken (as is proposed). It should be noted that most deforestation is not linked to charcoal. Further improvements could be achieved, especially through developing sugarcane electrical generation, downplaying coal, increasing natural gas, and more improvement in transportation energy use. The indirect impact on deforestation related to hydro development in the humid forest regions of Amazonia would be reduced by the slower rate of exploitation, among other measures.

LINES OF ACTION

Some key lines of action to conciliate better energy development and environmental objectives are:

- Continue to strengthen the policy of incorporating the socio-environmental dimension throughout the energy-planning cycle, from development of strategies to operation of plants. While progress is being made, it is necessarily slow. Greater participation of actors from outside the energy sector (from affected communities to other government agencies) is required, as is greater transparency of planning and essential data. The Brazilian energy sector has a tradition of excessive secrecy.
- Incorporate "externalities" in the price of energy. These "externalities" are the cost of mitigating environmental impacts as well as the value of the remaining "residual" impact—such as CO_2. This implies, as a prerequisite, a basic change in the pricing philosophy of the economic authorities which often does not even reflect the "conventional" costs of energy.
- Rationalize energy use. The potential is large and brings both economic and environmental benefits. The greatest potential is in end-use. The case of commercialized biomass is a separate case with large potential for savings from the transformation of raw biomass to end-use.
- Invest in technologies and regimes of operation which reduce impacts. The norms and analytical procedures orienting these investments should be adapted rather than simply copied from abroad. There are distinctive conditions and priorities in Brazil. On the other hand, strong protectionism of relevant equipment manufacture should be avoided.
- Seek synergisms between energy efficiency, environmental and other objectives. Some important interventions, as in the structure of urban transport, cannot be adequately justified by one or only two objectives. More generally it should be

recognized that major technological advances usually occur because they bring various kinds of improvement.

- Move away from emphasizing energy self-sufficiency as a goal. The emphasis should be more on with whom and under what terms the energy trade occurs than on how much. Self-determination is the goal.
- Seek international support for the transition to greater efficiency and reduced impacts. From a world-wide point of view such support may be worthwhile, due to the high return, It can also help break traditional barriers within the country. "Greenhouse-gas quota-related" investments would be an interesting case to examine. It may be much cheaper to produce a ton of CO_2 in Brazil than in Europe.

REFERENCES

1. World Bank—"Brazil: Energy Strategy and Issues Study: Princing and Investment Policy, Washington DC, May 9, 1990.
2. J. R. Moreira, A. D. Poole, M. A. Serrasqueiro, "Alternativas Energeticas e Amazonia," Instituto de Eletrotecnica e Energia/Universidade de Sao Paulo, 1990. See also J. R. Moreira, "Goals and Means of a Sustainable Development in Brazil and South America," Conference on Global Collaboration on a Sustainable Energy Development, Copenhagen, April 25–28, 1991.
3. Personal Information from Wilson Marques, PROCEL Coordinator, ELETROBRAS.
4. Relatorio sobre Matriz Energetica. Comissao de Reexame da Matriz Energetica Nacional, Ministerio da Infraestrutura, May, 1991.
5. Plano Nacional de Energia Eletrica —1987-2010, Relatorio Geral, ELETROBRÁS, Rio de Janeiro, 1987.
6. J. M. Ogden, R. H. Williams, and M. E. Fulmer, "Cogeneration Applications of Biomass Gasifier/Gas Turbines Technologies in the Cane Sugar and Alcohol Industries," Conference on Global Warming and Sustainable Development, São Paulo, Brazil, June 18–20, 1990.
7. J. Goldemberg, "Ecologia no Brasil—Mitos e Realidade—Energia e Meio Ambiente," Secretariat of Science and Technology, Federal Government, Brasilia, Brazil, 1991.

World Population Growth as a Scaling Phenomenon and the Population Explosion

S. P. Kapitsa

Institute for Physical Problems

Moscow, Russia

ABSTRACT

World population growth is treated and portrayed as a self-similar process of development, now entering a critical phase due to the global population explosion. A scaling law is established for all time well before and beyond the critical period. A universal expression valid everywhere is derived, treating the world population as a complex evolving system. These results indicate a limit for the world population of 13–15 billion inhabitants, to be reached by a rapid transformation in terms of history.

World population growth is seen to be the main global problem now facing humanity. In describing and studying this process one can pursue the detailed approach usually practiced in demography.[1,2] In this case the global population is obtained by summing up the partial contributions from all nations and parts of the world, each of which follows its own path of development. For future projections these partial contributions are extrapolated beyond the present. In this paper an attempt is made to treat the global growth as the development of a single system, an entity of its own, exhibiting and following its inherent fundamental properties for growth. Of these properties, we will consider the following three:

The first and, at present, most noticeable feature is the very rapid rise which the global population is now experiencing. This growth is so fast that it has been described as the global population explosion, an explosion whose rise time is remarkably short as compared to the duration of human history and is commensurable with the individual human life span itself.

The second feature or principle involved is that at all times sufficiently removed from this critical period, growth follows a self-similar pattern. In other words, there is no specific time constant involved and the growth is uniform and continuous. We do not in any way take into account the temporary rises and falls in population that happen in various regions and at different times. We consider only the gross overall population expansion—that is, a persistent character of our development right from the origins of humankind.

Finally, as we enter the critical period of rapid change we shall introduce a limit to the global growth rate as imposed by the reproductive nature of the human being and of society.

Let $N = K\psi(t)$ be a function that is to satisfy these conditions. Self-similarity may be expressed as an invariance principle by a functional equation,

$$\psi\left(\frac{t_2}{t_3}\right) = \frac{\psi(t_2)}{\psi(t_3)} \ , \tag{1}$$

for there is no specific time scale to which to refer as long as we are sufficiently far from the critical time t_1.

It may be shown that the function $N(t)$ then necessarily appears as a power law,

$$N = K(t_1 - t)^p \ , \tag{2}$$

expressing the fundamental property of scaling, of the temporal self-similar nature of growth. The invariance to a shift in time is expressed by the choice of t_1. The appropriate constants K and t_1 and the power index p should be determined by world population data.

It was first noticed by von Foerster[3] that world population growth may be described by an empirical expression

$$N = \frac{179 \cdot 10^9}{(2027 - t)^{0.99}} \ , \tag{3}$$

indicating the constants for Eq. (2). Later, von Horner[4] suggested a similar formula,

$$N = \frac{K}{t_1 - t} = \frac{200 \cdot 10^9}{2025 - t} \ , \tag{4}$$

that serves as a good approximation for world population growth as far as one does not approach the critical region near 2025. As Eqs. (3) and (4) diverge at t_1, this has led to consider t_1 as doomsday, after which the population becomes negative! Well before that, as the critical period for the population explosion is entered, we have to modify Eq. (4) to take into account that the growth rate is finite. The growth rate,

$$\frac{dN}{dt} = \frac{K}{(t_1 - t)^2} \ , \tag{5}$$

may be limited by introducing a characteristic time of change τ, eliminating the divergence and making the description beyond t_1 meaningful:

$$\frac{dN}{dt} = \frac{K}{\tau^2 + (t_1 - t)^2} \ . \tag{6}$$

The maximum growth is then attained at $t = t_1$,

$$\frac{dN}{dt_{\max}} = \frac{K}{\tau^2} \ . \tag{7}$$

On integrating Eq. (6), global growth will be described as

$$N - \frac{K}{\tau}\left[\frac{\pi}{2} - \tan^{-1}\left(\frac{t_1 - t}{\tau}\right)\right] , \tag{8}$$

approaching a limit $N_\infty = \pi K/\tau$ for $t \gg t_1$ and indicating the meaning of K. Expanding $\tan^{-1}$ into a series,

$$N = \frac{K}{t_1 - t}\left(1 - \frac{\tau^2}{3(t_1 - t)^2} + \frac{\tau^4}{5(t_1 - t)^4} - \cdots\right) , \tag{9}$$

the first term leads directly to Eq. (4), being independent of τ as long as $t_1 - t \gg \tau$. This asymptotic behavior of Eq. (8) is a significant feature of these solutions, distinguishing them from other models as the logistic ones. The total number of people M who ever lived may be obtained by integrating Eq. (8) from t_0 to t_1 and then dividing by an average age T;

$$M = \frac{K}{2T}\ln\left[\left(\frac{t_1 - t_0}{\tau}\right)^3 + 1\right] \simeq \frac{K}{T}\ln\frac{t_1 - t_0}{\tau} . \tag{10}$$

The relative growth rate halfway to saturation at $t - t_1$,

$$\frac{1}{N}\frac{dN}{dt}\bigg|_{t=t_1} = \frac{2}{\pi\tau} , \tag{11}$$

is less than its maximum value reached at $t_{max} - t_1 - 0.43\tau$ when

$$\frac{1}{N}\frac{dN}{dt_{max}} = \frac{2.3}{\pi\tau} . \tag{12}$$

Applying Eq. (8) to world population data, the values for K, t_1, and τ leading to a limit N_∞ may be calculated (Table I), where other pertinent results are given. These world population growth models are well grouped, and the best-fitting model III leads to $N_\infty = 15 \cdot 10^9$. Before the transition period, all models follow practically identical paths. As K is approximately the same for all cases, N_∞ is mainly determined by τ. For model III, $\tau = 40$ years and the maximum relative growth rate [Eq. (12)] is 1.81% per year to be expected at 1993. This rate is in good agreement with the present estimates based on world demographic data,[5] supporting model III as the optimal case, although model II, where $N_\infty = 13 \cdot 10^9$, gives comparable numbers.

The grouping and stability of these results attest to the robustness of these models. They describe the world population growth on a remarkable time scale within the limits of accuracy imposed by the data available. At best, even in this century, the accuracy

TABLE I.

Model	$10^{-9}N_\infty$	$10^{-9}K$	τ	t_1	$10^{-6}\left(\frac{dN}{dt}\right)_t^{max}$	$\frac{1}{N}\left(\frac{dN}{dt}\right)^{max}$, %
I	10	180	55	1988	60	1.31
II	13	185	45	2005	92	1.60
III	15	190	40	2010	119	1.81
IV	18	195	33	2017	180	2.18
V	25	200	25	2022	320	2.88
VI	∞	200	0	2025	200	2.86

Model	t_{max}	$10^{-9}N_{1990}$	$t_{0.9N_\infty}$
I	1964	5200	2173
II	1986	5135	2140
III	1993	5259	2130
IV	2003	5230	2117
V	2011	5306	2100
VI	(1990)	(5713)	2025

is a few percent, although in demography traditionally more digits are provided for than are realistically valid by their inherent accuracy [see Fig. 1(b).]*

In Table II, data for model III ($N_\infty = 15 \cdot 10^9$) are given in greater detail. The world population losses between World Wars I and II come out in a striking way, although on the scale of Fig. 1(a) they are barely noticeable. Beyond 1990, model III fits well within the forecasts for world population growth estimates by UN,[6,7] tending closer to the upper limit of these projections.[8]

The expression Eq. (8) may be considered as a universal phenomenological formula describing world population growth, following the principles stated. Behavior of this type, described by a scaling law, is often encountered in studies of complex systems, when a large number of processes of a similar nature occurring simultaneously are added up. Even in the case of exponential laws, we finally arrive at a power law, valid over a finite, but large, time span.

---

* Reference 1, p. 4.

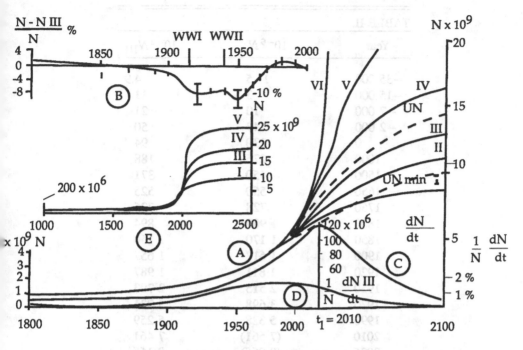

Fig. 1. (a) World population from 1800 to 2100 and projections based on models I to VI with UN maximum and minimum projections (Refs. 5–7); (b) difference between model III and world population data (%); (c) population growth rate for model III; (d) relative population growth rate for model III; and (e) population growth from 1000 to 2500 for models I, III, IV, and V.

A good example is the case, hopefully rather remote from demography, of the decay of radioactivity after fission of uranium, when the effective activity changes as $t^{-1.2}$. This typical scaling law is established shortly after the occurrence of fission, as it is divergent at $t = 0$. It is applicable from 0.1 sec to 100 years with an accuracy of 20%.[9] Thus over ten orders of magnitude in time we are dealing with a power law, although each of the more than a hundred contributing processes are all strictly exponential in their decay. One can also mention the expansion of the universe as described by a scaling law, taking us back to the very beginning of everything. As in the case of fission, we are receding from the critical events of the Big Bang. In the case of the population explosion, we are proceeding in the opposite direction in time and are now entering the critical region.

Applying Eq. (8) to the millennia of the distant past, not unreasonable results are obtained, up to the very origin of humankind some $t_0 = 2 \div 3$ million years ago when $N_0 = 6 \div 9 \cdot 10^4$ is an estimate for the size of the primordial tribe that corresponds

TABLE II.

Year	$10^{-6}N$	$10^{-6}N_{III}$
−35 000	1–5	4.9
−15 000	3	11
−7 000	10	21
−2 000	47	50
0	230	94
1000	275	188
1500	450	371
1650	550	525
1750	728	725
1800	907	894
1850	1 170	1 160
1900	1 617	1 657
1920	1 811	1 987
1950	2 515	2 793
1970	3 698	3 731
1990	5 328	5 259
2010	(7 561)	7 461
2025	(9 065)	9 165
2050	(11 162)	11 192
2130		13 394
2500		14 536

well with the estimate $\sim 10^5$ given by Coppens.[10] The total number of people who ever lived[10] is $M \simeq 100$ billion, if we assume $T = 20$ years.

Representative results are obtained for the period after the neolithic revolution (Table II). Since then not much has changed in the biological evolution of the human being, and a certain continuous pattern of social development has been established. As a global interactive entity, the human system really began evolving since the 16–17th centuries, when we can expect the principles of self-similar development to apply best.

The pattern of development after t_1 may be different from that which preceded t_1. The world population growth rate can decline symmetrically in time following Eq. (6) or a new regime could be set up when, for example,

$$\frac{dN}{dt} \simeq (t - t_1)^{-1} \text{ for } t > t_1 ,$$

leading to a logarithmically expanding world population model. In this case, the change in

$$\frac{1}{N}\frac{dN}{dt}$$

would be less rapid. Finally, we may note that well beyond t_1, in the case of a limited world population growth, the way in which N_∞ is approached also exhibits its self-similar nature.

In the immediate future, the world is heading for a period of rapid changes. At such a time instabilities may develop similar to those that led to World Wars I and II. To what extent can systemic instabilities be expected now? This is certainly beyond the agenda of these calculations, but they can hint at this possibility.

The transition to a limited world population, as it is to take place in the next few decades, would be a great change in the whole pattern of world population growth. It will involve profound changes in the habits, mentality, and life style, even values, of billions of people. The developed countries have evolved into the limited growth mode,[2] a change that can now be expected to happen worldwide in historically a very short time. In the framework of the description of world population growth, this process may seem to be happening independently of economic, technological, social, and environmental factors; their influence is taken into account indirectly by the concept of self-similar growth.

At present, interactions on a global scale by travel, trade, and communications have become a powerful unifying factor. This is leading to a certain synchronization of development for different parts of the world.

Such coherence in change shortens the time over which we can expect the transition to a limited global population to take place, partially explaining the rapidity of the population explosion.

With the onset of the world population explosion and the envisaged limit to growth at the same time, the environmental carrying capacity of our planet may present a limiting factor to development. Thus the population explosion and the approaching global environmental crisis are both happening at the same time, although the changes in population are going on at a greater rate and should be considered as independent, making them the most significant of all global issues.

The author wishes to thank Professor Nathan Keyfitz of IIASA for profitable discussions, Professor G. Barenblatt of the Institute for Oceanology, Academy of Sciences of the USSR, for elucidating some basic properties of scaling, and Dr. Sebastian von Hoerner for drawing my attention to Eqs. (3) and (4) after some of the present results were reported at the 41st Congress of the International Astronautical Federation in Dresden in 1990.

REFERENCES

1. N. Keyfitz and W. Flieger, *World Population Growth and Aging* (University of Chicago, 1990).

2. N. Keyfitz, *Population Change and Social Policy* (Abt Books, Cambridge, 1982); *Scientific American* **235** (2) (1976).
3. H. von Foerster, P. M. Mora, and L. W. Amiot, *Science* **132**, 1291 (1960).
4. S. von Hoerner, *Journal of the British Interplanetary Society* **28**, 691 (1975).
5. "World Population Prospects 1988," Population Series No. 106, New York, UN (1989).
6. A. G. Vishnevskij, "Vosproizvodstvo Naseleniya i Obshchestvo" (in Russian), Moscow: *Finansy i Statistika* (1982).
7. N. Sadik, "World Population," UN Population Fund, 1990.
8. "World Population Prospects 1990," New York, UN, 1991.
9. K. Way and E. P. Wigner, *Physical Review* **73**, 1318 (1948).
10. Y. Coppens, private communication, 1991.

Technology Options to Meet Environmental Requirements: Insight from MARKAL Modeling

John G. Hollins, Environment Canada (OSA)
Ottawa, Ontario K1A OH3, Canada
Tom Kram, Netherlands Energy Research Foundation ECN
P.O. Box 1, 1755 ZG Petten, The Netherlands
Richard Loulou, GERAD, École des Hautes Études Commerciales
5255, Avenue Decelles, Montréal, QC, H3T 1V6 Canada
Shigeru Yasukawa, Japan Atomic Energy Research Institute
Tokai-Establishment, Tokai-mura, Naka-gun, Ibaraki-ken, Japan
Douglas Hill
15 Anthony Court, Huntington, New York 11743, USA

ABSTRACT

The MARKAL model is being used by eleven nations in the IEA Energy Technology Systems Analysis Programme (ETSAP) to examine technology options to meet the threat of global climate change. MARKAL models the possible evolution of national, regional, or provincial energy systems over the next four or five decades. The energy system is represented by alternative sets of typical existing and potential technologies that extract, convert, transport, and use energy. Environmental requirements are established as constraints or objectives that, together with technical feasibility and economics, mold the evolving energy system. Rather than forecasting the future, the model evaluates various means to influence it. Now available in a user-friendly personal computer version, MARKAL offers the possibility of consistent analyses among many countries as energy and greenhouse questions are debated and negotiated in international fora. The initial findings presented here from Canada, Japan and the Netherlands show that technology could play a substantial role as human society seeks to come to terms with global environmental challenges. They also reveal the practical limitations and costs of technological solutions.

MARKAL

MARKAL (for MARKet ALlocation) is computer software designed to represent national, regional, or provincial energy systems as a set of technologies that extract, convert, transport, and use energy. In large systems, the technologies are chosen to be typical of a class of technologies, for example, a typical steam power plant fueled by natural gas. In smaller systems, individual technologies can be identified, for example, a specific proposed hydroelectric project.

Candidate energy technologies introduced into MARKAL are described by their costs and technical characteristics, including their effiency in using energy, and their environmental effects. Table I, for example, shows the coefficients of pollutant emissions by electricity generating technologies used in a MARKAL model of the Ontario energy system.

The model chooses among these technologies to define an optimal energy system as it evolves over time, usually a period of 40 to 50 years. In making the choice, the model must satisfy constraints that define the bounds of feasibility and other limitations. Inputs required by the model are projections of energy demands and the price of imported fuels. Ordinarily, the optimal energy system is defined as the one that meets the requirements at least total cost with future costs discounted at a specified rate, but other objectives can be introduced.

TABLE I. Example of Pollutant Emission Data for Electricity Generating Technologies in Quebec MARKAL Model (10^3 tonnes/ PJ except CO_2 10^6 tonnes/ PJ).

Technologies	SO_2	NO_x	CO_2
Lakeview power plant[a]	4.444	0.661	0.239
Lambton power plant[a]	4.444	0.661	0.239
Nanticoke power plant[a]	4.444	0.403	0.239
Atikokan-Thunder Bay power plants[a]	4.444	0.661	0.239
Lennox power plant	0.431	0.269	0.478
Natural gas steam power plant	0.000	0.069	0.119
Lennox power plant (conversion to coal)	0.444	0.069	0.239
Coal subcritical power plant	0.444	0.069	0.239
Coal supercritical power plant	0.444	0.069	0.239
Coal atmospheric fluidized bed	0.472	0.069	0.256
Coal integrated gasification combine cycle	0.131	0.069	0.224
Western Canada coal supercritical	0.639	0.069	0.253
Natual gas combine cycle	0.000	0.069	0.119
Lignite subcritical power plant	0.267	0.042	0.143
Natural gas combustion turbine	0.000	0.250	0.168
Light fuel oil combustion turbine	0.444	0.250	0.239
Natural gas cogeneration plant	0.000	0.069	0.119

[a]These plants are suitable for life extension and retrofitting processes, and the SO_2 emissions can be reduced by 90% by installing scrubbers. NO_x emissions can be reduced by 50% by using selective catalytic reduction equipments.

Although the resulting energy system is optimal, in the sense that no other system could meet the same requirements at less cost, it is perhaps equally significant that it is found to be technically feasible. All of the constraints on the energy system must be satisfied, and all the environmental requirements must be met. All of the numbers have to add up.

Environmental requirements may be specified in various ways. Maximum allowable annual emissions of carbon dioxide, sulfur dioxide, and nitrogen oxides, for example, can be specified as a constraint. In successive model runs, these constraints can be made more restrictive to determine how these limits affect the choice of energy technologies and the resulting costs.

Figure 1, for example, shows the change in the total annual carbon dioxide emissions from the Province of Quebec with different assumptions as to the allowable environmental emissions. With no constraints on environmental emissions, annual carbon dioxide emissions in 2020 would be double 1985 levels in a high economic growth scenario. A base case is defined in which acid rain restrictions on sulfur dioxide and nitrogen oxide

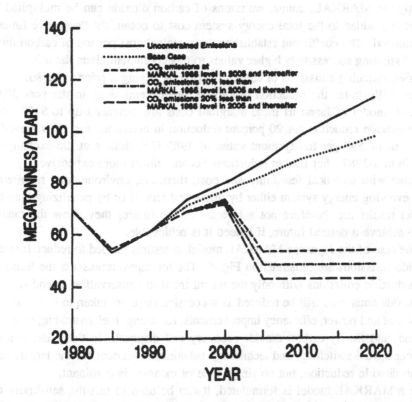

Fig. 1. CO_2 emissions from Quebec energy system with various constraints.
Source: C. Berger et al. (1991).

emissions are found also to restrict carbon dioxide emissions somewhat. To these acid rain restrictions is added a series of carbon dioxide restrictions which are to take effect in 2005 and subsequent years: to stabilize at 1985 levels, or to reduce them by 10 or 20 percent.

Figure 1, the result of a series of MARKAL runs, shows that there is an energy system that can meet each of these restrictions. Figure 2 shows the effect on the need for hydroelectric projects in Quebec in the base case, below, and with 20 percent carbon dioxide reduction, above. With the 20 percent carbon dioxide reduction, all the specific hydroelectric projects described in the model must be developed, approaching the limits of Quebec's economically feasible hydroelectric potential. It is remarkable that Quebec, which uses almost no fossil fuels in the generation of electricity, can still rely on its hydroelectric power to satisfy the carbon dioxide emission limits, even in this high economic growth case. Of course, there may well be other considerations that may alter this state of affairs, among them the current controversy around the flooding of large areas in Quebec North, with its adverse effects on human habitat, flora, fauna, and the release of mercury in the lakes and streams.

A second way to consider restricting emissions of carbon dioxide is by making them costly. In MARKAL, annual emissions of carbon dioxide can be multiplied by a coefficient and added to the total energy system cost to obtain the "objective function" to be minimized. The coefficient establishes the marginal cost per ton of carbon dioxide reduced. Assigning successively higher values to this coefficient alters the mix of energy technologies, reducing emissions of carbon dioxide by paying a price to do so.

Figure 3 illustrates the reduction in carbon dioxide emissions in the year 2030 in a MARKAL model of Japan as these marginal costs are increased up to $100 per ton of carbon dioxide reduction. A 20 percent reduction in emissions can be achieved at a marginal cost of $30 per ton (present value of 1985 U.S. dollars at the exchange rate with yen in mid-1987), but further reductions become much more expensive.

Together with technical feasibility and cost, therefore, environmental requirements mold the evolving energy system either by direct restrictions or by penalizing emissions. The model results are therefore not a forecast of the future; they show the measures needed to achieve a desired future, if indeed it is achievable.

For the case of the Japanese MARKAL model, measures needed to reduce future carbon dioxide emissions are illustrated in Fig. 4. The top curve represents the future path of carbon dioxide emissions with only the recent trend in conservation extended. Future carbon dioxide emissions will be reduced as successive steps are taken to introduce rational use of heat and power, efficiency improvements, recycling, fuel switching, technology substitution, and finally carbon dioxide recovery and disposal. In this case, efficiency improvements, fuel switching, and technology substitution account for the largest part of the carbon dioxide reduction, but no single type of measure is dominant.

Once a MARKAL model is formulated, it can be used to test the sensitivity of its findings to any variation in its input assumptions. MARKAL results are best reported as "If . . . then . . . " findings. Varying the assumptions can address uncertainties like the

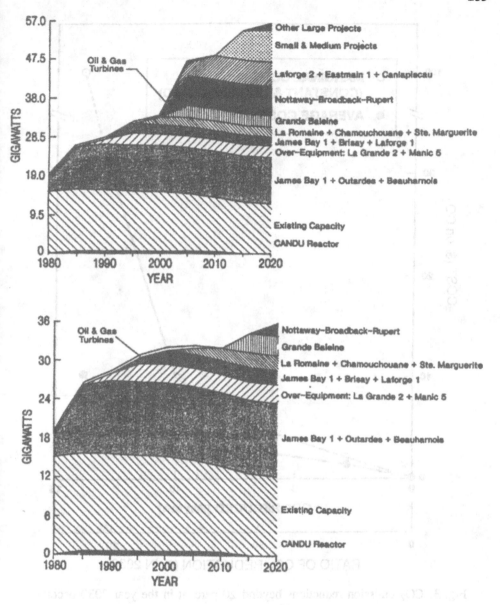

Fig. 2. With greater restrictions on emissions of CO_2, more of Quebec's potential hydroelectric capacity needs to be tapped. Source: C. Berger et al.
(1991).

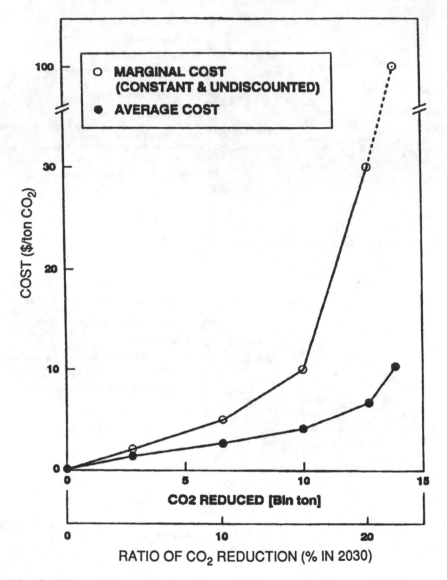

Fig. 3. CO_2 emission reductions beyond 20 percent in the year 2030 become much more expensive in the Japanese energy system model. Source: Yasukawa et al. (1991).

path of future oil prices and energy demands. It can also evaluate deliberate changes in energy policy.

The Netherlands, for example, has examined the implications of two qualitatively different energy futures in the TREND and GREEN scenarios shown in Fig. 5. Following the TREND means moving toward an all-electric system based on nuclear power. A

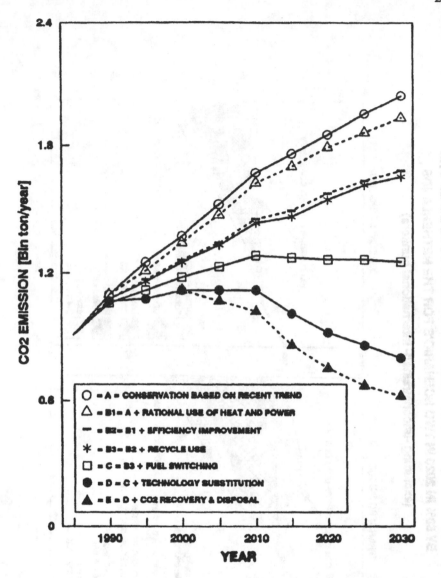

Fig. 4. A variety of measures contribute to reduction in CO_2 emissions in the Japanese energy system model. Source: Yasukawa et al. (1991).

GREEN scenario in which nuclear power is prohibited, on the other hand, moves toward a high-efficiency gas-based system.

In both cases, the Toronto goals of a 20 percent reduction from 1988 carbon dioxide emissions by the year 2005 with a long-term goal of 50 percent reduction can be met. The energy system costs may be as much as 50 percent higher following the TREND, however, where greater savings in energy end use are required.

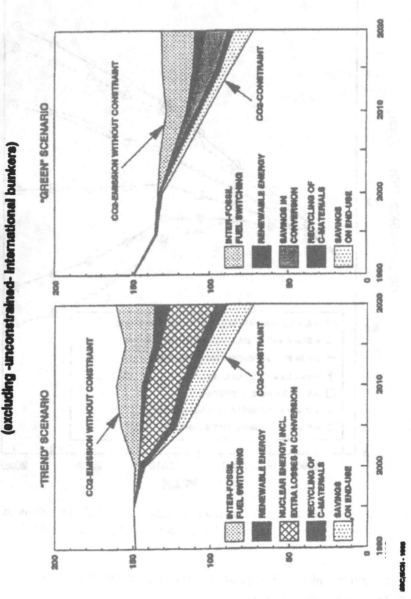

Fig. 5. Contribution of various options to reduce CO_2 emissions by 50 percent in 2020 in two scenarios for The Netherlands. Source: Okken et al. (1991).

In the TREND scenario, an abundant supply of CO_2-free nuclear electricity fosters the use of electric heat pumps and electric vehicles. Savings on heat demand are favored, but not savings on electricity use. The GREEN scenario, on the other hand, favors electricity savings and the use of high-efficiency gas equipment, gas combined cycle power plants, and combined generation of heat and electricity. In both cases, switching fossil fuels from coal to natural gas is a main contributor to the reduction of CO_2 emissions.

The technological options differ considerably, however. Only a limited number of "robust" technologies serve both futures: energy conservation in industry and residential space heating including low-energy buildings; renewables such as biogas, geothermal and offshore wind turbines; methanol production from natural gas especially for buses and trucks; electric buses and trolleys; and technologies using natural gas such as fuel cells, heat pumps, condensing gas boilers, and buses and trucks powered by compressed natural gas.

ETSAP

These examples of the application of MARKAL result from the work of the IEA Energy Technology Systems Analysis Programme (ETSAP). Participants in ETSAP presently include Belgium, Canada, Germany, Italy, Japan, the Netherlands, Norway, Sweden, Switzerland, the United Kingdom, the United States, and the Commission of the European Communities. The program is presently in the second year of the three-year Annex IV, "Greenhouse Gases and National Energy Options: Technologies and Costs for Reducing Emissions of Greenhouse Gases."

In addition to the national case studies, there is under way the standardization in a common model of a number of improvements in MARKAL in a user-friendly, highly automated personal computer version. This is expected to facilitate an outreach program in which a number of the participating countries are aiding non-IEA countries to establish MARKAL models of their energy systems. Among these are China, Indonesia, South Korea, and Tunisia.

MARKAL is now readily accessible to many users, leading to the possibility of consistent analyses from many countries as energy and greenhouse issues are debated and negotiated in international fora. The ETSAP participants provide a cadre for the expansion of cooperative international analysis of technology responses to global environmental challenges. ETSAP itself is a model for energy collaboration for the 21st century.

CONCLUSIONS

Although the final report of Annex IV of ETSAP is a year away, the ongoing national analyses permit some general observations on technology responses to global environmental challenges.

- Energy conservation and efficiency improvements are fundamental to reducing emissions that lead to acid rain and global warming, but they are not a panacea.

They are not the only source of emission reductions, and for large reductions in some countries perhaps not even the major source.

- Natural gas emerges as the critical fuel in the next few decades in those places where it can replace coal and oil.
- Renewable energy is likely to be justified only when a premium is to be paid for reducing carbon dioxide emissions, and its contribution is generally quite limited.
- Nuclear energy plays an important role in reducing carbon dioxide emissions in those countries where it is not too costly and otherwise acceptable. MARKAL studies reveal the cost to society of choosing whether or not to allow the use of nuclear power in a national energy system.
- While energy conservation, efficiency improvements, and fuel switching are the most important steps towards carbon dioxide emission reduction in the near term, technology substitution becomes a significant source in the longer term.
- MARKAL studies show that technology could play a substantial role as human society seeks to come to terms with global environmental challenges. They also reveal the practical limitations and costs of technological solutions.

The most suitable technology responses depend upon the energy-environmental-economic goals of individual countries and their circumstances: their climate, their natural resources, and their existing energy systems. These may defy the generalities. Switzerland, where electricity is now generated primarily with hydroelectric and nuclear power, has nothing to gain by switching to natural gas. Quebec, with its enormous hydroelectric potential, will depend heavily on renewable energy. The Netherlands, anticipating depleted natural gas fields, may find it economical to sequester carbon dioxide.

Some "robust" technologies can benefit a number of countries and will deserve international collaboration in R&D to accelerate their development. Cooperative MARKAL modeling studies in the international community will continue to help identify which technologies in which countries.

REFERENCES

Berger, C., R. Loulou, J. Soucy, and J. P. Waaub, "CO$_2$ Control in Quebec and Ontario," presented at the ETSAP/IIASA meeting, Laxenburg, Austria, May 21–22, 1991, GERAD, Montreal, Canada.

Okken, P. A., J. R. Ybema, D. Gerbers, T. Kram, and P. Lako, "The Challenge of Drastic CO$_2$ Reduction: Opportunities for New Energy Technologies to Reduce CO$_2$ Emissions in the Netherlands Energy System up to 2020," Netherlands Energy Research Foundation ECN, Petten, The Netherlands (March 1991).

Rowe, M. D., and D. Hill (eds.), "Estimating National Costs of Controlling Emissions from the Energy System: a Report of the Energy Technology Systems Analysis Project,"

International Energy Agency, BNL 52253, Brookhaven National Laboratory, New York, USA (September 1989).

Yasukawa, S., O. Sato, M. Konta, Y. Shimoyamada, T. Kajiyama, H. Shiraki, and Y. Tadokoro, "Preliminary Analysis on CO_2 Emission Reduction in Japan by MARKAL Model," presented at the ETSAP/IIASA meeting, Laxenburg, Austria, May 21–22, 1991, Japan Atomic Energy Research Institute, Ibaraki-ken, Japan.

Integrated Assessment of Energy Technologies
for CO_2 Reduction in the Netherlands

Tom Kram, Peter A. Okken, and J. Remko Ybema*
Netherlands Energy Research Foundation, ESC-Energy Studies
P.O. Box 1, 1755 ZG Petten, The Netherlands

ABSTRACT

Many energy technology options for reduction of CO_2 emissions are available. The contribution of individual options to CO_2 reduction is often calculated in a simple way starting from a constant reference situation and without considering interactions with other technological options. However from the policy viewpoint there is more interest in the contributions of options if they are evaluated in relation with other reduction options, fuel price paths and future energy demand projections and patterns. This paper reports on calculations performed with a process-oriented dynamic national cost-minimizing LP-model of the Netherlands energy system. Several scenarios are calculated with drastic (up to 70%) reductions of national CO_2 emissions in 2020, in order to identify cost-effective CO_2 reduction strategies. CO_2 reducing options in all scenarios include: energy conservation, renewables, recycling, fuel switch, nuclear energy, CO_2 removal and efficiency improvements. The role of electricity in the energy system is highlighted. Sensitivity analyses have been performed to assess the consequences of exclusion of some key technologies, like CO_2 removal, electric vehicles and electric heat pumps. Cost-emission trade-off curves for the total energy system are constructed.

1. INTRODUCTION

Recently a large and increasing number of new energy technologies is being proposed to decrease emissions of CO_2. These technologies include options on the supply side of the energy system as well as demand side options. Among energy supply side options are, e.g., clean coal technology, gas-fired fuel cells, combined heat & power, second and third generation nuclear power plants, solar PV, biofuels and wind energy. Demand side technology includes insulation, heat pumps, efficient appliances, conservation in industry and alternative transportation fuels.

Frequently large contributions to the limitation of CO_2 emissions are being claimed for each of the above mentioned individual technologies. However, quite often such

* This study was performed within the Dutch national research programme on global air pollution and climate change, co-sponsored by the Ministry of Economic Affairs, project 7007 and project 7061. Results will be included in multinational analysis in the IEA Energy Technology System Analysis Programme.

estimates are biased, e.g., by performing comparisons with some out-of-date reference technology, ignorance of developments on the other side of the energy system (either demand side or supply side) and not by considering variations in energy demand with time (e.g., diurnal and seasonal variations).

From a policy point of view there is not much interest in the contribution of one single technology to the reduction of CO_2 emissions. In developing future strategies an approach is more useful which considers all readily available and new technologies in one system, keeping in mind both the characteristics of demand technologies as well as those for supply options. A system approach is based on one or more scenarios (reflecting the evolution of population, economy and life style) and projections for fuel prices.

This paper reports on a study which was performed with an energy system computer model characterized for the situation of the Netherlands. This model is an optimization model that chooses mixes of technologies and fuels based on the costs of these. In Sec. 2 an introduction will be given to the model that has been used (MARKAL). In Sec. 3 the main characteristics of the Netherlands, its energy system, and two scenarios are briefly addressed. Section 4 discusses the role of various energy technologies under various CO_2 emission constraints. In Sec. 5 the costs related to limitation of CO_2 emissions are discussed briefly. Throughout the whole paper, and more explicitly in Sec. 6, the role of electricity in the energy system will be highlighted. Section 7 presents some conclusions.

2. THE MARKAL MODEL

The energy system model used in this study, the MARKet ALlocation (MARKAL) model, is a standard linear programming software package which is used to represent energy systems.[2] The model has been developed under the International Energy Agency (IEA) and is currently operated in 12 countries including the United States, Japan and the European Community. These users are organized in the Energy Technology Systems Analysis Programme (ETSAP).

Using detailed compilations of data characterizing available and prospective energy technologies, and incorporating projections and assumptions about the costs and availability of fuels, MARKAL configures an optimal mix of technologies to satisfy the specified useful energy demands. The model includes detailed data for the Dutch energy system: fuel prices; investment costs, availability, efficiency of several hundreds of different energy technologies; emission coefficients and emission abatement techniques. The model is used in a national costs minimizing mode with exogenous national maximum allowable emissions ('bubble' concept).

The model optimizes the energy system for the period 1980 to 2020 simultaneously in steps of 5 years each. This dynamic simultaneous optimization is referred to as perfect foresight, reflecting the activities of policy-makers as they are supposed to 'know the future.' Dynamic modeling is of vital importance for testing structural changes in the energy system under environmental constraints. Several new energy technologies are projected to be available at fixed points in the future. This reflects the impact of ongoing energy and environmental R&D. The model enables testing of such new technologies in a future energy system under various constraints ('technology assessment'). Few institutional and market barriers are assumed, to reflect the maximum potential of (new) energy technologies. On the other hand bounds are imposed on the speed of market penetration for new energy technologies to prevent unrealistic solutions, and there are lower bounds to ensure that older technologies will not be phased out too rapidly. Although, in real life, interest rates may vary among sectors (e.g., the time preference and discount rates within industry and the private sector are expected to be higher than within governmental sectors), from a national costs-minimizing point of view, only one discount rate should be applied. The model optimizes with a 5% discount rate; undiscounted cost figures are reported to allow for comparison with current cost figures.

A general MARKAL structure for the Netherlands energy system is given in Fig. 1. Here the different parts of the Dutch model are explained.

Sectoral demand / Upper right part of the general structure. The demand for energy is disaggregated into four major sectors: industry; residential and commercial; transport; and international bunkers, export and feedstocks. In these sectors specific energy needs are given exogenously for the various periods of the time-path considered. This demand is generated from scenarios.

For industry, specific electricity needs and 5 qualities of heat (greenhouse heat, low, medium and high temperature heat and process heat) have been specified. In the residential and commercial sector the energy demand is disaggregated into the needs for hot water, food preparation (cooking), electricity and space heating for various building types. The transport sector includes all main inland transportation modes. International ship transport and airplanes are excluded.

End-use technologies / Matrix on right part of structure. The matrix on the right part of the model structure connects the specific energy demands with the secondary energy carriers (presented in the central column). End-use technologies are the relations between energy demands and secondary energy carriers (presented by circles in the matrix). Only where a circle (technology) is given, demands and energy carriers are connected. Some energy carriers and demands are connected by more than one technology. A number within the small circle refers to this number of technologies. End-use technologies are specified by their respective energy conversion efficiencies, costs (investments and operation), lifetime and emission coefficients for NO_x and SO_2. The characteristics of a specific technology may change with time as a result of ongoing technology development. Extra energy conservation was also modelled as if it were a secondary energy carrier.

The model can choose the mix of secondary energy carriers and technologies to satisfy exogenous energy demand in the different years. For example, truck transportation (see demand sector) the model will choose some mix of fuels, consisting of diesel, compressed natural gas (CNG), methanol and ethanol (see matrix). Depending on fuel prices, technology costs, and (environmental) constraints the choice is made. This choice

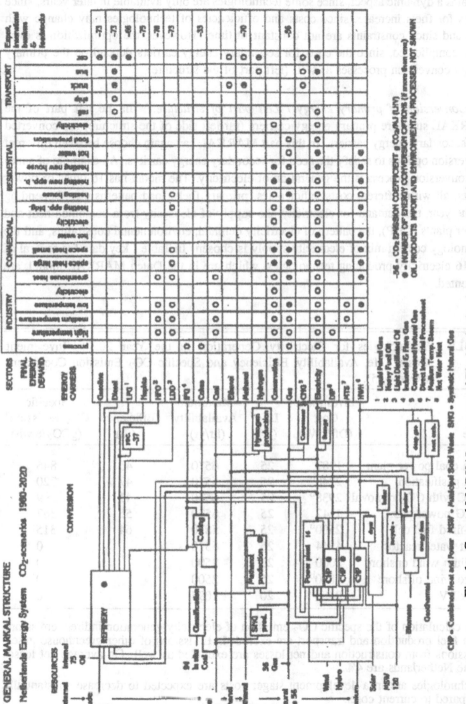

Fig. 1. General MARKAL structure of The Netherlands energy system, 1980 – 2020.

contains a dynamic aspect, since some technologies are only available in later years, since prices for fuels increase, since costs and efficiencies of technologies may change with time and since constraints are not constant in time. However, the optimization is even more complicated, since the costs for secondary energy carriers depend on the primary energy conversion processes as well (left part of the structure).

Conversion of primary energy / Left part of structure. In the left part of the MARKAL structure primary energy carriers (far left side of the structure) are converted into secondary energy carriers. In this part MARKAL can again choose between different conversion options to satisfy the need for secondary energy carriers. An example of a major conversion process is the generation of electricity. The model has 16 types of power plants, all with different costs, efficiencies, primary fuels, load patterns etc. Depending on the year, the demand for electricity, the supply of electricity from combined heat and power plants (CHP), the pattern of electricity demand, environmental constraints, and the technology costs a mix of electricity supply is chosen. In Table I key data for several of the 16 electricity producing technologies, which are in the Dutch MARKAL model, are presented.

TABLE I. Technologies for Electricity Generation in the Year 2020: Investment Costs, Life, Availability, Efficiency and Specific CO_2 Emission Coefficient (Refs. 4, 5).

Type	Investment Cost (Dfl/kWe)	Life (yr)	Maximum Availability (hr/yr)	Efficiency (%)	Specific CO_2 Emission[a] (gCO_2/Kwh)
Hard coal power plant	1987	25	6570	40	846
Coal gasification	2248[b]	25	6570	47	720
IGCC with CO_2 removal	2938[b]	25	6570	41	99
STAG power plant	1142	25	6570	55	367
Gas-fired fuel cell	2500[b]	25	5870	64	315
Light water reactor	4544	25	6570		0
Medium wind onshore	2000	20	2500		0
Large wind offshore	3800[b]	25	2500		0
Solar PV	2400[b]	20	1095		0

[a]For calculation of the specific CO_2 emission of electricity generation indirect emissions from fuel production and transport are excluded. Emissions of other greenhouse gases, emissions from construction and net losses are excluded as well. On average net losses in the Netherlands are 4%.

[a]Technologies are in a development stage; costs are expected to decrease substantially compared to current costs.

System borders

The domestic primary energy carriers are within the borders of the MARKAL model. In the case of wind energy (onshore and offshore), natural gas, oil, hydropower, solar, geothermal, biomass, and municipal solid waste (MSW),the resources for the Netherlands are characterized. The prices of imported fuels (crude oil, coal, etc.) have been set exogenously.

The CO_2 emission coefficients (gCO_2/MJ) can be found on the borders of the general MARKAL structure. Domestic extraction of gas and oil and import of fossil fuels accounts for CO_2 emissions. On the other hand, export of fuels, feedstocks, and bunkers are modeled as negative CO_2 emissions. Also specific recycling processes and CO_2 removal technologies are characterized by a negative CO_2 emission coefficient. By summing all CO_2 emissions (including subtraction of 'negative' CO_2 emissions), the total CO_2 emission from the net energy system is calculated.

3. THE NETHERLANDS, ITS ENERGY SYSTEM AND TWO SCENARIOS

The Netherlands is a densely populated and highly industrialized country in Western Europe. The area is $33\,000$ km^2; this is about one-tenth of the size of the State of New Mexico. Domestic natural gas is the most important energy source. The winter climate is moderate; 3000 heating degree days per year, with no significant cooling demand in summer. Main commercial energy users are the petrochemical, fertilizer, steel and aluminium industry, greenhouses and services. Approximately half of present oil consumption is for international bunkers and non-energy use (petro-feedstocks); the other half is for inland transport. The CO_2 emission from the Netherlands energy system is currently around 160 Mton per year. The Netherlands has large natural gas resources; there is no potential for hydropower.

At present electricity generation is mainly based on coal and natural gas. The average specific CO_2 emission coefficient of produced electricity is 603 gCO_2/kWh (see Table II). Coincidentally this value is almost the same level as for the United States. Compared to the U.S. situation, a larger share of fossil fuel in electricity production in the Netherlands is compensated by a larger contribution of gas relative to coal and high electrical-generation efficiencies. Note also the high CO_2 emission coefficients of Australia, Denmark and Greece and the low emission coefficients of Austria, France and Norway.

Scenarios for the period 1980–2020 are used in the model calculations, reflecting different world-wide socio-economic developments. These include a 'TREND' and a 'GREEN' scenario (see Table III).

The average annual economic growth rate is nearly 2.8% in the TREND scenario, and 2% in the GREEN scenario. In the GREEN scenario nuclear energy is not allowed and energy use decreases. By contrast in the TREND scenario there is no limit on the use of nuclear energy; industrial production and transport energy use are assumed to continue to increase following the next wave of international growth and trade (see also Fig. 1).

TABLE II. Electricity in Selected OECD Countries in 1988: Production, Efficiencies, the Role of Fossil Sources and the Average Specific CO_2 Emission Coefficients (Ref. 6).

Country	Electricity Production (TWh/yr)	Share of Fossil in Elec. Prod. (%)	Efficiency of Fossil Elec. Production[a] (%, LHV)	Average CO_2 Emission Coefficient[b] (GCO_2/Kwh)
OECD total	6140	58	38.3	470
Norway	97	0	—	2
France	363	4	34.7	38
Austria	42	20	45.2	123
Canada	465	23	37.1	201
Belgium	63	29	39.4	241
Spain	134	37	35.1	343
Japan	667	60	42.8	363
Germany (West)	367	56	35.8	504
Italy	179	78	39.5	530
Netherlands	58	94	41.7	603
United States	2872	72	37.3	611
Yugoslavia	84	64	35.0	625
United Kingdom	289	77	37.3	686
Denmark	27	100	45.2	806
Australia	130	88	34.7	827
Greece	33	92	34.0	878

[a] Values have been corrected for combined heat and power generation; the reference technology for heat production is assumed to be the gas-fired boiler with a heat efficiency of 100% (LHV).

[b] Average specific CO_2 emission coefficient of electricity production is calculated with emission coefficients for coal, oil and natural gas of respectively 94, 73 and 56 gCO2/MJ. For other solids local cofficients from [IEA/OECD] have been applied. Emission coefficients for nuclear and hydro are assumed to be 0 gCO$_2$/MJ. Indirect emissions and other greenhouse gases have not been included. Values have been corrected for combined heat and power generation; the reference technology for heat production is assumed to be the gas-fired boiler with an heat efficiency of 100% (LHV).

Useful energy demand projections and fuel prices are specified exogenously. All economically attractive ('no regret') energy conservation options have already been included in the energy demand projections, resulting in a fairly constant ('stabilized') CO_2 base line despite significant economic growth rates. In the base case scenarios the crude oil (and gas) price is supposed to increase from 8.6 Dfl/GJ in 1990 ($24/bbl at the dollar exchange rate of September 1990, $ = 1.7 Dfl) up to 18.5 Dfl/GJ by the year 2020. The

TABLE III. General Scenario Characteristics.

Scenario	Trend/Trend-R	Green/Green-R
International context	Cosmopolism	Solidarity
Industrial development	Export-oriented	Self-sufficiency
Transport	Increasing	Decreasing
Nuclear energy	Unlimited	Phase out
Energy price increase	Moderate	Moderate
Economic growth rate	2.5 to 3%	2%
NO_x/SO_2 constraints	Mild	Severe
National CO_2 constraint	Variable	Variable
CO_2 removal optional	No/yes	No/yes

price of natural gas is linked directly to oil prices. Coal prices in Dfl/GJ are about half the crude oil prices.

In the base case, the case without constraints on the emission of CO_2, national CO_2 emissions remain relatively stable. The TREND scenario shows a modest increase from 150-Mton CO_2/yr to 165 Mton/yr in 2020. In the GREEN scenario CO_2 emissions decrease to 130-Mton CO_2/yr in 2020. The evolution of the primary energy mix, calculated by the model in the TREND and GREEN scenario base case is presented in Fig. 2. Natural gas will become less dominant; coal and nuclear energy (only in the TREND scenario) will take over the role of gas.

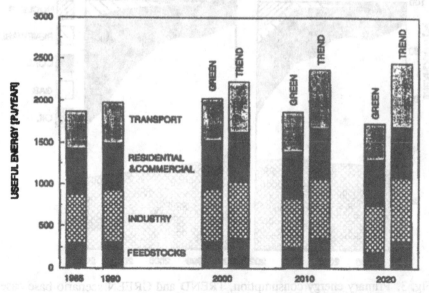

Fig. 2. Energy demand by sector, base case scenarios (Green: left; Trend: right).

268

Table IV shows that the average CO_2 emission coefficient decreases in 2020 in the TREND scenario while remaining constant in the GREEN scenario. In the TREND scenario gas is mainly replaced by nuclear energy. In the GREEN scenario the role of coal becomes dominant (see Fig. 3). The contribution of renewables to electricity production increases but remains modest in both scenarios. Coupled production of heat and power is attractive in both scenarios, resulting in higher net efficiencies for electricity generation.

TABLE IV. Average CO_2 Emission Coefficient in 2020 in the Base Case of the TREND and GREEN Scenario.

Scenario		Average CO_2 Emission Coefficient (gCO$_2$/Kwh)
1990		603
2020	TREND	349
2020	GREEN	601

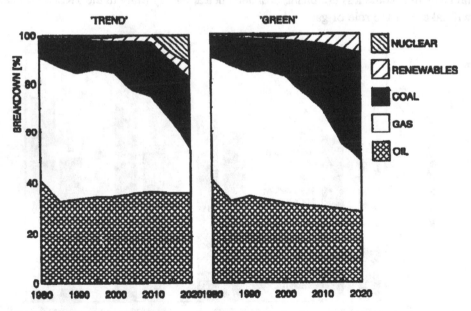

Fig. 3. Primary energy consumption, TREND and GREEN scenario base case (not CO_2 constrained).

4. OPTIONS FOR CO_2 REDUCTION UNDER
NATIONAL EMISSIONS CONSTRAINTS

The upper lines in Fig. 3 show the evolution of the annual CO_2 emissions in the TREND and GREEN scenario base case. In order to assess CO_2 reduction options, these scenario calculations were repeated with variable CO_2 constraints. The resulting CO_2 emission reduction in 2020 ranged from 20% to 70%, compared to the current (1990) CO_2 emission rate of 160 $MtCO_2$/y. In the high-growth TREND scenario more CO_2 reduction is needed, compared to the GREEN scenario, to meet these CO_2 emission constraints (dashed lines in Fig. 3). The NO_x and SO_2 constraints applied in previous scenario calculations (40% and 80% reduction by the year 2020) remained unchanged.

The options for CO_2 emission reduction chosen by the model to reduce CO_2 emissions can be aggregated:
- Savings on energy end-use
- Improvements in energy conversion efficiency
- Renewable energy
- Nuclear energy
- Intra-fossil fuel switch (coal, gas, oil)
- Recycling of fossil fuel derived carbonaceous materials;
- CO_2 removal.

Their respective contributions to reduce CO_2 emissions in the year 2020 are displayed in Fig. 4. The horizontal axis represents the various CO_2 reduction percentages from Fig. 3.

Energy end-use conservation is the most important option to stabilize CO_2 emissions in short term. At present, building insulation, public transport, luminescent lighting, appliance efficiency standards, and industrial energy conservation are being promoted. These 'no regret options' have already been included in the unconstrained baseline calculation. Additional energy conservation technologies are called upon for longer-term drastic CO_2 reduction. Savings on the use of electricity appear to be attractive in the GREEN scenario at lower CO_2 constraints than in the TREND scenario, due to the higher CO_2 intensity of electricity in the GREEN scenario.

Improvements in energy conversion efficiency decrease CO_2 emissions. At severe CO_2 constraints new high efficiency technologies, like fuel cells, are introduced. This contributes to lower CO_2 emission coefficients for electricity production (see Table I and Fig. 6).

Renewable energy is considered as a zero CO_2 option. The potential for renewable energy (wind, biomass, geothermal, solar energy and hydropower) in the Netherlands is restrained due to climatic and geographical conditions. Wind energy and solar help in achieving a low CO_2 intensity for electricity production (see also Table I and Fig. 6).

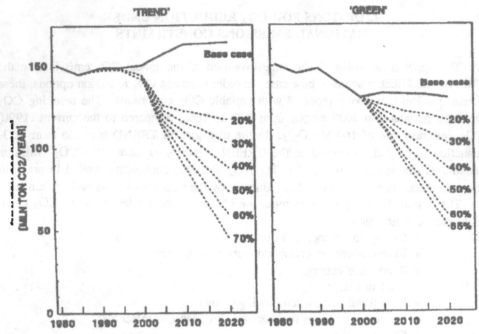

Fig. 4. National CO_2 emission (excluding international bunkers and export of petrochemical feedstocks). Trend and Green scenario base-case and CO_2 emission constraints.

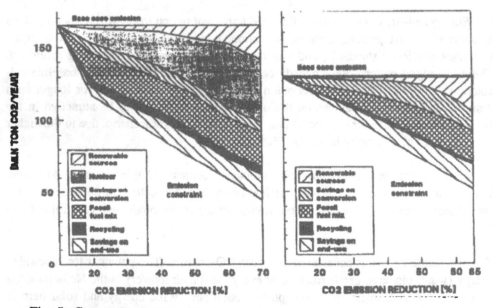

Fig. 5. Cost-optimal CO_2 reduction strategy mix by the year 2020, TREND, GREEN and GREEN-R scenario (scenario including CO_2 removal), impact of CO_2 constraints.

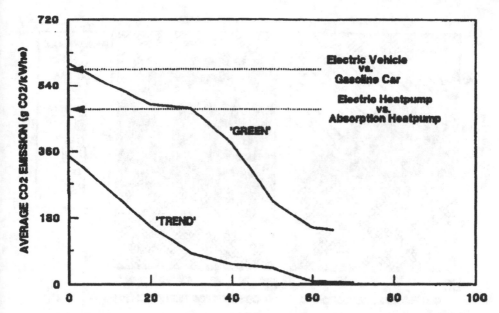

Fig. 6. Average CO_2 emission coefficient for electricity generation in 2020 under CO_2 constraints.

Recycling of fossil fuel derived carbonaceous materials such as waste plastics and lubricants, instead of incineration, reduces CO_2 emissions.

Fossil fuel switching affects CO_2 emissions because of differences in the carbon content of various fuels: the specific CO_2 emission from natural gas (56 gCO_2/MJ) is less than that for oil and coal (respectively 73 and 94 gCO_2/MJ). Substitution of natural gas for coal in power plants reduces CO_2 emissions (Table I). In addition fossil fuels will be substituted by electricity (electric heat pumps and electric vehicles) at more severe CO_2 constraints, reducing the average CO_2 emission (Fig. 6) far enough to create conditions which allow the introduction of electric alternatives in transportation and space heating. The introduction of those electric end-use options is strongest in the TREND scenario. In the GREEN scenario there is a small decrease of final electric use at modest emission reductions (see also Fig. 9).

In the TREND scenario a shift occurs from coal to nuclear power to meet the CO_2 constraints. Because nuclear energy is not allowed in the GREEN scenario, here a switch from coal to natural gas is induced by CO_2 constraints. In both scenarios the contribution from renewable energy increases (Fig. 6).

CO_2 removal is a new technical option. CO_2 removal will not be practised unless a CO_2 reducing policy is implemented. In this study CO_2 removal was added as a separate option, denoted by R (Removal). Cost figures for CO_2 removal are based on technologies which have been practised for many years by the chemical and oil industries.[3] CO_2

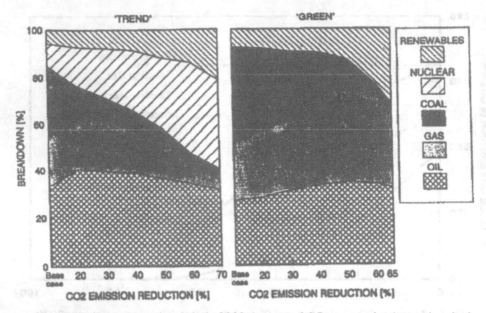

Fig. 7. Primary energy supply in 2020, impact of CO_2 constraint (petrochemical feedstocks included in oil share).

removal from Integrated Coal Gasification Combined Cycle (IGCC) power stations could significantly contribute to reduction of CO_2 emissions (see specific CO_2 emissions in Table I), consequently a reintroduction of coal in the primary fuel mix, under stringent CO_2 constraints results in the GREEN-R scenario (see Fig. 7). The availability of CO_2 storage capacity is a condition for this technology. In the Netherlands CO_2 removal is not expected to be restrained by lack of storage capacity in depleted natural gas reservoirs. Enhanced gas recovery could in fact become an important side benefit.

5. CO_2 REDUCTION COSTS

The model calculates the direct costs of CO_2 emission reduction. In Fig. 8 the undiscounted system costs are presented as a function of the cumulated emission from the Netherlands in the time period 2000–2020 under various CO_2 emission constraints. From Fig. 8 the marginal cost curves can be calculated. Initially marginal costs rise slowly to reach a level of around 50-Dfl/ton CO_2. Starting from 60% emission reduction, these costs rise sharply to 500 Dfl/ton CO_2 and more. To achieve such high emission reductions relatively cheap options are no longer sufficient and more expensive options are called upon, e.g., some forms of renewable energy and alternative automotive fuels. Such developments can trigger substantial economic restructuring, this might conflict with the original demand assumptions.

Due to the dynamic calculation in a network model, it is impossible to assign individual CO_2 reducing technologies in Fig. 8. For individual technologies in the LP-network

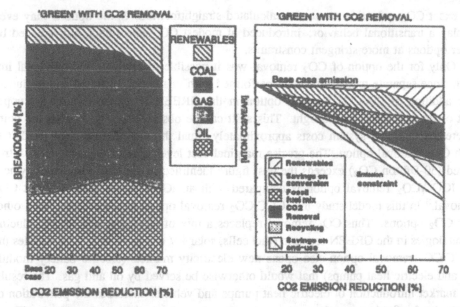

Fig. 8. Primary energy supply and cost-optimal CO_2 reduction strategy mix in 2020, impact of CO_2 constraint, GREEN-R scenario.

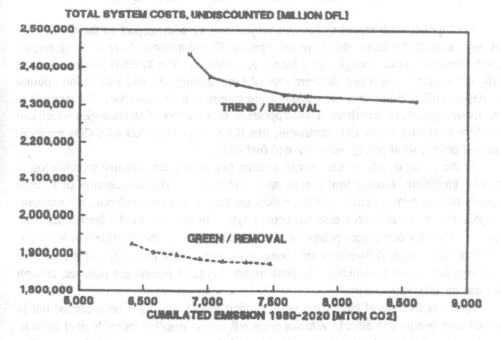

Fig. 9. Undiscounted system costs versus CO_2 emission reduction.

the exact CO_2 reduction cannot be calculated straightforward. Some options may even display a transitional behavior: introduced at modest CO_2 constraints, but replaced by other options at more stringent constraints.

Only for the option of CO_2 removal was it possible to calculate its individual impact, since separate model runs were performed with this option included and excluded. The addition of the CO_2 removal option in the GREEN-R scenario shifts the upper part of the cost curve to the right. This shift can be observed in Fig. 8 from the point where the marginal system costs approximately equal the marginal costs of the IGCC with CO_2 removal option. The precise marginal cost level where CO_2 removal is introduced (50-Dfl/ton CO_2) exceeds the cost figure identified in the feasibility study, where the IGCC/CO_2 removal option is compared with an IGCC power plant without CO_2 removal.[3] In this model study the IGCC/CO_2 removal option has to compete with other low CO_2 options. Thus CO_2 removal replaces a mix of electricity and heat producing technologies in the GREEN scenario (fuel cells, solar PV and gas fired CHP). Besides the IGCC/CO_2 removal option also opens new electricity markets (electric battery module car and electric heat pumps) that would otherwise be served by oil and gas. The resulting market introduction of electric heat pumps and vehicles increases the penetration of electricity for final energy demand.

6. CO_2 CONSTRAINTS AND ELECTRIFICATION

Assumptions with regard to demand projections, or with respect to the availability of large scale technologies that have low specific CO_2 emissions, have a strong impact on the optimal mix of strategies to reduce CO_2 emissions. The TREND scenario and the GREEN scenario reflect two different sets of such assumptions and indeed the optimal strategies differ. However some options seem robust in both scenarios. An example is the increasing role of electricity. There appears to be a number of technologies which can produce electricity at low CO_2 emissions, like IGCC in combination with CO_2 removal, nuclear power, wind energy, solar PV and fuel cells.

At the demand side of the energy system two new electricity-using technologies can be identified: electric heat pumps and electric cars. The introduction of electric cars is still uncertain; such is still dependent on the success in developing the natrium-sulphur battery or any other efficient battery type. In the case of the Netherlands the introduction of electric heat pumps is also uncertain due to the dominant role of gas (98% of the individual dwellings are connected to the natural gas grid) and the absence of a significant cooling demand. The Netherlands has large natural gas reserves, enough for use far into next century.

To assess the role of electricity, more explicit model runs have been carried out in which heat pumps and electric vehicles were left out as possible technological options. At drastic emission reduction the composition of the energy system was changed. The production of electricity is decreased, mainly in decentral coupled production and nuclear power. Gas fired absorption heat pumps and additional savings take over the share of

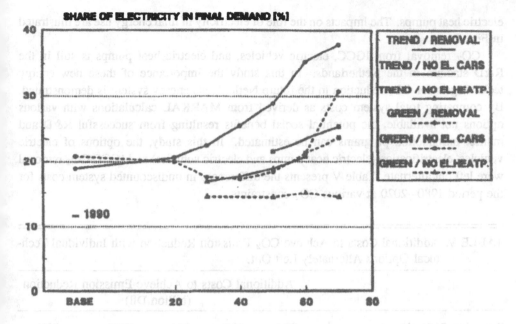

Fig. 10. Share of final electricity in final energy use under various CO_2 emission constraints.

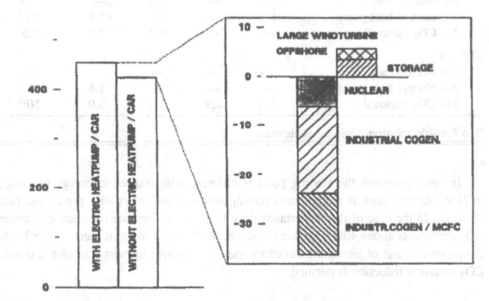

Fig. 11. Share of final electricity in final energy use under various CO_2 emission constraints.

electric heat pumps. The impacts on the role of electricity in final energy use are illustrated in Fig. 9.

CO_2 removal from IGCC, electric vehicles, and electric heat pumps is still in the R&D stadium in the Netherlands. In this study the importance of these new energy technologies for CO_2 reduction in the future Netherlands energy system is demonstrated. By comparing total system costs as derived from MARKAL calculations with various options not available, the potential social benefits resulting from successful R&D and market introduction programs can be estimated. In this study, the options of electric vehicles, electrification (electric heat pumps and electric vehicles) and IGCC/CO_2 removal were left out alternate. Table V presents the differences in undiscounted system costs for the period 1980–2020 at various CO_2 constraints.

TABLE V. Additional Costs to Achieve CO_2 Emission Reduction with Individual Technical Options Alternately Left Out.

Emission Reduction	Additional Costs to Achieve Emission Reduction (billion Dfl)				
	30%	40%	50%	60%	70%
TREND					
No electric cars	—	—	0.0	—	8.8
No electric heat pump	0.7	—	7.7	47.8	NF[a]
No CO_2 removal	—	—	6.3	9.8	25.9
GREEN					
No electric cars	—	—	—	—	0.03
No electric heat pump	—	—	—	1.8	21.5
No CO_2 removal	0.8	2.9	1.1	3.0	NF[a]

[a]No feasible solution could be achieved.

In other countries the starting point is different with respect to energy use (e.g., in New Mexico there is a significant cooling demand) and electricity production (see Table II). In the case of the Netherlands there is an increasing role for electricity under CO_2 constraints above 40% emission reduction. With this in mind, it seems rather likely the observed trend of increasing electricity use will continue in most countries if drastic CO_2 emission reduction is pursued.

7. CONCLUSIONS

The most important conclusion is the confirmation that a drastic reduction of CO_2 emissions with technical options is possible in the longer term at acceptable direct costs.

Such reduction will be achieved with a balanced mix of options. Energy conservation alone is not sufficient to reach stringent CO_2 emission reductions targets; structural changes in the supply side of the energy system are needed, like the switch from fossil fuels to renewables and nuclear, the change from coal and oil to natural gas, and CO_2 removal.

The interactions between measures on the demand side and the supply side of the energy system play an important role in the design of long term emission reduction strategies. The role of electricity will become increasingly important.

9. REFERENCES

1. Okken, P. A., J .R. Ybema, D. Gerbers, T. Kram and P. Lako, The challenge of drastic CO_2 reduction, opportunities for new technologies to reduce CO_2 emissions in the Netherlands energy system up to 2020, ECN-C—91-009, March 1991.

2. Rowe, M. D. and D. Hill (Eds.), Estimating national costs of controlling emissions from the energy system—a report of the Energy Technology Systems Analysis Project—IEA, NTIS, U.S. Dep. of Commerce, Springfield, VA 22161, September 1989.

2. Hendriks, C. A., K. Blok, W. C. Turkenburg. The recovery of carbon dioxyde from power plants. In: Climate and Energy—the feasibility of controlling CO_2 emissions. (eds. P. A. Okken, R. J. Swart, S. Zwerver), pp. 127–142, Kluwer Academic Publishers, Dordrecht, Netherlands, 1989.

The Accelerator Transmutation of Waste (ATW) Concept for Radioactive Waste Destruction and Energy Production

Edward D. Arthur

Los Alamos National Laboratory

I would like to describe a new concept that we are investigating here at Los Alamos which is called Accelerator Transmutation of Waste. Some very important possibilities are explored in this concept. First we are looking at the possibility of devising a workable system for destroying long-lived radionuclides in high-level nuclear waste. This is one of the main issues associated with nuclear power as it now stands: What do we do with the nuclear waste produced by the 100 or so reactors operating in the U.S. and the much larger number operating around the world? Our efforts in this area, particularly those of Charles Bowman at Los Alamos, led to some very interesting ideas for producing a long-term nuclear energy source with improved safety features compared to present-day reactors, with a very long-term energy supply that would essentially minimize or eliminate the long-lived high-level nuclear waste generated by the system.

We have set some challenging goals for ATW. We are pulling together several technologies that have been explored over the last several years at Los Alamos, including the development of high-current accelerator technology. A goal is to devise a system that can destroy all major components of long-lived nuclear waste. Our goal is different from the goals that programs elsewhere have set, in that we want to destroy not only the higher actinides that have received major emphasis in many transmutation concepts, but also other components that really dominate long-term risk scenarios, namely the products produced during fission. If we are successful in doing this, we can reduce the time scales for storing high-level waste in a repository from several tens of thousands of years to something perhaps less than a few hundred years. We think this system additionally provides some very important operational safety features.

We are looking at two systems, one for radioactive-waste destruction and one for energy production. They are coupled together, and this leads to some additional goals. The first thing one wants is to produce electricity efficiently. The fuel supply that we would use is natural thorium or natural uranium. Thus one has several thousand years of supply of fuel that could be used in this system.

One also wants to eliminate the long-lived, high-level waste produced during operation of that system. The approach we have been investigating centers around use of an accelerator to produce an intense source of spallation neutrons. Why use an accelerator coupled with a fissioning system? Figure 1 describes the fundamental impact of this coupling. In the figure, 500 fission events are used to drive an accelerator neutron source, and the impact that has on the basic neutron economy of the overall system is described. So if one has 500 fission events, one produces about 100 000 MeV of heat

Why Use an Accelerator?
Fundamental Impact

Example: 500 fission events used to drive an accelerator-spallation neutron source system

Heat energy	100,000 MeV
Electricity conversion(44%)	44,000 MeV
Beam power (45%)	19,800 MeV
# 800 MeV protons produced	24.8
@ 22 neutrons/proton	
Number of neutrons produced	545
@ 2.5 neutrons/fission (U233)	1250
For 500 source fissions	1794 neutrons TOTAL

The effective neutrons/fission is 3.6 versus 2.5 in a standard reactor system - a 40% increase

Fig. 1. Impact of an Accelerator Spallation Neutron Source on a Fission System.

energy. In ATW, we are looking at systems that would utilize molten salt for the conversion of heat to electric power where conversions on the order of 44% are possible. Thus, one gets about 44 000 MeV of electricity from the system. This electricity is then converted to beam power and using efficiencies appropriate for current accelerator technology, one gets about 20 000 MeV of energy. A medium-energy particle accelerator is utilized, so this electrical energy produces about twenty-five 800-MeV protons. Each of those protons produces about 22 neutrons, so one gets about 550 neutrons created from this spallation process. These neutrons are added to those produced in fission (about 2.5 neutrons per fission). Thus, for 500 source fissions one produces about 1800 total neutrons as compared with the 1200 that a standard reactor would produce. The effective number of neutrons per fission is about a 40% increase over that occurring in a reactor system. Reactor systems are constrained by nature to operate with 2.5 to 2.9 neutrons per fission. Coupling an accelerator leads to a significant increase in the parameter space that one has for a nuclear system. All this power does not have to be put back into the accelerator. About 10 or 20% of it can be used to run the accelerator, with the balance available for use on the commercial grid. This is why the coupled accelerator-fission system seems so attractive.

Let's return to the question of radioactive reactor waste. A 3000-megawatt power reactor system has a yearly discharge of about 33 metric tons of fuel. In that 33 metric tons of fuel one finds a spectrum of radioactive products, shown in Fig. 2. There are plutonium isotopes, higher actinides, such as neptunium and americium, and then there are fission products. Most of these have very long half-lives, and most are produced in rather significant quantities. These radionuclides have led people to consider alternatives for dealing with high-level nuclear waste that either don't involve repository storage or involve a minimal amount of repository storage capacity. (I should mention the country is in the process of evaluating Yucca Mountain as a reactor high-level waste storage site. If the process is successful and Yucca Mountain begins operation, then about 30 years later there may be need for another repository in the U.S.) Investigating alternatives to repository storage, or alternatives that can reduce the amount of material that has to go into repository storage, is clearly very important.

The radioisotopes problem drives storage of high-level waste. As mentioned above, most storage concepts have looked at the actinide portion of that inventory: Recycling or burning actinides is the best solution to the risks of long-term storage in a geologic repository. However, Fig. 3 provides a summary of some results by T. Pigford at Berkeley that indicates that the fission products, because of their yields in nuclear waste and because of their mobility in geologic storage environments, are more risky than actinides. Thus, from the very beginning, our goal was to design a concept that could destroy all major long-lived components. Our approach uses an accelerator to produce intense fluxes of low-energy neutrons or thermal neutrons. This is a departure from conventional wisdom concerning the best type of neutron spectrum to transmute or destroy actinides. Most approaches utilize a fast (harder) spectrum of neutrons. But we found that a system based on a high-intensity thermal neutron flux can destroy all of the components described

Reactor Radioactive Waste Production

3 GW Thermal PWR Yearly Discharge

Nuclide	Half-Life (yrs)	kg
238Pu	88	4.5
239Pu	24000	166
240Pu	6600	77
241Pu	14	25
242Pu	380,000	16
237Np	2,000,000	15
241Am	432	17
243Am	7400	3
99Tc	200,000	25
129I	16,000,000	6
135Cs	3,000,000	10
93Zr	1,500,000	23
107Pd	6,500,000	7
90Sr	30	13
137Cs	30	31

Fig. 2. Long-lived Radionuclides in High-level Nuclear Waste.

ATW's Intense Thermal Neutron Flux: Optimal Transmutation Performance

- Capability to effectively handle all long-lived radwaste components

- Optimized system (high flux, large destruction cross sections)

- Rapid conversion - effective half life in the system is short

Isotope	Natural Half Life (yrs)	Intense Thermal Neutron Environment
Pu239	24,000	0.003 yrs
Np237	2,000,000	0.06
Tc99	200,000	0.15
Cs137	30	2.9

Fig. 3.   Effective Radionuclide Half-Life in a Neutron Flux Environment.

above. The system operates in what seems to be an optimized fashion. Very high fluxes of neutrons do the destruction, and the cross sections or probability for transmutation at low energies are very large. This results in a very rapid conversion of material inside the system, as shown in Fig. 3. The figure compares the effective half-life in a system (which is essentially a product of flux times the probability for transmutation) to the natural half-life for a representative list of radionuclides in nuclear waste. In this high thermal flux system the effective half-life is very very short, translating into a very fast burn-up, which means the system comes to equilibrium rather quickly and operates much more efficiently than other types of reactor systems. Even for cases like cesium-137, which is a 30-year problem, the effective half-life is a factor of ten smaller than that associated with natural decay.

The basic components of the system are shown in Fig. 4. We begin with a particle accelerator that is similar to the Los Alamos Meson Physics (LAMPF) facility in terms of energy (800 MeV) but with a current that is about a factor of 100 larger. We believe that the prospects are good for achieving that current difference in light of certain developments that have occurred in the SDI neutral-particle-beam program. A beam of protons is taken out out of the accelerator and directed into a heavy-metal target. For each proton, about 22 neutrons are produced. This leads to a very intense flux of neutrons in a heavy-water moderator, about two orders of magnitude higher than the intensity that exists in a thermal reactor. The previous figure demonstrated the effective half-life of representative radionuclides, indicating that they burn up rapidly in this heavy-water blanket. One cannot, therefore, utilize the conventional methods of putting fuel into a nuclear system through solid rods. Instead, a flowing system must be used, and we have been exploring several ways of effecting such carrier loops within a system. Another feature is that this system has very low inventory of materials resident in it. If one chooses molten fluoride salts as the carrier material (this was a material that was used by Oak Ridge in the late '60s and early '70s in their design of a molten salt reactor experiment), one can extract heat from the system and convert it with very high efficiencies to electric power.

So let's look at what this environment that we have created does. In a waste-burning system one wants to achieve some rate of transmutation or rate of destruction per unit time (basically a product of the mass of material, the probability for transmutation, and the flux of neutrons). This system has about a factor of 10 higher cross section compared with other fast systems and about a factor of 10 to 100 times higher flux (so that one can reduce the amount of material that one has to have in the system by similar amounts). Figure 5 compares the inventory in this system to other types of concepts for waste destruction. For a fast-reactor system similar to the Integral Fast Reactor concept being developed at Argonne, or an accelerator driven system that uses fast neutrons (such as the Japanese Omega Program), very large amounts of material have to be loaded to get a desired performance rate. To achieve the same rate in the ATW system, we require less than a hundred kilograms of material. This inventory has some very important impacts in terms of the chemistry that one would use for separations—the capacity of the chemical separations systems are going to be small. At the end of life of the system

284

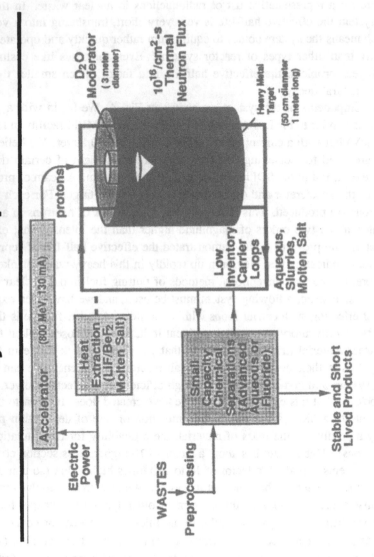

Fig. 4. ATW System Schematic.

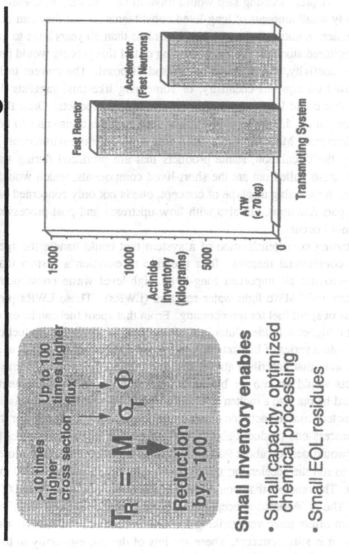

Fig. 5. Material Inventories in ATW and Other Transmutation Systems.

a residue will remain: conventional fast-neutron systems will have residues in the range of 1000 kilograms or so, while the ATW residue will be less than a kilogram.

This is the general concept and why it differs from other types of transmutation concepts that have been posed over the past 20 or 30 years. Figure 6 is a schematic of a radioactive-waste destruction system based on transmutation. High-level waste is brought into the system. A preprocessing step would have to operate very efficiently to separate out the relatively small amounts of long-lived constituents so that they can be introduced into the transmuter. Radionuclides with half-lives less than 30 years have to be pulled out and sent to monitored storage. The waste coming out of this process would have minimal amounts of radioactivity, and it would go to land disposal. The current technology for separations based on aqueous chemistry, or something like this, suggests that one can achieve goals that could lead to land disposal of effluent products. Once the long-lived components are removed, they would be introduced into the transmuter to be burned up to very high fractions. Material is circulated into this neutron environment, and at each step, or during the circulation, stable products that are produced during transmutation are pulled out. Also pulled out are the short-lived components, which would go to land disposal. Thus, in assessing this type of concept, one is not only concerned with how the transmutation part functions, but also with how upstream and post-processing chemical separations would occur.

We are starting conceptualization of a system that could handle the spent fuel discharged from commercial reactors. It is possible to envision a system that would be capable of destroying all important long-lived high-level waste constituents from the discharge of ten 1000-MWe light water reactors (LWRs). Those LWRs produce about 300 metric tons of spent fuel for reprocessing. From that spent fuel can be extracted a few metric tons of a higher actinide, plutonium, and a long-lived fission product mix that can be introduced into a series of blankets driven by one accelerator operating at a somewhat higher energy than that described above, but at a current that's in the same range. The fission products would go into the blanket's neutron target close to the neutron source. Actinides would be put into a molten salt part of this loop, and long-lived fission products would be extracted from the actinide fissions and sent into the inner part for their destruction. In the process one produces about 2000 megawatts of thermal power per blanket. Each blanket would destroy about 900 kilograms of the radionuclide mixture. Energy is also fed back to run the accelerator; this system could produce about 4000 megawatts of electric power. The reactors that would feed the ATW system produce 10 000 megawatts of electricity. The ATW-based concept could provide an additional 40% of electricity.

This system offers some very striking features in terms of radioactive-waste destruction. However, it is still a concept. There are lots of details, especially in the chemistry, that have to be worked out before it can play an important role in the management of high-level nuclear waste.

I would like to spend the last few minutes talking about the energy production system mentioned earlier. It is a system that we believe could safely provide a long-term energy supply because the accelerator drives a subcritical assembly, and it

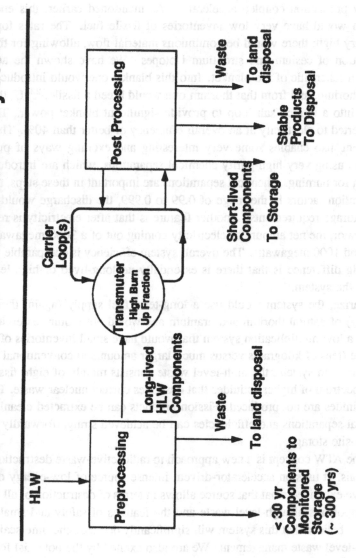

Fig. 6. Steps and Components of an HLW Management System Based on Separation and Transmutation.

would produce essentially no long-lived, high-level waste. This sounds a lot like the goals of the fusion program. However, unlike that program, ATW has a very short lead-time for development. We think that the individual parts of such a system are currently in existence. ATW could be demonstrated in a period of only 10 or 15 years.

Figure 7 illustrates the coupled accelerator-fission-blanket system that forms the basis for this energy production coupled accelerator. As mentioned earlier, this energy production system would have very low inventories of fissile fuel. The rates for actinide burn-up are very high; there would be continuous material flow, allowing for the natural isotope separation of cesium and strontium isotopes. We have shown the accelerator efficiency and a schematic of the blanket. Into this blanket one would introduce into the outer regions thorium, and from that thorium one would breed a fissile, ^{233}U, that would be introduced into a molten salt loop to provide significant blanket power. The power could be converted to electricity at an overall efficiency of better than 40%. This molten salt environment also enables some very interesting and exciting ways of pulling out fission products using very high-purity chemical separations, which are introduced back into the system for burning. Chemical separations are important in these steps, and if one achieves separation factors in the range of 0.99 to 0.999, the discharge would probably meet on-site storage requirements. Another feature is that after electricity is returned to run the accelerator, the net amount of electricity coming out of a 3000-megawatt blanket would be around 1000 megawatts. The overall system efficiency is comparable to current LWRs. The big difference is that there is essentially no long-lived or high-level waste coming out of the system.

To summarize, the system would use a long-term fuel supply (again, thousands of years of supply) of natural thorium and uranium and would not require external fuel enrichment. It is a low-multiplication system that would have small inventories of actinides at any one time (tens of kilograms versus much larger amounts in conventional systems). In a thorium/uranium system, the high-level waste consists mainly of eight fission products, not the spectrum of higher actinides that describes current nuclear waste. Plutonium and higher actinides are not produced. Fission products can be extracted cleanly, and required chemical separations and efficiencies can be achieved straightforwardly to enable our goal of on-site storage.

In short, the ATW concept is a new approach to radioactive-waste destruction and energy applications. It uses an accelerator-driven, intense source of low-energy or thermal neutrons. I have explained what that source allows in terms of destruction of all important long-lived components of high-level waste and the features of safety and small material inventories. We believe that this system will significantly decrease the time-scales associated with high-level waste management. We are also excited by the potential for creating a long-term energy source with a minimal legacy of long-term, high-level waste.

289

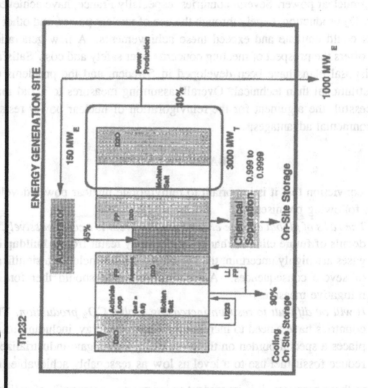

Fig. 7. An Energy Production Concept Based on ATW.

Implementing the Nuclear Option

David Bodansky

Department of Physics, University of Washington, Seattle, WA 98195

ABSTRACT

In view of the potentially serious nature of climate changes resulting from increasing production of carbon dioxide, it is important to pursue all practical mitigative measures, including nuclear power. Several countries, especially France, have achieved large reductions in CO_2 production, largely through the use of nuclear power, and other industrialized countries could emulate and exceed these achievements. A new generation of nuclear reactors offers the prospect of meeting concerns over safety and cost. Satisfactory nuclear waste disposal plans have been developed in Sweden, and the problems elsewhere are more institutional than technical. Overall, assuming measures to avoid major accidents are successful, the argument for the reinvigoration of nuclear power rests primarily on its environmental advantages.

BASIC PREMISES

The conviction that it is important to reinvigorate nuclear power development stems from the following premises:

- *The risks of global climate change are too great to accept passively.* Although the details of future climate changes which may result from a buildup of greenhouse gases are highly uncertain, the spectrum of risks includes a significant possibility of severe consequences. A responsible world should therefore take forceful mitigative measures.
- *It will be difficult to restrain increasing global CO_2 production.* The developing countries have a need to increase their use of energy, including fossil fuels. This places a special burden on the more energy profligate industrialized countries to reduce fossil fuel use to a level as low as reasonably achievable (ALARA).

Given these premises, it is imprudent to ignore any potentially effective measures for reducing CO_2 production. The most important energy measures are conservation, substitution of natural gas for coal, and increased use of solar and nuclear power. Alone among these, nuclear power faces substantial opposition. However, one cannot be confident that the alternatives will suffice without it. More particularly:

- *Conservation alone is not a solution.* Although conservation offers opportunities for large reductions in energy consumption, conservation is not a *source* of energy. Over the next few decades, additional sources will be required to supplement and replace coal and oil.
- *Natural gas represents only a short-term half-measure.* Although substituting natural gas for coal or oil would reduce CO_2 production, some CO_2 production

remains along with methane leakage. Further, resources are of uncertain magnitude and may not be adequate to sustain expanded natural gas use for more than several decades.

• *The potential of solar power is not clearly established.* There are severe environmental constraints on the expansion of hydroelectric power, the potential of biomass is not open-ended, and the world has little experience with the large-scale generation of electric power from wind, photovoltaic, solar thermal, or ocean sources. While development of solar power should be pursued vigorously, it is too soon to count upon its full success.

Nuclear power has already proved its effectiveness in reducing CO_2 production. While many industrialized countries talk about stabilizing or reducing CO_2 production during the next decade, several large users of nuclear power have already accomplished large reductions. In particular, France, Sweden, and Switzerland all reduced CO_2 production by more than 20% in the period from 1973 to 1989.

Despite these accomplishments, there are many barriers to an expansion of nuclear power. In the discussion below, we will consider the past record and current status of nuclear power, its potential contributions, and the difficulties standing in the way of a nuclear revival.

THE STATUS OF NUCLEAR POWER

As of June 30, 1991, there were 412 operating reactors in the world, with a combined net generating capacity of 320 gigawatts (GWe).[1] They were distributed among 25 countries (see Table I). Nuclear power accounted for 17% of world electricity generation in 1990[2] and for more than 40% of electricity generation in six countries: France (74.5%), Belgium (60%), Hungary (51%), South Korea (49%), Sweden (46%) and Switzerland (43%).[3]

The growth of nuclear power, which was very rapid in the 1970s and early 1980s has slowed markedly in recent years. World nuclear capacity at the end of 1990 was only 1% higher than a year earlier and generation was only 3.7% higher in 1990 than in 1989.[3] This may be contrasted with a 14% average annual growth in generation during the 1976–1986 decade.[4]

There is no prospect for an immediate resumption of rapid nuclear power growth. A comprehensive compilation in *Nuclear News* indicates that there were 101 additional reactors nominally on order or under construction as of June 30, 1991, with a total capacity of 87 GWe. Were all of these completed by mid-1999, this would represent an addition to capacity at a rate of about 3% per year. However, only 58 of these reactors (45 GWe) were listed as scheduled to be in operation before the year 2000. Although some changes are likely to occur, it is unlikely that the total will change greatly, and estimates of a 3% growth rate for the remainder of the decade must be viewed as highly optimistic. Even a 2% growth rate may not be achieved. Consistent with this pessimistic picture, the International Atomic Energy Agency now estimates that nuclear energy production

TABLE I. World nuclear status, by countries: number of units as of June 30, 1991, their net capacity (gigawatts electric), 1990 net nuclear generation (gigawatt-years), and percent of electricity from nuclear power in 1990.

| Country | Status 6/30/91[a] | | Net Generation (GWyr) | | Percent Nuclear 1990[c] |
	Number of units	Capacity (GWe)	1976[b]	1990[c]	
United States	111	99.7	21.8	65.9	19[d]
France	55	54.6	1.7	34.0	74.5
USSR	42	33.7	2.7	24.1	12
Japan	41	30.9	4.0	21.9	27
Germany[e]	21	22.4	3.3	16.4	28
Canada	19	12.8	2.0	7.7	14
United Kingdom	37	11.9	4.0	6.8	20
Sweden	12	9.9	1.7	7.5	46
South Korea	9	7.2	0.0	5.7	49
Spain	9	7.1	0.8	5.9	38
Belgium	7	5.5	1.1	4.6	60
Taiwan	6	4.9	0.0	3.8	38
Czechoslovakia	8	3.3	0.0	2.6	28
Switzerland	5	2.9	0.9	2.5	43
Bulgaria	5	2.6	0.5	1.5	36
Finland	4	2.3	0.0	2.1	35
South Africa	2	1.8	0.0	1.0	6
Hungary	4	1.7	0.0	1.5	51
India	7	1.4	0.3	0.6	2
Argentina	2	0.9	0.3	0.8	20
Mexico	1	0.7	0.0	0.3	3
Yugoslavia	1	0.6	0.0	0.5	5
Brazil	1	0.6	0.0	0.2	1
Netherlands	2	0.5	0.4	0.4	5
Pakistan	1	0.1	0.1	0.0	1
WORLD TOTAL	412	320	46.0[f]	218	17

[a]Data from Ref. 1.

[b]Data from Ref. 4.

[c]Data from Ref. 3.

[d]This is fraction from all sources; nuclear power accounted for 21% of 1990 generation by electric utilities.

[e]Generation in East Germany included in 1976 figures.

[f]Includes 0.4 GWyr of generation in Italy in 1976 (none in 1990).

will grow at an average rate of only about 2% or 3% per year for the period from 1990 to 2010.[5]

Some details of announced plans are exhibited in Fig. 1, for the ten largest nuclear power users.[1] In addition to displaying actual capacity as of June 30, 1991, plants then on order or under construction are shown, divided between those scheduled for completion before 2000 and those with later or indefinite completion dates. Figure 1 brings out the small magnitude of future development plans. There appear to be reasonably healthy nuclear programs in only three countries:

- *France.* France obtains about 75% of its electricity from nuclear power and in 1990 exported 12% of its total generation.[6] It is nearly saturated in terms of nuclear power use, especially as a large fraction of its non-nuclear power is inexpensive hydroelectric power. For the next half-dozen years, 6 new reactors are expected to go into operation at a rate of roughly one per year, including 3 of the new 1450-MWe N4 series. Overall it has been a very successful program, although not totally without difficulties. The greatest failure has been with the Superphenix breeder reactor, which has not operated reliably and, given the present and prospective uranium supply situation, has no clearly urgent mission.

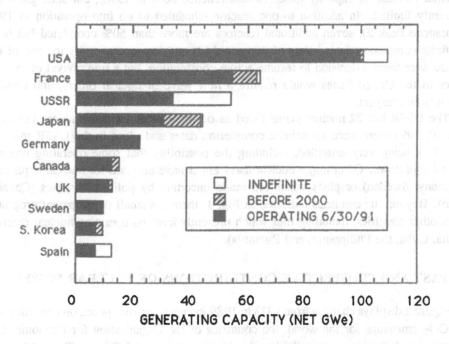

Fig. 1. Nuclear power generating capacity: actual capacity as of June 30, 1991, added capacity scheduled for operation before the year 2000, and additional plants on order or under construction with more distant or indefinite operation dates.

- *Japan.* Japan has the largest program for new construction, with the ambiguous exception of the USSR. As of June 30, 1991, 14 plants were on order or under construction (14 GWe), mostly conventional light water reactors, but including two 1315-MWe advanced boiling water reactors (ABWR), and one 280-MWe liquid metal fast breeder reactor. All but one are scheduled for completion by the year 2000.[1] Considering a broader time span, the Japanese government in October 1990 approved a plan which called for the construction of 40 reactors between then and 2010, roughly doubling its number of reactors and raising nuclear power's share of electricity generation to 43%.[7]

- *South Korea.* South Korea had 9 reactors in operation as of June 30, 1991 and 3 on order. Orders for 2 additional plants were announced in August 1991. All of these plants are scheduled for operation by 1999.[8] As with Japan, there are plans for large further expansions in the near future.

United States nuclear power generation more than doubled in the decade following a post Three Mile Island lull in 1980, continuing with an 8% rise from 1989 to 1990.[9] However, this expansion is coming to the end, with the completion in 1990 of all but one of the reactors remaining in the active pipeline. There is a prospect of gains from a further increase in capacity factor (which reached 66% in 1990), but such gains are inherently limited. In addition to one reactor scheduled to go into operation in 1993 (Comanche Peak 2), seven additional reactors are more than 50% completed but have indefinite completion dates.[1] Any of these could be put into operation by the end of the decade were there a decision to resume active construction, but a true revival of nuclear power in the United States would require a new wave of nuclear orders, and none is presently in prospect.

The USSR has 22 nuclear plants listed as on order or under construction, but as of June 30, 1991 there were no definite completion dates and plans in the USSR must be viewed as being very unsettled, including the possibility that some operating reactors may be shut down. Other major nuclear users either have no plans for nuclear expansion (Germany, Sweden) or plans which are made uncertain by political disputes (Canada, Spain). Beyond the countries indicated in Fig. 1, there are small development programs in ten other countries, including four which presently have no operating nuclear reactors (China, Cuba, the Philippines, and Rumania).

PAST AND PROSPECTIVE CONTRIBUTIONS OF NUCLEAR POWER

Figure 2 displays changes from 1973 to 1989 in energy consumption, oil consumption, and CO_2 emissions for the world, the countries of the Organization for Economic Cooperation and Development (OECD), the United States, and France (Refs. 10–12).*

---

* The CO_2 data are based on totals calculated by Marland and collaborators,[12] but subtracting the small contribution from cement production. I am indebted to Gregg Marland and the CDIAC for their help in providing these data.

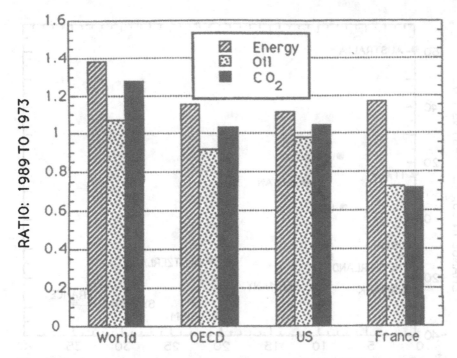

Fig. 2. Ratio of energy consumption, oil consumption and CO_2 production in 1989 to values in 1973, for selected regions and countries (see text).

Energy use rose in each case. For the world as a whole, oil consumption rose 8% and CO_2 emission increased 28%. For France, on the other hand, there was a 28% decrease in oil consumption and a 28% decrease in CO_2 production. The United States and the OECD (including the United States and France) were intermediate. In general, increases in all categories were higher for the developing countries than for the much more industrialized OECD countries, and the large increases for the world as a whole are largely explained in terms of "catching up" by countries with relatively low per capita energy use rates.

The success of France is primarily due to use of nuclear power. The role of nuclear power is brought out in Fig. 3 where for the major OECD countries, the change in CO_2 emissions from 1973 to 1989 is plotted against the fraction of total 1989 primary energy from nuclear power. (Countries with very low population or very low per capita GNP are omitted.) * The greatest reductions were achieved by large nuclear power users: France, Belgium and Sweden. The correlation is not perfect, but the trend is clear. Denmark was

* The nuclear fraction is based on OECD tabulations,[11] with the Total Primary Energy Supply (TPES) data modified to correspond to a 33% efficiency for conversion of "primary" renewable energy to electricity, rather than the 100% efficiency assumed in the original TPES tabulations. (This change has the effect of reducing the calculated nuclear fraction of primary energy for those countries which use hydroelectric power extensively.)

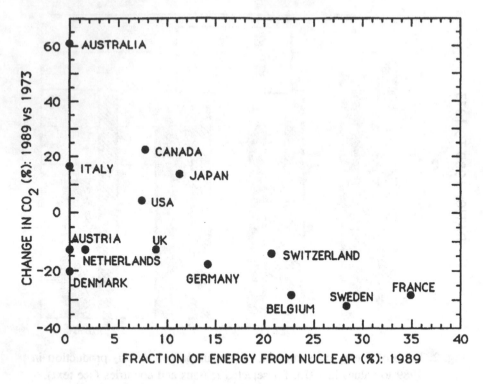

Fig. 3. Decrease in CO_2 production from 1973 to 1989 for major OECD countries, vs fraction of primary energy obtained from nuclear power (see text).

a striking exception to this trend, achieving a large CO_2 reduction without any generation of nuclear power. In large measure this was achieved by substituting electricity imports for electricity from coal in 1989,[11] with the imports coming mostly from Norway and Sweden.

To see how the example of France could be emulated and considerably extended, we consider energy consumption patterns in the United States. For the United States, CO_2 emissions come roughly one-third each from electricity generation, from varied direct uses of fossil fuels (mostly heating), and from transportation. It is easy to develop a scenario where the total emissions are halved over a period of thirty or forty years. This would correspond to a reduction in the neighborhood of 2% per year. The largest and most straightforward reduction could come by using nuclear power to replace virtually all fossil fuel use in electricity generation. The remainder could come from use of electricity to displace a portion of the fossil fuels used for heating purposes and, with more difficulty and probably a lower near-term impact, by the use of electricity for increased mass transportation and for vehicles powered with batteries or with hydrogen from electrolysis. Some of this displacement could be indirect, with natural gas freed by electricity in one application used to displace oil or coal in another.

In sketching this scenario, we tacitly assume that vigorous conservation programs are implemented so that total energy consumption does not increase over this period, despite population increases. The electrification program is not nuclear-specific, and the more that solar power can contribute towards electricity generation, the less would be the demand on nuclear power. Even at the extreme, however, the demands upon a nuclear power expansion would not be excessive. It would require something in the neighborhood of 500 or 600 GWe of new capacity, or a building rate of about 20 GWe per year over a period of thirty years. Normalized to the size of the national economy, this is less than the pace of France during the 1975–1990 period, when over 50 Gwe of nuclear capacity were added.

Other industrialized countries should be able to follow a path similar to that sketched above, although perhaps at not so rapid a rate because their potential for conservation may be less. Of course, in the long run the actions of the developing countries will be decisive. They now have about three-quarters of the world population, often with a high birth rate and a rapid increase in energy consumption. But the industrialized countries are still responsible for about three-quarters of global CO_2 emissions, and the steps they take over the next few decades will be very important in terms of actual emissions and in the example they offer.

NUCLEAR REACTORS: SAFETY AND COSTS

Reactor Safety

One of the main obstacles to nuclear power development is concern about reactor safety. In fact, outside the USSR, civilian nuclear power has had an excellent safety record. Even in the United States, which with the Three Mile Island accident has the most marred record among "western" countries, the overall record is very good. By 1990, a nuclear power industry had been created with about 110 operating reactors. They provided roughly 20% of total electricity generation in 1990 — more electricity than had been provided by all U.S. sources in 1955. This entire enterprise has been developed with no known deaths to the civilian population and a relatively low number of "ordinary" industrial casualties among workers.

TMI aside, the failings which have attracted most attention, including contamination at sites such as Hanford and Savannah River, are attributable to the weapons program, not to the quite separate civilian power activities. The mean exposure from the nuclear fuel cycle for people living within 50 miles of nuclear facilities was about 0.0005 mSv per year (0.05 mrem per year) in the early 1980s, less than 0.02% of the average radiation exposure from all natural sources.[13] Unlike fossil fuel plants, there are no noxious chemical emissions from nuclear reactors and the amount of land preempted is very small. In short, if accidents can be avoided, nuclear reactors are environmentally benign.

The Chernobyl accident provided a reminder that the consequences of an extreme accident can be severe. Given the poor design and operating practices, the Chernobyl accident sheds no light on the probability of an accident elsewhere. However, the TMI experience in 1979 showed that there was an appreciable chance of a major accident for U.S. reactors, although at TMI the concrete containment (absent at Chernobyl) kept external contamination low. Subsequently, the Nuclear Regulatory Commission and the nuclear industry have made extensive efforts to reduce the likelihood of an another accident of comparable, or greater, severity.

It is possible to monitor progress in reactor safety through application of probability risk analysis (PRA). A calculated accident probability is found through study of the network of conceivable individual malfunctions. Using the PRA formalism, the Nuclear Regulatory Commission has undertaken a program of analyzing reported "events" (i.e., reactor malfunctions at all levels), in order to identify precursors of sequences leading to severe core damage and to calculate an "inferred mean core damage frequency." This index, based on actual operating experience, has dropped from 3×10^{-3} per reactor year for 1969–1979 to 5×10^{-5} per reactor year for 1986–1987, an improvement of roughly a factor of 60.[14]

These numbers suggest that U.S. reactors are now much safer than they were before TMI. However, an even greater margin of safety would be desirable prior to a major nuclear expansion because, taken literally, they imply that, for example, with a worldwide total of 1000 reactors there still would be a 5% chance per year of core damage. In most cases, as illustrated by TMI, core damage will not lead to a major release of radioactivity outside the concrete containment which surrounds the reactor, but nonetheless risks at this level may be too high for comfort. As discussed below, new reactors are expected to provide a significant additional margin of safety.

Nuclear Power Costs

Costs of nuclear power vary widely between countries and within countries. France, generating 75% of its electricity from nuclear power, has lower electricity costs than most European countries, excluding those which obtain a large share of their electricity from hydropower.[6,15] In the United States, generating costs in 1989 varied among nuclear plants from 1.4 ¢/kWh to 22.9 ¢/kWh, with a median of 4.85 ¢/kWh.[16] Median costs for coal plants in 1989 were 2.73 ¢/kWh, with a range from 1.8 ¢/kWh to 4.2 ¢/kWh. Overall, nuclear costs had a broader range than coal costs, but on average were about twice as high. This reverses the situation of a decade earlier, when nuclear costs were lower than those of coal.

There was rapid rise during the 1980s in both construction costs and in operation and maintenance costs for nuclear plants. Revised standards adopted in the 1976–1980 period greatly increased the bulk and complexity of new reactors.[17] Adjusted for the same total capacity, the amount of concrete almost doubled, the amount of steel roughly doubled, and the length of electrical cable more than doubled. The number of hours of craft labor and construction times were greatly increased. Crucial to future economy is the reversal of the trend of increases, for both construction and operation costs.

Projecting future nuclear power costs, a joint 1989 study of the OECD Nuclear Energy Agency and the International Energy Agency made estimates for nuclear power and power from coal-fired plants. The study concluded:[18]

> Using the reference assumptions, nuclear plants are projected to have a
> significant economic advantage over coal-fire plants, for lifetime base-load
> power production,[sic] in Japan, a majority of countries in OECD Europe
> and in those regions of North America distant from coalfields.

The variation among countries depends strongly on the availability and cost of coal. Although the nuclear advantage is fairly large in some countries under the "reference assumptions," this advantage is highly dependent upon detailed assumptions of the models. It is substantially reduced and often reversed, for example, if the assumed real discount rate rises from 5% to 10%.

Especially in view of the past difficulties of predicting nuclear costs, it would appear unwarranted to claim any large future cost advantage for nuclear power in most countries. Usually, the balance between nuclear power and coal is close enough that the choice should be based on environmental and safety considerations, not considerations of direct cost. However, it may be noted that France, Japan and South Korea, the three countries with the healthiest nuclear programs, have reputations for economic pragmatism.

Next-Generation Reactors

One goal for the nuclear future, especially in the United States, is cheaper and safer reactors. Although safety and cost are often in conflict, in this case they are complementary. The key to future safety and economy for nuclear reactors is simpler, standardized units, in which the design incorporates desired safety features from the first, avoiding the need for subsequent changes. As part of good design, total demands for materials and labor used in construction are to be reduced and the construction times shortened, reducing costs.

In most countries, safety is being pursued simultaneously with the development of large evolutionary reactors. For example, the 1188-MWe Sizewell B reactor under construction in the United Kingdom and the two 1296-MWe advanced BWRs in Japan have lower calculated accident probabilities than do present reactors, but they are not radically different in design.

An alternative and much discussed approach is to move towards substantially different reactors which make extensive use of passive safety features. These are features which depend upon physical phenomena, rather than good response by the reactor operators and proper functioning of complex equipment. Examples include reliance on gravity and thermal properties of materials to insure safe operation and safe response in unusual situations. The reactors are sometimes termed "passively safe" (see, e.g., Ref. 19). Such features are incorporated in two variants of light water reactors: several advanced LWR designs for intermediate-size reactors (about 600 MWe) and the radically innovative PIUS reactor. Other serious contenders are small modular high temperature gas cooled reactors and liquid metal fast breeder reactors. None of these reactors has been built, but advanced design and testing has been carried out.

There is as yet no worldwide consensus as to whether it is preferable to proceed with large evolutionary reactors, taking advantage of both economies of scale and the benefits of accumulated experience, or to proceed with smaller reactors of more novel design. The latter place greater emphasis on passive safety and lend themselves to modular construction, but lack as extensive a base of experience.

With either path, the next generation of reactors can be expected to be still safer than the present generation. A suggested target for the next generation is a probability of less than 10^{-5} per reactor-year for a core damaging event (five times better than the 1986–87 record cited above) and, taking credit for the reactor containment, of less than 10^{-6} per reactor year for a large external release of radioactivity.[19] These numbers should not be considered to be quantitatively precise estimates or to be limits on appropriate safety ambitions for reactor designers. They are, however, indicative of the belief that an already "safe" technology (i.e., reactors in the Western world) can be expected to become even safer in the future.

NUCLEAR WASTES

No final repositories for spent fuel or high-level wastes are available in any country, and none is expected to be completed within the next decade or two. However, satisfactory solutions exist, at least in the form of detailed plans. In particular, in response to demands for a "solution," a plan was developed in Sweden during the late 1970s and early 1980s for the disposal of spent fuel in thick-walled canisters to be placed deep underground in holes in granite.

At all times, the "waste" is in the form of solid pellets of spent fuel. The total volume of the fuel extracted annually from a reactor is very small, of the order of 10 cubic meters allowing for spacing between fuel elements. Thus, this fuel could be held in about ten canisters, each with an internal volume of roughly 1 cubic meter. With so little volume, it is feasible to make the canisters very sturdy and corrosion resistant. Swedish plans call for using copper cylinders with 10-cm thick walls. There is little opportunity for water to reach these buried canisters or for wastes to travel far from them should any water reach the canisters and attack them. The designers of this system have concluded that it "has the capacity to isolate the waste from groundwater over periods longer than 1 million years."[20] The plan has undergone a number of reviews, which have concluded that it is sound, including one under the auspices of the U.S. National Research Council.[21] Most importantly, from an operational standpoint, the Swedish Government has found the plans to be satisfactory.

In contrast, the United States appears incapable of formulating a serious policy for the handling of high-level radioactive wastes. After several decades of inconclusive attempts, a presumably final waste handling policy was enunciated in the Nuclear Waste Policy Act of 1982. This set forth a sequence of steps which would lead, with due deliberation, to the selection of a site for deep geological disposal of the wastes, with operation of the site to begin in 1998.

The mandated program, which fell well behind schedule, was revised by 1987 amendments. These identified a site at Yucca Mountain in Nevada as the sole candidate for "site characterization," i.e., studies to determine suitability as a waste repository site. The State of Nevada has vigorously opposed and to some extent has succeeded in impeding these studies. Recognizing the slow pace of progress, a new schedule was enunciated by the DOE in 1989 which, if the site is found to be acceptable, would lead to the beginning of operations by "approximately 2010." Given the record, it may be reasonable to expect that the site will, at best, not be put into operation before 2020.

A common thread in the failure of U.S. efforts to implement a waste disposal program is an apparent lack of serious purpose. This is exhibited, for example, in the disparate treatment of waste hazards. Spent fuel is now accumulating in cooling ponds at the reactor sites. The spent fuel is very hot, both radioactively and thermally, and not particularly well isolated from the environment. However, there is virtually no expressed public concern, and very little safety concern in engineering or scientific circles. Were the spent fuel placed in protective canisters and transferred to a deep underground repository, they would be far better isolated from the environment as they decay to lower levels of radioactivity. But it is this later phase that elicits most of the expressed concern. Were radioactive wastes a real hazard, a serious society might be expected to try to isolate them promptly.

The approach is related in part to the issue of responsibility to future generations. Of course, each generation has a responsibility which is not to be treated lightly. That is the motivation, for example, for the present concern over global climate change. In the case of waste disposal, however, the professed concern is carried to extremes which find few counterparts. Suggested standards for exposures from potential repository leaks, extending for 10,000 years, are far more stringent than standards for exposure of people today. We appear to wish to shield future generations from exposures far smaller than we accept for ourselves today, most conspicuously the present risks from indoor radon. In effect, it is tacitly assumed that future generations will have lost the ability to detect and cope with radioactivity. It is beyond the pale to raise the possibility that instead of being technologically inept compared to ourselves, people in several thousand years will not only know how to detect radioactivity but will have found a way to avoid or cure cancer. The illegitimacy of such a speculation suggests that we are not dealing with substantive risks but with symbols.

Influencing this approach is the fact that most of the scientific community does not consider nuclear wastes to be a serious technical problem, while some of the environmental community may be comfortable in keeping the issue alive. Michael McCloskey,

Chairman of the Sierra Club, is quoted in Luther Carter's *Nuclear Imperatives and Public Trust* as saying:[22]

> I suspect many environmentalists want to drive a final stake in the heart of the nuclear power industry before they will feel comfortable in cooperating fully in a common effort at solving the waste problem. . . . Their concern would arise from the possibility that a workable solution for nuclear waste disposal would make continued operation of existing plants more feasible, and even provide some encouragement for new plants.

This is part of a breakdown of social cohesion and trust. The prestige of government and "experts" is low, and there is little willingness to rely on their assurances that the civilian waste disposal program poses little risk. In this suspicious climate, appeals to perceived local self-interest fall on fertile ground and attempts even to explore potential sites, much less develop them, face intense local opposition. The scenario of federal-local conflict is now being played out in Nevada, where the State authorities are doing all they can to thwart the site characterization program. The outcome is uncertain, but the process is not conducive to careful and expeditious study.

In response to public distrust and demands for a near-perfect system, the federal government has adopted an approach which demands full knowledge of all aspects of a 10 000-year program before proceeding to implement a solution. A Position Statement of the Board on Radioactive Waste Management of the National Research Council suggests that this approach may be a large part of the problem:[23]

> There is no scientific or technical reason to think that a satisfactory geological repository cannot be built. Nevertheless, the U.S. program, as conceived and implemented over the past decade, is unlikely to succeed . . .
>
> In the face of public concerns about safety, however, geophysical models are being asked to predict the detailed structure and behavior of sites over thousands of years. The Board believes that this is scientifically unsound and will lead to bad engineering practice.

Instead of the rigid approach of "demanding absolute certainty about the safety of the repository for 10,000 years" (*ibid.*, p. 1) the Board suggests a more flexible and experimental approach, in which detailed plans could be modified as increased knowledge is gained about the site being developed. This, as the Board points out, will require a change in the existing laws and regulations under which the waste program operates. More fundamentally, it would require a broader realization that the overall hazards of geological disposal are not very great, a belief in the soundness of the general plan, and a trust in the competence and good faith of the people managing the program.

NUCLEAR POWER AND NUCLEAR WEAPONS

Perhaps the most profound reason for hesitation about nuclear power is the linkage between nuclear power and nuclear weapons. The existence of a nuclear power infrastructure, even if initially devoted solely to peaceful commercial power generation, provides personnel, equipment, and materials which could ease the path to nuclear weapons. The problem would be made worse by the existence of breeder reactor programs which put plutonium in wide circulation.

In addition to these positive linkages, there are also negative ones. Competition for energy resources, in particular oil, has long been seen as having the potential of escalating into war, conceivably including nuclear war. Expanded use of nuclear power would lessen the pressures, through displacement of oil and other fossil fuels in electricity generation and by substitution of electricity for oil in other sectors. It may be noted that the one war in which nuclear weapons have been used appears to have been motivated in part by Japan's concern over access to oil supplies. More recently, the intervention of the United States against Iraq is widely thought to have been strongly motivated by oil considerations. Further, it was oil wealth which gave Iraq the resources to pursue its nuclear weapons program.

For states which admit to having nuclear weapons, there has been no positive linkage. They all developed weapons well before they developed civilian nuclear power. First operation of civilian nuclear power plants came in the United Kingdom in 1956, in the United States in 1957, in the USSR in 1958, and France in 1964. All had nuclear weapons well before then. China, which is scheduled to have its first commercial nuclear power in 1991 has had a bomb since the mid-1960s. The historical record is clear: if a country is determined to have nuclear weapons, it is simplest to go for weapons first and nuclear power later.

Among countries which are suspected of having nuclear weapons or of trying to develop them, currently or in the past, some have no commercial nuclear power generation (Iran, Iraq, Israel, and North Korea) and some generate a small amount of nuclear power (Argentina, Brazil, India, Pakistan, and South Africa). In addition, there are many countries with large commercial nuclear power programs which appear to have no weapons ambitions, for example Japan, Germany, Canada, Sweden, South Korea, and Spain.

The picture is not crisp. It may be that some of the states at the threshold of nuclear weapons, now or in the future, can derive benefit from the existence of a civilian power program. But the links are not clear in any instance, and the example of Iraq shows how a truly determined state can pursue a surreptitious program which places no reliance on civilian nuclear power.

Undoubtedly, a completely nuclear-free world would offer fewer immediate opportunities for states which decide to "go nuclear." But that would require the abandonment of all nuclear weapons, of civilian nuclear power, and of reactors used in research and in isotope production. Even so, this would not suffice to erase the memory and, to some, the hope of future nuclear weapons. Instead, we live in a nuclear world and will continue to live in one. It is difficult to see how the problems of weapons proliferation will be lessened if countries such as the United States, Germany and Sweden were to phase out their own nuclear power programs, because there is little prospect that all relevant countries would follow suit. Efforts would be better directed towards minimizing proliferation dangers in the context of a world with nuclear power.

OBSTACLES TO NUCLEAR POWER DEVELOPMENT

Presumably, opposition to nuclear power would diminish were it widely accepted that the next generation of reactors will be safe and inexpensive, that nuclear waste disposal problems can be responsibly handled, and that nuclear power would not greatly exacerbate proliferation dangers. However, there is not likely to be an abrupt, universal change of opinion. Instead, the increased acceptance of nuclear power, if it is to occur, will depend upon a less dramatic shift in views about nuclear hazards coupled with increased concerns about the hazards of doing without it.

We will consider these matters in the context of the United States, but many features of the pattern are repeated elsewhere. Already, there is much diffuse support among the public in the sense that it is widely believed that more nuclear power will eventually be necessary — some time, some place. However, at present there is effective opposition to a new nuclear facility in one's own vicinity, here and now. With few exceptions it is difficult to shut down an operating nuclear facility, and most ballot initiatives to achieve this have failed. With even fewer exceptions, it is difficult to build a new facility. This is a prescription for paralysis.

The division between many levels of government further complicates matters and contributes to the paralysis. Important roles are played by the President and a host of bodies including the Nuclear Regulatory Commission, the Department of Energy, and the Environmental Protection Agency. Congress and the Courts also have crucial roles. There is a confused division of initiative and authority between the federal government, the States, and in some cases individual cities or counties. Smooth progress requires near unanimity, with a single determined player often having the power to block progress. Overall, there is more to gain politically from opposing nuclear power than from urging it, and many of those in government who favor nuclear power in principle oppose federal activism in the economic realm.

It will take great political vigor and initiative to move out of this impasse. Some important efforts are now underway, such as efforts for reforms in the procedures for the licensing of new reactors. However, there is no very large supporting constituency for these and additional initiatives. The electric utilities, which once were strong supporters of nuclear power, are now reluctant to commit further to it. There are large financial risks if a project falls into disfavor, and the rewards for success are modest. For utilities, support of conservation is something of an economic and political free ride. Their return on investment will be assured by sympathetic rate-setting commissions and the onus of possible electricity shortages will in part be shifted to the consumers who have failed to conserve sufficiently. If the situation gets too bad, they can build gas turbines fairly quickly and with relatively small capital investments.

A potential pro-nuclear constituency lies in the scientific and engineering community. But for the most part there is little sense of urgency, with a general indifference to energy issues and a feeling that the fate of nuclear power is in other hands. Nevertheless, should energy issues again move to the fore of general consciousness, the technical community can play an important role in informing a skeptical and worried public. Of course, much will depend upon evolving perceptions of climate change dangers and the future successes of solar power.

Another potential source of pro-nuclear activism lies in segments of the environmental movement—environmental revisionists. Despite all past appearances, the main argument *for* nuclear power is an environmental one. If nuclear power operates as designed, it is clean and safe and places minimal demands upon transportation and land resources. Working flawlessly, it is close to an environmental ideal, an embodiment in terms of amount of materials, if not in size of units, of "small is beautiful." Given a period of accident-free operation, it is not unreasonable to expect that some environmentalists will take a fresh look at nuclear power, following the recurring impulse to question received wisdom. The moral fervor of such a neo-environmental movement could conceivably provide the spur to a nuclear revival.

In the meantime, even without this influx of new support, there are successes in improving the operation of existing reactors, in moving (albeit slowly) towards approved designs for a new generation of reactors, and in gradually overcoming the roadblocks facing waste disposal at Yucca Mountain. And, always, there is the example of the success of France and the steady progress of Japan.

REFERENCES

1. "World List of Nuclear Power Plants," *Nuclear News* 34 (10), 61–80 (August, 1991).
2. "Capacity of nuclear power plants edges up in 1990," *IAEA Newsbriefs* 6, No. 1(48), 1 (January/February 1991).
3. *Commercial Nuclear Power, 1991: Prospects for the United States and the World*, Report DOE/EIA-0438(91) (U.S. Department of Energy, Washington D.C., 1991).
4. *International Energy Annual 1986*, Report DOE/EIA-0219(86) (U.S. Department of Energy, Washington D.C., 1987).
5. *IAEA Newsbriefs* 6, No. 4(51), 5 (August/September 1991).
6. *Nuclear Power in France is Doing Well!*, Press Release, (Electricité de France, Paris, July 1, 1991).
7. "Japan: Nuclear need stressed in three government acts," *Nuclear News* 33 (15), 51 (December 1990).
8. "Late News in Brief," *Nuclear News* 34 (10), 25 (August 1991).
9. *Monthly Energy Review, August 1991*, Report DOE/EIA-0035(91/08) (U.S. Department of Energy, Washington D.C., 1991).

10. *International Energy Annual 1989*, Report DOE/EIA-0219(89) (U.S. Department of Energy, Washington D.C., 1991).
11. *Energy Balances of OECD Countries, 1980–1989*, International Energy Agency, Organization for Economic Co-operation and Development (OECD, Paris, 1991).
12. G. Marland, Oak Ridge National Laboratory (private communication).
13. *Ionizing Radiation Exposure of the Population of the United States*, NCRP Report No. 93 (National Council on Radiation Protection and Measurements, Bethesda MD, 1987), p. 53.
14. T. E. Murley, "Developments in Nuclear Safety," *Nuclear Safety* 31 (1), 1–9 (1990).
15. F. Nectoux, *Crisis in the French Nuclear Industry* (Greenpeace, Amsterdam, 1991).
16. *Electric Plant Cost and Power Production Expenses 1989*, Report DOE/EIA-0455(89) (U.S. Department of Energy, Washington D.C., 1991), pp. 11–12.
17. *A Review of the Economics of Coal and Nuclear Power*, Department of Energy draft report 9/30/81 (unpublished), Sect. III.A.1.
18. *Projected Costs of Generating Electricity from Power Stations for Commissioning in the Period 1995–2000* (OECD, Paris, 1989), p. 12.
19. C. W. Forsberg and A. M. Weinberg, "Advanced Reactors, Passive Safety, and Acceptance of Nuclear Energy," *Annu. Rev. Energy* 15, 133–52 (1990).
20. T. Papp, "The role of the canister in a system for the final disposal of spent fuel or high-level waste," in *The Disposal of Long-Lived and Highly Radioactive Wastes*, edited by A. S. Laughton, L. E. J. Roberts, D. Wilkinson, and D. A. Gray (Royal Society, London, 1986), p. 40.
21. *A Review of the Swedish KBS-3 Plan for Final Storage of Spent Nuclear Fuel*, National Research Council (National Academy Press, Washington D.C., 1984).
22. L. J. Carter, *Nuclear Imperatives and the Public Trust: Dealing with Radioactive Waste* (Resources for the Future, Washington, D.C., 1987), p. 431.
23. *Rethinking High-Level Radioactive Waste Disposal: A Position Statement of the Board on Radioactive Waste Management*, National Research Council (National Academy Press, Washington D.C., 1990), p. vii.
24. "Nuclear power status around the world," *IAEA Bulletin* 23, No. 1, 43 (1991).

The Modular HTGR: Its Possible Role in the Use of Safe and Benign Nuclear Power

Massoud T. Simnad
University of California
San Diego, CA 92093

ABSTRACT

The advantages of the modular high-temperature helium gas-cooled nuclear power reactor are discussed in terms of the basic design of this system and its inherent and passive features that provide the safety and protection of the plant investment. With its environmental compatibility and safety characteristics, the MHTGR offers a number of advantages that should assure it a significant role in contributing to energy production and mitigating climate change.

INTRODUCTION

The increasing awareness of the potential environmental consequences of burning fossil fuels and their rapid depletion has revived interest in nuclear power world-wide. The continued growth of the world's population produces a corresponding growth in energy demands. Recent studies predict a world population of ten to fifteen billion within the next fifty years. The corresponding increase in energy demand will require at least a tripling of current energy production, particularly when world-wide rising expectations in standards of living are considered. The results of several detailed studies, such as the IIASA study,[1] indicate that nuclear power in the range of 2500 to 4000 GWE will be required to meet this demand, in addition to the contributions from other sources of energy. This means that several thousand large power reactors will be in operation, compared to the approximately 400 power reactors currently in operation in about 26 countries.

The acceptance of nuclear energy for electric power generation is influenced by the requirements of safe and reliable operation, as well as competitive energy costs. However, the future generation of nuclear power plants must have the characteristics of "inherent safety" which has been defined by Alvin Weinberg as "safety which relies not upon the intervention of humans or electromechanical devices, but on the immutable and well-understood principles of physics and chemistry." The electric utility viewpoint, as expressed by Gordon Heins, is that the new generation of nuclear power reactor designs "must be smaller in unit size, simpler to operate and maintain, and incorporate passive safety features, yet still be competitive with fossil-fueled options."[2]

During the period 1944–1955, Farrington Daniels did pioneering studies at Oak Ridge National Laboratory and at the University of Wisconsin on the design of an inherently safe, high-temperature, helium gas-cooled power reactor (HTGR),[3] in which the basic characteristics of advanced HTGR reactors were established. These included helium gas as the coolant in a direct-cycle gas-turbine primary system, the selection of graphite as the moderator and core structural material, and the choice of uranium carbide and thorium carbide as the fissile and fertile materials, respectively, in a $^{235}U/Th/^{233}U$ fuel cycle.

THE DEVELOPMENT OF HTGRs

The HTGR was revived in the mid-1950s in the U.K. (DRAGON reactor), the U.S. (Peach Bottom and UHTREX reactors), and in the F.R. Germany (AVR reactor).[4-6] The prototype reactors operated very successfully and demonstrated the feasibility and advantages of the HTGRs. Larger commercial prototypes were then built and operated in the U.S. (300 MWe Ft. St. Vrain) and in the F.R. Germany (300 MWe THTR). All these prototype reactors have been shut down following their useful life. A small prototype HTGR for process-heat applications is under construction in Japan. Research and development studies on the modular HTGR are in progress in the U.S., Germany, Russia, Japan, and the P.R. China.[4-6]

THE MODULAR HTGR (MHTGR)

The development of the helium-cooled reactor has been directed toward a high degree of inherent safety.[4,5] The smaller (130 MWe), or modular, MHTGRs have been designed since 1984 to be so safe that, even under the most severe postulated accidents, they would not result in core damage or require the evacuation of the public from the site boundary. This goal is to be achieved within the framework of efficient and economic generation of electric power and process heat.

The design of the MHTGR addresses the changed utility requirements, the need for public acceptance, and the Congressional initiative for "inherently safe" power reactors. The utility goals include the following basic factors:
- reduction of uncertainties in costs and schedules
- reduction of financial commitments
- minimizing investment risks
- shorter planning cycles
- smaller power additions

These requirements led General Atomics and the U.S. Department of Energy since mid-1984 to the development of the smaller MHTGR power plants, with emphasis on improved safety, greater reliability, and competitive economics.

MHTGR PLANT DESIGN

The modular concept permits deployment of individual reactor modules to match the growth requirements or financial constraints of the utilities. Each reactor module in a power plant is housed in a vertical cylindrical concrete enclosure which is fully embedded in the earth below ground level (Fig. 1). The reactor enclosure serves as an independent confinement structure having a vented and filtered exhaust system (Fig. 2). Above ground auxiliary structures house common systems for fuel handling, helium processing, and other services.

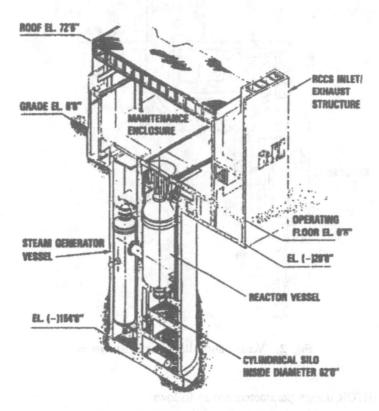

ROOF EL. 72'6"

GRADE EL. 8'6"

MAINTENANCE ENCLOSURE

RCCS INLET/ EXHAUST STRUCTURE

STEAM GENERATOR VESSEL

OPERATING FLOOR EL. 6'8"

EL. (-)20'0"

REACTOR VESSEL

EL. (-)154'6"

CYLINDRICAL SILO INSIDE DIAMETER 62'0"

Fig. 1.   MHTGR elevational view.

Each reactor module is contained within three steel vessels: a reactor vessel, a steam generator/circulator vessel, and a connecting concentric cross-duct vessel (Fig. 3). The reactor vessel, 22.5 ft in diameter and 72 ft in length, contains the core, reflector, and associated supports. A shutdown heat exchanger and a shutdown cooling circulator are located at the bottom of the reactor vessel. Top-mounted standpipes contain the control-rod drive mechanisms and hoppers containing boron carbide pellets for reserve shutdown. These standpipes are also the access ports for refueling and in-reactor inspection.

In a power plant it houses a vertical cylindrical concrete enclosure which is fully embedded in the earth. A region below grade level (Fig. 1) below the enclosure serves as an inlet/exhaust confinement structure, serving as intake and filtered exhaust system (Fig. 2). Above, ground level there are the common system with fuel handling, fuel reprocessing and other service...

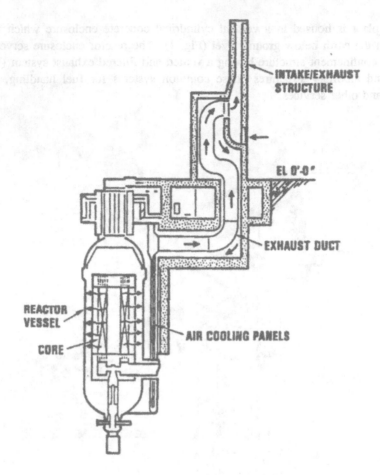

Fig. 2. Vented and filtered exhaust system.

The MHTGR design parameters are as follows:

Thermal power	350	MWt
Electric output	135	MWe
Net efficiency	38.4	%
Steam conditions	1005°	F/2515 psia
Helium coolant temperature	1268°	F outlet
	497°	F inlet
Helium pressure	925	psig
Core power density	5.9	W /cu cm

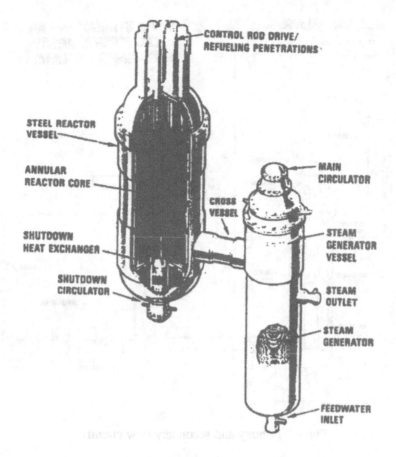

Fig. 3. Reactor, cross vessel and steam generator.

REACTOR SYSTEM COMPONENTS

A helical-coil steam generator and the electric-motor-driven main circulator are contained in the steam generator vessel, which is 14 ft in diameter and 85 ft in length. Feedwater enters the steam generator through the bottom-mounted header and superheated steam exits through a side-mounted nozzle. The main helium circulator is mounted on top of the steam generator vessel.

The primary and secondary flow circuits are shown in Fig. 4. The helium coolant flows down through the core, where it is heated by the nuclear reactions. The hot helium then flows through the inner cross-duct and downward over the steam generator bundle where its heat is transferred to the water to make steam. The steam is piped to a two-turbine generator unit.

312

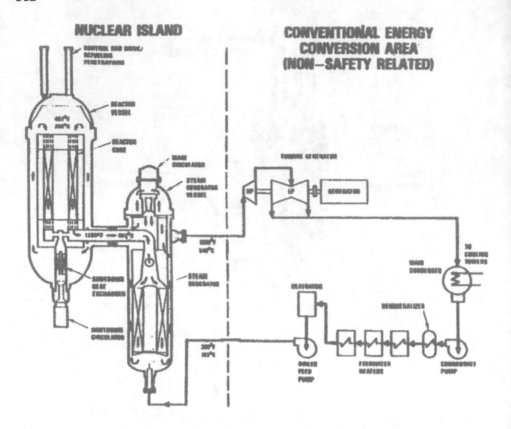

Fig. 4. Primary and secondary flow circuits.

The active core region in the reactor is composed of graphite fuel element blocks that are hexagonal in cross section, 14 in across the flats and 31 in long (Figs. 5, 6). The fuel is in the form of low-enriched (20% ^{235}U) fissile uranium oxycarbide (UCO) and fertile ^{238}U or ^{232}Th. The particles are bonded together in fuel rods, 0.5 in in diameter and 2 in long, which are contained within sealed vertical holes in the graphite fuel element blocks. Vertical coolant holes are also provided in the fuel-element blocks. The fuel elements are stacked in columns to make up an annular-shaped core having a radial thickness of about 3 ft, an average outer diameter of 11.5 ft and a height of 25 ft. Unfueled graphite blocks surround the active core to form replaceable inner and outer radial and upper and lower axial reflectors. Permanent reflector blocks are located at the outer periphery of the replaceable graphite blocks.

Reactor power is controlled by articulated control rods that travel in the vertical channels located in the inner and outer graphite reflector regions. A reserve shutdown system, when activated, permits small boron carbide spheres to enter channels located in the innermost row of fuel columns.

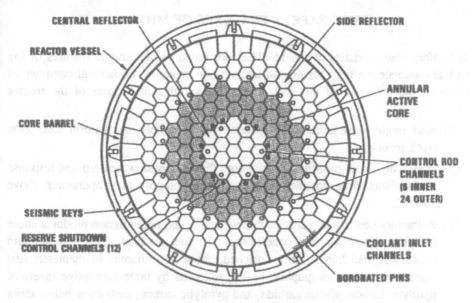

Fig. 5. Reactor cross section.

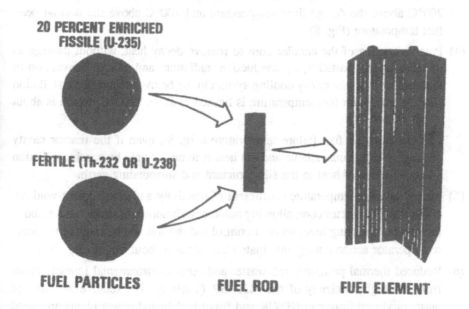

Fig. 6. Fuel element components.

PASSIVE SAFETY FEATURES OF MHTGR

The safety characteristics of the MHTGR are based on the unique features of the design that eliminate the need for human intervention or active mechanical components for safety of the public or of the investment. These inherent features of the reactor include:

(1) Inert single phase helium gas coolant, with extremely low neutron absorption cross section.

(2) Stable, high-heat-capacity graphite core that allows slow heat-up and response to transients, does not melt, and retains its integrity to temperatures above 3000°C.

(3) Refractory-coated fuel particles that are stable and retain fission products under the most severe accident conditions. The fuel particles consist of uranium oxycarbide fuel kernel (20% enriched, about 350 microns in diameter) first coated with a porous graphite buffer, followed by three successive layers of pyrolytic carbon, silicon carbide, and pyrolytic carbon, each layer being about 50 microns thick (Fig. 7). Coated particles are mixed with a graphitic material, formed into fuel rods, and then inserted in holes drilled in the graphite fuel-element blocks (Fig. 6). No failure of refractory coatings on the fuel particles occurs if the fuel is maintained below about 1800°C, which is approximately 200°C above the design limit temperature and 700°C above the normal peak fuel temperature (Fig. 8).

(4) Passive cooling of the annular core to remove decay heat, without damage to the core or fuel particles, by conduction, radiation, and natural convection to the reactor enclosure cavity cooling system in the below-ground silo installation (Fig. 9) Maximum fuel temperature is limited to about 1600°C, which is about

200°C below the fuel failure temperature (Fig. 8), even if the reactor cavity cooling system is unavailable and the heat is removed only by passive radiation and conduction of heat to the silo structure and surrounding earth.

(5) Strong negative temperature coefficient of reactivity and zero coolant-void co-efficient of the reactor core, allowing automatic shutdown at temperatures above the normal operating level without control rod motion and insensitive to incorrect operator action during anticipated transients without scram (Fig. 10).

(6) Reduced thermal pollution, rad waste, and other environmental impacts result from the high efficiency of the MHTGR (Table I). The relative amounts of waste produced from an MHTGR and fossil-fuel-fueled power plants are listed in Table II.

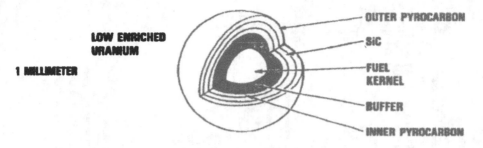

Fig. 7. Coated fuel particle (TRISO).

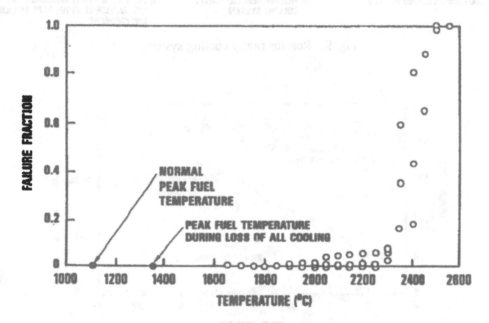

Fig. 8. Stability of TRISO-coated particles.

ECONOMIC AND LICENSING BENEFITS FROM THE MHTGR

Although traditional scaling laws favor larger plants, the MHTGR economics benefit from: (l) the elimination of excess engineered safety systems; (2) nonsafety classification of balance of plant; (3) pre-licenced, standardized design; (4) shortened schedules; (5) serialized factory fabrication; (6) multiple units on a site.

The results of extensive cost estimates indicate that a 4-unit plant of 1800-MWt total capacity will be competitive with modern coal-fired plants and advanced water-cooled reactors. The capital cost target (overnight) for such a plant is estimated to be

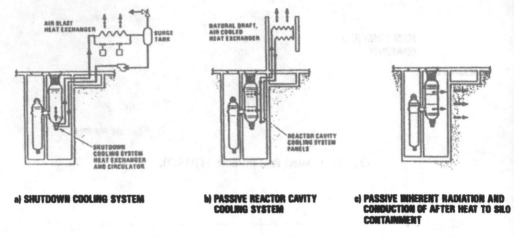

Fig. 9.   Reactor cavity cooling system.

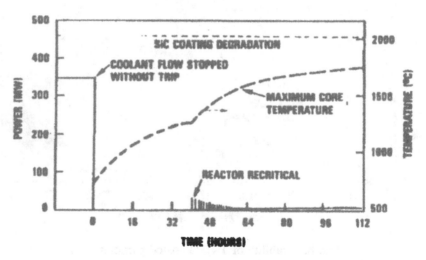

Fig. 10.   Core temperatures and reactor power.

$1500/KWe. The busbar cost of electricity appears to be 10 per cent less than the cost from an equivalent-sized modern coal-fired plant (Table III).

The smaller, simpler, easier to operate and maintain MHTGR is forgiving of mistakes. Also, it reduces the requirements to develop a large, expensive support infrastructure, thus permitting a more rapid deployment and beneficial return from nuclear energy in developing nations (Fig. 11).

The range of MHTGR applications is listed in Fig. 12. These include technologies that will have a major impact on reduction of air pollution, such as fossil fuel recovery,

TABLE I. Comparison of Reactors and Fossil Plant.

	MHTGR 4 × 350	Equivalent PWR 1 × 1670	Equivalent BWR 1 × 1600	Equivalent Fossil Plant 1 × 1540
Core thermal power [MW(t)]	1400	1670	1600	1540
Plant electric output [MW(e)]	540	540	540	540
Plant efficiency (%)	38	−32	−34	−35[a]
Core power density (KW/lℓ)	5.9	98	54	N/A
Design fuel burnup (MWd/Te)	83 000	32 000	23 400	N/A
Refueling cycle	1/2 core per 20 months	−1/3 core per year	−1/4 core per year	Continuous 2.0 million tonnes per year

[a]Power demands of sulfur dioxide scrubbers and other pollution-control equipment required on new coal plants reduce coal-to-electricity efficiencies to 35% or lower.

TABLE II. Waste Production from Nuclear and Fossil.

	Waste	Solid	To Atmosphere
Nuclear[a]	Radwaste		
High level		30	—
Low level		40	—
Coal[b,c]	Ash	230 000	—
	CO_2	—	7 500 000
	SO_2	—	3 400
Oil[b,c]	Ash	16 000	—
	CO_2	—	5 000 000
	SO_2	—	2 400
Natural gas[b,c]	Ash	—	—
	CO_2	—	3 500 000
	SO_2	—	—

[a]Without reprocessing.
[b]Assumes 100% fly ash removal and 95% SO_2 removal.
[c]Neglects NO_x production (most severe for natural gas).

TABLE III. Comparison of MHTGR and Coal Power Costs.

	4-Module MHTGR	Coal Plant[a]
Thermal rating, MW(t)	1400	1475
Net electrical rating, MW(e)	539	500
Capacity factor, %	80	80
Construction schedule, MO	36	48
Total capital costs, $M	1083	688
Specific capital costs, $/kW(e)	2009	1376
Fixed charge rate	0.095	0.095
Busbar costs,[b] mills/kWh		
Capital	27.4	18.7
Fuel cycle	9.8	23.g[c]
D&M	7.5	8.2
Decommissioning	0.5	—
Total	45.3	50.8

[a]Based on pulverized coal plant with precipitators and desulfurization equipment.
[b]Based on 30-year levelized costs.
[c]Based on $1.75/MBtu for 2005 startup and 1% real escalation after 1968.

coal gasification and liquefaction, steel production by direct reduction of iron ore with hydrogen, cogeneration systems, district heating, and hydrogen production by electrolysis or thermochemical water splitting. A study of MHTGR cogeneration desalination systems was carried out by General Atomics, Bechtel National, Inc., and Gas-Cooled Reactor Associates for the Los Angeles Metropolitan Water District. Fig. 13 shows the flow sheet for this process, which can produce large volumes of fresh water from saline waters with no impact on climatic change.[4]

Direct-cycle gas turbine power plants for near-term application have been under study for direct Brayton-cycle power plants based on Modular Gas-Cooled Reactor cores.[7,8] These conceptual designs are based on a 200 MWt pebble-bed reactor (MGR-GT), and on 350 MWt and 450 MWt prismatic cores. The outlet gas temperature is 850°C. Busbar efficiencies in the range of 45 to 50% appear possible, resulting in a lower-cost modular nuclear generating system. The conclusion from these studies is that the high performance, economic potential, and reliance on available technologies support near term development of this power system.[7,8]

The civilian MHTGR pre-licensing application technical review was completed in early 1988. The key licensing documents were submitted to the U.S. Nuclear Regulatory

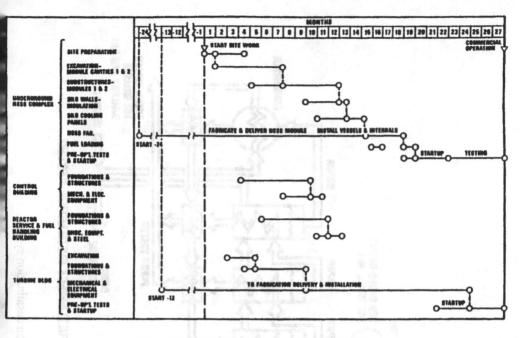

Fig. 11. MHTGR construction schedule.

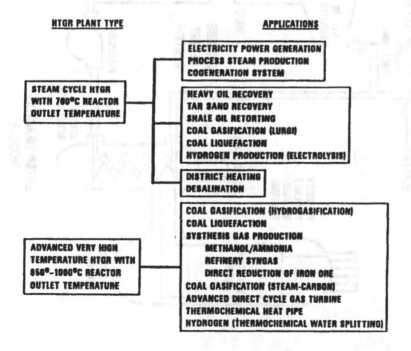

Fig. 12. MHTGR applications.

320

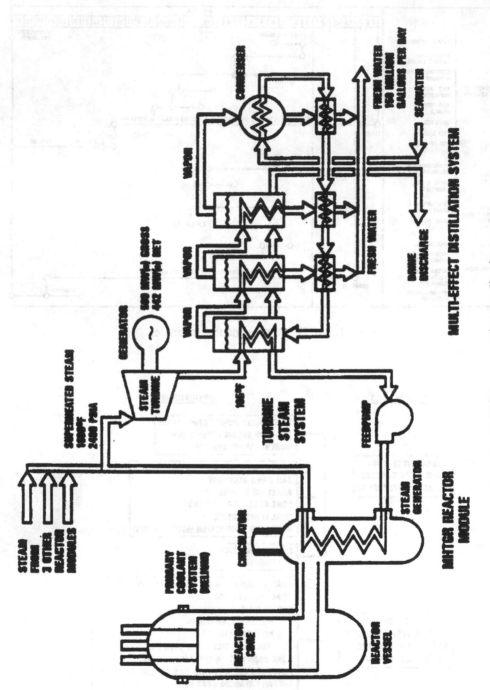

Fig. 13. MHTGR cogeneration desalination system.

Commission in 1987. The draft safety evaluation report (NUREG 1338) was issued in February 1989. The Advisory Committee on Reactor Safeguards (ACRS) agreed with the NRC staff endorsement in a letter dated October 13, 1988: "Neither the designers, the NRC staff, nor have the members of the ACRS been able to postulate accident scenarios of reasonable credibility for which an additional physical barrier to release of fission products is required in order to provide adequate protection to the public."

CONCLUSION

The MHTGR can contribute significantly to the mitigation of air pollution and climate change by serving diverse energy needs world-wide. The design of this passively safe nuclear power system can satisfy the concerns of the public, the government, and the utilities about nuclear safety and investment protection.

ACKNOWLEDGMENT

The author would like to record his appreciation to R. A. Dean, C. Hamilton, and W. Simon for helpful discussions and assistance in the preparation of this paper at General Atomics.

REFERENCES

1. W. Hafele, Energy in a Finite World (International Institute for Applied Systems Analysis, Salzburg, Austria, 843 pp., 1980).
2. G. L. Heins, *Energy: The Intl. J.* **16** (1/2), 1 (1991).
3. M. T. Simnad, *Energy: The Intl. J.* **16** (1/2), 25 (1991).
4. M. T. Simnad, A. J. Goodjohn, and J. Kupitz (eds.), High-Temperature Helium Gas-Cooled Nuclear Reactors: Past Experience, Current Status and Future Prospects, Special Issue of *Energy: The Intl. J.* **16** (1/2), 1–610 (1991).
5. International Atomic Energy Agency, Gas Cooled Reactor Design and Safety, Tech. Rept. Series No. 312 (IAEA, Vienna, 1990); and Status and Prospects for Gas-Cooled Reactors, TRS No. 235 (1984).
6. W. Steinwarz, N. Kirch and E. Rohler, (eds.), Topical Issue on the Status of the German High-Temperature Reactor Development, *Nucl. Eng. Design* **121** (2), 131–326 (1990).
7. L. M. Lidsky, D. D. Lanning, J. E. Staudt, X. L. Yan, H. Kaburaki, and M. Mori, *Energy: The Intl. J.* **16** (1/2), 177 (1991).
8. Proc. Intl. Workshop on the Closed Cycle Gas-Turbine Modular High-Temperature Gas-Cooled Reactor (MIT, Cambridge, MA, June 17–19, 1991).

CANDU Heavy Water Reactors and Fission Fuel Conservation

Duane Pendergast
AECL CANDU
Sheridan Park Research Community
Mississauga, Ontario, L5K 1B2, Canada

SUMMARY

The role of nuclear energy in the mitigation of climate change is examined. Recent papers which raise doubts about the potential of nuclear power are discussed. The magnitude of energy available from nuclear fission is restated and compared with current world energy consumption. It is concluded that the full development of fission power can provide world energy for thousands of years. An application of the CANDU heavy water reactor in a symbiotic relationship with light water reactors to conserve fission fuel is explained.

INTRODUCTION

Scientists have been predicting, for several decades,[1] that rising carbon dioxide levels in the atmosphere will lead to increased average world surface temperature. Early articles on this "greenhouse" effect projected that the warming would be beneficial as plant growth would be enhanced and the onset of an anticipated ice age could be delayed indefinitely. This initial optimism about the consequences of the greenhouse effect then gave way to concerns that human interference with the atmosphere might be harmful.[2] Concern about global warming reached a crescendo during the summer of 1989. This initial concern is dampening. Recently publicized estimates of the magnitude of warming[3] are on the low side of widely ranging predictions. Fault is found[4] with modelling simulations. Expectations that climate change can easily be avoided by conservation and efficient energy use are leading to complacency.

Nuclear power is acknowledged by most energy researchers to release very little carbon dioxide to the atmosphere. Nevertheless the large-scale implementation of nuclear power is questioned by some[5,6] as an appropriate response to the greenhouse "problem." Reasons given range from the ongoing debate about safety and nuclear proliferation to more recent doubts about the adequacy of nuclear fuel supplies. Some analyses suggest nuclear power simply can't be implemented quickly enough to turn around increasing carbon-dioxide levels.

This paper reviews the potential of nuclear power as a means of reducing carbon emissions. A role for the CANDU reactor in conservation of nuclear fuel supplies is explained.

NUCLEAR ENERGY AND CLIMATE CHANGE—THE SCEPTICS' VIEW

Fossil fuels are believed by most climatologists to be the major cause of observed increasing carbon dioxide levels in the atmosphere. Nuclear industry visionaries have identified the tremendous potential[7,8] to replace fossil-fuel energy with nuclear-derived energy.[9] The greenhouse issue has raised the possibility that the massive conversion to nuclear energy, envisaged by them, is essential now. The operation of nuclear power plants is acknowledged, even by those who believe nuclear power is not the way to achieve reduced global warming, to be relatively carbon dioxide free.[10] Why then would nuclear power be rejected as a solution?

The two studies cited in the introduction raise lingering doubts about the safety of nuclear power. I will sidestep the safety issue. That issue is ongoing and well debated elsewhere. They then go on to discuss other aspects of the large-scale deployment of nuclear power plants.

One study[11] postulates increasing global energy use scenarios with a substantial nuclear-energy component. Although nuclear power is nearly carbon-dioxide free, the models of energy use increase projected include a large component of fossil fuel energy. This inevitably leads to increasing carbon dioxide. Additional analysis indicates to the authors that increased efficiency of energy use is the quickest, most cost-effective way to reduce carbon dioxide emissions for the scenarios chosen.

Another evaluation[12] examines nuclear fuel supplies. The large-scale conversion to nuclear power using current reactor designs is evaluated. This would quickly exhaust projected reserves of high grade uranium ore. Breeder reactors are rejected as being too slow to develop to the levels of fossil energy displacement needed to combat the greenhouse effect. The potential to utilize low-uranium-content ores in the quest for greater uranium supplies is then considered. The analysis *assumes* that fossil fuels would be the source of energy used to retrieve uranium. The energy derived from the uranium becomes less than the energy which could be derived directly from the fossil fuel *assumed* to be used in mining the low-grade ores.

The ultimate alternatives to fossil-fuel energy envisaged by sceptics of the nuclear energy alternative are solar, wind and biomass energy. These are thought to be the source of energy for electricity production and, ultimately energy for transportation and other energy-consuming processes.

I believe the sceptics are excessively pessimistic about nuclear power's ability to play a role in providing low-carbon-dioxide energy. Two studies were selected for discussion. The first of these selects growth scenarios which preclude direct comparison of nuclear- and fossil-fuel-derived energy. Assumed growth in fossil-fuel use overwhelms the positive contribution of nuclear power. The second assumes continuing use of fossil fuel to recover low-grade uranium ores. In the long run nuclear energy would be used for uranium recovery. A common theme of both studies is that action taken to ameliorate the greenhouse effect must be undertaken on an urgent basis due to projections of large temperature increases in the next century. These projections warrant reexamination.

CLIMATE CHANGE AND THE NEED FOR URGENT ACTION

Recent scientific articles, based on results from computer modelling of the climate, are predicting temperature increases at the low end of the broad range of uncertainty (1.5°C to 4.5°C increase[13] for a doubling of carbon dioxide) in this science. Some of the differences in results arise from such simple changes in the models as a change from an assumed sudden doubling of carbon dioxide to a time-linear increase of carbon dioxide to the same end point.[14] Other major uncertainties include those of cloud[15] and ocean circulation modeling. In any case, these lower temperature-increase projections are beginning to be generally accepted as a genuine improvement in modelling. These new predictions allow time for a more careful and deliberate planned response to rising carbon-dioxide levels.

This current view of the lack of need to take quick action to compensate for the greenhouse effect is not very satisfying to those who want to undertake a massive response to reduce carbon dioxide emissions — right now! It appears that solid evidence of actual greenhouse-induced warming, perhaps combined with a tax on carbon emissions to the atmosphere, would be needed to initiate action in the near term. This isn't likely to be implemented in view of the developing consensus that climate changes are likely to be more modest than anticipated by the media a couple of years ago. The positive side of this is that we expect to have more time to implement efficiency improvements in energy use and production. This breathing space will allow for more measured development of the full potential of nuclear derived energy as well.

THE POTENTIAL OF NUCLEAR ENERGY

Climate warming has the potential to be the major incentive over the next two or three centuries for further development of nuclear energy as the major source of energy. Nuclear power plants, as they exist today, can be demonstrated to have minimal impact on the environment relative to fossil-fuel-derived energy. It is recognized by its friends and foes to produce very little carbon dioxide relative to other major means of energy production. As mentioned in the previous section, pioneers of the nuclear industry outlined the vast energy available from nuclear fission. This potential is worth reviewing. Newcomers to the industry, like me, haven't been exposed to this.

Nuclear-fission-derived electricity already makes a significant contribution to world energy supply. This proven technology substantially reduces the greenhouse effect relative to fossil-fuel use, which surely would have displaced it. Nuclear fission has the potential to provide a far greater fraction of world energy needs. Its possible widespread adoption raises additional questions of environmental and economic concern. It is the responsibility of the nuclear industry to ensure balanced understanding of these concerns related to the risks and benefits of alternative energy sources in the minds of the public. The industry will then be in a strong position for consideration as a major source of energy should climate change turn out to be as serious as many believe.

Are there sufficient nuclear-fuel reserves to support a massive conversion to nuclear power? Current global energy needs are roughly equal to the electrical output of seventy five hundred 1000 megawatt nuclear plants.[16] Quick calculations based on the fissile uranium recovery and consumption[17] of existing power reactors reveals that seventy five hundred 1000 megawatt plants would use up projected uranium reserves of 24 million tonnes[18] in one or two decades. That's not encouraging. However, existing reactors use uranium inefficiently, so there is a potential to extend the use of these reserves by a factor of about 50[19] through various forms of "breeding" additional fissile material. This would extend the energy content of these existing "reserves" to last about 500 to 1000 years at current energy consumption rates.

An extension of the human life style as we know it for another millennium may seem too short for some of us. Again, the nuclear industry is holding a trump card. Estimates of uranium reserves given above are based on ores which allow profitable recovery at a selling price[20] of $130 per kg. The current price is at an all time low[21] of $20 per kilogram. This price is worthy of some reflection. Existing reactors consume about 20 to 30 mg of natural uranium per kwhr of electricity. This works out to a current market value uranium cost of about 1/20th of a cent per kwhr. This is a very tiny fraction of the current electricity selling price of 5 to 15 cents per kwhr in Canada and the United States. An increase in the price of uranium to $4000 per kgm would increase the uranium share of electricity cost to about 10 cents per kwhr, effectively doubling its cost. We would complain, of course, but we could cope with such an increase easily through more efficient use and perhaps giving up just a little bit in our lifestyle. The incentive to producers to exploit lower-grade uranium deposits[22] would greatly expand usable reserves. Studies of deposits indicate that the earth's crust contains several orders of magnitude more uranium than is counted as a reserve or resource at current prices. This suggests uranium could serve as our sole source of energy for thousands of years, even if used in the current wasteful fashion. The sea also contains about 4 *billion* tonnes of uranium[23] although it is quite dilute. Japan has undertaken a pilot project to recover it and has found that it would cost in excess of ten times current prices.[24] Although this is a bit indefinite, it is well below the uranium cost of $4000 per kgm which we estimated would double electricity costs and suggests this source may be practical. Recovery of half of this uranium would thus supply all our current energy needs from existing reactor technology for about 2000 years.

So far we have discussed fission power systems as we know them now. We've demonstrated that nuclear fuel supplies are available for several thousand years for "once-through" fuel use as reactors are currently operated. All of the uranium in natural uranium (140 times more than the original fissile content) is potential reactor fuel if breeder reactors are used. In addition thorium, which is about 4 times as abundant in the earth's crust, can be used to create fission-reactor fuel material. Experience shows that inherent losses in the conversion process reduce the net gain[25] to about a factor of 50. We can thus expect fission power to serve us with abundant energy for many tens or hundreds of thousands of years instead of the 10 or 20 years suggested by energy pessimists. Fifty years of research and development has provided access to this bounty. Workable techniques to begin extracting it are in place.

CANDU REACTORS AND THE CONSERVATION OF NUCLEAR FUEL

It seems that the political and economic climate at present precludes any massive nuclear-power program to sharply reduce carbon emissions to the atmosphere. Nevertheless fossil fuels will become more expensive as supplies are depleted. Wind and solar power may not turn out to be an economic option. What can the nuclear industry do to preclude energy analysts' concerns that nuclear fuel supplies are insufficient to be a significant source of energy? Perhaps incremental improvement to existing reactors to reduce uranium use is an attractive alternative. The development and deployment of reactors which breed additional fissile material in excess of that used could be delayed. The extraction of uranium from dilute sources such as low-grade ores or seawater could be saved for the more distant future.

Existing reactor designs can be used to utilize the energy available from fissile material in used-fuel stockpiles. I will focus on the Canadian CANDU (CANada Deuterium Uranium) system. A basic feature of the CANDU reactor is the heavy-water moderator. Heavy water is particularly effective in conserving neutrons during moderation to the energy levels needed to generate fission. The fissile content of the fuel can thus be particularly low while an energy-producing chain reaction is sustained. It is this property which allows a CANDU reactor to operate effectively with natural uranium fuel of 0.72% fissile content. Light-water reactors are typically operated with the fuel enriched to about 3% fissile content. Since light water is poor at conserving neutrons, the fuel has to be discharged before the fissile material is depleted.

In fact, the "spent" fuel from a light-water reactor is a very good source of fuel for a CANDU reactor. Recent publications[26,27] explore several ways of reusing spent fuel from light-water reactors. These indicate that the energy from a given quantity of natural uranium can be nearly doubled by a second use of LWR fuel in a CANDU reactor. Most of the reuse-recycling processes described require some kind of chemical processing of the spent fuel. The goal is to extract the fissile from neutron-absorbing materials. The studies have indicated that most of the benefits can be obtained by direct[28] use of LWR fuel in a CANDU reactor. The direct-use process envisaged at present is to remove the fuel pellets from LWR fuel and reform them to a suitable form for fabrication into CANDU fuel bundles. (The fuel assemblies which have evolved for use in LWR and CANDU reactors are similar in the materials used. The shape and size of the fuel assemblies prevent direct transfer from current light water reactors to CANDU reactors.)

In addition to the greater energy extraction from uranium that would be realized from this re-use of spent fuel, another benefit is obtained. The volume of spent fuel to be stored is reduced.

Another scheme under study[29] at Atomic Energy of Canada indicates that a rearrangement of the geometry of the fuel and moderator by alternating close- and far-spaced fuel assemblies in the core has the potential to triple[30] the utilization of natural uranium in CANDU reactors without recycling. If proven practical, this would be even more effective than the synergistic relationship between the LWR and the CANDU as a conserver of uranium supplies. Since it uses natural uranium as fuel, the energy-intensive uranium-enrichment process would also be bypassed.

SUMMARY AND CONCLUSIONS

The enormous appetite for energy that man has developed in the past century has raised genuine concerns about climate changes postulated to result from increasing carbon dioxide in the atmosphere. A couple of years ago massive changes in energy use or supply were thought by many to be needed soon to reduce sharply carbon dioxide releases to the atmosphere. Some changes in modelling predictions towards lower warming have captured the heart of the media. The sense of urgency has worn off for now and more time is thought to be available for reasoned development of appropriate policy and response options. The nuclear industry that has developed over the past 50 years is a good candidate as an alternative source of energy that generates little carbon dioxide. Tremendous amounts of energy are available with full development of the nuclear fission option. Concerns have been raised by energy analysts about the possibility of running short of nuclear fuel. These concerns can be alleviated in the short term through ingenious use of existing reactor designs. The examples cited include reuse of LWR fuel in CANDU reactors and a modification of the CANDU design, which, studies indicate, would extend fuel supplies by a factor of three with a corresponding reduction in spent fuel.

The youth of the nuclear-power industry suggests fertile ground for considerably more innovations and improvements in reactor development. We can expect that many more will be discovered as the full potential of nuclear-fission energy supply is realized in the decades to come.

REFERENCES

1. Callendar, G. S., "The Artificial Production of Carbon Dioxide and Its Influence on Temperature," *Quarterly Journal of the Royal Meteorological Society* 64, 223–240 (1938).
2. Plass, G. N., "Carbon Dioxide and Climate," *Scientific American* 201 (1), 41–47 (June 1956).
3. Washington, W. M., and G. A. Meehl, "Climate Sensitivity Due to Increased CO_2: Experiments with a Coupled Atmosphere and Ocean General Circulation Models," *Climate Dynamics* 4, 1–38 (1989).
4. Brookes, W. T., "The Global Warming Panic," *Forbes Magazine* (December 25, 1989), pp. 96–102.
5. Mortimer, N., "Aspects of the Greenhouse Effect," Public Enquiry, Proposed Nuclear Power Station Hinkley Point C, FOE-9, Friends of the Earth, 26–28 Underwood Street, London, N1 7JQ (June 1989).

6. Keepin, W., and G. Kats, "Greenhouse Warming: Comparative Analysis of Nuclear and Efficiency Abatement Strategies," *Energy Policy* (December 1988), pp. 538–561.

7. Weinberg, A., "Continuing the Nuclear Dialogue, Selected Essays," American Nuclear Society, La Grange Park, Illinois (1985).

8. Lewis, W. B., "Nuclear Energy and the Quality of Life," *IAEA Bulletin* 14 (4), 2–14 (AECL–4380, December 1972).

9. Scott, D. B., "The Coming Hydrogen Age: Preventing World Climatic Disruption," World Energy Conference, Div. 2, Session 2.3, Paper 2.3.3, Montreal, September 17–22, 1989.

10. Loc. cit. 5, Sections 3.1 and 3.2. Figures 1 and 2.

11. Loc. cit. 6, p. 552.

12. Loc. cit. 5.

13. Hare, F. K., "The Global Greenhouse Effect," Proceedings of the Toronto Conference on the Environment, World Meteorological Association, WMO–710, Toronto (1988), pp. 59–68.

14. Loc. cit. 3, Abstract, p. 1.

15. Cess, R. D. et al., "Interpretation of Cloud-Climate Feedback as Produced by 14 Atmospheric General Circulation Models," *Science* 245, 513–516 (August 4, 1989).

16. Pendergast, D. R., "The Greenhouse Effect: A New Plank in the Nuclear Power Platform—Part 1," *Engineering Digest* 36 (6) (December 1990).

17. Based on "burnups" of 6500 and 33 000 megawatt-days/tonne for CANDU and LWR's, respectively, and recovery of 72% of the fissile uranium-235 from natural uranium for use in LWR fuel.

18. Weinberg, A. M., "Are Breeder Reactors Still Necessary?" *Science* 232, 695–696 (May 9, 1986).

19. Stevens, G. H., "Plutonium: A Fuel for the Future?" *The OECD Observer* (October–November 1989), pp. 22–25.

20. Loc. cit. 18.

21. Robinson, A., "Rio Algom Plans to Close Two Mines," *Globe and Mail* (Toronto, January 27, 1990), p. B1.

22. Deffeyes, K. S., and I. D. McGregor, *Scientific American* 242 (1) 66–76 (January 1980).

23. Tabushi, Iwao, and Yoshiaki Kobuke, *Mem. Fac. Engg., Kyoto Univ.* 46 (1), 56–60.

24. *Uranium Information Newsletter,* No. 5 (May 1989).

25. Loc. cit. 19.

26. Hastings, I. J., et al., "Synergistic CANDU-LWR Fuel Cycles," 6th Korea Atomic Industrial Forum/Korea Nuclear Society Joint Annual Conference, Seoul, Korea (April 15–17, 1991).

27. Pasanen, A., et al., "Recent Advances in the LWR/CANDU Tandem Fuel Cycle," ENS/ANS-Foratom Conference Transactions, Vol. IV, pp. 2100–2104.

28. Loc. cit. 27, pp. 2101 & Table 1.

29. Dastur, A. R., and A. C. Mao, *Canadian Nuclear Society Bulletin*, Technical Supplement (May/June 1989), pp. 1–6.
30. Dastur, A. R., A. C. Mao, and P. S. W. Chan, The Use of Subcritical Multiplication to Improve Conversion Ratio in Heavy Water Lattices, International Conference on the Physics of Reactors: Operation, Design and Computation, Marseille, France (April 23–26, 1990).

Future Contributions of Renewable Energy Technologies And Impacts on CO_2 Emissions

Robert A. Stokes

National Renewable Energy Laboratory

Golden, Colorado 80401-3393

Just over two years ago, President Bush directed the Department of Energy (DOE) to develop a National Energy Strategy (NES). During the past two years, 18 public hearings were conducted throughout the U.S. to develop inputs to this planning process. As part of the overall NES effort, five of the DOE laboratories developed an interlaboratory white paper estimating the potential contributions of renewable energy.* I will begin by summarizing some of the results of the NES with particular emphasis on the potential contributions of renewable technologies to the nation's energy mix over the next 40 years. I will discuss three examples of renewable energy technologies which have achieved remarkable progress in efficiency of energy conversion and reduction of costs associated with their use during the past decade. I will conclude with a brief discussion of the impacts on U.S. emissions of carbon dioxide resulting from different levels of renewable energy utilization over the next 40 years.

The NES base-case projection for primarily energy consumption by fuel type over the next 40 years is shown in Fig. 1. This base-case scenario assumes a 2.4% annual growth rate for U.S. GNP over the 40-year planning period. Energy use is expected to increase at only 60% of the GNP growth rate, reflecting continued energy efficiency improvement in all end-use sectors. Even though U.S. economic growth is projected to be significantly lower in the next 40 years than in the previous 40, the NES business-as-usual case projects a 60% increase in primary energy use over today's levels by 2030. In this base-case projection, renewables' contributions to primary energy more than doubles. With no change in policy, the contribution of nuclear-generated electricity to U.S. energy supply is assumed to drop almost to zero as nuclear plants are assumed to be retired at the end of their design lives. Perhaps the most significant result from the perspective of global climate in the base-case NES scenario is the large increase in projected utilization of coal over the planning period.

Figure 2 provides a graphic summary of the flow of primary energy to the end-use sectors of buildings, industry, and transportation. Figure 2 also clearly shows the significance of electricity as an intermediate energy carrier. Electricity consumption in recent years has grown at approximately twice the rate of overall primary energy consumption.

---

* Idaho National Engineering Laboratory, Los Alamos National Laboratory, Oak Ridge National Laboratory, Sandia National Laboratory, Solar Energy Research Institute, *The Potential of Renewable Energy—An Interlaboratory White Paper,* Solar Energy Research Institute, SERI/TP-260-3674 DE90000322, Golden, Colorado, March 1990.

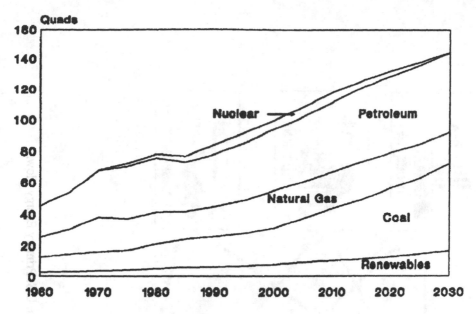

Fig. 1. Primary Energy Use by Energy Source
NES Current Policy Base Case.

This preference for electricity is expected to continue, reflecting its substantial amenity value. Unfortunately, the processes of conversion of fossil fuel to electricity and its subsequent distribution to end-use consumers have the potential to reduce overall energy system efficiency. The likely continuation of the trend to electrification will therefore provide an even greater challenge to the nation in its attempt to achieve greater and greater improvements in energy efficiency. The trend toward electrification can, however, encourage the transition from fossil fuels to renewable and nuclear generation because substitution of non-CO_2 emitting sources of electricity for fossil fuels is essentially transparent to the eventual consumer of the electricity. Figure 2 also indicates that the transportation end use is essentially decoupled from the electricity sector in today's energy system.

Figure 3 illustrates the various energy forms for renewable technologies. Even though many of the renewable technologies are best suited to producing electricity, they are also capable of producing process or space heat, chemical energy, or even high-value fuels and chemicals. The Interlaboratory White Paper* concluded that renewable energy technologies have the potential to expand their contribution from the current level of nearly 7 quads (approximately 8% of U.S. energy) to nearly 1/3 of total U.S. energy requirements by 2030. This level of penetration by renewable technologies could be achieved despite the assumed limitation of supplies from intermittent electrical technologies, such as wind and photovoltaics, to no more than 20% of the nation's electric generation mix. Breakthroughs in electrical storage technology based on advanced batteries or even

---

The Potential of Renewable Energy, op. cit.

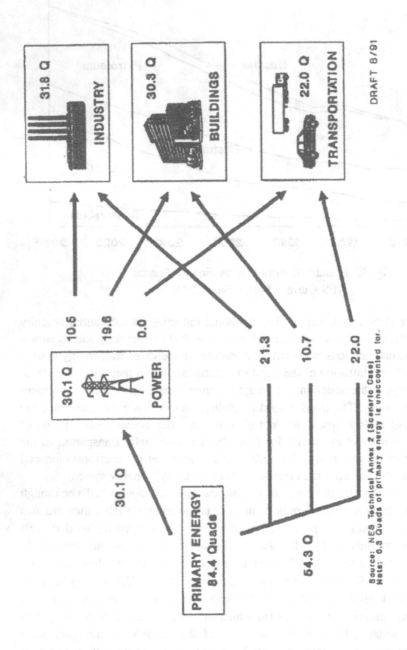

Fig. 2. [Energy Flow Diagram from Meridien. U.S. Primary Energy Flows in 1990 (Includes Electricity Losses).

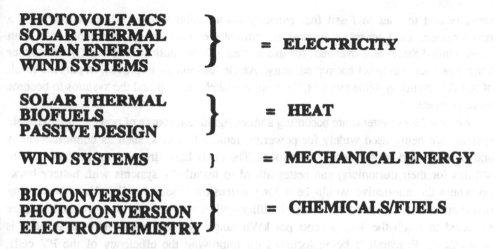

Fig. 3. List of Types and Uses of Renewable Energy.

superconducting magnetic energy storage (SMES) during the 40-year period could increase renewable contributions to the electrical sector significantly.

As an indication of the kind of progress achieved in renewable energy development during the last decade, I will summarize results in three areas: wind-generated electricity, photovoltaics, and production of ethanol for transportation fuel from lignocellulosic biomass feedstocks. The remarkable progress in each of these areas suggests that the projections for contributions by renewable energy summarized in the paper are achievable.

Wind energy has expanded its role in electricity generation in the past two decades, particularly in California. Windmills have long been used for pumping water, grinding grain, and other tasks, and more recently for generating power for such tasks as powering radios at remote locations. The intensive R,D&D effort of recent years has brought wind-generated power to the commercial stage, particularly for larger remote stand-alone applications and for electric utility supply in locations where wind resources are particularly good. The researchers have identified improvements in blade airfoil design which will improve efficiency, allowing greater output of power from better wind sites and also making lower intensity wind resources economical to develop. In addition, improvements in gear and bearing mechanisms, blades made of composite materials, and support structures are expected to reduce costs and increase reliability. As a consequence, the cost of wind-generated electricity is expected to drop from about 8 cents per kWh today to about 3 cents per kWh in 2030, assuming an acceleration of R,D&D efforts. At such a cost level, wind power would cost enough less than oil- or gas-generated electricity so that utility systems might be designed to utilize substantial wind-generating capacity and to shut in fossil fuel capacity before wind when loads dictated.* On a smaller scale, wind generator systems are currently economical and are being sold, both in the United States and overseas, for use in stand-alone applications, usually with either battery or package-diesel backup capacity. As the technology improves, in part the result of R,D&D on utility-scale systems, the costs are likely to drop and the systems to become more competitive.

Photovoltaics systems are becoming a more significant source of power. Smaller scale systems are being used widely for powering remote facilities, such as communications and lighting, and in developing countries. The costs have dropped to the point where utilities (or their customers) can better afford to install PV systems with battery back-up where the alternative would be a long extension (about 1 mile) from an existing distribution line. Meanwhile, costs of utility-scale systems have also dropped and are expected to reach the 4 to 5 cents per kWh range by 2030, particularly if R,D&D is accelerated. Research is being focussed on improving the efficiency of the PV cells, reducing the costs of the cells and extending their life, and in reducing the cost of the supporting systems. Multijunction cells provide a particularly attractive option. A number of alternative technologies are being investigated; success in meeting program cost objectives is not dependent on success in any one pathway.

PV systems, in a number of potential forms, might be used to provide power generation at the outskirts of utility systems, reducing the transmission and distribution investments otherwise required. PV modules integrated in roof tiles, for example, might provide the bulk of a household's power, with the utility's distribution system as back-up.**

Liquefaction of *biomass* for use as transportation fuels may be the largest single potential use of renewable energy. Significant progress is being realized in research in several areas. Today the production of liquid fuels from biomass is primarily via fermentation of corn constituents to produce ethanol, entailing the use of high cost raw

---

* In an operating situation, the highest marginal cost plant is shut in first when loads decline. Wind energy, having very little out-of-pocket variable costs, would be among the last to be shut in if available. At 3 cents *total* cost, however, it may even be more economical to plan facilities with the expectation that oil, gas, and coal plants would be shut in when wind-derived energy is available. The out-of-pocket costs of oil- and gas-fired plants will exceed 3 cents per kWh by 2000 or shortly thereafter. Some coal plants burning low-sulfur coal and having SO_2 scrubbers may have out-of-pocket fuel and maintenance costs in excess of 3 cents per kWh by 2030, particularly those plants having high coal transportation costs or those located remotely from markets and that transport power instead. (The coal plants in Arizona and Nevada serving the Pacific Coast may be examples.)

** If remote wind and PV systems become sufficiently numerous, direct current household and commercial appliances and lighting systems may be commercialized, reducing or eliminating the DC/AC/DC conversions now required, even in stand-alone systems, which "penalize" the renewable technologies.

materials. The advanced technologies will utilize lower cost materials — wood or other cellulosic materials, oil seeds, or microalgae — as feedstocks. Methanol is projected to be producible at a cost of 55 cents per gallon from biomass by 2000 ($10 per million Btu), assuming the feedstock cost goals are achieved by then or waste materials are available. Ethanol is projected to be available at a cost of 60 cents per gallon (or $7 per million Btu) shortly after 2000, the timing largely dependent upon whether R,D&D is accelerated, using an enzymatic hydrolysis process and an improved fermentation system. Increased yields, reduced residence times, and higher viable concentrations of ethanol in the fermentation reactors are among the key R,D&D targets. Fast pyrolysis of biomass to produce gasoline via a syncrude is expected to be commercialized by 2020, producing gasoline at a cost of 90 cents per gallon ($7.00 per million Btu). Accelerated R,D&D would move the date up to 2005 to 2010. R,D&D on producing diesel fuels from plant seeds or from microalgae is also progressing, offering alternative sources of renewable liquid fuels; these technologies have similar cost objectives, but it may take a decade longer to reach them.

Biomass Feedstock may be the most important area where acceleration of R,D&D is needed. The achievement of the cost goals for liquid fuels is highly dependent on the availability of feedstock at a cost of $2.00 per million Btu. If a substantial portion of the nation's liquid fuels is to be replaced with biofuels, the feedstock has to be available over wide areas at the low costs indicated. More than in the case of the other renewable technologies, wide-spread, large-scale demonstrations must be made to farming and forestry interests to convince them that land should be withdrawn from other crops and animal production and dedicated instead to energy crops.

As part of the National Energy Strategy studies, the nation's national energy laboratories undertook an analysis of the potential contributions that renewables might make to the nation's energy supplies under two scenarios, in addition to the base case.*

The scenario shown in Fig. 4 is based on the assumption that research, development, and demonstration efforts in support of renewable energy are expanded immediately by the federal government and that support continues through the next several decades. The level of support is assumed to be sufficient to permit the national laboratories to conduct maximum efficient and effective levels of research and development, and undertake demonstration projects as needed to bring emerging technologies into commercial readiness and acceptance promptly. Renewable energy was assumed to capture markets when the projected economics were favorable.** The substantial increase in renewable energy utilization projected would not reverse the expansion of fossil fuel usage.

---

* Idaho National Engineering Laboratory, Los Alamos National Laboratory, Oak Ridge National Laboratory, Sandia National Laboratories, Solar Energy Research Institute, *The Potential of Renewable Energy*, SERI/TP-260-3674, DE90000322, Golden, Colorado, 1990.
** No counter-response of fossil fuel prices was assumed nor were any assumptions made regarding accelerated R,R&D or other incentives for conservation or for nuclear power development.

Another study was made of the possible impact of pricing renewable energy in a manner that would reflect its values in minimizing adverse environmental impacts. "Premiums" were assigned to renewable energy based on the fossil energy type that might be replaced.* The "market pull" effects of the premiums accelerated renewable energies' achievement of competitiveness and market expansion, as reflected in Fig. 5.

Achievement of these projections of renewable energy's potential contribution to the nation's energy supply would *reduce* the increase in carbon dioxide (CO_2) emissions, but *not avoid an increase*. Figure 6 shows the carbon emissions which would occur in the base case and the reduced levels provided in the R,R&D and premiums cases. Moreover, the economic growth rate (2.4% per year) projected for the 1987–2030 is much lower than recent (post World War II) precedent; faster economic growth would lead to greater CO_2 emissions as additional fossil fuels would likely be used. Figure 6 shows the estimated carbon emissions from fuel consumption in the three future scenarios compared to recent levels. (The CO_2 and methane emissions from fossil fuel production are *not* included in this analysis.) If the objective is sought of avoiding increases or achieving reductions of CO_2 emissions, the option of accelerating development and deployment of renewable energy is unlikely to be adequate by itself. Much greater penetration would have to be achieved than appears likely even under favorable circumstances.

A *target* of a 20% reduction in CO_2 emissions has been proposed. Figure 5 indicates that a dramatic expansion of renewable energy use would be unlikely by itself to permit the nation to achieve that goal. In addition, it has been estimated that a 50% reduction in CO_2 emissions may be required to arrest the trend toward global warming. Such an achievement is clearly outside the range of possibility using renewable technologies alone. The magnitude of such a reduction is indicated in Fig. 7.

Obviously then, if carbon emissions are to be minimized, a combination of efforts on many fronts will be required. For illustrative purposes, I have constructed a simple scenario for 2030 which assumes "heroic" levels of efforts to avoid the use of fossil fuels by conservation or substitution. I do not suggest that this precise scenario is feasible; the simplistic methodology used may "double-count" some of the reductions in fossil fuel use, the costs may be unacceptable, and investments may have to be made in additional fossil fuel facilities before the alternative technologies—nuclear and conservation as well as renewable energy—are commercially available. The point of the scenario is to demonstrate that *major* changes must be made in energy use patterns and investments in all consuming sectors if CO_2 emissions are to be reduced in the United States. First, conservation efforts and investments are assumed to expand and have similar impact as in the 1970 to 1987 period, with the result of reducing energy demand growth to half that projected in the base case. (This might imply additional shifting of steel, aluminum, auto, and other energy-intensive goods production to foreign sources, thereby reducing U.S. CO_2 emissions but increasing them elsewhere.) Nuclear power plants, which were

---

* No corresponding premiums were assigned to conservation investments. No attempt was made to evaluate nuclear energy with respect to environmental values or costs.

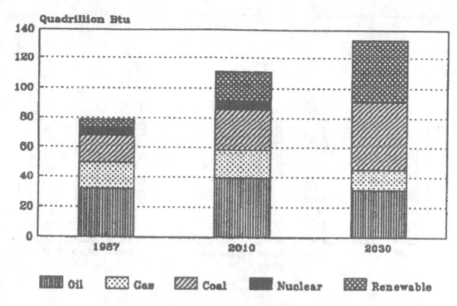

Fig. 4. U.S. Energy Supply, Accelerated Renewables R,D&D Case.

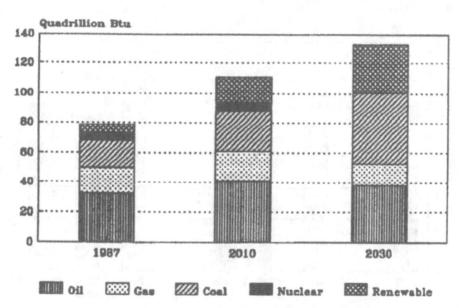

Fig. 5. U.S. Energy Supply, National Premiums Scenario.

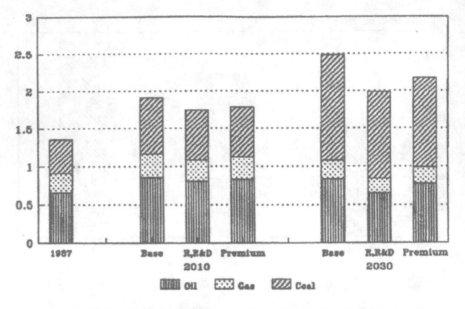

Fig. 6. Carbon Emissions from U.S. Energy Use
(Gigatons of Carbon per Year).

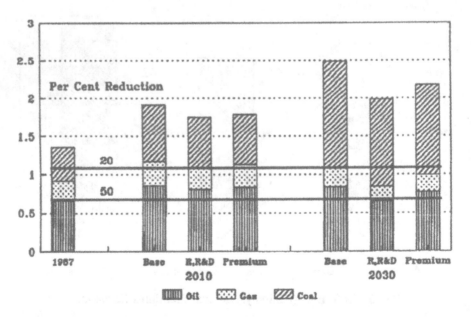

Fig. 7. Carbon Emissions from U.S. Energy Use
(Gigatons of Carbon per Year).

assumed in the base case to be retired when their current licenses expire, were projected to be life-extended (or replaced) and an equal amount of new capacity added, effectively doubling current nuclear capacity. Finally, renewable energy was projected to penetrate all markets where it appeared physically feasible. All auto and truck fuel, after additional vehicle efficiency improvements, would be supplied by renewable energy fuels — alcohols or biomass synthetics. No new coal-fired power plants would be commissioned; those now under construction would replace existing capacity. New base load plants would be nuclear, hydro, geothermal, or wood-fired (or other renewable energy if economical batteries were to be developed). Peaking power would be supplied by renewables and oil and gas use reduced. Buildings would be retrofitted for active and passive solar technologies. The costs of the transformation, however high they might be, were ignored. The modified energy use in this hypothetical scenario is shown in Fig. 8.

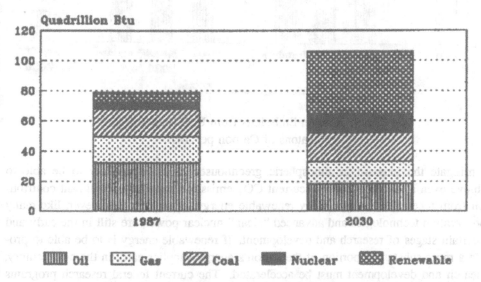

Fig. 8. U.S. Energy Supply, Maximum Fossil Fuel Replacement Scenario
(Includes Accelerated Renewables R,R&D).

Despite the herculean efforts described in the scenario, only a 20% reduction in CO_2 emissions is achieved, as indicated in Fig. 9. If a 50% reduction is needed, steps such as a complete replacement of all coal-fired power plants with nuclear or renewable energy plants might be required. Moreover, there is no indication that other countries, particularly those such as India, China, and Russia that have large coal resources, would be willing to invest in higher-investment-cost alternatives. The developing countries would also be a question. (I set aside the question of self-interest; some countries may see an individual advantage in global warming, even if most of the world would be adversely affected.)

Renewable energy can, nonetheless, make a substantial contribution to reduction of CO_2 emissions from levels they might otherwise achieve. Moreover, other steps

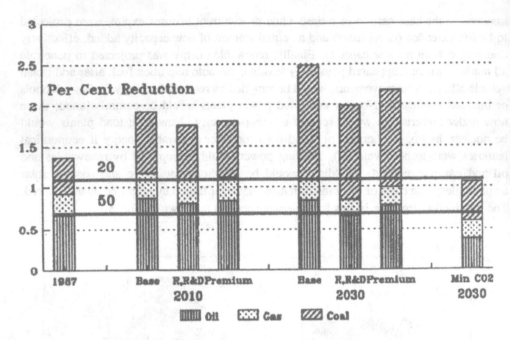

Fig. 9. Carbon Emissions from U.S. Energy Use
(Gigatons of Carbon per Year).

to mitigate the increase in atmospheric greenhouse gases are unlikely to be able to achieve even a 20% reduction in current CO_2 emissions without a significant contribution from renewable energy. Many renewable energy technologies, however, like many conservation technologies and advanced "clean" nuclear power, are still in the early and uncertain stages of research and development. If renewable energy is to be able to provide a more substantial portion of the nation's energy supplies early in the next century, research and development must be accelerated. The current federal research programs have identified a wide variety of promising technological pathways which offer promise of providing *competitive* renewable technologies in several sectors, in addition to the currently economic hydro, geothermal, wood combustion, and small-scale photovoltaics and wind technologies.

Energy Conservation Versus Renewables and the Threat of Global Warming

J. A. Laurmann

Marine Science Institute

University of California, Santa Barbara

1. INTRODUCTION

Steps to reduce anthropogenic emissions of greenhouse gases are now being actively pursued, with a number of industrialized countries committed to future cuts in fossil fueled carbon dioxide releases. The rationale for action and its pubic acceptance rests on an appeal for avoidance of a long-term. future-generational loss of environmental quality and of habitability, coupled with claims that costs of mitigative steps are minimal. These qualitative assertions are just now being put to quantitative test, and it seems inevitable that controversy will ensue. We see the reinvigoration of an old energy growth debate that first achieved international notice with the Club of Rome's presentation of the "limits to growth" hypothesis. The language used in discussion has changed — the central theme is now "sustainable development," and the long-standing argument between pro-industrial growth advocates and low growth conservationists has yet to rise in prominence in argument on the global warming issue, but will surely come.

We anticipate that the debate will be heated, so that it will be especially difficult to maintain the detachment of scientific research agenda development that is necessary if the uncertainties in our understanding of the problem are to be rationally addressed and if the logic for dealing with the potential climate change is to be properly handled. The argument of the sixties between the limited, soft energy path and the large energy growth believers devolved largely on social science issues concerning future possibilities in such matters as potential technology development. The threat of global warming has added a new dimension, in as much as extensive analysis in the physical sciences is now part of the problem. Projection of the timing and nature of the climatic effects of increasing greenhouse gas emissions is, at the moment, arguably as unsure as is the social sciences' prediction of the future. However, we do hope and expect that the former shortcoming can be redressed, given enough time and attention. It is our intent in this paper to try to illuminate the energy debate by relating it to some important characteristics of the anticipated climate change that we have yet to understand properly. One feature that exhibits particular sensitivity to the latter is the choice between reducing energy use through energy conservation and through enhanced energy efficiency versus moves towards renewable energy to replace fossil fuels.

2. THE POLITICAL MILIEU

The last three years have seen a rapid escalation of public attention being paid to a long-standing global environmental issue. The role of carbon dioxide in maintaining an equitable climate for the planet has been understood for over a century, and predictions of global warming from increasing industrial use of fossil fuels have been made for over fifty years. The order of magnitude of the projected change has not changed from estimates made twenty years ago, at the start of attention by organized scientific research, and it is not clear what has caused the recent remarkable increase in public and political concern with this environmental problem. Understanding the social forces at work in effecting this change is important, since it may be that the prominence of the topic could fall as rapidly as it rose—an eventuality that might be worse than over-reaction.

Proposed steps for mitigation of potential climatic change have followed the tentative suggestions made at the June 1988 Toronto Conference: "The Changing Atmosphere: Implications for Global Warming," reductions of greenhouse gas emissions of some 25% in a decade or two being called for. As in all pronouncements made by various national and international entities, cuts are initially to be restricted to developed countries, or transfer payments to be made to developing countries to pay for their emissions reduction efforts. Table I lists the commitments made by a number of countries as of the end of 1990, although none of these are internationally binding. The latter could change with development of a Framework Convention on Climate Change, planned for ratification at the UN Conference on Environment and Development in 1992.

Global reduction of greenhouse emission rates by a quarter are recognized as being insufficient to notably diminish the rise in temperatures predicted by current climate models if fossil fuel energy use continues at its projected growth rate. Stabilization of atmospheric concentrations is often cited as a goal for avoiding deleterious climatic change, and, for carbon dioxide—the most important greenhouse gas—this requires a cut in emission rates of between 60% and 80%, at current emission rates (IPCC, 1990). Note that stabilization of carbon dioxide emissions, in contrast to atmospheric levels, by, for example, substituting renewable energy for all newly planned fossil fuel based energy sources, will not stop atmospheric concentrations nor global temperatures from rising.

Figure 1 presents examples of the effect of four suggested emissions reductions policies on atmospheric concentrations and globally averaged temperatures, as calculated by general circulation climate models and reported by the IPCC (1990). Shown are the median "best guesses" for global temperature increases to the year 2100 for the four scenarios. The uncertainty is very large (c.f Sec. 3), with an estimated range of equilibrium temperature rise of between $1.5°C$ and $4.5°C$ above the preindustrial, at CO_2 doubling levels (IPCC, 1990), though these limits have been assigned no quantitative statistical measure. It should be noted that Fig. 1 plots temperature rise above the current value, about $3/4°C$ higher than the preindustrial (1850) temperature. Applying the lower climate sensitivity limit to the BaU (business as usual) scenario of Fig. 1 results in

TABLE I(a). Current Known Targets and Plans for Carbon Emission Reductions in Developed Countries (from Bowler et al. (1990).

	Present CO_2 (%)	Policy
USA	22.0	Against emission controls.
USSR	18.4	Against emission controls at present.
Japan	4.4	Bring emissions to 1990 level by 2000.
Germany	3.2	30% reduction on 1987 level by 2005.
Britain	2.8	Bring emissions to 1990 level by 2005.
Canada	2.0	As a first step, get emissions to 1990 level by 2000.
France	1.9	Recommends 20% cuts by 2005, up to 50% by 2030.
Italy	1.8	Bring emissions to 1990 level by 2000. Parliamentary resolution for 20% cut by 2005.
Australia	1.6	20% cut by 2005.
Netherlands	0.65	No increase after 1995; 3 to 5% reduction by 2000, followed by substantial cuts.
Belgium	0.5	Bring emissions to 1988 level by 2000.
Denmark	0.3	20% reduction.
Finland	0.26	Bring emissions to 1990 levels by 2000 at least.
Sweden	0.22	Bring emissions to 1988 level by 2000.
Norway	0.22	Bring emissions to 1990 level by 2000.
Switzerland	0.2	Supports getting to 1990 level by 2000, 20% cut proposed.
Ireland	0.14	Supports getting to 1990 level by 2000.
New Zealand	0.1	20% reduction by 2000.

diminution of the temperature rise trajectory to one that approximately matches the case B scenario.

The temperature responses shown in Fig. 1 include a (very uncertain) allowance for the non-steady thermal transient effect of rising carbon dioxide concentrations on the climate system, based upon an upwelling/diffusive mixing model of ocean-atmosphere coupling. In these depictions, the realized temperature increase is about 1°C below the equilibrium estimate at doubling of equivalent carbon dioxide concentrations.

A recent set of calculations (Schlesinger, 1991) allowing for thermal non-equilibrium and for a reduced climate sensitivity to greenhouse-gas forcing has been made to inquire

TABLE I(b). Examples of Worldwatch Institute's Proposed Carbon Dioxide Emissions Targets (Flavin 1990), aimed at equalizing emissions per capita. Net reduction of 12% below 1990 level by 2000, 38% below projected value.

	Current (tC/capita)	Targets (% cut)
Kenya, India, Niger	0.5	+3.0
China, Nigeria, Philippines	0.5–1.0	+1.5
Indonesia, Mexico, S. Korea	1.0–1.5	0
Italy, France, New Zealand	1.5–2.0	−0.5
Japan, Thailand, Peru	2.0–2.5	−1.0
Britain, W. Germany, Brazil	2.5–3.0	−2.0
Australia, USA, USSR, Colombia, Côte d'Ivoire	3.0	−3.0

into the importance of immediate reductions in emissions as called for in the national programs cited above. Figure 2 illustrates the results. Shown is the effect on temperature rise in delaying initiation of the IPCC reduced emission strategies (depicted in Fig. 1) from 1990 to 2000. In every case the difference in warming trajectory is minimal, under $0.5°C$ in temperature change.

The above presentations, which do not include depictions of possible climate changes, were actions to reduce emissions to be delayed for several decades. Such late response scenarios are not popular for policy analysis at the moment, but we believe they are in fact important to consider, since they represent an outcome that we see as quite probable. Thus, as we discuss below, the known facts of global warming and, above all, the realpolitik of the situation, suggest that no significant actions to mitigate potential climate change are likely until the threat becomes obvious to all of the world community that has a stake in preserving the present climate. This means a delay in taking action, at least until quantification in the climatic sciences is on a much securer footing, or perhaps until a warming trend is clearly identified and established to be caused by anthropogenic emissions.

3. IMPACTS OF CLIMATE CHANGE

Projections of possible climatic changes as illustrated in Fig. 1 by themselves do little indicate to the seriousness of the threat. Adding information on past climatic changes, of the type plotted in Fig. 3 provides a reference level that can be used to argue for preventive action. For example, the IPCC "business as usual" projected temperature increase

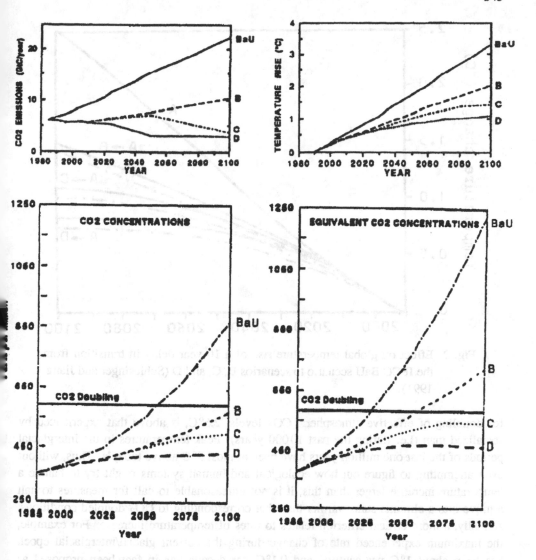

Fig. 1. IPCC policy response scenarios.

BaU: Coal intensive, modest efficiency increases, modest CO controls, continued deforestation, agricultural CH_4 and NO_2 emissions uncontrolled. Montreal CFC Protocol
partially implemented (Ref. 4).

B: Lower carbon fuels emphasized, notably natural gas. Large efficiency increases,
stringent CO controls, deforestation reversed, Montreal Protocol implemented.

C: Shift towards renewables and nuclear energy post 2050. CFCs phased out, agricultural emission limited.

D: Renewables and nuclear in first half of the century. CO_2 emissions stabilized early
in industrialized countries with modest growth in developing countries. CO_2 emissions reduced to half of 1985 level by 2050 (thereby stabilizing CO_2 atmospheric
concentrations.

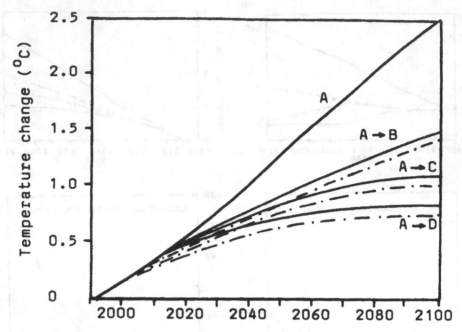

Fig. 2. Effect on global temperature rise of a 10-year delay in transition from the IPCC BaU scenario to scenarios B, C, and D (Schlesinger and Jiang, 1991).

for doubling of effective atmospheric CO_2 levels, 2.5°C, is above that experienced by socialized man (i.e. during the past 10000 years). Peak temperatures in the interglacial periods of the last one million years have been at most 3°C above today's. Thus, without even attempting to figure out how ecological and human systems might try to handle a temperature increase larger than this, it is not unreasonable to call for measures to halt a move into a climatic state warmer than that corresponding to CO_2 doubled conditions. Actually, a more critical criterion relates to rates of temperature increase. For example, the maximum experienced rate of change during the current glacial/interglacial epoch has been about 1°C per century, and 0.1°C per decade has in fact been proposed as an upper limit for anthropogenically induced allowable climate change (Jaeger, 1987), though decadal scale climatic variations have been somewhat larger. Support for this much more severe restriction is provided by an ecosystem sustainability limit argument (actually maximum boreal forest migratory adjustment rates). Climate impact damage estimates support the claim that rate, rather than the level of temperature change, is critical for setting climate-change toleration limits (Sec. 5).

In spite of the apparent strength of this argument, we need to note that contrary opinions exist. Some Soviet scientists assert that a return to Cretaceous climate conditions (some 5°C warmer that today) is desirable, at least in temperate and northern climes (Budyko & Sedunov, 1988). Even if this is not the case, it is sure that the initial effects

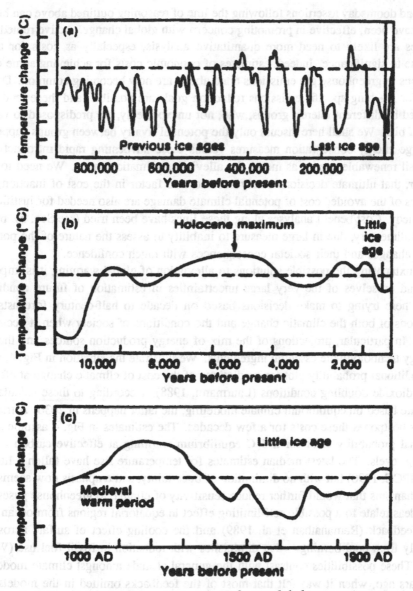

Fig. 3. Paleoclimatalogical reconstruction of past global temperature averages
relative to present mean (15°C) (IPCC, 1990).

of a gradual heating of the globe will have mixed results, improving living conditions
in some regions and worsening them in others. We can also expect that initially some
economic interests will benefit rather than lose. This mixed result implies difficulty in
global acceptance of the portended climate change as a negative to be avoided.

Broad doomsday assertions following the line of reasoning outlined above can be, and in fact have been, effective in promoting concern with global change. Active remediative measures are likely to need more quantitative analysis, especially as costs for taking them rise to significance. Indeed, estimates of economic costs for achieving some of the reductions in greenhouse gas emissions cited above are now becoming common. Diverse means for reaching specific emissions reduction goals are possible, and these tend to be supported by different interest groups, who, not unexpectedly, are predisposed to various forms of bias. We shall here discuss only the potential rivalry between groups supporting very large energy conservation measures versus those advocating rapid introduction of non-fossil renewable energy as means for alleviating climatic change. We need to note, however, that ultimate decisions surely need to also factor in the cost of inaction; then estimates of the avoided cost of potential climate damage are also needed for justification of a remedy. Cost/benefit analyses along these lines have been tried (see Sec. 5), but are as yet rudimentary, due in large measure to inability to assess the nature of the potential climate changes and their societal consequences with much confidence.

In juxtaposing the possible solutions to alleviation of global warming it is important to remind ourselves of the very large uncertainties in estimation of future conditions. We are here trying to make decisions based on decade to half-century forecasts, and predictions of both the climatic change and the conditions of society when it occurs are needed. In particular, projections of the mix of energy production sources and the state of energy technology are essential ingredients. We illustrate the situation in Fig. 4, which is a conditional probability plot for predictions of the cost of climatic change at effective carbon dioxide doubling conditions (Laurmann, 1988). According to these calculations, which are based on equilibrium climate modeling, the latter happens in 2030. A transient analysis postpones these costs for a few decades. The estimates in Fig. 4 assume a 50% ln-normal probability error for a 3°C equilibrium warming at effective carbon dioxide doubling levels. The latest median estimates for temperature rise have fallen a little (to 2.5°C, IPCC, 1990) but with no diminution of uncertainty, and emphasis now seems to be on mechanisms that would further reduce sensitivity of climate to greenhouse gases. The latest ideas relate to a possible self-limiting effect in equatorial regions from ocean water vapor feedback (Ramanathan et al. 1989) and the cooling effect of sulfate aerosols — primarily from coal burning — that diminishes with reduction in fossil fuel use (Wigley, 1991). These possibilities contrast with the general attitude amongst climate modelers a few years ago, when it was felt that most of the feedbacks omitted in the models were positive and would enhance greenhouse gas warming.

Greenhouse gas warming is a global problem requiring global action for its mitigation. However, its impacts are heterogeneous and disparately felt in different parts of the world and by different economic sectors. Different nations place different stress on their needs for fossil fuel use and for industrialization. The description of the problem given above excludes such important features, and these can lead to further reasons for variance in attitude toward the need for remediative measures and their cost acceptability. The following section, which deals with technological possibilities for reducing emission

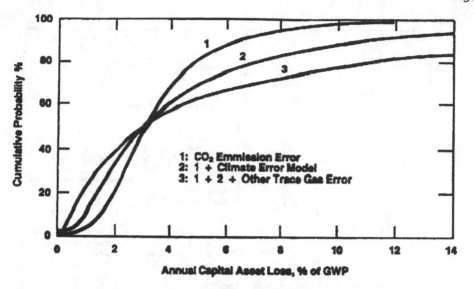

Fig. 4. Effect of projection error on climate damage estimation.

rates, treats the issues in single-measure global terms, without further elaboration of such geopolitical problems, thus no doubt oversimplifying some important controlling factors.

4. THE COST OF REDUCING GREENHOUSE GAS EMISSIONS

The greenhouse gas problem has seen fruition of the long nurtured seed of the low energy use enthusiasts. There is certainly an abundance of evidence that, even at today's energy costs, much can be done to improve the cost efficient production and utilization of most forms of energy. Arguments are based first, on technical engineering analyses, and, second, on the presence of artificial (non-market driven) governmental energy subsidies. Obviously also excluded to date in energy market pricing are the exogenous environmental costs of future greenhouse gas warming (as well as, in most jurisdictions, other environmental costs). There is little dispute as to the logic and need for a "least cost energy" approach, but there certainly is disagreement as to the economic savings that it could bring. Figure 5 illustrates the latter; it shows two sets of engineering analyses on the costs and benefits of reduction in electricity use through various energy efficiency/conservation steps. A recent Academy of Sciences evaluation, Fig. 6, on the same theme, projects the possibility of notably low costs for the US in achieving reductions in emissions: the cost of a 10% reduction in CO_2 equivalent emission rates varies from about \$10/tC (ton of carbon equivalent) to a net benefit of \$80/tC. These optimistic results apply only under today's energy use conditions. Marginal costs will increase after presently existing inefficiencies are eliminated. It should also be mentioned that the technical rationale for paying to implement this means for reduction of energy use can be supplemented by an energy security rationale argument.

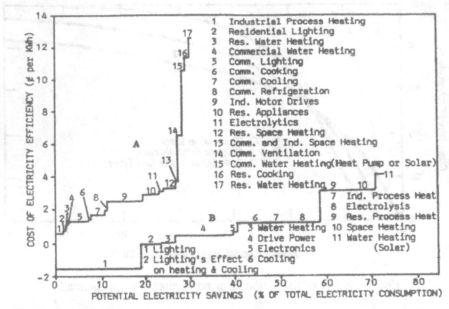

Fig. 5. Estimates of costs of improvements in electric and use efficiencies. A: Electric Power Research Institute; B: Rocky Mountain Institute (Ficket et al., 1990).

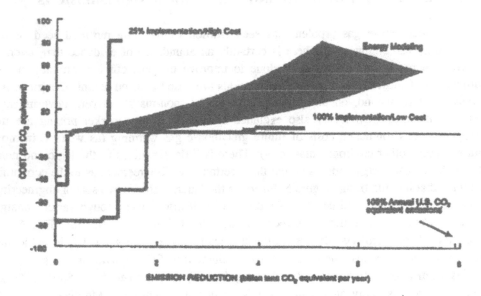

Fig. 6. Engineering and energy/economic model estimates of US costs for greenhouse gas emissions reductions (NAS, 1991). Note that 3.667 tCO_2 contains 1 tC.

There are grounds for disagreement on the efficacy of the conservation approach for emissions reduction, and these become more critical as longer term, larger reductions are sought. We draw no conclusions, but summarize the issues as follows:

- *Future energy intensity and price elasticity of demand.* The optimistic projections assume substantive decorrelation of economic from energy growth in industrialized countries, based in part on empirical evidence since the first OPEC oil crisis. However, there is room for disagreement over the causes of this shift and its permanence, one item being growth in the service sector at the expense of manufacturing industry.

- *"Real" pricing of energy supply.* Issues entering include marginal cost pricing by electric utilities, allowing for reduced demand (and hence cost) of energy by user energy conservation, price support for home use by energy producers, economically correct allowance for resource depletion, Third World energy distribution inefficiencies and pricing anomalies, and of course the inclusion in the energy price of exogenous social costs, such as those from climatic change. The "least cost" rationale should eliminate all these possible sources of difficulty, but we note that political impediments abound. We also note that many countries now impose a tax on energy as a means for raising general revenue, and this may or may not fit into a least cost rationale, depending on their social costs of energy use.

- *Engineering versus econometric estimation of the energy efficiency contribution to greenhouse gas reduction.* We have already cited the differences that can occur in projections made in these ways (as depicted in Fig. 6). An omission made in most engineering analyses that is likely to become critical if very large scale reductions in net energy use are anticipated, is the additional secondary industry costs that arise with major restructuring of the industrial suppliers of high energy efficient technologies. This type of resource limitation can affect not just costs but also the rate at which emissions can be reduced, an important matter if a rapid response to the threat is needed. The lacunae between micro and macro analysis capability is particularly critical for a rapid response scenario, such as that called for in the Toronto target (a 20% cut in CO_2 emissions by 2005), or the reductions needed to stabilize atmospheric CO_2 concentrations (60% to 80% at today's emission rate).

- *Greenhouse-gas reduction implementation.* The energy conservation/efficiency approach requires either taxation or emissions regulation if fossil fuel use is to be reduced by more than that achievable through a simple least cost rationale. Observe that economic restrictions on fossil fuels should favor renewable energy equally with conservation. However, the former response is de-emphasized in most of the plans for greenhouse warming mitigation now proposed. This affects, not only immediate specific measures for reducing energy consumption, but also allocations for national R& D expenditures. We also point out that if the "least cost" approach results in a fall in energy prices or a fall in its anticipated growth rate — as it will if environmental costs are excluded in energy pricing — we should expect a rise and not a fall in total energy use (and hence greenhouse gas emissions), according to simple supply/demand argument. From this point of view it is clear that the fiscal modality used for implementing a greenhouse gas reduction policy is crucial.

Also excluded from all discussion to date is the critical role of OPEC in setting world energy prices in a fossil fuel use constrained scenario. For example, if the industrialized countries were to successfully implement a large carbon emissions reduction program, without cooperation from the developing world, an OPEC policy of reducing oil prices in order to maintain sales could run counter to the objective of reducing global fossil fuel use. Indeed, it seems quite likely that developing countries would be eager to enhance their economic development by taking advantage of a lowering of energy prices and increasing their energy use (Laurmann, 1992).

Most emissions-reduction cost calculations allow for the introduction of non-carbon based sources of energy, as the costs of fossil fuels increase. In energy/economy modeling they are introduced through assumed "backstop" prices for nuclear, photovoltaic and biomass energies. A sample set of results from a recent calculations made with one such model (Manne & Richels, 1989, 1990) is given in Fig. 7. Figure 7(a) shows the effect on US GNP of a carbon tax policy aimed at reducing US CO_2 emissions 20% beneath their present level by 2020, and stabilizing them thereafter. Figure 7(b) shows costs for stabilizing world CO_2 emissions at today's level. In neither case are mitigation costs overwhelming, even though they are much higher than energy conservation enthusiasts would have us believe (see Sec. 7 below). An instructive comparison of these findings can be made with estimates of damages that would be incurred from unabated climate warming, Fig. 7(c). The latter shows damage estimates based on both a linear and a quadratic dependence of climate damage on temperature rise, and assumes damages to be either 3% of GNP, an early historic estimate (Laurmann, 1980), or 1.3% of GNP, a recent value from Nordhaus (1991), at CO_2 doubling. No temporal allowance for climate non-equilibration is included in Fig. 7(c).

Intercomparison of the results shown in Fig. 7 suggests that stabilization or modest reduction in emissions would not be cost effective until late in the next century, though the enormous uncertainties in all these attempts at quantification should not be forgotten in trying to draw conclusions. Moreover, Mann and Richels may be overpessimistic regarding future energy efficiency improvement potential. Adding 1%/yr autonomous energy efficiency improvement rate (the non-price induced rate) to the zero value assumed in the results shown in Fig. 7, cuts the costs in half, as Fig. 8 illustrates. As shown there, if it is further assumed that carbon dioxide removal from coal burning and advanced renewable energies become available at low cost, emission reduction costs can be cut by a factor of four. Energy efficiency improvement rates even larger than 1%/yr are commonly claimed to be possible (see Sec. 7), with reduction in costs by as much as another factor of

353

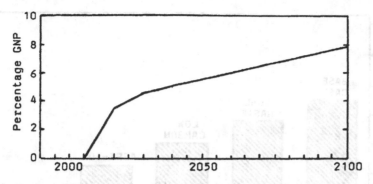

Fig. 7(a). Projected US GNP loss for a 20% carbon dioxide emissions reduction by 2020, followed by emissions stabilization.

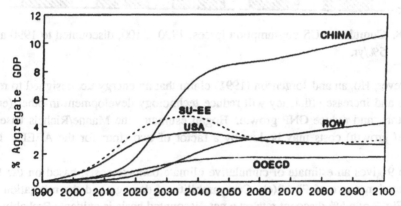

Fig. 7(b). Projected GNP losses for limiting global CO_2 emissions to current net global value.

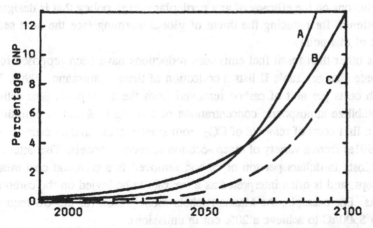

Fig. 7(c). Climate damage cost projections. A: quadratic damage function, 3°C sensitivity; B: linear damage function, 3°C sensitivity; C: quadratic damage function, 1.3° sensitivity.

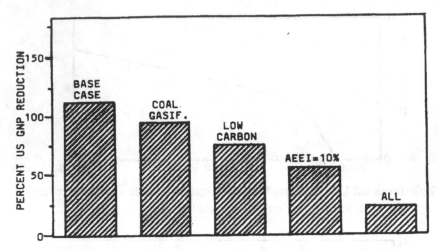

Fig. 8. Cumulative US consumption losses, 1990–2100, discounted to 1990 at 5%/yr.

four. However, Hogan and Jorgenson (1991) claim that an energy tax designed to reduce energy use and increase efficiency will reduce technology development in other (energy using) sectors, and reduce GNP growth. Hogan estimates the Manne/Richels base case (zero AEEI growth) costs may be low by a factor of two (four for the AEEI = 1%/yr case).

Figure 9 gives an estimate of cumulative climate damage costs, based on the linear damage function curve of Fig. 7(c). Comparability of damage and the mitigation costs shown in Fig. 8 at a 5% discount rate on a net discounted basis is evident. Probably more important are the large variations arising from uncertainty in damage estimate (shown in Fig. 9) and from the choice of discount rate, suggesting an extreme delicacy in our ability to draw conclusions on the efficacy of any particular energy policy that is designed as an economic optimum for reducing the threat of global warming (see the next section for amplification of this point).

Measures other than fossil fuel emissions reductions have been proposed for alleviation of climate change. Table II lists a collection of these (Laurmann, 1990). We have included both costs per unit of carbon removed from the atmosphere plus estimates of the cost to stabilize atmospheric concentrations of CO_2 (a 60% cut in emissions rate). The table also lists costs of removal of CO_2 from smoke stacks and an estimate made by Nordhaus (1991a) from a variety of macro-economic energy models. The latter is plotted in Fig. 10. Costs in dollars per ton of carbon removed is a common cost measure for mitigation steps, and is often interpreted as a tax rate to be levied on the carbon content of fossil fuels. For example, the Manne-Richels model result for the US requires a tax rate rising to $500/tC to achieve a 20% cut in emissions.

A recent detailed economic assessment based on the MARKEL model (ETSAP, 1991) is given in Fig. 11, showing marginal as well as average emissions reductions costs.

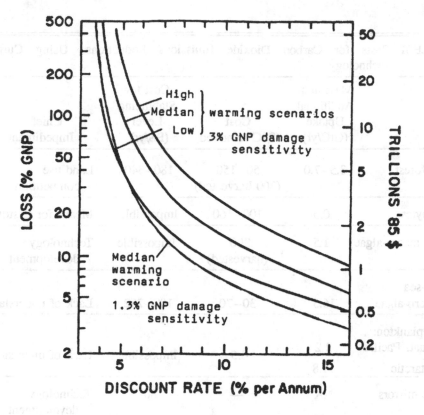

Fig. 9. Cumulative climate change damage costs, discounted to 1985; linear
damage function; warming assumed to continue indefinitely into the
future.

Others (e.g., Chandler 1989) also call for high tax rates to reduce fossil fuel and net energy use, but claim that this can be balanced by reduction in other taxes so as to have no net effect on GNP.

The results quoted in Table II serve to re-enforce the observation made earlier concerning the difficulty of making policy choices. They indicate that costs for a variety of possible means for alleviation of climate warming are of the same order of magnitude. The large and inevitable error in all of these make optimum choice of remedy difficult, and a strong tendency to postpone taking any action. Moreover, climate damage costs can also be of the same order of magnitude (Fig. 4). Reduction of uncertainty in this situation has high pay-off, and an economic hedging strategy, in which funding research to reduce uncertainty plays a major role, becomes a logical choice (Manne & Richels, 1991).

TABLE II. Costs for Carbon Dioxide Emissions Reductions, Using Current Technology.

	Maximum Additional Uptake (GtC/yr)	Cost ($/tC removed)	Cost for Stabilizing Level (B$/yr)	Chief Impediment
New forest	3.5–7.0	50–150 (300 harvested)	180–540	Land use competition
Halopytes	0.5	100–160	Impossible	Means for C storage
Coast macro-algae	1.5	300 (harvested)	Impossible	Technology development
Open-sea macro-algae	16.5	30–70	108–252	Lack of understanding
Phytoplankton: Equat. Pacific	1.2	0.5	Impossible	Lack of understanding
Antarctic	1.8			
Space mirrors	NA	—	80[a]	Technology development
Stack scrubbing	NA	133[a]	Impossible	Stationary sources only
Emissions limitation	NA	10% cut: 10 100% cut: 190	277	Time to take effect

[a]Capital costs amortized at 8%/yr.

5. COST/BENEFIT ANALYSIS

The science of climate change damage estimation is in its infancy, and it is difficult to justify its use in a quantified assessment of climate mitigation policy. However, comparisons of the type made in Sec. 4 are possible, and a cost/benefit analysis has been made by Peck and Teisberg (1991a) that confirms the broad conclusions described there. Peck and Tiesberg have used the Manne-Richels Global 2100 model, and added a climate damage function to formulate a consumption loss mitigation approach for optimizing greenhouse gas emission reduction policy. Figure 12 shows the optimum emissions trajectory and the associated carbon tax requirement. The climate damage function is

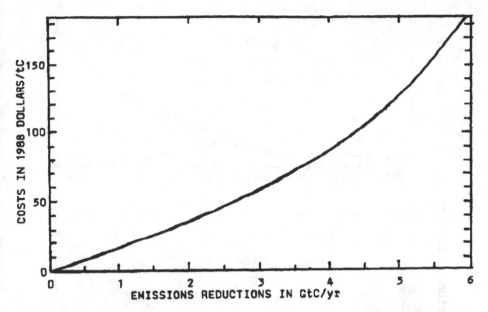

Fig. 10. Equilibrium analysis of average annual costs of CO_2 emissions reduction. Multiply ordinate by abscissa to get costs in billion \$/yr.

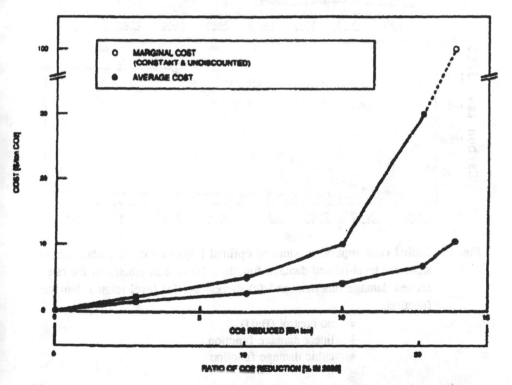

Fig. 11. MARKAL model estimation of global costs of carbon emissions reductions.

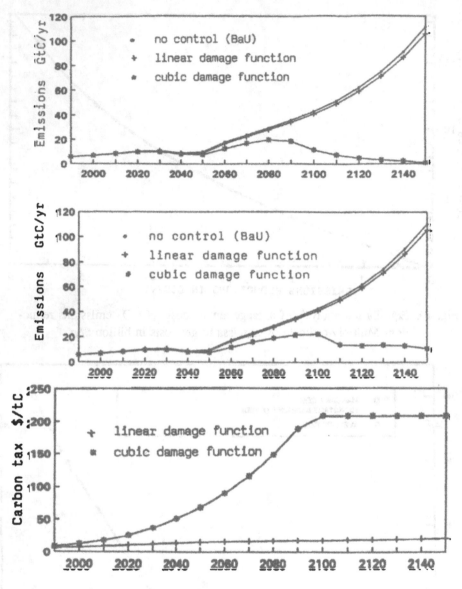

Fig. 12. Global consumption maximized optimal trajectories: (a) carbon emissions for level-related damage function; (b) carbon emissions for rate-related damage function; and (c) carbon tax for level-related damage function.

- no control (BaU)
+ linear damage function
* cubic damage function

assumed to be either linear or cubic in temperature rise, and to equal 2% of GNP at a temperature corresponding to CO_2 doubling (c.f. with Fig. 4). The major conclusions of this analysis confirm some points made earlier in this paper: firstly, with a linear damage function, costs with and without mitigation are comparable and small, until the mid next century; secondly, strong non-linearity can completely change this result, and the optimum solution then calls for a large carbon tax increase in the next few decades, with large emissions reductions. Non-linearity of the climate damage function is recognized as an indisputable property of the biosphere—sufficiently large warming will eventually destroy ecosystems and habitability of the earth. Unfortunately, the form of the non-linearity is completely unknown, and assumptions made in the damage estimates we have cited are arbitrary, so that the non-linear Peck/Tiesberg conclusion that early action is actually needed is not demonstrable from our present understanding of the interdependence of climate and nature. The only evidence that may be gleaned for quantification could perhaps come from estimating climate conditions that might define the boundaries of stability of ecosystems and of social systems; these could then serve to define the limits which greenhouse warming would be allowed to reach. We do know that high (i.e., non-linear) costs and absence of early action, imply a likely need for drastic and rapid reductions of emissions at a late date that are bound to be difficult to consummate.

Nordhaus (1991b) has developed an analogous optimum growth model, and drawn similar implications. With a quadratic climate damage function and 1.3% GNP loss for CO_2 doubling, he has concluded that there is little incentive to take mitigative actions.

Peck and Teisberg (1990b) have also calculated an optimum trajectory for a climate damage function that is rate of temperature change dependent. With a linear damage function, conclusions are as before—the optimum path is essentially unchanged from a no mitigation policy, Fig. 13. With third order damage dependence, emission reductions could be required early, dependent on the magnitude of rate change cost; the three curves in Fig. 13(b) correspond to a 2% GNP loss arising from either 0.25°C [the value used in Fig. 13(a)], 0.20°C, or 0.15°C per decade rates of increase of temperature. The first value corresponds roughly to the "business as usual" IPCC scenario (Fig. 1). If appeal is made to the paleoclimatalogical record to set a rate limit (Sec. 3), the last two lower values may be appropriate, and, according to Fig. 13, these require early and large emissions reductions.

6. THE CRITICAL PHYSICAL UNCERTAINTIES

The emissions reductions possibilities outlined above lead us to the following enumeration of critical features of the climate system that should influence choice amongst the policy options.

The need for better understanding of the atmospheric retention characteristics of greenhouse gas emissions is most critically felt for estimating the rate of diminution of atmospheric concentrations (and the associated temperature changes) accompanying

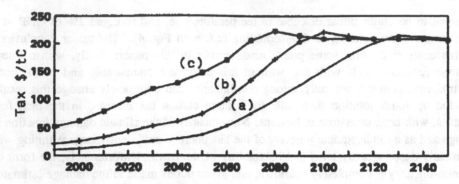

Fig. 13. Carbon tax for level-related damage function; three cost insensitivities; cubic damage function.

a policy of fossil fuel use and emissions reduction. The long-lived gases are of most concern, amongst which carbon dioxide is the largest contributor. For the latter, complex and poorly understood properties of the bio-geochemical cycling of carbon are involved. The major sink for carbon is the deep oceans, into which it slowly diffuses. It turns out that the fractional CO_2 uptake by the oceans increases as emissions fall, an important but poorly quantified property that tends to ease the mitigation task. It turns out that the chief source for a large delay in the heating response to gradually increasing atmospheric concentrations of carbon dioxide also comes from the ocean. The larger the ocean thermal capacity, the slower will global temperatures rise as emissions increase, and even if emission increases halt, temperatures can still keep on rising until equilibrium is reached.

Climatic response to altered levels of greenhouse gases under equilibrium conditions has been modeled extensively, though even the globally averaged figures for warming that these models yield are suspect and have large probable error (cf. Fig. 4). Policy analysis has to allow for the uncertainties in climate change projection, and we have techniques for doing this. More cumbersome to handle are the additional uncertainties in calculating the transient climatic behavior, and the thermal inertia role of the oceans has to be better understood for proper treatment of this form of uncertainty. As mentioned in the previous paragraph, it is this property of the climate system that is especially important for assessing the climatic effects of emissions reductions, especially if these are massive, as might be needed in a late response strategy in which rapid replacement of fossil fuels would be called for.

Extensive improvements in the simulation of the climate system are needed if quantitatively reliable, regional climatic change effects are to be derived. For these, not only is better understanding of the physical mechanisms controlling climate required, but the numerical models must be refined so as to be able to describe conditions on a much finer scale than by the typical 300-km scale now used. Even given these improvements, the impacts of climate warming are unlikely to be accurately assessable for conditions markedly different from previously experienced climatic extremes. But, as the sample cost projections described above indicate, it appears that only the prospect of much larger climatic changes can justify remediative action; the prospects for being able to derive damage cost estimates for these climatic conditions are remote, the best we can hope for are qualitative descriptions of possible future states. We are therefore faced with the uncomfortable prospect of having to make decisions on taking greenhouse gas remediative actions based on very loosely definable knowledge, with little chance of improving on the latter.

7. EFFECTIVENESS OF THE EFFICIENCY ENHANCEMENT SOLUTION

The central issues can most simply be illustrated by considering the following three progressively more severe criteria for mitigation of the greenhouse effect that have been put forward:

(A) stabilization of greenhouse gas emissions;

(B) reduction of emission rates by 20%, followed by stabilization;

(C) stabilization of atmospheric concentrations of greenhouse gases at their present levels.

Figure 14 is a rough portrayal of the energy efficiency increase rates needed to achieve various levels of emissions reductions by increasing efficiencies without the introduction of renewable energies.

Selection of one (or none) of these depends not just on assessment of the nature of the climatic changes that the policy option seeks to avoid and on the means for doing so, but on normative attitudes concerning risk-taking and discounting the future. The latter are central attributes of the global warming problem that are likely to be the prime determinants in policy choice, although they are rarely identified as such. This being said, we go on to less argumentative, more easily discussed factors that distinguish these three possibilities.

Provided the political will exists, the first option should be fairly easily applicable to the industrialized countries; there could be secondary non-greenhouse environmental benefits, and even a modest economic gain for small emissions cuts. However, costs are likely to be high for a number of developing nations, and subsidization from wealthy countries would probably be essential to reach the global goal. Figure 15 gives estimates by the Japanese Institute for Energy Economics for increases in energy efficiency through the year 2025 needed to keep global emissions at the 1988 level. These values are additional to increases in efficiency that are projected to occur without an emissions reduction policy.

There is disagreement on whether conservation and improved efficiency can alone achieve target (A). The annual improvement rate of 2% in energy efficiency required to do so (Fig. 14) is beyond anything achieved globally in the past, though energy intensity reductions of this size occurred in the OECD during the 1980s. A more generally acceptable figure is 1% per year as economically justifiable for industrialized countries

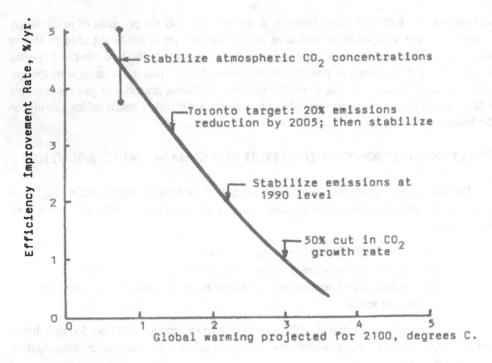

Fig. 14. Approximate energy efficiency improvement needed to reduce global warming.

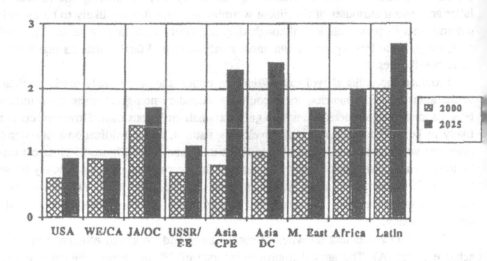

Fig. 15. Additional energy efficiency improvement rates needed to stabilize global carbon dioxide emissions at 1988 level (Institute for Energy Economics, 1990).

(OTA, 1991; Carlsmith et al. 1990), and perhaps also for developing countries under present conditions of extremely inefficient operations (Rose et al, 1984, Torok et al. 1990). A 2% rate could require subsidization — the DOE Interagency Task Force (1990) cites a 1% GNP loss for the US to stabilize emissions by 2005; yet others (e.g., Chandler et al. 1988) claim no cost, and even a net savings for the US in achieving carbon dioxide emissions stabilization. Maintaining constant emissions requires continuation of this efficiency increase rate to the middle of the next century through improvement in technologies for energy production and use, and it is difficult to predict whether this can in fact occur.

The criterion (B) approximates the Toronto target, and is one cited in many national plans (Table I). Chandler (1989) estimates that an annual energy efficiency improvement of 3.3% can meet it at no net cost, a claim disputed by the majority of analysts. A 200% tax on coal is needed to bring about the reduction in carbon energy intensity, but (so it is claimed) with no net loss in world GNP. An Office of Technology Assessment report (1991), restricted to looking at the United States, estimates that a 20% emissions reduction can be achieved by 2015 with an energy intensity increase of 2.4% per annum (though less than 2% is from improved efficiency), with uncertain economic effects — estimates ranging from a net savings to a 1.8% GNP loss.

Criterion (C) is the most difficult to effectuate, requiring between 60 and 80 percent cut-backs in emissions rates. Krause (1990) estimates that an energy efficiency improvement rate of between 3.3 and 5 percent a year is needed to halve present day CO_2 emissions, and may still not suffice to stabilized atmospheric levels. This should be compared with a 1979–1986 OECD rate of 2.6%/yr, and may not be feasible at any cost on the global scale.

In summary, it seems probable that up to a 1% global improvement rate in energy efficiency can be met with no net cost, at least at present. Note that this figure includes price induced energy conservation as well as technology improvement effects (cf. the model results of Mann and Richels, discussed above). There is dispute as to whether this will be true in the future, several analysts believing that efficiency improvement will become more difficult with time. Even greater argument concerns the cost-effectiveness of efficiency improvement rates of 2–3%/yr. Such rates are achievable for the industrialized countries, at arguable cost, but it is not settled whether global rates as high as this are possible at any price. Issues such as those discussed above in Sec. 4 are involved.

8. CONCLUSIONS

The degree of warming mitigation stemming from greenhouse gas reduction of a few tens of percent yields only modest cost savings from avoided climate damages — of the same order as median estimates of the costs of emissions reduction. This is the range in which energy conservation can play an important role, though for rapid action, its effectiveness relative to the introduction of renewable energies is in dispute. Allowing for the large uncertainties in all economic calculations that we know how to make, cost/benefit analysis in this situation is incapable of giving a clear directive for action. Hedging strategies that allow for the uncertainty can be developed; the major conclusion from these is that research to reduce uncertainty has high pay-off. The most important question to determine is the transient climatic response with rising, or modestly falling, emissions.

Delay of action, with stabilization at higher atmospheric concentrations, could require stronger cut-backs, as well as yield higher equilibrium temperature increases. However, as best we can tell at the moment, delays of a decade or two will not exacerbate the timing of warming very much, nor make mitigative actions much more expensive. In fact, we can expect time delay will assist in the development of more efficient and cheaper non-polluting energy sources.

To better assess the actual path of temperature changes in such delayed response scenarios, we critically need more understanding of the transient response of the carbon cycle and climate systems. We expect that too long a delay will result in warming magnitudes that it may be wise (and cost-effective) to avoid. At issue here is the size of climate damages associated with a doubling of effective carbon dioxide levels (two to three degrees C). From a cost/benefit viewpoint, there appears to be little to choose between allowing the change to occur and paying to stop it — within the uncertainty of our knowledge, costs are comparable. However, non-linearity of the climate damage function past doubling, can radically change the balance, and dictate mitigative action. Unfortunately, we have very little information on the shape of the climate damage function beyond effective carbon dioxide doubling temperatures.

In such late response scenarios the speed with which emissions can be reduced becomes critical for avoiding too large a temperature increase. Market penetration time for the non-polluting technology may well require early start of mitigation. Determining how early requires interactive analysis of the dynamics of climate response and the means and costs of rapid reductions in emissions.

If a non-economic based warming reduction target is set, such as one defined to avoid exceeding past natural climatic variations, much more severe emissions limitations can be required. In fact, it would seem that the $1/10°C$ per decade limit suggested by some as the maximum allowable rate of temperature rise could be reached and surpassed soon from greenhouse warming. If this criterion is accepted, immediate actions, going beyond energy conservation, are needed. However, the basis for choice of a particular limit is weak. We know that intercentennial natural climate variability is $1°C$, thereby leading to the $0.1°C$ per decade figure, but natural rates of temperature change are higher on shorter time scales. In order to place this methodology on a secure footing, proper statistical analysis of the spectrum of climate variations over a ten to one hundred year period is needed, and made over geographical scales appropriate to ecosystem and human system response. It will also be important to discover what temporal variations of climatic conditions set thresholds of sustainability. In many cases, simply specifying maximum allowable temperature or temperature rate increases will not suffice. For instance, a sequence of seasonal extremes over a specific number of years, or exceeding of a given rise rate over a specific number of decades, could be critical determinants for survivability of particular systems or species.

Overall, we judge that highest priority should be given to assessing the nature of and means for mitigation of climate changes that could occur during the later stages of a continued increase in fossil fuel as the world's primary energy source. Viewed in the context of the large uncertainties that are indigenous to the greenhouse problem, shorter term (say, prior to effective carbon dioxide doubling) greenhouse gas induced climatic change impacts are both too small and too far in the future to be stressed. Realpolitik suggests that no major mitigation actions will take place in this situation. We recommend attention be turned to the much more serious possibility of major climatic impacts occurring in the second half of the next century and the need to take early actions to avert these. The latter would require large-scale replacement of fossil fuels or elimination of their emissions; energy conservation and improvements in energy production efficiencies would have a secondary role. It is important that evaluation of minor short term gains that might be possible from near term mitigation steps not dominate consideration of the global warming problem, to the detriment of adequate analysis of a longer term more critical threat.

The large uncertainty in our ability to project future global warming means that increasing research on the subject has high insurance value. Maintenance of a resilient energy system, prior to commitment to major remediative steps, is another obvious hedging tactic. But we should also note that extreme risk aversity under high risk, highly uncertain conditions implies a different conclusion. It is thus quite possible for particular attitudinal and normative standards to lead logically to a different conclusion—one that calls for immediate, and even drastic action to be taken now to ensure the avoidance of a future catastrophic, although low probability event.

REFERENCES

Bowler, S., M. Cross, J. Gribbin, C. Joyce, D. MacKenzie & T. Thwaites, 1990. The Politics of Climate: a Long Haul. *New Scientist*, October 27.

Budyko, M. I., & Yu. S. Sedunov, 1988. Anthropogenic Climate Change Conference on Climate and Development, Hamburg FRG, November.

Carlsmith, R., W. U. Chandler, J. E. McMahon & D. J. Santini, 1990. Energy Efficiency: How far can we go? Oak Ridge National Laboratory, ORNL/TM-11441.

Chandler, W. U., H. S. Geller & M. R. Ledbetter, 1988. Energy Efficiency: A New Agenda. American Council for an Energy-efficient Economy, Washington D.C.

Chandler, W. U., 1989. Assessing Carbon Emission Control Strategies: The Case of China. Workshop on Energy and Environmental Modeling and Policy Analysis, MIT Center for Energy Policy Research.

Energy Technology Systems Analysis, 1991. Newsletter #5, August.

Fickett, A. P., C. W. Gellings & A. B. Lovins, 1990. Efficient Use of Electricity. *Scientific American*, September, 65–74.

Flavin, C., 1990. Slowing Global Warming. *Environmental Science & Technology*, February.

Goldemberg, J., T. Johansson, A. Reddy & R. Williams, 1988. Energy for a Sustainable World. Wiley Eastern Ltd. New Delhi.

Hogan, W. W. 1990. Comments on Manne & Richels: Manne, A. S., & R. G. Richels, 1990. CO_2 Emission Limits: An Economic Analysis for the USA." *Energy J. 11* (2), 75–85.

Interagency Task Force, 1990. The Economics of Long-Term Global Climate Change; A preliminary assessment. US Department of Energy, Office of Policy, Planning and Analysis. Washington D.C.

Intergovernmental Panel on Climate Change, 1990. Scientific Assessment of Climate Change. World Meteorological Organization/U.N. Environmental Program. Cambridge University Press, Boston.

Jaeger, J., 1987. Developing Policies for Responding to Climatic Change. Summary of workshop held in Villach 28 September–20 October. Beijer Institute, Stockholm.

Krause, F., 1990. Energy Policy and Climate Stabilization. International Project for Soft Energy Paths. San Francisco, CA.

Laurmann. J. A., 1980. Assessing the Importance of CO_2 Induced Climate Change Using Risk-Benefit Analysis. In "Interactions of Energy and Climate," Reidel, Boston, MA.

Laurmann, J. A., 1985. Market Penetration of Primary Energy and its Role in the Greenhouse Warming Problem. *Energy 10* 6, 761–775 .

Laurmann, J. A., 1988. Future Energy Use and Greenhouse Gas Induced Climatic Warming. In "Carbon Dioxide and Other Greenhouse Gases: Climatic and Associated Impacts." Ed: R. Fantechi & A. Ghazi Kluwer.

Laurmann, J. A., 1989. Evaluating Costs of Global Climate Change. Marine Science Institute, University of California, Santa Barbara.

Laurmann, J. A., 1990. Countering Climatic Change. Proceedings of a "Symposium on the Climatic Effects of Increased Fossil Fuel Burning and Energy Policy Implications for the Asia-Pacific Regions." Economic and Social Commission for Asia and the Pacific, Tokyo, December 12–14.

Laurmann, J. A., 1992. World energy prices, geopolitics and global warming. Accepted for publication as guest editorial in *Int'l. J. Energy Environment Economics*.

Manne, A. S,. & R. G. Richels, 1990. CO_2 Emission Limits: An Economic Analysis for the USA. *Energy J.* **11** (2), 51–74.

Manne, A. S., & R. G. Richels, 1991. Buying Greenhouse Insurance. International Energy Workshop, Laxenburg Austria, June 18–20.

National Academy of Sciences, 1991. Policy Implications of Greenhouse Warming. Committee on Science, Engineering, and Public Policy. Washington D.C.

Nordhaus, W. D., 1991a. A Survey of Costs of Reduction of Greenhouse Gases. *The Energy Journal*, March.

Nordhaus, W. D., 1991b. An Optimal Transition Path for Controlling Greenhouse Gases. International Energy Workshop, Laxenburg Austria, June 18–20.

Office of Technology Assessment, 1991. Changing by Degrees, Steps to Reduce Greenhouse Gases. Congress of the United States, Washington D.C.

Peck, S. C., & T. J. Teisberg, 1991a. CETA: A Model for Carbon Emissions Trajectory Assessment. International Energy Workshop, Laxenburg, Austria, June 18–20.

Peck, S. C., & T. J. Teisberg,·1991b. Temperature Related Damage Functions: A Further Analysis with CETA. Draft, Electric Power Research Institute.

Ramanathan, V., R. D. Cess, E. F. Harrison, P. Minnis, B. R. Barkstrom, E. Ahmad & D. Hartmann, 1989. Cloud-Radiative Forcing and Climate: Results from the Earth Radiation Budget Experiment. *Science* **243**, 57–63, January 8.

Rose, D. J., M .M. Miller & C. Agnew, 1984. Global Energy Futures and Carbon Dioxide Induced Climate Change. MITEL 83-015, MIT Energy Laboratory, Cambridge MA.

Schlesinger, M.E., & Xingjian Jiang, 1991. Revised projection of future greenhouse warming. *Nature* **350**, 219-221, March 21.

Torok, S., W. Foell, T. Siddiqi, A. Keesman & T. Nagao, 1990. Climatic Effects of Increased Fossil Fuel Burning and Energy Policy Implications for Asia and the Pacific. International Energy Workshop. Honolulu, Hawaii, June 7–8.

Wigley, T. M. L., 1991. Could reducing fossil-fuel emissions cause global warming? *Nature* 349, February 7, 503–506.

Hot Dry Rock

David V. Duchane
Earth and Environmental Sciences Division
Los Alamos National Laboratory

ABSTRACT

The earth contains a vast quantity of thermal energy stored in hot rock beneath its surface. While in a few places this energy can be obtained by bringing naturally occurring hot water or steam to the surface, the general case is that hot rock at depth is essentially dry. Over the past two decades, techniques have been developed to extract the thermal energy found in hot dry rock (HDR) in a manner that is both practical and environmentally benign.

THE HDR PROCESS

Geothermal energy in the form of hydrothermal resources such as water or steam is utilized in a number of places either to provide direct heat or to produce electricity.[1,2] Worldwide, however, geothermal energy is found most often in the form of hot dry rock (HDR). During the past two decades work has been underway at the Los Alamos National Laboratory and at other institutions to develop techniques to extract the energy from HDR.

In the HDR process,[3] a well is drilled to reach rock which is sufficiently hot to be useful. Water is then pumped down the well under pressures high enough to open up natural joints in the rock and create a man-made reservoir consisting of a relatively small amount of water dispersed in a large volume of rock. One or more wells are subsequently drilled to intercept the reservoir some distance from the first, and the system is operated by circulating pressurized water down one well (the injection well), across the reservoir, and up the other wells (the production wells). As the water flows across the hot reservoir, it becomes heated by contact with the hot rock. At the surface, this thermal energy is extracted by a heat exchanger and the water is recirculated to repeat the process. The same water thus flows repeatedly around the system in a closed loop to mine the heat from the depths of the earth.

The depth at which useful hot rock occurs varies from place to place with the local geothermal gradient—the rate at which the temperature of the earth increases with depth. Figure 1 is a geothermal gradient map of the United States. It shows that temperatures generally increase much more rapidly with depth in the western part of the country than in the east. Drilling costs are a major factor in determining the feasibility of developing HDR resources.[4] Table I summarizes the estimated cost of electricity from HDR as

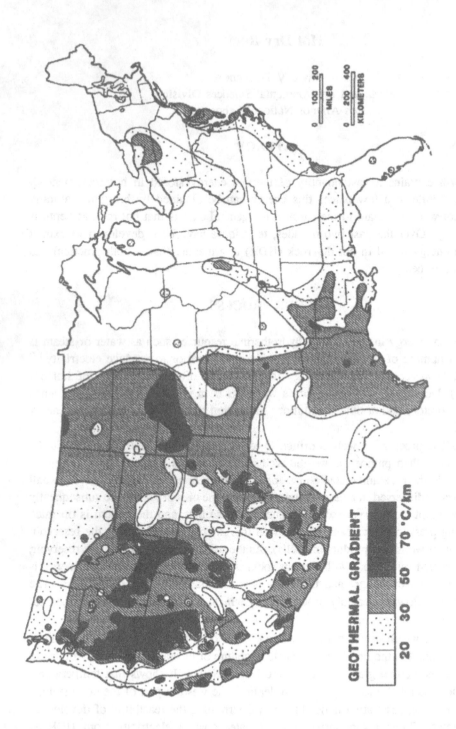

Fig. 1. IA geothermal gradient map of the United States.

TABLE I. Costs of Electricity from HDR Resources.

Geothermal Gradient of Resource (°C/km)	Break-even Electricity Price ($/kWh)
80	0.05–0.06
50	0.08–0.09
30	0.16–0.18

a function of geothermal gradient. It is clear that, at present, economics favor early development of HDR in the high gradient areas of the western United States.

The Los Alamos National Laboratory has been investigating HDR technology for about the past 20 years. In the late 1970's, a small HDR reservoir was developed. It was operated periodically over a two-year period to demonstrate that it is possible to extract the thermal energy from HDR. Over the past decade a larger, Phase II reservoir has been built. Figure 2 shows the Phase II HDR system. The reservoir is centered at about 3.6 km in rock at a temperature of 240°C. It is ellipsoidal in shape with an estimated volume of about 20 million cubic meters. The locations of numerous microearthquakes generated during the formation of the reservoir form the basis for establishing its location, orientation, and shape as illustrated in the figure. Two wellbores penetrate into the reservoir. They average about 110 meters apart in the reservoir region.

ENVIRONMENTAL CHARACTERISTICS

Because of the closed-loop design, the HDR process has extremely favorable environmental characteristics. Only heat is removed from the earth under normal operating conditions. There are no significant operational by-products or long-term wastes accumulate. In addition, shut down of the system at the end of its useful life can be accomplished by simply closing in the wells which access the underground reservoir. Techniques to accomplish this have been developed and proven in the petroleum industry.

Conventional hydrothermal power sources have been found to emit only small amounts of greenhouse gases such as carbon dioxide and hydrogen sulfide. In an HDR system, small quantities of greenhouse gases may also be picked up in the circulating fluid. These gases may either be retained in the fluid circulating within the loop, since the water is always kept under relatively high pressure and is not vaporized to steam, or they may be periodically released under controlled conditions.

HDR systems are generally developed in tight, crystalline rock many thousands of meters below potable water resources. For this reason, there is no danger of ground or surface water pollution. Some water is used in operating an HDR system. Current work

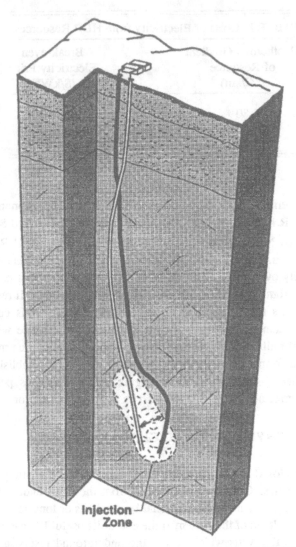

Fig. 2. The Phase II HDR system at Fenton Hill, NM.

has indicated that this amount can be very small and that most of the water either fills microcracks within the reservoir or slowly diffuses from the reservoir at a very low rate.[5]

Figure 3 shows the results of an experiment in which the HDR reservoir at Fenton Hill was continuously pressurized, but not flowed, over a two-year period. In this experiment, water was simply pumped into the injection well to maintain pressure on the system, while the production well was shut in. It is clear from the figure, that the amount of water required to keep the system pressurized declined with time. Undoubtedly, this water loss is due to permeation from the boundaries of the reservoir which decreases with time. At

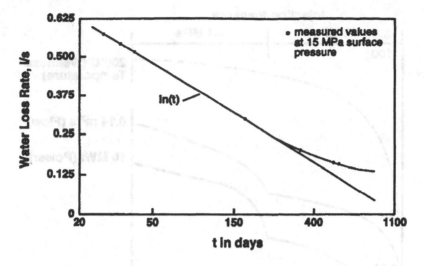

Fig. 3. Water consumption over time at elevated pressure.

a mean reservoir pressure of 15 MPa above hydrostatic, water consumption eventually declined to less than 0.2 ℓ/s, an extremely small amount in comparison to the large size of this reservoir.

The land requirements for a HDR power plant may be very small in comparison to other energy sources. Because the heat is stored naturally underground, no provision need be made for fuel storage on the surface. The wells themselves occupy only a few square meters each. Space will be required for the power plant, of course, but by analogy to conventional geothermal facilities,[6] these are envisioned to be of a modest scale rather than the extremely large units required for economic energy production from coal or nuclear sources.

These promising environmental characteristics of HDR have been supported by results of experimental operations conducted to date. As mentioned above, many of the environmental features can be substantiated by direct analogies to commercial operations in related industries. Long-term testing of the Phase II HDR facility at Fenton Hill is expected to further confirm that this energy source is among the best in the world from an environmental standpoint.

THE LOS ALAMOS LONG-TERM TESTING PROGRAM

A 30-day flow test of the Phase II HDR system was conducted in 1986.[7] Some important results of that experiment are illustrated graphically in Fig. 4. As shown, the production temperature and flow rate continually increased over the term of this test and thermal power production levels as high as 10 megawatts were achieved. This test was conducted using an improvised surface facility and rented equipment. The encouraging

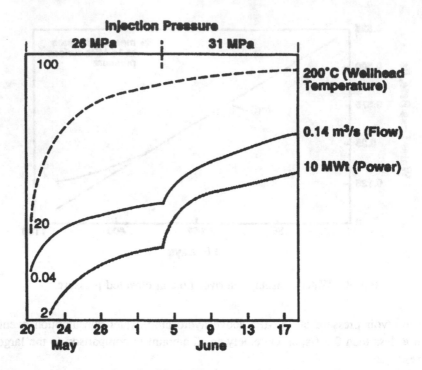

Fig. 4. Some results of a 30-day flow test of the Phase II HDR reservoir.

results, however, provided an incentive for developing a facility suitable for long-term operations.

Construction of a surface plant at Fenton Hill was completed late in 1991. This plant provides the means to extract thermal energy from the Phase II HDR reservoir on an extended basis. Figure 5 is a schematic drawing of the important components of that plant. The entire facility has been built to industrial power plant standards and is highly automated.[8]

Final preparations are now underway for a long-term flow test (LTFT) of the Phase II reservoir. If adequate funding is obtained, the duration of this test will be for at least one year and perhaps longer. The primary goal of the LTFT is to demonstrate that thermal energy can be obtained from the Phase II reservoir in useful quantities on a sustainable basis. A variety of experimental data will be collected over the duration of the test to address environmental and economic issues related to HDR energy production.

Several scientific and technical studies will be conducted during the test. Seismic surveillance will be maintained to assure that reservoir growth is not being induced under the usual operating scenarios anticipated during the test. In a few cases, the injection pressure may be raised to a level designed to cause reservoir expansion (as indicated by microearthquakes) in order to more severely test the reservoir or to obtain more seismic information about reservoir characteristics.

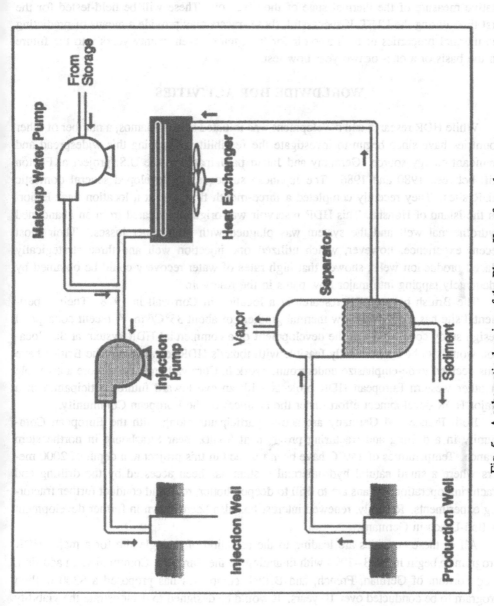

Fig. 5.   A schematic drawing of the Phase II surface plant.

Tracer tests will be employed periodically during the LTFT. Radioactive tracers will be used to assess changes in the hydraulic characteristics of the reservoir as the test proceeds. In addition, temperature-sensitive tracers have been developed to give a quantitative measure of the thermal state of the reservoir. These will be field-tested for the first time during the LTFT. If successful, these tracers may provide a means of predicting the thermal properties of the reservoir for five, ten or even twenty years into the future on the basis of a one- or two-year flow test.

WORLDWIDE HDR ACTIVITIES

While HDR research and development was initiated at Los Alamos, a number of other countries have since begun to investigate the feasibility of tapping this widespread and abundant energy source. Germany and Japan participated in the U.S. project at Fenton Hill between 1980 and 1986. The Japanese subsequently developed several domestic HDR sites. They recently completed a three-month flow test at a location near Hijiori on the island of Honshu. This HDR reservoir was originally created from an abandoned hydrothermal well and the system was plagued with high water losses. Their most recent experience, however, which utilized one injection well and three strategically placed production wells, showed that high rates of water recovery could be obtained by adequately tapping into major flow paths in the reservoir.

The British began HDR research at a location in Cornwall in 1978. Their experimental site has a relatively low thermal gradient of about 35°C/km. A recent conceptual design study concluded that the development of a commercial HDR system at that location would not be economically feasible with today's HDR technology. The British have thus decided to de-emphasize underground work in Cornwall and take a more active role in other western European HDR projects with an eye toward future participation in a major HDR development effort under the auspices of the European Community.

Both France and Germany are active participants along with the European Community in a drilling and fracturing program at Soultz, near Strasbourg in northeastern France. Temperatures of 140°C have been reached in this project at a depth of 2000 meters where a small natural hydrothermal system has been accessed by the drilling and fracturing operations. Plans are to drill to deeper, hotter, rock and conduct further fracturing experiments. Recently, renewed interest has also been shown in further development at Bad-Urach in Germany.

All of these activities are leading to the selection of a single site for a major HDR Program to begin in 1993–1994 with financing by the European Community. In addition, a consortium of German, French, and British companies has proposed a $300 million program to be conducted over 10 years. It would be designed to demonstrate the viability of HDR in the areas of low thermal gradient that are typical of northern Europe.

Significant HDR work has also been conducted in Russia during the past year. Drilling and fracturing operations were carried out at Tirniaus near Elbrus in the Caucasus Mountains. The experimental work followed the Fenton Hill model with original drilling to 3.6 km followed by fracturing operations at pressures up to 60 MPa (8700 psi). Mechanical problems led to abandonment of the deepest portion of the original wellbore, but sidetracking and redrilling to an adjacent location at the same depth and further fracturing are planned. The Russians could greatly benefit from some of the advanced technologies developed at Fenton Hill. Under the right conditions, joint U.S.-Russian cooperation could lead to benefits for both parties.

SUMMARY

HDR can provide a vast source of energy for mankind without contributing to the increase of greenhouse gases in the atmosphere. The technology can be employed with minimal environmental impact upon aquatic and terrestrial environments and does not produce long-term wastes and their attendant disposal problems.

Research and development work has been underway to develop means for extracting useful energy from HDR for the past two decades. Pioneering work was done at the Los Alamos National Laboratory. Their Phase II HDR system is ready for long-term testing to demonstrate that sustainable heat delivery is possible under practical operating conditions. A number of other countries are also investigating HDR as an abundant, indigenous, and clean energy source. As technology development moves forward, it seems certain that HDR will come to play a significant role in the energy picture of the environmentally-conscious world of the twenty-first century.

REFERENCES

1. D. H. Freeeston, *Geothermal Resources Council Bull.* 19 (7), 188–198 (1990).
2. G. W. Huttrer, *Geothermal Resources Council Bull.* 19 (7), 175–187 (1990).
3. J. W. Tester, D. W. Brown, and R. M. Potter, Los Alamos National Laboratory Report LA-115114-MS (1989).
4. J. W. Tester and H. J. Herzog, Massachusetts Institute of Technology Energy Laboratory Report MIT-EL 90-001 (1990).
5. D. H. Brown, Proceedings of Geothermal Review IX, U.S. Department of Energy, Geothermal Division (1991), pp. 153–157.
6. J. W. Rannels and D. V. Duchane, *Power Generation Technology* (Sterling Publications, London, 1990), pp. 177–180.
7. Z. V. Dash, Ed., Los Alamos National Laboratory Report LA-11498-HDR (1989).
8. R. F. Ponden, Proceedings of Geothermal Review IX, U.S. Department of Energy, Geothermal Division (1991), pp. 149–151.

Calculating the Net Greenhouse Warming Effect of Renewable Energy Resources: Methanol from Biomass

Rex T. Ellington and Mark Meo
Science and Public Policy Program, University of Oklahoma
Norman, OK 73019 U.S.A.

ABSTRACT

We present a quantitative method for accounting for all greenhouse gases (GHGs) emitted from technical systems for full system life to give an unambiguous basis for comparing the total greenhouse warming forcing of alternative projects or plans. It captures all GHG emissions from activities possibly separated in location or time, and thus characterizes more completely the technical or policy trade-offs available to decision makers also interested in energy efficiency or reduction of costs. The method is timely as it affords opportunity for defining more accurately the key criteria necessary to make informed decisions at national and international levels.

The method employs a systems perspective by using it to include the entire supporting infrastructure, not just the costs, emissions, etc. of a primary process. Also, a life-cycle-plus perspective is used. For comparison of alternative systems, analysis of each should include GHGs emitted during the construction of the system and those released throughout its operating life and salvage period plus final decay of residual GHGs. We demonstrate the fundamental soundness of systems and life-cycle analysis by presenting a practical approach for assessing directly the total greenhouse warming effect of a biomass system via the Warming Forcing Factor.

1. INTRODUCTION

As debate continues about an international agreement for stabilizing and/or reducing GHG emissions, concern has been raised about the imperative for the more developed and industrialized nations to acknowledge the need for developing nations to use cleaner fuels, such as renewable resources, in the most efficient manner. Should flexibility for meeting agreed upon criteria be built into a global market for GHG emissions trades, for example, compliance among nations and development sectors could be subject to a variety of constraints including the availability, cost, and feasibility of adopting appropriate technologies. Under these circumstances, decision making within individual nations, which is likely to follow a "bottom-up" pattern of policy change, may benefit from the active transfer of proven climate-benign advanced techologies from industrialized nations.[5,6] Yet, for such an international effort to succeed, clear agreement will have to be reached on a verifiable method for calculating the net global warming effect of alternative technical systems, particularly energy and energy-dependent systems.

Since industrial and agricultural activities release GHG emissions in a continuous manner that varies in quantity over the lifetime of their operation, strategies developed to aid planning and decision making must take these characteristics into account. Various measures developed for this purpose which relate the warming potential of all GHGs to an equivalent gas, usually carbon dioxide, exhibit one or more critical flaws in failing to account explicitly for continuous releases of GHGs.[7-10] They may fail to account completely for the entire suite of GHGs released as a result of project construction, operation, and decommissioning activities, or oversimplify the absorption traits of specific trace gases to calculate greenhouse forcing equivalency to CO_2, or assume single-pulse outputs. In view of the pressing need to develop a common framework for comparing the effectiveness of alternative global warming mitigation actions, the adoption of imprecise or misleading global warming potential measures for use in technology assessment could create unforseen problems.

We have developed a quantitative method for accounting for all GHGs emitted from technical systems for the full life of each activity to give decision makers an unambiguous basis for comparing the total greenhouse warming forcing of alternative projects or plans.[11,12] This method can capture all source emissions of GHGs from activities that may be separated in location and/or time more fully than current methods, and thus characterize more completely the technical or policy trade-offs available to decision makers who may seek to maximize other goals such as energy efficiency or a reduction of total costs while minimizing risks associated with global warming. As an illustration, the net global warming effect of a biomass-based methanol fuel production system is summarized in this paper. The example is timely as it affords an opportunity for defining more accurately the key criteria necessary to make informed decisions at national and international levels.

2. ENTIRE EFFECT OF ANY PATH

A single defect which characterizes many of the approaches to mitigating GHG effects is the factor of incompleteness. Evaluations of fuels which concentrate on end-of-pipe emissions often ignore all previous steps in the production chain, which could change the evaluation markedly. Industrial process evaluations often focus primarily on point-source emissions and ignore distributed emissions in other sectors with which they may be connected. In addition, discussions of agricultural systems often ignore the many fossil fuel subsidies existing through direct energy uses and chemical inputs. These are all parts of the correct definition of the basic problem.[13]

A second area in which some analyses are found wanting pertains to the documentation of all GHG emissions associated with a system, particularly from sources imbedded in subsidiary processes and operations. Third, a fault of some analytic methods is that time-saving shortcuts are attempted, such as expressing the behavior of other GHGs as equivalent CO_2.[7-9,14] Accumulating evidence suggests that the warming effect of each GHG should be calculated by itself, and the effects added to get the total warming forcing.[15] Fourth, some methods try to characterize the total effect of a system by use of response relationships which are based on the injection of single pulses of GHGs into the atmosphere.[16] In truth, important systems will inject GHGs into the atmosphere for many years, often at varying rates. Also, effects will continue to persist years after the systems are shut down. Thus, a complete analysis must include the whole system lifetime plus after-effects.

We argue first, that a systems perspective is a more powerful way to achieve completeness than most current methods by using it to include, in the analysis, the entire supporting infrastructure of a primary process, not just its own costs, emissions, etc. Second, we argue that a life-cycle perspective must be used to include the complete effects of each element from birth to well after its demise. For comparison of alternative systems, analysis of each should include all GHGs emitted during the construction of the entire system (invested emissions) and those released throughout its operating life and salvage period (direct emissions) plus the final decay of residual GHGs, i.e., its full life-cycle. Third, we seek to demonstrate the fundamental soundness of systems and life-cycle analysis by presenting a practical approach for assessing directly the total greenhouse warming effect of a system via the Warming Forcing Factor. This approach utilizes the absorption effectiveness and atmospheric lifetime of each and every GHG emitted by a system, with all the rigor available for describing each.

2.1 Systems Concept

The systems concept is used widely in engineering and scientific analysis. When a system includes a process or operation of interest that has global impact and elements that may be at different locations around the globe, it must be identified completely and carefully to prevent oversights and errors. With each primary process, industrial or agricultural, there is a supporting infrastructure which sustains it. Conversely, without the primary process, there would be no need for each infrastructure element or a *pro rata* share of that element to exist. This combination of elements constitutes the GHG-emitting system; every operation from raw material production to final waste treatment, through all intermediate steps including transportation, should be accounted for whether or not any of the operations are separated in time or location from the primary process.

The approach is especially helpful when considering replacement of, for instance, fossil fuel systems with biomass systems (Fig. 1). Growing wood for combustion to replace coal is often cited as having zero net CO_2 generation, i.e., photosynthesis fixes an amount of carbon equal to that burned and respired.[17,18] However, the labor and energy used to plant and raise seedlings and the labor, fuel, and other energy expended in chemicals, logging, transport, and processing result in GHG emissions. Much of this support represents a fossil fuel subsidy that would necessitate planting substantially more trees than were burned to attain a zero sum balance. A systems viewpoint discloses fully that many apparently benign energy sources can, in fact, have hidden GHG emissions associated with raw material mining and processing, construction and operation. In general,

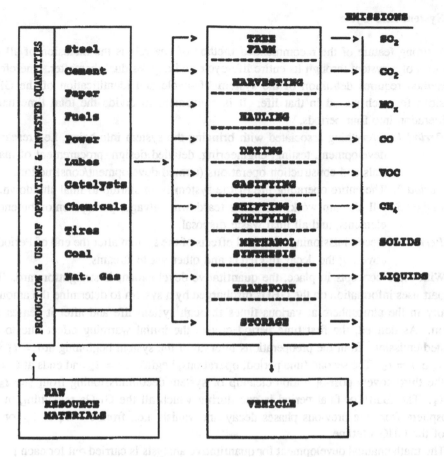

EMISSIONS

Steel	→ SO₂
Cement	→ CO₂
Paper	→ NOₓ
Fuels	→ CO
Power	→ VOC
Lubes	
Catalysts	→ CH₄
Chemicals	→ SOLIDS
Tires	
Coal	→ LIQUIDS
Nat. Gas	

TREE FARM — HARVESTING — HAULING — DRYING — GASIFYING — SHIFTING & PURIFYING — METHANOL SYNTHESIS — TRANSPORT — STORAGE

PRODUCTION & USE OF OPERATING & INVESTED QUANTITIES

RAW RESOURCE MATERIALS — VEHICLE

Fig. 1. System Definition for Methanol from Woody Biomass.

the more complex the system, the more opportunities there are for energy consumption and GHG emissions.

The next requirement imposed for method development is to include any and all GHGs that might be released from the system to the atmosphere. Further, the method should provide a measure of the total warming effect of the system throughout its entire life and any post-operation decay of the GHGs emitted. This approach requires that the instantaneous radiation absorption efficiency of each GHG must be considered as well as its total lifetime in the atmosphere. This method differs from the authors' Emissions Index in which the integrations were calculated over the lifetimes of the species.[7] It also differs from the IPCC Global Warming Potential measure, which is the instantaneous radiative forcing times concentration in the atmosphere integrated over an unspecified time divided by the corresponding values for carbon dioxide.[8] Both have been criticized for their dependence on linearity assumptions and neither ties directly to use in life-cycle phased analysis.

2.2 System Life-Cycle

A strong feature of the recommended method of analysis is the inclusion of all the emissions of a system through its entire life-cycle and any residual thereafter. Therefore, the method requires definition of the system life-cycle and identification of the GHG emissions for each period in that life. It is convenient to divide the total time under consideration into four periods.

Period 1. Everything associated with bringing the system into being, i.e., research, development, testing, engineering, detailed design, procurement of materials and construction operations (termed development/construction).

Period 2. The entire operating life of the system from startup to final shutdown.

Period 3. All cleanup work, including teardown, salvage, perhaps entombment of elements, and all final waste disposal.

Period 4. Decommissioning: All final effects of the system after the end of Period 3, covering the decay of GHGs and other waste streams.

With these concepts in place, the quantitative development is straightforward. The first part uses information on the GHGs discharged by a system to determine the amounts actually in the atmosphere at various times through system life and after it ceases operation. As defined, the first time period covers the initial warming effect due to the invested emissions from the preoperations interval of the system beginning at $t = t_1$ and ending at $t = t_2$. The second time period, operations, begins at $t = t_2$ and ends at $t = t_3$, and the third covers post-operation cleanup or system decommissioning from $t = t_3$ to $t = t_4$. The fourth or final period is one during which all the GHGs remaining in the atmosphere from the previous phases decay and vanish, i.e., from $t = t_4$ to t_∞ or the end of the GHG lifetime.

The mathematical development for quantitative analysis is carried out for each period and then the relationships are added to cover the entire life-cycle-plus time effect of the system. A simple material balance is written for a given period, i.e.,

$$\frac{dN(t)}{dt} = P(t) - \lambda(t)N(t) , \qquad (1)$$

where $N(t)$ is the number of moles of the GHG emitted by the system existing in the atmosphere at any time, t, $P(t)$ is the rate of the GHG emission by the system as a function of time, and $\lambda(t)$ is the rate of decay of any mole of the GHG in the atmosphere. Thus, while the GHG is being injected into the atmosphere, part of that released earlier in the period has been removed by one or several processes. Integration of this equation with the proper relationship for $P(t)$ and $\lambda(t)$ gives the number of moles which are present in the atmosphere at any time, with rigor limited only by the information on $P(t)$ and $\lambda(t)$.

For purposes of illustration and preliminary analyses, we made several simplifying assumptions: 1) the rate of injection, P, is constant at an average value for each period, as shown in Fig. 2; and 2) the rate of decay for each GHG is constant, suggesting that

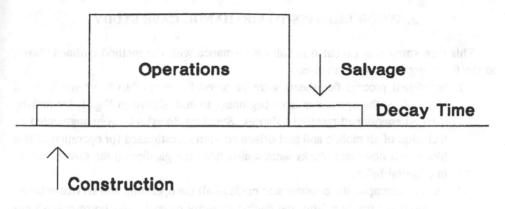

Fig. 2. Simplified Profiles for System Periods.

simple first-order decay occurs. Thus, a simplified material balance can be written as follows,

$$\frac{dN(t)}{dt} = P - \lambda N(t) , \qquad (1a)$$

where P is the rate of injection, moles/unit time, and λ is the rate of decay, in moles per unit time per mole present. Dickinson and Cicerone[19] applied this balance to constant emissions of CFCs. Integration for real values of λ yields

$$N(t) = \frac{P}{\lambda}(1 - e^{-\lambda t}) + N(t_0)e^{-\lambda t} \qquad (2)$$

In the approach discussed here, the total greenhouse warming forcing is calculated for the system for each active gas alone, as a function of the concentration at any instant, times its absorption efficiency, integrated over the duration of the period or the lifetime of the GHG for the decay phase. The concentration value is the ambient atmospheric value plus the emission of the system. Similarly, the absorption efficiency value is for that portion of the response curve applicable at the time. Then, the total system warming forcing is calculated by summing the effects of the individual gases.

Finally, the total warming forcing can be related to the total lifetime useful output of the system (e.g., kwh) to yield an index of performance: the Greenhouse Warming Effect Index. As noted, essentially all the data needed for calculations should be available from careful material and energy balances normally carried out for each block of the technical flowsheets needed to describe the system. As a footnote to the procedure, GHGs which decay into GHGs, such as methane and carbon monoxide to carbon dioxide, can also be treated by slight modification of the material balance equation.[12] Only simple integrations are needed to convert GHG emissions and lifetimes of gases to warming effects for spreadsheet presentation. These are easily carried out by table or on personal computers.

3. WOOD BIOMASS-TO-METHANOL CASE STUDY

This case study was executed in full conformance with the method outlined above, so the following steps were involved.

1. Individual process flowsheets were prepared for every block, or sub-block if required, of the operations from beginning to end, shown in Fig. 1, for making detailed energy and material balances. Standard flowsheets were augmented by inclusion of all mobile and fuel-driven equipment estimated for operation of that block. All flowsheet blocks were scaled from the gasifier so the flowsheets are in material balance.

   a. For example, the estimate was made of all the operating and service vehicles used for the tree farm, the engine capacity of each, and horsepower-hours of operation per cycle. Then, based on fuel use per horsepower hour, total direct fuel consumptions were developed. To these consumptions were added the consumptions in the fuel chain to prepare each gallon of fuel going into an engine, i.e., indirect consumptions. These represented a fossil fuel subsidy often overlooked for energy use and emissions analysis.

   b. The flowsheets were also detailed to show consumables such as fertilizers, pesticides, herbicides, water, building heating fuel, and electric power. Both direct and indirect energy consumptions were estimated for each.

   c. For transportation blocks, primary equipment needs were determined and consumptions estimated. Auxiliary vehicles were included. Other consumables such as lubrication oils, shop heat and power were estimated. Then both direct and indirect energy consumptions were estimated.

   d. For wood-chip gasification and conversion to methanol, a contractor's study[20] was selected and examined in detail. Additions were made as necessary to cover auxiliary vehicles, and oversights as regards owner's costs not considered by the contractor. Add-on direct and indirect fuel energy and chemical consumptions were estimated.

2. Detailed lists of all equipment in every block were prepared which included all commodity materials used to bring every block of the operating flowsheet into being. This was done by disaggregating equipment, machinery, buildings, services, and civil installations into units of basic materials, such as steel, concrete, copper, aluminum, etc.

   a. This listing was employed to sum up the total amount of each basic material invested in plant and machinery to bring the system into being. Then, information on the energy use and emissions of the system which produced the material, e.g., cement, was used to convert the amount of material installed into "invested" energy and emissions of a direct nature.

   b. Secondary information on the system used to produce the commodity material was employed to estimate the invested energy and emissions due to commodities used to create this system, e.g., steel used to build the concrete plant. A reaction to this calculation might be that it deals with second-order infinitesimals. This may be true, but drmi á priori rejection of these quantities is not warranted until several detailed analyses prove the case.

    c. Finally, to add to the operating component of invested energy and emissions, estimates must be made of human inputs to R&D, engineering, design, and construction. Also, fuel and power usage for construction and construction equipment are estimated to derive energy use and emissions.

3. The operating life of the system was set, along with the stream factor, turnaround and maintenance philosophy, to yield the useful output per unit time. This defines the energy output, which can be compared with the total operating and invested energy inputs to yield energy efficiency.

4. Energy balances were prepared by use of the flowsheets and material balances. These quantities were aggregated to determine total operating inputs and invested inputs for calculation of simple ratios or efficiencies. The integrated lifetime useful output divided by the integrated operating inputs yields a simple operating efficiency. The integrated output divided by the total of the operating and invested inputs yields the total system energy efficiency.

5. Emission quantities were more difficult to determine and aggregate. Direct fuel usages were converted to emission quantities based on fuel composition and NO_x generation characteristics of engines, i.e., gasoline versus diesel. Similar summations were carried out for all other combustion processes. Finally, all emissions were aggregated, invested emissions for Period 1 and operating emissions for Period 2. For this first analysis, the salvage phase (Period 3) was ignored. Use of new information from the Electric Power Research Institute[21] regarding errors in nitrous oxide values obtained from grab samples markedly reduced the estimated nitrous oxide emissions from the anticipated values. Emissions were converted to average values for both the construction and operating phases and the relationships for constant-value releases were used for calculating the moles of each GHG in the air at any instant.[11]

6. Greenhouse warming forcing calculations were carried out using the basic relationship for warming forcing.[11] This relationship uses the moles present times the instantaneous absorptivity of the GHG divided by the volume of the atmosphere. The expression is integrated over each period of project life, plus the additional time required for the residual concentration of the GHG to decay to zero.

7. Abbreviated results from the study are presented to provide a picture of the end-to-end values for a complete system (all values are subject to slight revision in final calculation rechecks). An input of 310 tons of green wood per operating day yields 170 tons of dried wood and a product of 77 tons of fuel grade methanol. An operating energy efficiency of 41 percent was calculated, including external energy subside. The primary GHG emission, of course, is carbon dioxide. Some 160 kg of CO_2 are emitted for every 10^9 J of methanol produced by the system

386

(390 lb CO_2/million Btu) and another 69 kg of CO_2 are emitted for every 10^9 J of methanol burned in an engine. The profile of CO_2 existing in the atmosphere, above background, due to the system is shown in Fig. 3.

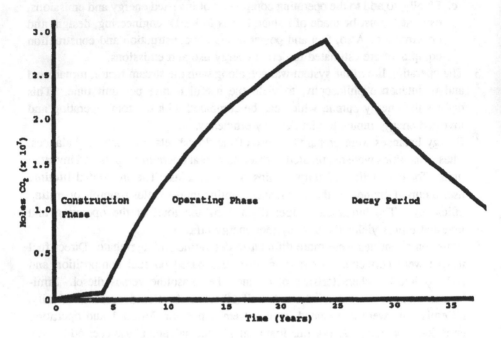

Fig. 3. Moles CO_2 in Atmosphere from Project vs Time.

4. IMPLICATIONS FOR GREENHOUSE POLICY

The method of analysis presented above is directly applicable to comparison of alternative technological paths to the same end use, i.e. kwh, vehicle miles, pounds of plastic, etc., even though the paths are markedly dissimilar. This result obtains because the complete warming forcing effect of each path is calculated for a given system life. The associated warming forcing is then normalized by relating it to the total useful output of the system, i.e., kwh per unit warming or the reciprocal thereof. This approach provides a way to compare systems of markedly different components, for example biomass-to-methanol (agricultural plus industrial) with conventional petroleum-to-gasoline (industrial fossil fuel) for vehicular fuel. It would be equally applicable to electric car systems.

For comparing different systems, the method facilitates identification and quantification of pollution conversions and intersectoral transfers. For example, if methanol from wood is substituted for gasoline, nominal improvement in urban smog may be obtained. However, in essence, that pollution is moved from the city and converted to a doubling of the CO_2 discharged to the atmosphere elsewhere. Identification of the intersectoral transfer enlarges the scope of the problem; should the rest of the nation have a voice in approving a local solution which will affect the nation and the world?

Information gathering, sharing and compilation are important elements in the Administration's global environmental-change research effort.[22] Our proposed method fits this effort like a glove. It should be clear that the same system definition can be used to calculate the economic cost, net energy use, and net global warming forcing. All three measures can be related to units of useful output to obtain a common base. All can be developed with slight extension of the energy, material, and money balances normally carried out.[23,24] Sensitivity analyses can be made in the normal manner. With cataloging of the data, on consistent bases and on generic systems, approaches with the same output can be rank-ordered in each category according to their desirability.

The federal administration appears to favor development of a market-based approach, as in the 1990 Clean Air Act Amendments, to attain compliance with possible atmospheric stabilization objectives. Thus, it is implied that a "bottom-up" approach resulting from the aggregation of many individual voluntary decisions would be preferred to a small number of "top-down" mandates. In a practical sense, these decisions would be made by many smaller, decentralized groups based on prevailing conditions at the time.

Even with the possible adoption of national targets for GHG emissions reduction which could reorient entire industries, decisions would be made at the margins of existing conditions. Thus, individual companies or groups of investors would be faced with questions of the type outlined above; should a certain amount of crude oil production and refining be shut down and a matching amount of biomass-based methanol production be constructed to take its place to meet vehicular fuel needs? The proposed analytic method is easily applied here.

The "sunk cost" practice from economic analysis is a concept that has definite applicability to GHG warming forcing analyses.[25] A sunk cost is one that was incurred in the past and cannot be altered by present or future action. Economic decisions related to future action should not be permitted to be affected adversely by sunk costs.[26] The same rationale should be applied to decisions regarding GHG emissions and the warming forcing of individual systems; what happened in the past should not influence decisions regarding the future. A particularly important area is that of choosing between continued operation of an existing system and shutting it down and building and operating an alternate system.

The first analysis of all systems should be on the basis of "grassroots," start-from-scratch creation of the system. This is obviously correct for situations such as the wood-biomass-to-methanol system for auto fuel discussed herein. However, where the basic decision is a choice between building the biomass system, continue with gasoline, or await reformulated gasoline, comparison of these alternatives as grassroots systems would be inappropriate. The relevant warming forcing for the gasoline system, where adequate or over-capacity already exists, does not include construction of the system. Since it exists already, prior emissions have long since entered the ecosystem and cannot be recalled.

Therefore, the warming forcing of the gasoline system will come from future operations, salvage, and decay, plus the use of the gasoline. For reformulated gasoline, the analysis should be based on the prorated effect of the current system, plus the prorated effect of construction and operation, etc., of facilities to produce new blending agents (e.g., MTBE). Otherwise, the gasoline system is falsely burdened, except in those countries where the gasoline system does not exist. Applicability to this type of situation is a strong attribute of our method of analysis.

The method outlined above is particularly applicable to the assessment of basic technologies. Others have recognized the need to include more components in general economic analysis[27] and are gradually broadening the scope of analysis toward the system definition outlined above. Yet great difficulty exists in utilizing general, sectoral economic data from many different sources, with unknown gaps and assumptions, to continually include new sectors and increase the accuracy of the overall result. Careful studies of generic systems using the proposed method can provide benchmarks for adjusting major sectoral studies and rank ordering basic approaches. Further, with the responsibility the national laboratories have regarding technology transfer,[28] they must be careful to make the correct technology available. This means that any of the three scenarios typified above could apply with respect to a given developing country. The transfer agent should be able to defend quantitatively the choice for any transfer. With our method, the analyses to do so are relatively simple.

5. CONCLUSIONS

In any program or agreement for controlling greenhouse warming forcing in which transfer of clean technologies to developing countries would be central, the entire effect of each candidate technology should be defined clearly. This requires a systems perspective, consideration of lifetime-plus effects, and a determination of the effect of each GHG separately. A method of analysis which meets these requirements has been described. When applied to trade-offs such as biomass-based methanol fuel versus gasoline, effects such as the intersectoral transfer of pollution and increased greenhouse warming forcing become evident. We believe this method shows real promise and should be used as the basis for the collection and sharing of information between public and private agencies to provide detailed data for rank-ordering generic systems with respect to their respective greenhouse warming forcing, energy usage, and dollar cost.

6. ACKNOWLEDGMENTS

This research was supported by a grant from the Sarkeys Energy Center, University of Oklahoma. We thank Dawlat El-Sayed for her assistance with the analysis and Thomas James, Director, Science and Public Policy Program for his contributions. Lennet Bledsoe and Kathie Ewing helped prepare the manuscript.

7. REFERENCES

1. Benedick, R. E., A. Chayes, D. A. Lashof, J. T. Mathews, W. A. Nitze, E. L. Richardson, J. K. Sebenius, P. S. Thacher, and D. A. Wirth. *Greenhouse Warming: Negotiating a Global Regime* (World Resources Institute, Washington, DC, 1991).

2. Lashof, D. A. and D. A. Tirpak, eds. *Policy Options for Stabilizing Global Climate*, Report to Congress (U.S. Environmental Protection Agency, Washington, DC, 1990).

3. Sand, P. H. "International Cooperation: The Environmental Experience." in *Preserving the Global Environment*, ed. J. T. Mathews, (W. W. Norton & Co., New York, 1991).

4. Victor, D. G. "How to Slow Global Warming." *Nature* 349 (1991) 451.

5. DeCanio, S. J. and K. N. Lee. "Doing Well by Doing Good: Technology Transfer to Protect the Ozone." *Policy Studies Journal* 19:2 (1991) 140.

6. Intergovernmental Panel on Climate Change. *Climate Change: The IPCC Response Strategies* (Island Press, Washington DC, 1991).

7. Ellington, R. T. and M. Meo. "Development of a Greenhouse Gas Emissions Index." *Chemical Engineering Progress* 86:7 (1990) 58.

8. Houghton, J. T., G. T. Jenkins, and J. J. Ephraums, eds. *Climate Change: The IPCC Scientific Assessment*. Report prepared for the Intergovernmental Panel on Climate Change by Working Group I (Cambridge University Press, Cambridge, England, 1990).

9. Lashof, D. A. and D. R. Ahuja. "Relative Contributions of Greenhouse Gas Emissions to Global Warming." *Nature* 44 (1990) 529.

10. Hammond, A. L., E. Rodenburg, and W. Moomaw. "Accountability in the Greenhouse." *Nature* 347 (1990) 705.

11. Ellington, R. T. and M. Meo. "Greenhouse Gases, Utilities and Climate Policy: Performance Measures for Assessing Trade-offs." Paper 91–122.1, 84th annual meeting, Air & Waste Management Association. Vancouver, B.C. June 16–21, 1991.

12. Ellington, R. T., M. Meo, and D. Baugh. "The Total Greenhouse Warming Effect of Technical Systems: Analysis for Decision Making." submitted to *J. Air & Waste Manage Assoc.*, 1991.

13. Ellington, R. T. and M. Meo. "Rational Guidelines for National and International Decision Making about Global Warming." in *Earth Observations and Global Change Decision Making 1990: A National Partnership*, Vol. 2, ed. I. Ginsberg and J. Angelo (Kreiger Publishing Co., Melbourne, FL, 1991).

14. Rodhe, H. "A Comparison of the Contribution of Various Gases to the Greenhouse Effect." *Science* 248(1990) 1217.

15. Wang, W. C., M. P. Dudek, X. E. Liang, and J. T. Kiehl. "Inadequacy of CO_2 as a Proxy in Simulating the Greenhouse Effect of Other Radiatively Active Trace Gases." *Nature* 350 (1991) 573.

16. Rhoades, E. A., S. B. Holbert and C. W. Berish. "Method for Assessing Greenhouse Gas Emissions from Proposed Projects." Paper 91–128.3, 84th annual meeting, Air and Waste Management Association, Vancouver, B.C. (1991).

17. Brower, M. *Cool Energy—The Renewable Solution to Global Warming* (Union of Concerned Scientists, Cambridge, MA, 1990).

18. Nierenberg, W. A. "Atmospheric CO_2: Causes, Effects, and Options." *Chem Eng Prog* **85**: 8 (1989) 27.

19. Dickinson, R. E. and R. J. Cicerone. "Future Global Warming from Atmospheric Trace Gases." *Nature* **319** (1986) 109.

20. Stone and Webster Engineering Corporation. *Economic Feasibility Study of a Wood-Gasification Based Methanol Plant*, Publication SERI/STR-231-3141 (Solar Energy Research Institute, Golden, CO, 1987).

21. Kokkinos, A. "Measurement of Nitrous Oxide Emissions." *EPRI Journal* **15**: 3 (1990) 36.

22. Bromley, D. A. "Keynote address." in *Earth Observations and Global Change Decision Making, 1990: A National Partnership*, Vol. 2, ed., I. Ginsberg and J. Angelo (Kreiger Publishing Co, Melbourne, FL, 1991).

23. Fava, J. A., R. Denison, B. Jones, M. A. Curran, B. Vigon, S. Selke and J. Barnum, eds., *A Technical Framework for Life-Cycle Assessments* (SETAC Foundation for Environmental Education, Inc., Washington, DC, 1991).

24. Mortimer, N. D. "Energy Analysis of Renewable Energy Sources." *Energy Policy* **19**:4 (1991) 375.

25. Gurney, K. R. "National Greenhouse Accounting." *Nature* **353** (1991) 23.

26. Stermole, F. J. *Economic Evaluation and Investment Decision Methods* (Investment Evaluation Corp., Golden CO, 1984).

27. DeLuchi, M. A. "Emissions of Greenhouse Gases from the Use of Gasoline, Methanol, and Other Alternative Transportation Fuels." in *Methanol as an Alternative Fuel Choice: An Assessment*, ed. W. L. Kohl (The Johns Hopkins Foreign Policy Institute, Washington DC, 1990).

28. Schriesheim, A. "Toward a Golden Age for Technology Transfer." *Issues in Science and Technology* **7**:2(1991) 52.

Making Better Use of Carbon:
The Co-Production of Iron and Liquid Fuels

John H. Walsh
Energy Advisor

This paper is published by Abstract only, at the request of the author, who will furnish edited copies to requestors.

ABSTRACT

It is becoming increasingly probable that emissions of carbon dioxide from the fossil fuels will have to be reduced to deal with the threat of global warming. This issue is likely to be resolved one way or the other in the period 1995–2000 which does not leave much time to develop responsive options. In the intervening period, the policy of the Government of Canada is to restrict emissions of greenhouse gases to their 1990 level by 2000 by encouraging greater efficiency in the use of energy. Should further reductions prove necessary, economic instruments would probably be employed but these would not be applied before a comprehensive international agreement is reached.

There are three main themes in this paper: it is assumed that measures to control carbon dioxide emissions will in fact be necessary by the turn of the century; that there will be a rising need for liquid fuels to supply the world's steadily growing fleet of motor vehicles for many years to come: and that the future niche for Canada's industry in such a constrained world will be the operation of the energy-intensive industries with efficiency and sensitivity.

The steel industry, as a large user of coal and energy in other forms, would be significantly affected by a need to reduce carbon dioxide emissions. This paper first examines what might be called the conventional responses open to the industry in such a situation and then explores a new possibility for making better use of carbon by the co-production of iron and liquid fuels.

Solid Fuel Conversion for The Transportation Sector

Alex E. S. Green

Clean Combustion Technology Laboratory

Department of Mechanical Engineering

University of Florida, Gainesville, Florida

ABSTRACT

The conversion of solid fuels to cleaner-burning and more user-friendly solid, liquid, or gaseous fuels spans many technologies. Here we consider coal, residual oil, oil shale, tar sands, tires, municipal solid waste, and biomass as feedstocks and examine the processes which can be used in the production of synthetic fuels for the transportation sector. Mechanical processing can lead to potentially usable fuels such as coal slurries, micronized coal, solvent refined coal, vegetable oil, and powdered biomass. The thermochemical and biochemical processes considered include carbide production, liquefaction, gasification, pyrolysis, hydrolysis, fermentation, and anaerobic digestion. The products include syngas, synthetic natural gas, methanol, ethanol and other hydrocarbon oxygenates, synthetic gasoline, and diesel and jet engine oils. We discuss technical, economic, energy, and environmental aspects of synthetic fuels and their relationship to the price of imported oil. This work is a summary of a recent collection of papers entitled "Solid Fuel Conversion for the Transportation Sector" published by the Fuels and Combustion Technology (FACT) Division of the American Society of Mechanical Engineers (ASME).

The full paper is available from the author.

Waste-to-Energy, Municipal—Institutional

Alex E. S. Green

Clean Combustion Technology Laboratory

Mechanical Engineering Department

University of Florida, Gainesville, FL

THE CONTINUED ROLE OF FIRE

Fire has been a civilizing force for mankind since the dawn of civilization [see Fig. 1(a)], sometimes compared with the discovery of speech, writing and agriculture.[1] However, carbon dioxide and other emissions from fire are now viewed by some as the doom of civilization [see Fig. 1(b)]. Figure 2 provides an overview of local, regional and global anthropogenic emissions to the atmosphere largely associated with the use of fire for energy purposes.[2] Which one of these problems is the most urgent at this time is a matter of considerable debate. This conference is primarily concerned with the greenhouse issue whereas the recent technical work of our laboratory relates to the issue of toxics. Nevertheless, as we shall show, there are close relationships between the global greenhouse, stratospheric ozone depletion and toxic emission problems.

MUNICIPAL WASTE AND THEIR GREENHOUSE EMISSIONS

Municipal waste is now regarded as a significant source of fuel for electricity generation. Figure 3 gives the breakdown by weight of the 1988 municipal waste output in the United States.[3] Of the approximately 180-million tons generated, one might conservatively project the use of 100-million tons for waste-to-energy use. Granting an average of 5000 Btu/lb, one readily calculates the energy potential,

$$10^8 \text{ tons} \times 2.103 \text{ lb/ton} \times 5.10^3 \text{ Btu/lb} = 10^{15} \text{ Btu/lb} = 1 \text{ quad}$$

If one adds waste biomass from forestry operations or from agriculture, this source of energy might readily reach about 3 quads. Cultivated biomass energy crops could provide some 8 extra quads according to the pessimists or as much as 12 quads according to the optimists.[4] In either event the sum would be a significant part of our national energy use, which currently totals about 80 quads.

The combustion of municipal waste and biomass, of course, will contribute CO_2 to the atmosphere and to this extent would exacerbate the greenhouse problem. However, for disposables, such as newspaper made from biomass, the uptake of CO_2 in growing the pulp would largely balance the emissions of CO_2 during combustion. Accordingly, from the CO_2 standpoint the use of energy from the combustion of biomass and much of municipal waste are natural components of a solution to the greenhouse problem.

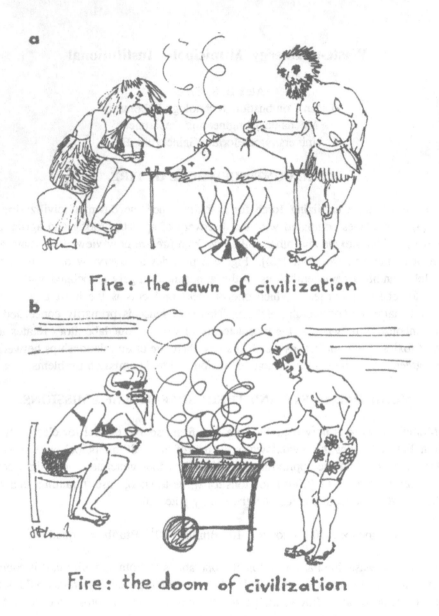

Fig. 1. The Fire Dilemma (courtesy Jean Francois Louis).

Some of the other gaseous emissions of municipal waste facilities also absorb in the long wave earth radiation region, which controls the global atmospheric temperature. The 800 to 1200 cm^{-1} region, or 12.5 to 8.3 micron region, is of particular importance to the anthropogenic greenhouse problem since this region is near the peak of the earth-atmosphere long-wave radiation that is not blocked by H_2O and CO_2 bands. Many

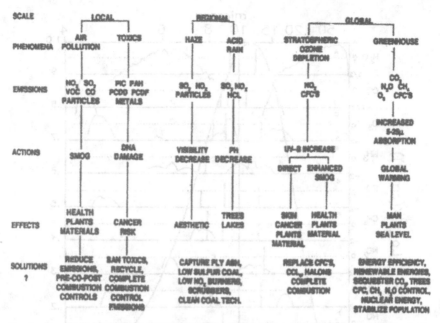

Fig. 2. Anthropogenic emissions to atmosphere: transportation, utilities, industrial, commercial, residential, landfills, incineration, agricultural (from Green, Ref. 2).

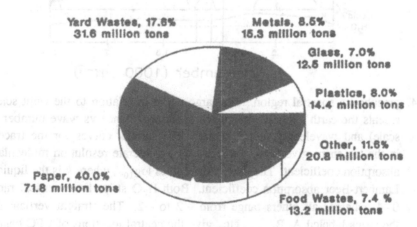

TOTAL WEIGHT • 179.6 million tons

Fig. 3. Materials generated in MSW by weight, 1988 (from EPA, Ref. 3).

products of incomplete combustion (PIC) absorb in this region so the eventual use of municipal waste, biomass, and fossil fuels, must be considered carefully in this context. Figure 4 displays the Planck spectrum for a 255 K black body which calculates to be the

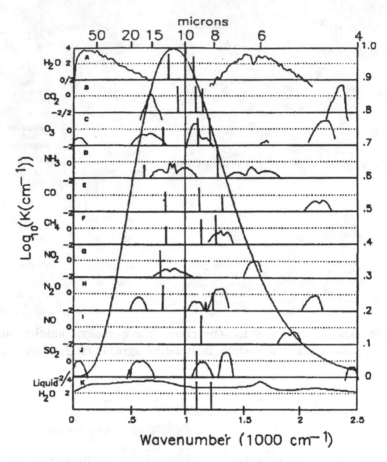

Fig. 4. greenhouse spectral region. The large curve in relation to the right scale represents the earth's relative longwave spectral radiance vs wave number (lower scale) and wavelength (upper scale). The smaller curves for the trace gases in the ten strips give $\log_{10}k$, where k is a moderate resolution molecular band absorption coefficient. The lowest strip gives $\log_{10}k$, where k is the liquid water Lambert-Beer absorption coefficient. Both H_2O strips have vertical ranges of 0–4, whereas all others range from −2 to +2. The straight vertical lines in the strips labeled A, B, . . . , etc., give the central locations of CFC bands with heights represented by $\log_{10}I$, where I are the integrated band intensities all on vertical scales from 0–4. Here, A – $CFCl_3$(F11), B – CF_2Cl_2(F12), C – CF_3Cl(C13), D – CF_4(F14), E – $CHClF_2$(F22), F – C_2F_6(F116), G – CCl_4, H – $CHCl_3$, I – CHF_3, J – CH_2F_2, and K – $CBrF_3$.

average global temperature in the absence of a greenhouse effect. Also shown on this diagram are the active absorption bands of H_2O, CO_2, O_3, NH_3, CO, CH_4, NO_2, NO,

SO$_2$, and liquid water and the positions and approximate band strengths of important chloro-flourocarbons (CFCs) in the long wavelength region.[5,6]

Table I is a list of the spectral features of C$_1$ to C$_3$ products of combustion including chlorinated PICs which have emissions or absorption bands in this range.[7-10] The

TABLE I. List of Spectral Features in Chlorinated Hydrocarbon Combustion.[a]

Wavelength (cm^{-1})	Species		Spectral Type	
793.9	Carbon tetrachloride		Broad peak	
795.8	1,1-Dichloroethylene	Q		PQR
801.4	Tetrachloroethylene		Broad peak	
824.6	Ethane		Broad peak	
826	trans-1,2-Dichloroethylene		Broad peak	
850	Trichloroethylene		Double peak	
857	cis-1,2-Dichloroethylene		Double Q branch	
858.5	trans-1,2-Dichloroethylene	Q		PQR
869	1,1-Dichloroethylene		Double Q branch	
896.6	Vinyl chloride	Q		PQR
898.6	trans-1,2-Dichloroethylene	Q		PQR
917	Tetrachloroethylene		Broad peak	
937.3	1,1,2-Trichloroethane	Q		PR
940	Trichloroethylene		Twin peaks	
942.3	Vinyl Chloride	Q		PQR
949.6	Ethylene	Q		PQR
973.2	Ethyl chloride	Q		PQR
1020.2	Methyl chloride			
1033.5	Formaldehyde	Q		PQR
1086.2	1,1-Dichloroethylene			
1087.4	1,1,1-Trichloroethane		Double peak	
1095.2	1,1-Dichloroethylene		First of double peak	
1097.4	1,1-Dichloroethylene		Second of double peak	
1200.6	trans-1,2-Dichloroethylene	Q		PQR
1209.3	1,1,2-Trichloroethane	Q		PQR
1220.7	Chloroform		Twin peaks	
1232.2	Ethylene dichloride	Q		PQR

[a]Adapted from Hall et al., Ref. 7.

carbon-chlorine stretching region from 500 to 750 cm^{-1} is sufficiently different from the CO_2 band from 630 to 720 cm^{-1} to permit identification of virtually all chlorinated hydrocarbons by spectra in this range. Thus PICs might play a significant role in the anthropogenic greenhouse problem. Table II presents some data on long-lived industrial solvents which are also PICs.[11-13] Most of these lifetimes are determined by reactions with OH. Emissions of carbon monoxide a major product of incomplete combustion can lower the natural OH concentrations which would lengthen the lifetimes and hence increase the equilibrium concentrations of the greenhouse gases in Table II. Not enough attention has been given to the relationship between PICs, particularly chlorinated PICs, and the greenhouse problem. Chlorinated PICs have been detected in the Kuwaiti oil fires.

TABLE II. Atmospheric Data for Industrial Solvents Which Are Emitted from Combustion (Ref. 9).

Compound	b.p. (°C)	Half-life		Decomposition
Benzene	118.5	26.7	days	OH reaction
Carbon tetrachloride	76.5	30–50	years	Photolysis, OH
Chloroform	61.7	80	days	OH reaction
1,1-Dichloroethane	57.3	62	days	OH reaction
Pentachloroethane	161	1.8	years	OH, photolysis
Tetrachloroethylene	121	2	months	OH reaction
Toluene	110.6	1	day	OH reaction
1,1,1-Trichloroethane	74.1	0.5–25	years	OH, photolysis
Trichlorofluoromethane	23.7	52–207	years	Photolysis

Long-lived chlorinated PICs should diffuse into the stratosphere where they would be photodissociated by the harder ultraviolet radiation of the sun. In these cases the chlorine atoms created can catalytically destroy ozone molecules. Thus chlorinated PICs are greenhouse and ozone depletion threats.

Many chlorinated PICs are carcinogenic, mutagenic, teratogenic, neurotoxic, cause reproductive dysfunction or are acutely or chronically toxic.[9] The larger municipal incinerators when properly operated expose chlorinated PICs to long times at high temperatures which helps break down these compounds. However, institutional combustion facilities, particularly medical waste facilities, are much smaller and, when operated in the conventional starved-air mode, can be a large source of toxic PICs.[10-13]

VOLATILE ORGANIC COMPOUNDS

The Clean Combustion Technology Laboratory (CCTL) has been engaged in interdisciplinary energy-environmental studies related first to the combustion of coal and co-firing of coal and natural gas,[14,15] then to cellulosic biomass (CB) and non-hazardous waste (NHW),[16-21] and more recently to the measurement and minimization of toxic products of institutional waste incineration.[22-24] Our incinerator facilities at Tacachale (formerly named Sunland at Gainesville) includes a 1972 500-lb/hr (227-kg/hr) Environmental Control Products (ECP) incinerator. Figure 5 shows this facility which was reconstructed to facilitate emission measurements.

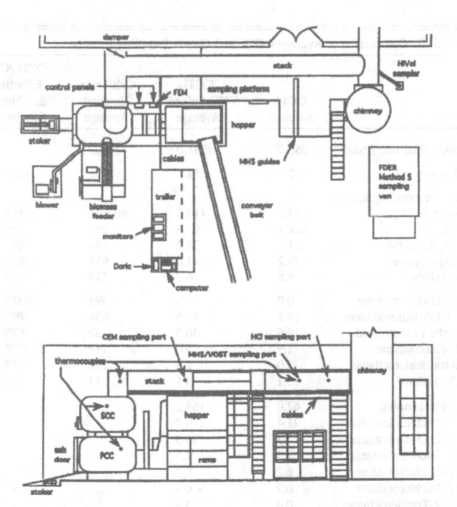

Fig. 5. Clean combustion technology laboratory research incinerator (from Green et al., Ref. 22).

We have compared our volatile organic compound (VOC) emission factors with corresponding emission factors given in reports of the California Air Resources Board (CARB) on medical incinerators without post combustion controls.[10-12] Table III shows these comparisons. Note that in almost all cases our emission factors are substantially lower than the average concentrations of the same VOC measured by CARB when normalized to the same O_2 concentrations.

Table III also shows a comparison of averages of emission factors of volatile chlorinated hydrocarbons when burning non-hazardous waste (NHW) with emission factors from burning PVC-spiked NHW. Note the large increase in the emission of

TABLE III. Comparison of Average CCTL and CARB Sampling Results.

	CCTL Average	CCTL PVC-Spiked Average	California Hospital Average	CCTL/Cal Effective Scrubbing Efficiency
Average feed rate (lb/hr)	356.8	351.0	558.3	
HCl (g/kg)	2	12	17	88%
Emission rates (μg/kg)				
Benzene	143.0	419.2	80 501	100%
Toluene	128.1	93.8	891	86%
Ethylbenzene	21.4	38.2	146	85%
m&p-Xylene	19.2	17.0	1 614	99%
o-Xylene	4.6	6.3	315	99%
1,2-Dichloroethane	0.0	0.0	593	100%
1,1,1-Trichloroethane	14.1	14.6	634	98%
Carbon tetrachloride	18.6	30.2	29	35%
Trichloroethene	1.9	5.0	184	99%
Tetratrichloroethene	2.0	2.8	272	99%
Chloroform	16.1	65.5	17	5%
Chlorobenzene	62.0	169.5		
1,2-Dichlorobenzene	21.4	110.4		
1,3-Dichlorobenzene	17.7	76.5		
1,4-Dichlorobenzene	2.1	7.3		
1,1-Dichloroethane	6.2	0.0		
1,1-Dichloroethene	0.9	0.3		
1,1,2-Trichloroethane	0.0	3.8		

several compounds, particularly carbon tetrachloride, chlorobenzene, chloroform, and dichlorobenzene. Emissions of nonchlorinated volatile aromatics (except benzene) stayed about the same.

We attribute the lower toxic emissions indicated in the CCTL columns of Table III to:

(1) our protocols of separating (a) toxics and recyclables and bagging only non-hazardous waste and (b) avoidance of chlorinated plastics by the Tacachale purchasing agent. This measure also explains our lower HCl levels and further assists in lowering chlorinated organic emissions.

(2) our retrofit measures promoting good combustion: (a) installations of a stoker, (b) the incorporation of a strong blower for extra underfire air, (c) the installation of an extra tangentially directed overfire air blower (originally for use as an educing agent for our biomass feeder) (d) our operational protocol of running at higher temperatures (1800° primary, 2100° secondary) rather than those of conventional starved-air incinerators ˙1400° primary, 1700° secondary).

In addition to the VOC measurements of the CCTL, the work of Bulley in New Zealand[25-27] has demonstrated that reduced PVC input and good combustion conditions substantially lowers the chlorinated dioxin (PCDD) and chlorinated furan (PCDF) output from medical waste combustion. Figures 6(a) and 6(b) illustrate his recent results. Between the two efforts it has become clear that good combustion and avoidance of chlorinated plastics will greatly reduce emissions of toxic chlorinated PICs.

CONCLUSIONS

Municipal waste, agricultural waste and cultivated biomass can contribute significantly to our national energy requirements with minimal net contributions to the atmospheric CO_2 burden. However other emissions from these incinerators can present potential smog, toxic haze, acid rain, ozone depletion and greenhouse problems (see Fig. 2). Chlorinated organic products of incomplete combustion (ClPICs) represent a triple threat to the atmospheric environment in that they are strong greenhouse gases, can foster ozone depletion and are often toxic at very low concentrations.

Clarification of these potential negative impacts of ClPICs are mostly coming out of studies of emissions from medical waste incinerators since PVC is more strongly represented in the current medical waste stream than in the MSW stream. However, these institutional situations should be more manageable since the chief executive officer or purchasing agent can tightly control the institution's plastic purchases. This situation does not yet prevail, however, and medical waste facilities are often the major source of toxic emissions in many communities. However, we are hopeful that this situation will change before much longer.[28] Municipal waste to energy problems will take some strong public education measures plus some national regulations or an international agreement of a similar nature to the Montreal Protocols to reduce the amounts of harmful materials. Establishing a "pollution prevention" ethic, in which products are evaluated as to all their

402

environmental impacts from the "cradle to the grave," is the key to the eventual solution
of this problem.

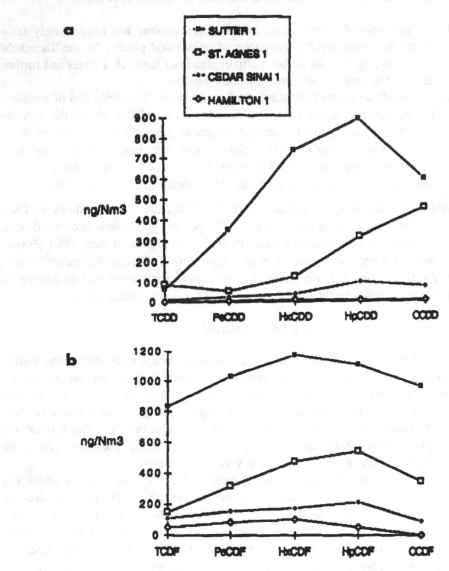

Fig. 6. PCDD and PCDF comparisons in ng/Nm³ @ 12% CO₂ (adapted from
Bulley, Ref. 26).

ACKNOWLEDGMENT

I would like to thank Dr. Jean Francois Louis for his encapsulation of the issue of *FIRE* in Figs. 1(a) and 1(b).

REFERENCES

1. "Fire," in the *Encyclopaedia Britannica*, Vol. 9, 1958, p. 265.
2. A. Green, "Foreward (sp CK??) to Advances in Solid Fuel Technologies ," A. E. S. Green and W. E. Lear, Jr., Eds., Published by Fuels and Combustion Technology (FACT) Division of the American Society of Mechanical Engineers, *FACT* 9 (1990).
3. "Characterization of Municipal Solid Waste in the United States: 1990 Update," United States Environmental Protection Agency, EPA/530-SW-90-042, June 1990.
4. D. L. Rockwood and G. M. Prine, "Alternative Production Systems: Wood Crops," in *Methane from Biomass, a Systems Approach*, W. H. Smith and J. R. Frank, Eds. (Elsevier Science Publishing Co., New York, New York, 1988), p. 27.
5. A. Green, Ed., "Greenhouse Mitigation," American Society of Mechanical Engineers ASME-H00513 (ISBN No.)-7918-0379-1) *FACT* 7 (1989).
6. A. Green "Ozone Depletion and Greenhouse Warming," Proceedings of Woods Hole Conference, October 23, 1989.
7. M. J. Hall, D. Lucas and C. P. Koshland, "Measuring Chlorinated Hydrocarbons in Combustion by Use of Fourier Transform Infrared Spectroscopy" *Environ. Sci. Technology* 25 (2) 260–267 (1991).
8. A. Green, C. Batich, and J. Wagner, 11Advances in Uses of Modular Waste to Energy Systems," *FACT* 9, *ibid*, Ref. 2.
9. House of Representatives Report 101-952, *The Clean Air Act Amendmentsm* U.S. Gov't. Printing Office, Washington, DC, 1990.
10. A. Jenkins, "Evaluation Test on a Hospital Refuse Incinerator at Saint Agnes Medical Center, Fresno, CA," ARB Test Report No. SS-87-01, 1987.
11. A. Jenkins, "Evaluation Test on a Refuse Incinerator at Stanford University Environmental Safety Facility, Stanford, CA," ARB Test Report No. ML-88-025, 1988.
12. A.Jenkins, "Evaluation Retest on a Hospital Refuse Incinerator at Sutter General Hospital, Sacramento, CA," ARB Test Report No. C-87-090, 1988.
13. J. C. Lauber and D. A. Drum, "Best Controlled Technologies For Regional Biomedical Waste Incineration," Paper 90-27.2 in Proc. 83rd Annual Meeting of Air & Waste Management Association, Pittsburgh, Pennsylvania, June 24–29, 1990.
14. A. Green, Ed., *Coal Burning Issues*, University Presses of Florida, Gainesvillle, Florida, pp. 1–380, 1980.
15. A. Green, Ed., *Alternative to Oil, Burning Coal with Gas*, University Presses of Florida, Gainesville, Florida, pp. 1–140, 1981.
16. A. Green, G. Prine, B. Green, et al., "Coburning in Institutional Incinerators," in Proc. of Air Pollution Control Association International Specialty Conference on

Thermal Treatment of Municipal, Industrial, and Hospital Wastes, APCA. Pittsburgh,Pennsylvania, pp. 61–81, 1987.

17. A. Green, D. Rockwood, J. Wagner, et al., "Co-Combustion in Community Waste to Energy Systems," FACT Division of the American Society of Mechanical Engineers, New York, New York.,*Co-Combustion* **4**, 13–27 (1988).

18. A. Green, "Clean Combustion Technology," in *Proc. of Conf., There is No Away,* University of Florida Law School, Gainesville, Florida, Available through EPA-NTIS, 1986.

19. A. Green, "Community Waste-to-Energy System Technologies," *Florida Scientist* **52** (2), 104–118 (1989).

20. A. Green, J. Wagner, B. Green, et al., "Co-combustion of Waste, Natural Gas and Biomass," *Biomass* **20**, 249–262 (1989).

21. A. Green, H. VanRavenswaay, J. Wagner, et al., 1990, "Co-feeding and Co-firing Biomass with Non-Hazardous Waste and Natural Gas," *Biomass* (1990).

22. A. Green, C. Batich, D. Powell, et al., "Toxic Products From Co-Combustion Of Institutional Waste," in *Proc. of 83rd Annual Meeting of the Air & Waste Management Association*, Pittsburgh, Pennsylvania, paper 90-38.4 (1990).

23. A. Green, C. Batich, J. Wagner, et al., "Advances in Uses of Modular Waste to Energy Systems," Presented at the Joint Power Generation Conference ASME-IEEE, Oct. and published in Green, A., and Lear, W. (Eds), 1990, Advances in Solid Fuels Technologies, Fuels and Copmbustion Technology (FACT) Division of ASME, New York, New York, 1990.

24. A. Green, J. Wagner, C. Saltiel, J. Blake, and Xie Qi Ma, "Medical Waste Incineration with a Toxic Prevention Protocol," Proc. of 84th Annual Meeting of the Air & Waste Management Association, Vancouver, British Columbia, June 16–21, 1991, Paper 91-33.5.

25. M. M. Bulley, "Medical Waste Incineration in Australasia," Paper 90-27.5 in Proc. of the 83rd annual meeting of the Air and Waste Management Association in Pittsburgh, Pennsylvania, June 24–29, 1990.

26. M. M. Bulley, "Medical Waste Incineration in Australasia," Paper 90-27.5 in Proc. of the 83rd annual meeting of the Air and Waste Management Association in Pittsburgh, Pennsylvania, June 24–29, 1990.

27. M. M. Bulley, "Incineration of Medical Waste: Treating the Cause Rather than the Symptons," *Clean Air*, May 1991.

28. A. Green, Ed., *Pollution Prevention and Medical Waste Incineration* Van Nostrand-Reinhold, in press.

CFC and Halon Restrictions: Impacts on Energy

Robert E. Tapscott
Center for Global Environmental Technologies, New Mexico
Engineering Research Institute, University of New Mexico
Albuquerque, NM 87131-1376

ABSTRACT

Owing to concerns about depletion of stratospheric ozone, the global community is facing restrictions on the consumption and use of some of its most useful chemicals — the chlorofluorocarbons (CFCs) and halons. CFC uses include refrigeration, air conditioning, foam blowing, cleaning, and aerosols. Halons have only one major application, as fire extinguishing agents. The amended Montreal Protocol, an international treaty, calls for a complete phaseout by the year 2000 of regulated halons (for all but "essential" uses), CFCs, and some other chemicals. The 1990 Amendments to the U.S. Clean Air Act also call for phaseouts of these chemicals for all but certain specified uses and even then only with special approval and for limited quantities. Technologies to decrease or eliminate the consumption of regulated chemicals include both alternative engineering approaches and replacement chemicals. Alternative engineering approaches substitute new processes and equipment to decrease dependence on ozone-depleting materials. Replacements are chemicals having decreased ozone depletion potentials. In most cases, alternative technologies will involve tradeoffs in economy, effectiveness, safety, and energy requirements. The energy consumption changes associated with some alternative technologies provide a feedback to affect energy requirements. In many cases, attempts to solve one global environmental problem through regulatory action can exacerbate another.

INTRODUCTION

Much of this Conference has dealt with feedback and coupling — the effect of one phenomenon on another. Dr. Bengtsson has spoken of the role of positive and negative feedbacks, from clouds, on global warming[1] and Dr. Kapitsa has presented an excellent overview of feedback between population, energy, and climate.[2] I would like to use a slightly modified version of a figure that Dr. Kapitsa used in his talk (Fig. 1), being presumptuous enough to add another arrow (the arrow shown in black). This short paper discusses the relationship between climate, energy needs, and technologies.

EFFECT OF CLIMATE CHANGE ON ENERGY

Climate change can affect energy needs in two ways, natural and anthropogenic. Here, "natural" denotes energy requirement changes that occur in response to climate

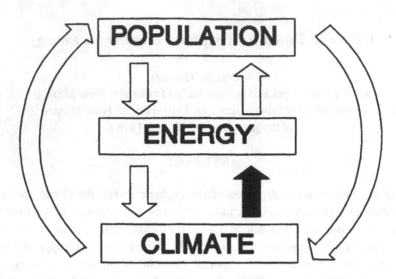

Fig. 1.   Feedback between population, energy, and climate.

changes. One example is, of course, the increased requirement for air conditioning as temperatures increase. "Anthropogenic" feedback refers to the effect on energy options of manmade efforts to prevent, mitigate, or postpone a climate change. This brief presentation covers only the latter—the effect on energy options of regulatory measures to counteract climatic change.

The term "climate change" is being used here in its broadest sense to include not only temperature and precipitation, but also such phenomena as acid rain and stratospheric ozone depletion. This paper looks at only the last phenomenon.

REGULATIONS TO LIMIT STRATOSPHERIC OZONE DEPLETION

In 1974, F. Sherwood Rowland, a chemistry professor at the University of California, and Mario J. Molino published a revolutionary paper.[3] Their calculations indicated that very stable halogenated halocarbons, the chlorofluorocarbons (CFCs), could drift into the stratosphere and catalytically deplete stratospheric ozone.

To meet this still unproven threat, the Vienna Convention for Protection of the Ozone Layer was adopted in March 1985, establishing a framework for development and adoption of an international treaty to protect stratospheric ozone. In September 1987, 24 countries signed the Montreal Protocol on Substances that Deplete the Ozone Layer, a treaty to limit the consumption of CFCs and halons. The Montreal Protocol became effective on 1 January 1989. Today, over 70 nations have ratified the Protocol.

The Montreal Protocol restricts the consumption of the most widely used ozone-depleting materials. Here, consumption is defined as production minus exports plus imports. The 1990 London amendments to the Protocol call for an eventual phaseout of the originally regulated CFCs and (except for "essential uses") halons. Additional chemicals — in particular, carbon tetrachloride and methyl chloroform — were also added. The 1990 Amendments to the U.S. Clean Air Act call for an even more rigorous phaseout schedule for CFCs. The Act also calls for a total phaseout of halons. The phaseout schedules are shown in Figs. 2 and 3.

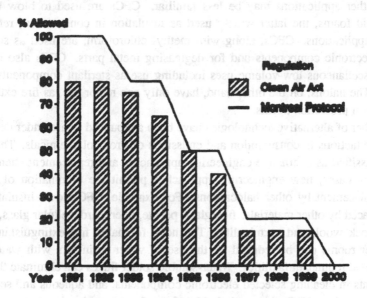

Fig. 2. Mandated CFC consumption phaseout.

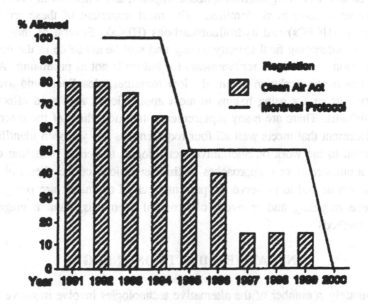

Fig. 3. Mandated halon consumption phaseout.

CFC AND HALON ALTERNATIVES AND REPLACEMENTS

Owing to their low toxicity, high stability, and ideal physical properties, CFCs have a variety of uses. Their use in aerosols, refrigeration, and air conditioning is well known; however, other applications may be less familiar. CFCs are used to blow both flexible and rigid foams, the latter widely used as insulation in construction, refrigerators, and other applications. CFCs, along with methyl chloroform, are used as solvents for cleaning electronic components and for degreasing metal parts. CFCs also have a variety of miscellaneous low-volume uses including use as sterilant components and food freezants. The halons, on the other hand, have only one major use, as fire extinguishing and explosion protection agents.

A number of alternative technologies have been announced or are under development to permit reductions in consumption and emissions of Protocol chemicals. These can be roughly classified as alternative engineering approaches and replacement chemicals.

In many cases, new engineering approaches permit the elimination of chemicals without replacement by other halocarbons. For example, CFC-blown insulating foams can be replaced by other materials: fiberglass, perlite, fiberboard, cellular glass, insulating concrete, rock wool, and vermiculite. The need for halon fire extinguishing systems in computer rooms can be reduced by the use of water sprinklers with water-resistant computers and redundant facilities. Water-soluble solder fluxes can eliminate the need for CFC solvents in cleaning selected electronic components, and aqueous and semiaqueous solvents can replace CFCs and methyl chloroform for many cleaning applications.

Despite these alternative engineering approaches, there is a wide search for alternative halocarbon chemicals having decreased ozone impacts, and a number of ozone-protective CFC replacements have been identified. The most important of these are hydrochlorofluorocarbons (HCFCs) and hydrofluorocarbons (HFCs). Several of these alternative chemicals are undergoing final toxicity testing and will be available in the near future.

The situation for chemical replacements for halons is not as promising. Alternatives must have low ozone depletion potentials, low toxicities, cleanliness and low volatility (the primary reasons for using halons in most applications), and good effectiveness as fire extinguishants. There are many replacements that meet three of these requirements, but no replacement that meets well all four requirements has yet been identified.

In addition to the work on alternative technologies to permit reduction of Protocol chemicals, a number of new approaches to limit emissions of the chemicals to protect stratospheric ozone and to preserve the present stock of chemicals are being developed. These include recycling and recovery of Protocol chemicals, bank management, and destruction methods.

NEGATIVE IMPACTS ON ENERGY

Unfortunately, a number of the alternative technologies involve negative impacts on energy consumption. We only have time to look at a few of these. Thus, for example, some alternative refrigerants have a lower efficiency and, therefore, a larger power consumption than do the existing refrigerants. In general, foams blown with replacement chemicals have poorer insulating abilities than do foams blown with CFCs, as do alternatives to foams. Note that this places a severe restriction on meeting energy consumption requirements for home refrigerators, which will use alternatives in both the refrigeration equipment and in the foam insulation. Aqueous solvents often require greater energy consumption for drying as compared to CFCs.

In some cases, the impacts are indirect. An excellent example is the lower effectiveness of many of the halon replacements, which will increase the problem of fire protection for nuclear power plants and Alaskan North Slope oil production facilities. Additional energy consumption will be expected to meet the expected requirement for destruction of excess CFCs and halons. In many cases, larger and heavier compressors are required for the new refrigerant replacements, causing increased fuel consumption by automobiles and aircraft.

CONCLUSIONS

We have only had an opportunity to look at a very few negative energy impacts due to regulatory measures to address stratospheric ozone depletion. However, this has been sufficient to show that in many cases, attempts to solve one global environmental problem can exacerbate another.

REFERENCES

1. L. 0. Bengtsson, "Model Simulations of Greenhouse Warming — Any Observational Evidence?," Global Climate Change: Its Mitigation Through Improved Production and Use of Energy, Los Alamos, NM, 21–24 October 1991.
2. S. Kapitsa, Global Climate Change: Its Mitigation Through Improved Production and Use of Energy, Los Alamos, NM, 21–24 October 1991.
3. M. Roland and F. S. Molina, Nature 249, 810 (1979).

Supply Price Analysis of Power Sector
Demand Management and Supply Options

Steven G. Diener
Steven G. Diener & Associates Limited
123–21 Vaughan Road
Toronto, Canada, M6G 2N2

BACKGROUND AND OBJECTIVES

More than any other energy supply industry, the electric power sector embodies several interrelated dimensions of energy analysis. These issues include, among others, the need to develop long-term forecasts of electricity demand; the magnitude of investment requirements (in Ontario alone, these can amount to over $20 billion over a 25-year planning period); the dizzying array of competing and rapidly changing supply and demand management measures and technologies; the significant and diverse socio-economic and environmental impacts attached to these options; and the multi-faceted role of utilities serving as power producers, buyers of independent power and promoters of demand management.

This note is intended to address the principles of supply price analysis, a tool for the techno-economic assessment of alternative supply and demand management options, and to discuss some of the applications of this analytic tool in assessing utility financial incentive programs and electricity-related environmental issues. To state my thesis at the outset: I consider supply price analysis the foundation and single most important aspect of power sector planning. Without a disciplined analysis of cost-competitiveness, technology choices and investment decisions will be made without regard for the societal benefits of economic efficiency and with undue regard for the metaphysical views of those with simple solutions, whether they belong to the "small is beautiful," "big is beautiful," or other schools of dogmatic belief. This is not to deny the importance of considering and, where possible, measuring both the socio-economic and environmental consequences of these options. Indeed, in certain cases, these so-called "external costs" (which can be positive or negative) may be "internalized" or added to the more narrowly defined costs initially estimated in the analysis of cost-competitiveness.

CONCEPT OF SUPPLY PRICE AND SUPPLY CURVE

Alternative supply and demand management options generally vary in terms of such characteristics as investment cost, construction or implementation period, economic or service life, and the annual quantities of energy output or savings. A useful tool for the consistent assessment and comparison of these potentially dissimilar options is based on the concept of "supply price" (also referred to as "resource cost" or "levelized unit energy cost"). Supply price is defined as the constant or levelized production cost per unit of energy output, using real (inflation-adjusted) costs and discount rates, and excluding transfer payments like taxes and subsidies. Put another way, given a particular real discount rate, the supply price of an option represents the real price of electric energy it needs in order to earn sufficient revenues to cover its capital and operating costs. It is calculated as the present worth (PW) of capital and operating costs divided by the PW of electrical energy output or savings. For those interested in delving behind the formula, Table I provides a step by step derivation.

In the simplest case where a quick and preliminary comparison is adequate, an option will be shown as economically feasible if its expected supply price is lower than that associated with a benchmark bulk supply technology. It is this type of analysis that is frequently used to compare demand management and alternative energy to conventional bulk supply options. For example, a cogeneration or demand management project may be shown to carry a favourable supply price when compared to a coal-fired thermal power plant. Table II provides illustrative supply prices for selected supply and demand management technologies.

But life is not this simple. As already suggested by Table II, all supply and demand technologies other than selected bulk supply options generally have a broad range of supply prices depending on site specific and other considerations. In these cases, instead of using average or a narrow band of supply prices, the analyst should develop a "supply curve" showing the quantities of electric power that an option could provide at various supply prices. It is this supply curve that helps define the economic as opposed to the technical potential for alternative technologies. The economic potential for a technology (measured in terms of power or energy) is the point where its supply curve intersects that of a competing bulk supply alternative such as nuclear or coal-fired thermal power. Figure 1 provides an example of a cogeneration supply curve for Canada developed by the author several years ago.

Similar supply curves can be developed for small hydro, wind energy, biogas, energy from waste, demand management and other technologies in order to identify their particular economic limits. Note that there is no need to carry out a supply price analysis of each and every potential site relevant to a technology. For each technology, it is sufficient to estimate only the number of sites in each of several (5 to 20) market categories and perform a supply price analysis on one case study per category using the following data:
- average or typical project capacity
- average capacity factor and electric energy output
- capital and operating cost
- value of non-electric output (for example, tipping fees earned by an energy from waste facility).

Using the example of energy from waste, it may be sufficient to develop three to five case studies representing three to five categories (ranges) of municipal solid waste (MSW) generation. For each range, an average MSW tonnage is determined and used

TABLE I. Derivation of the Resource Cost Formula.

Let:

I_t = capital investment made in year t;

O_t = operating cost incurred in year t;

Q_t = energy output quantity in year t;

$P1$ = per energy unit product price to meet investment costs;

$P2$ = per energy unit product price to meet operating costs;

P = real per unit resource cost or total per unit product price;

K = cost of capital (discount rate);

T = project life;

(I_t, O_t, $P1$, $P2$, and K are all expressed in real terms).

For economically viable projects, the relationship between the cost and value of a capital good is given by:

$$\text{Capital cost} \equiv \sum_{t=0}^{T} \frac{I_t}{(1+K)^t} = \sum_{t=0}^{T} \frac{P1 \times Q_t}{(1+K)^t} \equiv \text{Capital value} \quad ,$$

hence,

$$P1 = \left(\sum_{t=0}^{T} \frac{I_t}{(I+K)^t} \right) \Big/ \left(\sum_{t=0}^{T} \frac{Q_t}{(1+K)^t} \right) .$$

The product price must also cover the operating costs, O_t. This component of the total product price is defined analogously:

$$\sum_{t=0}^{T} \frac{O_t}{(1+K)^t} = \sum_{t=0}^{T} \frac{P2 \times Q_t}{(1+K)^t} \quad ,$$

hence,

$$P2 = \left(\sum_{t=0}^{T} \frac{O_t}{(1+K)^t} \right) \Big/ \left(\sum_{t=0}^{T} \frac{P2 \times Q_T}{(1+K)^t} \right) .$$

The total per unit price or real resource cost P, is $P1$ plus $P2$, or:

$$\left(\sum_{t=0)}^{T} \frac{I_t + O_t}{(1+K)^t} \right) \Big/ \left(\sum_{t=0}^{T} \frac{Q_t}{(1+K)^t} \right) .$$

That is, the per unit real resource cost is the ratio of the present valued total costs to the present valued quantity of energy output.

TABLE II. Illustrative Supply Prices of Electricity
Supply and Demand Options.

	$ per MWh (1991 Cdn. $)
Nuclear (CANDU)	35–45
Coal	45–50
Combined Cycle	60–65
Cogeneration (Industrial)	30–45
Cogeneration (Non-industrial)	40–60
Small Hydro	50–90
Demand Management	
Low-cost measures	25
Mid-cost measures	40
High-cost measures	70

Sources: Ontario Hydro (1990); Acres International (1987); Diener (1988); Marbek (1987); Diener and Dupont (1984).

as the basis of estimating case study data shown above. The technical and economic potential calculated for each case study is then multiplied by the number of sites in each category to determine the total technical and economic potential for each category and finally, for each technology.

Although a truism, the notion of "garbage in, garbage out" cannot be over-emphasized in conducting supply price analysis. Advocates of particular technologies and mixes of power generation are adept at making assumptions and estimating input data in a way that enhances the cost-competitiveness of one or more preferred options. Understated capital and operating costs, ignored indirect costs, high capacity factors, low discount rates for capital intensive options, overstated economic lives, and as discussed later, selective accounting for environmental impacts, are some of the more common reflections of biased and self-interested analysis. To protect the public interest against such bias, those entrusted with energy policy should elicit third-party reviews of plans and analyses put forward by proponents, whether they be utilities, independent producers, advocates of demand management, environmental groups or any of the numerous "anti-something" groups. In Ontario, examples of such reviews include the recent inquiries on the costs of thermal and nuclear power and the current Environmental Assessment Board hearings on Ontario Hydro's 25-year expansion plan. These hearings, a particularly broad and

414

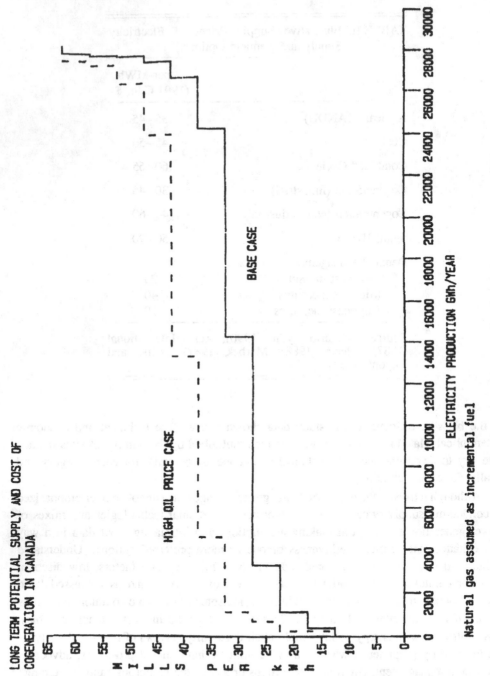

Fig. 1. Long-term potential supply and cost of cogeneration in Canada.

ambitious undertaking, cover all (not only the environmental) aspects of Hydro's plans and involve over 40 intervenors funded by Hydro to the extent of over \$20 million.

SUPPLY PRICE ANALYSIS AND FINANCIAL INCENTIVE PROGRAMS

The issue of financial incentive programs revolves around the appropriate level of financial assistance for particular demand management and energy supply alternatives. Note that the economic potential of an option may or may not exceed its financial potential, where financial potential is defined by the total capacity (power) or electric energy that meets the investment decision-maker's investment criteria (such as payback period or discounted cash flow rates of return), using inflation-inclusive cash flows and including taxes and subsidies. A useful policy compass may be developed if each demand and supply option is classified into one of four categories: (a) both economically and financially feasible options; (b) neither economically nor financially feasible options; (c) economically feasible but financially infeasible options; and (d) financially feasible but economically infeasible options. An illustrative categorization of selected demand management and alternative energy sources is shown in Fig. 2. Support in the form of financial incentives should go only to those options that show economic but not financial feasibility. Conversely, support should be withdrawn from those that exhibit only financial and not economic feasibility. More precisely, the level of financial assistance for any economic option should be set at an amount equal to that required to achieve financial feasibility.

SUPPLY PRICE ANALYSIS AND ENVIRONMENTAL IMPACTS

As the present conference demonstrates, the topic of environmental impact is vast and complex — one that this note can only address briefly and selectively. But that may be sufficient to make two points. First, no source of electric power, whether demand or supply based, is exempt from an array of environmental impacts, most though not all of which represent negative externalities. Second, any attempt to monetize these impacts through debits and credits applied to supply prices initially based on internal costs will of necessity be selective in terms of what impact(s) are considered, and generally subjective and arbitrary in terms of the valuation attached to the selected impacts. Let me underline that this is not a call to ignore environmental issues in making technology and investment choices in the power sector. It is rather a plea for expanded effort toward a more comprehensive enumeration and description of impacts without the false precision reflected in the monetization of selected impacts within the calculus of per kilowatt-hour supply prices. Utilities and their bureaucratic and political masters should then make the technology choices based on explicit and largely qualitative tradeoffs both among diverse environmental impacts attached to options with roughly equal cost, and between "high cost, low impact" and "low cost, high impact" technologies.

ECONOMIC FEASIBILITY

		No	Yes
COMMERCIAL FEASIBILITY	**Yes**	• Off-oil conversion to conventional natural gas furnace • Conventional electric heating in new homes • Propane for fleet vehicles • CNG for fleet vehicles	• Conventional gas furnace in new homes • Condensing gas furnace in new homes • All-electric heat pump in new homes • Industrial waste heat recovery to save oil or gas • Biomass for process heat, using 100% (MSW) to save oil or gas • Diesel fuel for fleet vehicles • Small hydro projects
	No	• Off-oil conversion to all-electric heat pump • Off-oil conversion to electric plenum heater • Off-oil conversion to solar hot-water system • Off-oil conversion to central wood furnace • Solar hot-water system in new homes • Central wood furnace in new homes • Biomass for process heat, using 50% MSW and 50% wood waste to save fuel oil • Biomass for process heat using 100% wood waste to save oil or gas • Methanol and methanol blends for fleet vehicles • Diesel, propane, CNG, methanol, and methanol blends in private cars	• Off-oil conversion to condensing gas furnace • Industrial cogeneration of electricity • Biomass for process heat, using 50% MSW and 50% off-site wood wastes to save natural gas

Fig. 2. Classification of energy technologies according to economic and commercial feasibility — Ontario.

These tradeoffs are especially important when a particular environmental impact is paramount concern, as is surely the case with emissions of greenhouse gases. In is situation, appropriate policy measures are required to encourage, through carbon xes and emission limits, a switch from more to less blameworthy technologies, that from coal-fired plants to bulk supply options such as: (i) nuclear power and, to a sser extent, hydraulic and gas-fired thermal power; (ii) independent power options such small hydro, wind power and cogeneration; and a host of (iii) demand management tions. Here too, knowledge of comparative supply prices and supply curves is valuable. e information helps in the selection of options that are not only benign in terms of eenhouse gas emissions, but also cost-competitive in terms of supply price and hence ngruent with goals of efficient resource allocation. The outcome of this selection ocess is unlikely to be the same for all utilities, as it depends on the region- and utility-ecific availability of technologies and their comparative supply curves. It is more likely at the selection process will point to a mix of supply and demand options, with no single hnology serving as a panacea for each and every economic and environmental concern. e challenge before us is to determine as objectively as possible the make-up of this ptimal" mix of electricity supply and demand technologies. The interaction of ideas ong the scientists, engineers, economists, and policy-makers present at the Los Alamos nference is an important step in meeting this challenge.

Part Three:

Climate-Change Policy and Decision Making

The Human Dimensions of Climate Change

Charles N. Herrick

NOAA, Office of the Chief Scientist

Washington, DC 20235

ABSTRACT

The social and environmental impacts of climate change are potentially enormous. Research is needed to reduce uncertainties and guide the development of rational response strategies. Global climate research activities need to include a strong component in the social sciences. Broadly, social science provides insight into the values and preferences that shape the "Human Dimension." Social science tools and methods can be used to integrate climate-change assessments, assuring that research activities are driven by social relevance rather than disciplinary paradigms. Recent and historical examples are discussed to illustrate some of the human dynamics associated with climate change.

CLIMATE CHANGE: WHAT IS THE PROBLEM?

Under the "conventional" understanding of scientific procedure, the definition of a problem is approached from an "objectivist" perspective, from the perspective of a disinterested researcher able to observe independent realities. The researcher is held to be directly aware of something given, something which has not been affected by human judgments. Under such a model, "our task is to be adequate to the description of objects [and situations] that exist independently of our activities; we may fail and we may succeed, but whatever we do the objects . . . will . . . be what they were before we approached them." (Fish, 1980)

On the other hand, one could adopt a more sociological perspective on the issue of problem definition. Problem definition is a process in which we *denote* the things on which we will focus and frame the context in which we will study them. For example, if we define something as a "rural poverty" problem, a whole array of concerns, players, analysis methods, and agencies orient themselves accordingly. If we define it some other way — as a carrying capacity issue, as an economic distribution issue, or as an agricultural or public welfare issue — then the arrangement of players and analysis methods would alter. To be blunt, the world does not denote and frame itself, and data do not on their own indicate that for which they serve as evidence. Problem denotation is imposed by humans engaged in particular activities. As Donald Schon writes, "problems do not present themselves as givens . . . When we set the problem, we select what we will treat as the 'things' of the situation, we set the boundaries of our attention to it, and we impose upon it a coherence which allows us to say what is wrong and in what direction the situation needs to be [analyzed]." (Schon, 1983)

This is not to deny the existence of an independent world, "out there," but to argue that human comprehension of that world is mediated through scientific conceptualization. As Helen Longino summarizes, "Once we have decided on a system for measuring movement, the speed of an object is not arbitrary. The sorts of things we measure, however, will change as our needs, interests, and understandings change." (Longino, 1990)

Consider an example. In 1980 Congress established the National Acid Precipitation Assessment Program (NAPAP) to investigate the causes and effects of acidic deposition and to recommend strategies for the control and mitigation of adverse effects. NAPAP's responsibility was to provide credible, well-reviewed technical findings and science-based recommendations to inform the public decision process. Lasting ten years and costing nearly $600 million, NAPAP included scientific, technological and economic studies focused on the following technical areas: emissions of acid precursor pollutants, atmospheric chemistry, atmospheric transport and dispersion, and atmospheric transformation dynamics, acid deposition regimes and air quality monitoring, and effects on surface waters, aquatic life, forests, agricultural crops, building materials and cultural artifacts, human health, and visibility. The principal objective of the Program was to produce an unbiased scientific understanding upon which to base policy.

The example of NAPAP is enlightening. Congress established NAPAP to reduce scientific uncertainty surrounding the issue of acid rain; specifically, Congress wanted to know if acid rain was a problem. The "problem" with such a charge is that acid rain — like global change — is not a tidy, uni-dimensional phenomenon; rather, acid rain can be validly characterized from numerous perspectives: chemical, meteorological, biological, ecological, social, and others. Within each of these perspectives, further specialization is not only possible, but from a research point of view, absolutely necessary. The "problems" of any of these sub-disciplines are not necessarily relevant to other substantive areas. They involve different spatial and temporal scales and levels of aggregation, different chemical species, various degrees of analytical detail due to different levels of expertise and/or funding, different assumptions concerning the relevant time-frame over which effects research should focus, broadly divergent modeling approaches and statistical tests, confusion regarding the relative importance of direct, indirect, or interactive effects, and so on.

From the perspective of pure research and technological development, NAPAP was fully successful. The Program produced an enormous amount of peer-reviewed data, findings and projections. However, NAPAP lacked an overarching, extra-disciplinary framework that would allow it to characterize acid rain as a problem, non-problem, or something in between. As we mentioned, problems depend upon perspective and perspective is a matter of disciplinary approach and/or public value. There is no such thing as a generic problem. Lacking a framework, NAPAP was all too often forced to (A) report findings in excruciating disciplinary detail, sparing no caveats; or (B) retreat into uninterpreted statistics. Neither approach is especially helpful to non-specialist decision makers.

It is thus a mistake to suppose that "good science" or "more science" can provide the *answers* for science-based policy disputes. A high-quality research program such as NAPAP can produce a *"banquet"* of findings, but cannot determine which banquet items must be considered by decision makers and weigh most heavily in the policy choice. Indeed, a wealth of high-quality findings may even accentuate a dispute by providing antagonists with grist for debate. (Ozawa, 1991)

Problem definition, and thus, factual relevance, can only be determined and evaluated from a particular frame of reference. NAPAP lacked such a framework; we defined our universe in terms of disciplinary science because we had no other fulcrum for integration and evaluation. This means that policy antagonists could pick and choose from the scientific banquet, collecting supporting facts or uncertainties, not building consensus.

ASSESSMENT FRAMEWORK: INTEGRATING FACTS AND VALUES

This, of course, begs the question, "What is a framework and why is it central to a talk on the 'Human Dimensions' of global change?" I think the answer to such a question is probably long, obtuse, and situationally variant. I will not try to answer it. I will, however, suggest that an assessment framework involves the formal conjunction of scientific information (data, findings, and uncertainties) and public values. Scientific information is important because it characterizes the physical world, while values put the information in context. The value side of the picture can be illuminated by social science. Social science thus provides the public value information necessary to assure that physical science is socially relevant.

Social science involves the systematic study of the behavioral processes of individuals, groups, and institutions. (NRC, 1991) Consider energy consumption. Energy consumption is crucially related to climate change, and hence, to technological development. But consider the extent to which energy consumption is a social issue: Energy consumption involves *geography* (distances between human settlements, production and consumption locations); *demography* (who lives where, what they do, etc.); *economics* (relative costs of labor and energy as factors of production); *political science* (influence of interest groups over research funding and policy formulation, regional and international political alignments); *sociology, anthropology, and social psychology* (information on value and preference development, risk perception, etc.).

In what follows, I provide four examples of how social problems involve an amalgam of science and value; how sometimes one can trump the other, and why the traditional bifurcation of fact and value leads not to objectivity but to a possible loss of relevance.

I. The Dust Bowl — Most Americans are familiar with the Dust Bowl, the Great Plains drought of the 1930s. Fewer are aware that the Great Plains were subject to another significant drought during the 1950s. Interestingly, the two drought events were nearly comparable in terms of meteorological and hydrological severity. However, the societal disruption and hardship were far more pronounced in the 1930s than the 1950s.

(Warrick, 1977) The physical effects of the two droughts were virtually the same, yet societal consequences were notably different. Why?

Between the 1930s and the 1950s a number of social adjustments were implemented, affecting societal response dynamics:

- emergence of an extensive structure of federal institutions for drought relief and rehabilitation,
- increase in the scope of agricultural practices aimed at soil and water conservation,
- increase in the use of financial and grain reserve systems, and
- large increase in irrigation.

This example illustrates that the transformation of things like temperature and humidity changes into "social and political changes is not necessarily direct. . . . the vulnerability of a country to climate change depends no less on its social, institutional, and technological structure then on the particular character of the change in question." (Meyer-Abich, 1980: 69)

Implication I. *The relationship between climate and socio-economic activity may be significantly different for different types of societies or economic sectors. Climate change may lead immediately to pronounced social adjustment or, at the other extreme, propagate slowly through the socio-economic system necessitating only limited adjustments.*

II. Perception Determines Consequence — In December, 1990, a predicted earthquake prompted residents of a small town in southeast Missouri to pack and secure valuable belongings, shore up flimsy structures, close schools, and generally prepare for a major disaster. The earthquake did not happen. Many scientists felt the earthquake prediction to be less than credible and, indeed, irresponsible. Nevertheless, "science" caused a significant social reaction.

Consider a second example: Forecasts of rainfall, temperature extremes, and snowpack can be useful for estimating seasonal water supply. This is especially true in areas of high variability in precipitation. In 1977, the Bureau of Reclamation issued a forecast for the Yakima Valley of Washington for less than half the average of total water supply.

The forecast prompted some farmers to invest heavily in groundwater wells and irrigation systems, others decided not to plant that season. In all, millions of dollars were either lost or invested in drought remediation measures. (Glantz, 1986)

As it turned out, the drought failed to materialize. The farmers argued that they had invested needlessly, based on information from an incorrect forecast. Forecasts of global climate change could have similar action-inducing consequences. It is a proposition of sociology that "If men define situations as real, they are real in their consequences." An implication of this proposition is that individuals and social groups can react to global climate change regardless of whether a significant change actually occurs. (Wimberley, 1991)

Implication II. *There may be significant human response to climate change even if no change occurs, raising questions of economic opportunity cost, societal risk perception and aversion, and the proper role of science-for-policy.*

III. Rain Follows the Plow — During the first half of the 19th century, the popular conception of the American West was one of vast wastelands capable of supporting only beasts and "savage" Indians. It was widely believed that the West could not support civilized life, which, for the most part, meant that Western lands were not amenable to the agricultural methods practiced in the East.

By the 1850s things had begun to change. Dry land farming techniques had been introduced, small-scale irrigation developed, and railroads were beginning to make access less problematic. Developments such as these provided the substance for a promotional campaign that depicted the West as a bountiful garden, as the incarnation of Jefferson's agrarian democracy. Further fueling the admonition to "go west" was an alliance between science and the promotion of Western commerce. Based largely on anecdotal evidence, it had long been suspected that settlement acted to change climate. After the war this association acquired the legitimacy of scientific documentation. The Government sponsored Geological and Geographical Survey of the Territories had reported increasing precipitation for several years in succession, resulting in the broad acceptance of a theory known as "Rain Follows the Plough." The reasons cited for this coincidence of increased precipitation and settlement were sketchy, but by the late 1870s it was conventional wisdom that Western rainfall was sufficient for agricultural development and expansion.

About the only disparagement expressed against this view was registered by John Wesley Powell, the second Director of the United States Geological Survey. Based on findings from a comprehensive survey, Powell argued that the West was not capable of supporting increased agricultural settlement, that the association between settlement and precipitation change was spurious, and that settlement practices needed to be reformed in light of documented climatological deficiencies. In sum, Powell argued that Western settlement needed to be carefully weighed against the carrying capacity of land and climate.

The implications of Powell's "deficiency argument" were far-reaching. Few Western institutions were left untouched. Most importantly, the prevailing system of gridiron property and jurisdictional boundaries needed to be superseded by "hydrographic basins" or watershed units in which settlement and water allocation would be determined by natural drainage contours. Farms would be clustered around water sources to maximize resource use, and the size of claims would be determined by the fertility of the soil and the amount of available water. In other words, Western land use would be controlled by a scientifically-based centralized planning system that was fundamentally at odds with individual initiative, private property, riparian law, gridiron townships, and the political sovereignty of states. Powell's approach was ultimately defeated.

Implication III. *Powell's scientifically credible approach was inconsistent with the society and culture of the West. Powell made the mistake of thinking that physical facts would be harder, more compelling than social facts. The opposite is often true.*

IV. Contextuality of Scientific Information — It is plausible to suggest that global change will be severe, extensive in its consequences, and very damaging and disruptive of social life. Mitigation and/or adaptation approaches could require significant changes in manufacturing processes, extreme regulation of transportation, centralized control over utilities, unprecedented restrictions on land and property use, and the implementation of controls over many aspects of home and family management. How well-suited are current governmental structures to conduct such broad-ranging programs of climate-change management? In particular, could a de-centralized, federalist, political-economic system adapt to such an increase in responsibilities, or in adaption would it change into something far less republican and more authoritarian? I pose this not as a threat, but rather as a legitimate research question.

In a classic study, Karl A. Wittfogel argues that deficiency societies tend to have much more authoritarian/totalitarian regimes than societies of relative abundance. Wittfogel was the architect of the controversial "hydraulic society" thesis. Put simply, there is an inverse relationship between the extent of water availability and the centralization of political power. Adequate precipitation and/or small-scale irrigation are consistent with individual or feudalistic initiative and a decentralized government structure. However, where the scale of water management escalates, when more and larger dams and water distribution networks are constructed, political power comes to be concentrated in the hands of an elite, usually a ruling class, of bureaucrats. As the environmental historian, Donald Worster, describes this

> the chief characteristic [of the hydrologic society] was that it interfered on a massive scale with the flow of the watershed, forcing water miles and miles out of the path of least resistance, running ever more complex risks of environmental degradation, requiring as a result of that danger a constant, intense vigilance. Reorganizing the fundamentals of nature in such a way demanded in turn the consolidation of the loose mosaic of villages into a broader, more powerful instrumentality . . . [resulting in] a state . . . with a bureaucratic organization to design and administer the water system.

World political systems range from industrialized democracies to single-party authoritarian regimes. Political-economic systems can be centralized or de-centralized, bureaucratic or charismatic, stable or revolutionary, pluralistic or elitist, laissez-faire or defacto socialist. These varying regimes are likely to differ profoundly in two key areas: 1) their acceptance and ability to use scientific information, and 2) how they are equipped to deal with a range of response strategies.

Implication IV. *The value and applicability of scientific information varies with socio-political context. The comparative analysis of analogous political/administrative situations is necessary to ascertain the strategic, tactical, and administrative possibilities for achieving various policy goals with respect to climate change. (Mann, 1983)*

CONCLUSION: NEED FOR PARALLEL INTEGRATION

Examples I–IV illustrate the partial understanding that could be generated by a global change research and assessment process that fails to recognize the "Human Dimension," or to be more specific, by a research and assessment program that fails to *incorporate* social science. In other words, science and technology tell only part of the story, or maybe they provide the elements of the story, but not the coherence. It is up to the social and economic sciences to provide the "glue" that binds the banquet of findings into a rational response strategy.

Let's make an heroic assumption; let's say that science "solves" all the interesting global change questions within the next five to ten years. We'll assume that uncertainties surrounding ocean/atmosphere/cryosphere interactions have been satisfactorily resolved; that the role of clouds has been incorporated in general circulation models, which, in turn, have been validated to the fullest possible extent. Pick your favorite issue or uncertainty and imagine it resolved or gone. Assume that we know when global change will occur, at what rate, and how it will be manifested on sub-grid scales. What do we do with this information? (Mahlman, 1991)

As we have already discussed, changes in the physical climate may not impose one state-of-affairs over another. From the perspective of particular societal activities, a change in climate might impose no significant change at all. At the other extreme, profound societal changes may come about from the mere belief that climate will change, even if such a change never materializes.

Here, then, is the problem: Science provides information on *change*, but change, considered alone, does not entail a particular social problem. The *translation* of physical change into something known as "social risk" cannot be accomplished without the integration of the value component. As Sheila Jasanoff writes, in normal science, "Fact-gathering . . . is directed toward areas that [disciplinary paradigms] predict are important or toward elucidating ambiguities in the paradigm. In policy-relevant science, however, the questions scientists ask (or fail to ask) are not guided by a scientific paradigm alone, but by more instrumental considerations arising from the policy process." (Jasanoff, 1986)

Science is the process of obtaining new or refined information through research. Research is an ordered, rule-governed process of inquiry aimed at understanding physical, chemical, biological and social phenomena. Applied research is inquiry aimed at increased understanding for specific societal goods and activities. Technology is the process whereby the results of applied research are utilized to construct or improve tools. Tools facilitate human activities (ORB, 1991).

Assessment is a process in which research, applied research, and technological information are marshalled to project the consequences of alternative courses of action. Alternative courses of action indicate different policies, or bundles of public value. Public value, or the human dimension, thus provides the lodestar of assessment relevance. A social research agenda needs to be integrated with the physical research agenda. This integration needs to be parallel rather than sequential. To be meaningful, social research cannot work from the "finished product" of physical science; social and physical science need to work in parallel to "define" a common problem.

REFERENCES

Fish, Stanley. 1980. *Is There a Text in This Class?* Cambridge, Massachusetts: Harvard University Press.

Glantz, Michael H. 1986. Politics, forecasts, and forecasting: Forecasts are the answer, but what was the question? In: R. Krasnow (ed.) *Policy Aspects of Climate Forecasting.* Washington, DC: Resources for the Future.

Jasanoff, Sheila. 1986. *Risk Management and Political Culture.* New York: Russell Sage Foundation.

Longino, Helen E. 1990. *Science as Social Knowledge.* Princeton, New Jersey: Princeton University Press.

Mahlman, Jerry D. 1991. *Understanding Climate Change.* Climate Research Needs Workshop, Joint Climate Project to Address Decision Maker's Uncertainties. Science and Policy Associates, Washington, DC.

Mann, Dean E. 1983. Research on political institutions and their response to the problem of increasing CO_2 in the atmosphere. In: R. S. Chen, E. Boulding and S. H. Schneider (eds.) *Social Science Research and Climate Change: An Interdisciplinary Appraisal.* Dordrecht, Holland: D. Reidel Publishing Co.

Meyer-Abich, Klaus M. 1980. Chalk on the white wall? On the transformation of climatological facts into political facts. In: J. Ausubel and A. K. Biswas (eds.) Climate Constraints and Human Activities. New York: Pergamon Press.

NAPAP. 1991. *The Experience and Legacy of NAPAP: Report of the Oversight Review Board of the National Acid Precipitation Assessment Program.* Washington, DC: NAPAP Office of the Director.

NRC. 1991. *Global Environmental Change: Understanding the Human Dimensions*. National Research Council, Committee on Human Dimensions of Global Change.

Oversight Review Board. 1991. *The Experience and Legacy of NAPAP: Report to the Joint Chairs Council of the Interagency Task Force on Acidic Deposition*. The Oversight Review Board of the National Acid Precipitation Assessment Program. NAPAP Office of the Director: Washington, DC

Ozawa, Connie P. 1991. *Recasting Science: Consensual Procedures in Public Policy Making*. Boulder, Colorado: Westview Press.

Schon, Donald. 1983. *The Reflective Practitioner: How Professionals Think in Action*. New York: Basic Books.

Warrick, Richard A. 1977. Drought in the US Great Plains: Shifting social consequences. In: K. Hewitt (ed.) *Interpretations of Calamity*. London: Allen & Unwin Inc.

Wimberley, Ronald C. 1990. Social Risks and Consequences of Global Climate Change. American Geophysical Union, 1990 Annual Meeting.

Worster, Donald. 1985. *Rivers of Empire*. New York: Pantheon Books.

What the Greenhouse Skeptics Are Saying

William W. Kellogg
Senior Scientist (Retired)
National Center for Atmospheric Research
Boulder, Colorado 80302

ABSTRACT

Both theory and observations now strongly suggest that a serious global warming due to human activities is probably already under way, and the cause must be the greenhouse effect, due mostly to the additional carbon dioxide released into the atmosphere by fossil fuel burning. Nevertheless, an articulate minority of respectable scientists is maintaining that the notion of a global warming is highly exaggerated, and therefore the leaders of governments and industry need not take any immediate action to slow the warming. Scientific uncertainty is usually invoked as the main reason for this skepticism, and other reasons include the observation that a few places on the Earth have actually grown cooler lately, that changes in solar activity may be the real cause of the observed global warming, and that there are hitherto overlooked negative feedback mechanisms in the climate system that could counteract the effects of a greenhouse warming. Although these kinds of arguments are probably wrong or at least highly questionable, they are frequently invoked by those in high places who would prefer not to take any strong action to stem the flood of fossil fuels here and abroad.

INTRODUCTION

The rationale for the conviction that a serious global warming due to the greenhouse effect is probably under way has been stated many times, both at this conference and elsewhere. An authoritative statement about greenhouse warming was published last year by the International Panel on Climate Change (IPCC).[1] It follows that most of those who subscribe to this notion feel prompted to take some kind of action to deal with the situation, action leading to a reduction in emissions of greenhouse gases — or, at the very least, to prepare for the climate change. These are not popular notions, though, since the actions that seem to be called for are, it is argued, generally unpleasant and probably expensive.

All this is greeted by skepticism by those who subscribe to another school of thought, one which argues that any draconian actions to avert the alleged climate change are uncalled for at this time. The arguments set forth are usually along the line that there is too much uncertainty in the picture that scientists are drawing of a future warmer Earth, and that it may be premature to try to do anything about the situation — especially since such action might be disruptive or even harmful.

Recently a number of scientists have provided ammunition for the second school of thought, either purposefully or unknowingly. These skeptics have brought forth arguments that tend to discredit the notion that mankind is warming the Earth — or that at least downplay its seriousness. The media have often given considerable coverage to such pronouncements; and, of course, those policy-makers who are reluctant to take action welcome the pronouncements of these "naysayers." Let's wait and see who turns out to be right, they say.

In a recent paper[2] I have attempted to analyze a number of the statements of the skeptical naysayers, and this will be a summary of the various arguments. Those who desire a more substantial discussion are invited to read my paper.

A LITANY OF THE SKEPTICS

The Uncertainty Principle: Scientists are trained to be skeptical of new ideas, and tend to profess that they are uncertain about the ultimate truth of any theory. The history of science shows that we have been fooled before, just when we believed that we had a theory well established. This sense of uncertainty is particularly prevalent when we are dealing with a system as complex as that which determines our climate, and indeed there are many things about that system which we do not yet understand. We fully expect, for example, that the oceans will provide some surprises in the future. To some extent, then, the skeptics of the greenhouse theory of climate change are justified in exploiting this deep-seated feeling of uncertainty, a feeling shared by all serious scientists. However, is it fair to utterly discredit the greenhouse theory of global warming because scientists admit that there are some gaps in their knowledge?

We Have Not Seen Any Global Warming Yet: This appeal comes in two parts. First, the observed global mean temperature rise in this century of about 0.6°C is claimed by the skeptics to be "statistically insignificant," in view of the large variations in global temperature that have occurred in the past. It could be just a temporary feature of a noisy record, they maintain. However, a simple signal-to-noise analysis shows that the probability of the 90-year rise in temperature being a product of the random fluctuations is considerably less than one percent. The second part of this argument depends on the observation that in fact there has been virtually no observed warming trend in the continental United States, and that the North Atlantic and the North Pacific have actually grown cooler in the last two decades or so. Thus, they say, there must be something wrong with the theory. The answer is not simple, but it must be pointed out that the Lower Forty-Eight states occupy less than 5 percent of the area of the world, so are not representative of the globe; and, as for the northern oceans, we can see that the changes in both atmospheric and oceanic circulations can account in large part for the regional cooling. After all, it would be surprising if the response to the greenhouse effect were a simple and uniform one in all parts of the world.

There Are Negative Feedbacks That Will Counteract the Greenhouse Effect: Climate modelers are well aware of the fact that the complex climate system has many positive (amplifying) and negative (damping) feedback loops, and the current climate models take as many of these into account as computer speed and human ingenuity will allow. So far, however, no convincing negative feedback has been identified that has been overlooked by the modelers and that would justify ignoring the messages of the climate model experiments. It must be acknowledged, however, that the way we deal with clouds in our models is still unsatisfactory, and the oceans are certainly poorly understood. Thus, we may expect the skeptics to advance further suggestions involving clouds as a possible negative feedback mechanism that would slow the greenhouse warming. The oceans, on the other hand, are more often linked to possible positive feedbacks, though this is still a controversial topic.

The Observed Warming Is Due to Changes in the Sun Rather Than the Greenhouse Effect: Could it be that the observed global warming in this Century is due to some other cause than the greenhouse effect? We cannot absolutely rule out that possibility, and a favorite surrogate has been changes in solar activity and output of radiation from the sun, which we now know is a slightly variable star. The observed 0.6°C rise could be accounted for if the so-called "solar constant" had risen by about 0.6 percent. There are two serious problems with this idea, however. First, the changes in sunspots and microwave emissions from the sun, which are our long-time indicators of solar activity, do not seem to be in step with the global surface temperature changes — though there do appear to be important but subtle changes in stratospheric circulations and temperatures that correlate well with solar activity changes. Second, now that we have more than ten years of accurate measurements from satellites of the solar output, free from interference by our atmosphere, we can see that the changes in output over a solar cycle are far too small to account for the observed global change. Thus, it seems that adherents of this view will have to search for another surrogate for the greenhouse effect.

Satellite Observations Show No Global Warming in the Past Ten Years: A NASA team of scientists and engineers has developed and flown a passive microwave radiometer on Nimbus-7 that can monitor the upward radiation by oxygen in the middle troposphere. This was announced as a new and "precise" survey of global temperature. The originally published record is less than ten years long, however, covering most of the decade of the 1980s, and the fact that it showed considerable variations but no clear trend over this short period hardly constitutes a denial of the 90-year trend. Nevertheless, this record was greeted by a few skeptics as further evidence against the reality of a global warming.

The Warming Will Be Good for Us: There is a third school of thought that acknowledges the validity of the greenhouse theory of global warming, and also points to the observed warming trend as real. However, it diverges from the consensus of the scientific community in maintaining that a global warming will generally be beneficial for humanity. Clearly, the message here is resonant with the message of the skeptical naysayers, since in effect it says to the policymakers: Do not attempt to slow the emissions of greenhouse gases — let the warming proceed! This argument, including the policy conclusions, has been most forcefully advanced by the distinguished climatologist, Mikhail Budyko, and some of his colleagues in Leningrad (now called Saint Petersburg again).

Budyko maintains that, while there may be a temporary drying trend in the centers of the temperate continents in summer as predicted by most of the climate model experiments, as the global warming continues and approaches 3 to 4°C there will generally be more rainfall, and both agriculture and natural ecosystems will prosper as never before. The evidence for this contention is the reconstruction of conditions during the late Pliocene Period of 3 to 4 million years ago, when (as claimed by Budyko) it was several degrees warmer than now and there were no major deserts anywhere in the world. "This casts doubt on the expediency of carrying out very expensive actions aimed at retarding or terminating global warming during the nearest decades," he said recently in a joint paper with Y. S. Sedunov.

FINAL COMMENTS

It would be hard to draw any very firm conclusions on the basis of what has been said above, but I believe that a jury listening to the arguments of the skeptical naysayers would conclude that they had not made a very good case against the generally accepted notion that a global warming was in store for the world—one that is probably actually taking place already. Furthermore, this warming must be due to the greenhouse effect, and that in turn is mainly a product of mankind's insatiable use of fossil fuels. That is what I have concluded, and that is the message (though cautiously worded) of the IPCC Report.[1]

No doubt there will be further discussions of all these issues in the years to come, and perhaps a decade from now the evidence will be so obvious that there will be little room for skepticism. Most of my colleagues would agree with that prediction.

However, this Conference has been dealing with the measures that might be taken to reduce greenhouse gas emissions by adopting a more farsighted energy strategy here and abroad (energy conservation and substitutions for fossil fuels), and the main question that keeps being raised is: How long can we wait before getting started? Those who would prefer to put off taking decisive action can continue to derive comfort from the skeptical naysayers and those who believe the warming will be beneficial.

REFERENCES

1. IPCC, Climate Change: The IPCC Scientific Assessment. Rept. prepared for the International Panel on Climate Change by Working Group 1, J. T. Houghton, G. J. Jenkins, and J. J. Ephramus, Eds. (Cambridge University Press., 1990).
2. W. W. Kellogg, *Bul. Amer. Meteorological Soc.* **72**, 499–511 (1991).

Climate Change: Is the Message to Politicians:
Just Wait and See?

Kjell Håkansson, Consulting Director, Studsvik AB
Nykoping, Sweden

ABSTRACT

This paper represents points of view given by me at the summary session of the Los Alamos symposium on "Global Change" as well as opinions expressed in discussions during the symposium and my interpretation of presentations.

It is not my intention to suggest a perfect global strategy but rather to present a way of thinking which takes into consideration that scientific evidence is insufficient at the starting point.

THE SCIENTISTS' MESSAGE

Analyzing global warming is a complicated task. Even if scientists have now in their hands very powerful tools in the form of computers and modeling techniques, the output from these is not better than what the data that are put in represents and the deficiencies are considerable. The models also need further development.

It is a fact, however, that the ongoing increase in emissions of greenhouse gases will result in global warming. How much is not verified and forecasts differ from $0.5-4.5°C$ for the next century. $1.5° \pm 0.5$ is a best guess with present state-of-the-art.

Another fact is the close correlation between the explosive population growth and the growth of greenhouse gas emissions. How long this can go on is questionable. Estimates based on global natural resources indicate a leveling off effect at 10, or maybe as much as 15, billion.

WORST CASE SCENARIO

If temperature should increase as much as $4.5°$, the impact on climate and sea level will be dramatic and may even supersede what human beings can adapt themselves to. The effect might well be concealed or damped for a period of time by the ability of oceans to accumulate CO_2. But when the oceans become saturated, temperature will most likely rise faster.

This should then be considered together with a world population of 10–15 billion.

GLOBAL CHANGE STRATEGIES

Lead Times

There is obviously a time lag between the time of emission and the time when effects can be registered. It means that even if we turned off emissions completely, temperature should continue to rise due to already emitted gases. It also means that continued emissions at an increasing rate before measures are taken will have an even higher impact.

Changing the energy mix will take a long time before it statistically has an influence, and new technologies have lead times of 40 to 75 years before they reach full effect in relation to their potential.

Changes in population growth can, however, happen rapidly in countries where economic, social and religious conditions are changed. Sweden is an example where such changes appeared during two decades.

Alternative Strategies

Even if there are great uncertainties, a growing share of politicians are of the opinion that evidence is sufficient to call for "Action now." This is specifically the case in Europe where earlier international agreements on SO_2 and CFC has opened the way for CO_2.

On the other hand, the uncertainties and the probability of moderate temperature increase as presented by the scientist at the symposium can be taken as a good excuse for doing nothing and "Just wait and see." This strategy or lack of strategy would be specifically adequate if those scientists are right who expect a global cooling and consider higher temperature from greenhouse gases a remedy and means of postponement of this much more serious development.

My opinion is that for the time being we cannot ignore the worst case scenario. The predictions from the scientists shall not be looked upon as reason for doing nothing but rather as an important option to have lead time available for powerful action programs to allow a structural change away from Fossils. Such a structural change may or may not be necessary but we need to have a "Preparedness for the worst case."

DEVELOPING A PREPAREDNESS STRATEGY

Prolonged Lead Time

The more we can postpone a rapid rise in temperature and other damages to air, the greater is the likelihood of having necessary structural changes in place when needed, with the understanding that the lead time really is utilized for adequate preparations.

In many countries we now see national energy strategies emerging with attempts to apply the concept of sustainability from the Brundtland Commission. But they all represent primarily an optimization within existing energy structures. There is considerable potential for reduction of greenhouse gases in energy conservation and complementary use of new technologies in certain narrow niches. It represents mainly a one-time correction of energy intensities and are as such a welcome contribution but they will rapidly be counteracted by rising increases in GDP and the demand from a growing population.

The basic idea national strategies have been built around in the past has gradually changed. After the war, reduced dependence on imported fuels was of great importance, hence the rationale for the development of nuclear power in countries like Sweden and Japan. This motive was highly enhanced during the OPEC crises and reduced dependence on oil got the highest priority. The discovery of huge natural gas deposits and the conversion from coal and oil to gas in Europe primarily has also become an important factor. One could say that even without any outlined policy for greenhouse gases, considerable reductions have taken place. In Sweden, CO_2 emissions have been reduced by 40% since 1970 as a result of the introduction of nuclear power and a consistent energy conservation after 1974.

The deterioration of the Communist system has opened up a new huge potential for energy conservation and improvement of environmental conditions which calls for international efforts and coordination. The result from activities in those countries will be much greater than a corresponding effort in, for instance, Western Europe.

Avoided Suboptimizations

An essential ingredient in a Preparedness Strategy will be to avoid ad hoc measures which will be doubtful or even detrimental to total systems solutions. A spectacular example can be taken from the coal power plants. It is indeed a mature business with an old-fashioned technology, principally the same as a hundred years ago. Fuel efficiency has gradually been increased by marginal technical improvements until lately when the curve has turned downwards again. The reason is that the SO_2 problem has been tackled by adding scrubbers which consume part of the power output. Solving SO_2 has made CO_2 worse!

The rapid growth of energy demand in Developing Countries (DCs) will require investments in new production capacity nationally. I believe that the technology needed has to be developed separately for the DC's based on their conditions which differ in many ways from ours in the industrialized countries. Technological transfer often fails because it is an export business complementary to the home market. A more rational approach is to work with technical support to a growing national industry or be prepared to establish subsidiaries in the new markets when suitable.

Preparing Future Structural Changes

Today the industrialized world can be sorted into divisions with regard to power production. The first division, with the lowest generating costs, uses hydro, nuclear and combinations. Here we find Norway, Canada and Sweden. Second come countries with nuclear and coal, like France and Germany; then countries with nuclear, coal and/or oil/gas, like USA and Japan; and finally those who depend only on fossils—coal and/or oil/gas, like Italy, Portugal, Denmark and most DCs.

For those who already are out of the fossil fuel trap the goal must be not to fall into it again. For those who have a fairly balanced mix it is maybe most clever to concentrate on improvements within that mix until the clouds clear up. Most exposed are, on the one hand, industrialized countries wholly dependent on fossils and the DCs for whom fossils are less capital intensive and thus the only affordable option for growth. These countries in the trap should be given first priority attention with regard to intermediary strategies and international support.

One major polluter is a joint global problem—the car and his driving force the Otto engine, another mature technology with vested interests so strong that they are hard to beat. And the transportation sector continues to increase its contribution to emissions when other sectors reduce.

This discussion leads to the conclusion that every country is unique in its structure and should analyze its own strategic options and that strong international support should be considered to those who are caught in the Fossil Fuel Trap—e.g., DCs, Eastern Block Countries—in order to open up the trap and take the first steps out of it.

Develop Adequate Technologies

It cannot be emphasized enough that the lead time available has to be utilized efficiently for the development of technologies requested for the conversion from non- or low-emitting fuels to different forms of end use energy.

The use of Biomass has to struggle with its inherent weaknesses: the low-energy density and the need for chemical conversion techniques. In order for biomass to supply long-term satisfactory energy, technologies for farming, harvesting and conversion to end-use energy have to find new and drastically improved solutions. Solar energy faces similar generic problems. The technical society should raise the ambition level considerably and scrutinize future options using a tougher set of criteria, ruling out wishful thinking.

Nuclear fission faces, on the other hand, the well-known problems connected to its high energy density. Irrespective of whether light water reactors are safe enough or not, this technology seems to have come to a dead end. Three Mile Island, Chernobyl, and remaining distrust about certain reactor designs (like the graphite-moderated), and the resulting regulatory bureaucracy syndrome in important countries like the USA have ruined public confidence and—as important—the confidence of the financial world. The first-generation nuclear levels off.

But it doesn't necessarily mean that nuclear fission is out of the picture. Why should a design concept primarily aimed at powering a submarine, developed by the military with defense money, and handed over free of charge to U.S. industry be the final word in fission for civil power generation? I am convinced that nuclear fission will be a necessary player in the future energy mix if the worst case scenario comes to fruition, and most likely it will be competitive from different aspects even if global warming will be more modest.

A new generation of nuclear reactors should be based on new engineering concepts which can meet enhanced criteria for safety and fuel efficiency. From a financial point of view, new reactors must promise improved economies of scale (smaller plant modules) and reduced lead times for construction. The fuel should be designed to prohibit use in weapons. This is a real challenge, very ambitious but not impossible. Like fusion, it will require international coordination and—I would say—a new generation of engineers who can look at the task without prejudice and restrictions created by vested interests. The size of the task is such that I doubt it can be carried out by the power equipment industry, specifically as it is badly hurt by the inconsistencies of the current market.

The third technological challenge is to change the transportation function away from its current dependence on fossils. It is not only a matter of substituting for the Otto engine but rather to tackle the overall transport system. Someone looking at our planet from outside must wonder how we use our brains when we allow ourselves to be trapped for a great share of our lifetime in a car running at minimal speed and poisoning the environment with emissions from inefficient combustion. And, we must admit, this is a disgrace for our civilization! And what will it look like when we are three times as many?

Basic Scientific Challenges

One message came across quite clearly at the symposium: The scientists are not in agreement. Many persuading facts of evidence were given that we can expect to enter a period of global cooling in a not too distant future. My reaction to the presentation is that it is necessary to have the global cooling and warming scientists to speak a compatible language and to bring their scientific results into the same frame of reference. Strictly logical the work made by "the global warming group" is of no real value as long as they sidestep the sciences which research the long-term changes of our planet. They have to be brought together in the same team "Climate Change," and the subject is just too serious to be ignored. Unfortunately I don't see that the symposium was successful in this respect. The "global coolers" were in such minority that their message hardly came through.

Is scientific evidence contradictory? Is it possible now to bring it into a conclusive format? Or is a coordinated research program necessary to give future answers? If yes, it should be part of a Preparedness Strategy.

MESSAGE TO THE POLITICIANS, SUGGESTED CONTENTS

1. Well, my most important point is that there is no reason to sit down, lean back and do nothing. Evidence is clear that we emit to the air in an increasing pace and that

it has an effect. There is, a worst case which will have threatening effects. But there is no reason to panic; scientists give us lead time.

2. The lead time must be utilized rationally which means
   - preparing for structural changes globally,
   - developing adequate non-emitting technologies,
   - increasing scientific efforts to verify whether we can expect a global cooling which overrides global warming.

   This will form a Strategy of Preparedness for the Worst Case.

3. We deal with a global problem which calls for international agreements about the strategy and coordination of actions.

   There's a growing awareness among politicians of what needs to be done. The USA, which contributes 25% of carbon to the air, equal to all DCs, cannot neglect its responsibility to participate without losing face.

4. This strategy will require a vast financial support which also will have to be based on international coordination and collaboration.

   Economic theory has to be adopted to a new view: Environment will be treated as part of our resource base. It should be the target for investments and maintenance in the same way as are capital resources. Environment is part national, part regional and part global. This should be considered when a financial structure for the environment is built up.

   It is important that environmental charges to end users will be designed in a way to support national as well as international components in the environmental resource mix. If, however, they will be regarded and used as an integrated part of national taxation, they will probably fail to fulfil their prospective purpose:
   - to give environment a financial tool for obtaining sustainability,
   - to restrict the use of the environment as it is no longer free of charge, and
   - to fund global as well as national investments in the environment.

Sequential Policies for Abating Global Climate Change

James K. Hammitt and Robert J. Lempert
RAND, 1700 Main Street
Santa Monica, CA 90407-2138

ABSTRACT

Rational policy analysis concerning potential global climate change must recognize the tradeoff between (a) delaying abatement measures until the links between policies, greenhouse gas (GHG) fluxes, climate, consequences, and welfare are better understood, and (b) the possible foreclosure of policy options and increased cost of abating climate change as GHG releases and investments in physical capital continue along their current paths. To analyze this tradeoff, we present a two-period model of sequential policy making. In the first period, climate sensitivity and the damages associated with climate change are unknown, and policy makers may choose to invest in generic abatement options — energy efficiency and low-emitting energy technologies. In the second period, uncertainties about climate sensitivity and damages are substantially reduced and policies must be adapted accordingly. We describe the optimal first-stage policies and characterize their dependence on parameters of the model.

INTRODUCTION

The possibility that significant and widespread climate change will be induced by anthropogenic increases in the atmospheric concentration of carbon dioxide, methane, chlorofluorocarbons, and other greenhouse gases (GHGs) creates a public-policy question of unprecedented scope. Climatic change and measures to mitigate and/or adapt to it could affect human societies world-wide for many decades. The appropriate response is currently a matter of international debate, where the central issues include the extent and timing of measures to limit anthropogenic emissions of greenhouse gases (cf. Morrisette and Plantinga, 1990).

In evaluating near-term policies, it is useful to recognize three fundamental characteristics of the issue: scientific uncertainty, anticipated learning, and long response times. These characteristics create a tension between competing policy recommendations: to take precautionary actions now to slow or limit the extent of climate change, or to postpone costly measures until we can more accurately predict their effectiveness, and then choose a response contingent on our improved understanding. Early action has the benefit of allowing a specified cumulative reduction in GHG emissions to be spread over many years, so that technologies with relatively low marginal cost per unit emission reduction may be employed. Postponing action has the benefit of shifting costs to the future, where their magnitudes are discounted and less costly abatement options may become available.

However, postponing action may limit our ability to prevent some degree of climate change.

We explore the appropriate timing of greenhouse-abatement policies using a simple model to simulate the evolution of climate change and abatement costs under specified policies. Our analysis addresses the extent to which the global community should begin actions to prevent or limit climate change, neglecting distributional questions of how abatement actions and costs should be distributed among nations. In the following sections, we describe the basic approach, present the simulation model, and characterize the conditions under which early abatement actions are preferred.

POLICY-ANALYTIC APPROACH

Consider the sequential decision problem illustrated in Fig. 1. In the first period (taken as 10 years), policy makers must select an interim abatement policy characterized by its cost, reduction in GHG emissions relative to a business-as-usual baseline, and effects on the capital stock (that influence the costs of subsequent decisions). This choice must be made despite the currently incomplete understanding of the links among anthropogenic GHG emissions, atmospheric composition, climate change, and effects on natural and human systems.

At the beginning of the second period, uncertainties about climate change and its effects are assumed to be resolved, and policy makers choose the least-cost abatement policy to limit climate change to the "optimal" level, described below. The second-period policy is contingent on the new information about climate change and consequences. It also depends on the first-period choice: the stringency (and costs) of the second-period policy are inversely correlated with the stringency of the first-period policy (conditional on the resolution of scientific uncertainties).

We use the increase in global annual-mean surface temperature above the preindustrial level ΔT as an index of climate change. The impacts of climate change will depend on the geographic and temporal distribution of temperature, precipitation, storms, and other factors. Nonetheless, ΔT is the climatic variable that can be projected with greatest confidence. We assume the net welfare effects of regional climate changes (incorporating the effects of climate change on economic activities, benefits and costs of adaptive measures, ecosystem damages, and political and ethical dimensions of the consequences) are closely correlated with global-mean warming. The unknown function relating net damages to ΔT is likely to be nonlinear and may be discontinuous at points where eco- and social systems exhibit threshold behaviors. For the set of climate trajectories we examine (for which ΔT increases monotonically to some peak value and subsequently declines), we assume that damages are equal for trajectories that achieve the same maximum value of

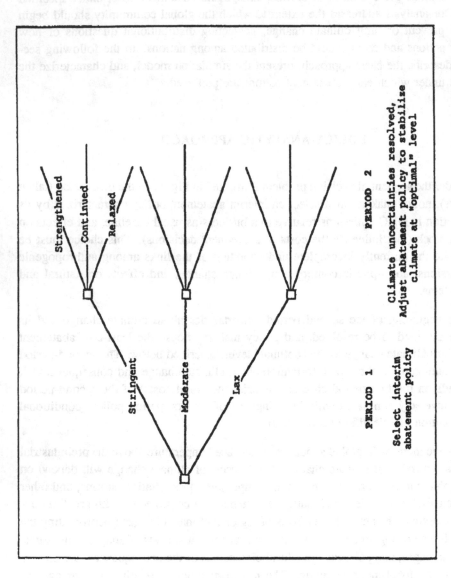

Fig. 1. Decision tree for sequential policy choice. At the beginning of period 1 (1992), an interim abatement policy is selected. At the beginning of period 2 (2002), uncertainties about climate sensitivity and the damages of climate change are substantially resolved and the policy is amended to achieve the optimal degree of climate change. A

ΔT, denoted $\Delta T_{\max}$, and that damages increase with $\Delta T_{\max}$. We compare the abatement costs of such trajectories.*

We aggregate all of the uncertainties about climate change and its consequences into two parameters. The climate sensitivity ΔT_{2x} indexes the magnitude of climate change as a function of GHG concentrations; it is defined as the equilibrium ΔT corresponding to the reference case of doubled CO_2 concentration. The climate target ΔT^* is the "optimal" level of climate change to accept, defined implicitly as the level which equalizes the marginal costs of more stringent abatement and the marginal costs of damages from greater change (after accounting for benefits and costs of efficient adaptation). Although data needed to estimate the relationship between the marginal cost of damages and ΔT are extremely limited, we can estimate the marginal abatement cost of limiting climate change to a specified value of ΔT^* and so infer the perceived marginal damages implicit in choosing any particular target.

The assumption that uncertainties about climate sensitivity and the optimal level of change are substantially resolved in a decade is extreme; however, the costs of the second-period policy can be interpreted as approximating the expected value of beginning the second period of a multiperiod dynamic program, conditional on the first-period policy and assuming optimal sequential decision making thereafter (Hammitt, 1990). We also evaluate the sensitivity of our conclusions to the assumed duration of the period before climate uncertainties are assumed to be resolved.

It is possible to assign prior probability distributions to the uncertain parameters ΔT_{2x} and ΔT^* and calculate the first-period policy that minimizes the expected sum of abatement and damage costs for these distributions, as well as the economic value of improved information about these parameters (Hammitt and Cave, 1991; Manne and Richels, 1991a). For the present, we do not propose prior distributions but explore the sensitivity of our conclusions to plausible values of the climate sensitivity and target.

SIMULATION MODEL

We simulate the evolution of climate and abatement costs as a function of policy using a simple model of the dynamics of anthropogenic GHG emissions, atmospheric concentrations, and ΔT. We describe the atmospheric-concentration and climate components first.

Atmospheric GHG Concentrations

Atmospheric concentration of GHGs at time t is modeled as a function of preceding anthropogenic emissions $F(\tau)$, $\tau < t$. For simplicity, we include CO_2 only. The difference between year t and preindustrial GHG concentrations

---

* A more general measure of total damages might incorporate the rate of climate change and its duration, e.g., damages $= \int_0^\infty D[\Delta T(\tau), d\Delta T(\tau)/dt]d\tau$, where the integrand measures the damage associated with climate change at time τ.

$$\Delta C(t) = \int_0^t F(\tau) \left[\alpha_0 + \sum_{i=1}^4 \alpha_i \, e^{(\tau-t)/\lambda_i} \right] d\tau \; , \tag{1}$$

with weights α_{0-4} = 0.13, 0.20, 0.32, 0.25, 0.10 and time constants λ_{1-4} = 363, 74, 17, 2 years. Equation (1) is Maier-Reimer's and Hasselmann's parameterization to the results of their analysis of CO_2 concentrations using an ocean general circulation model, which includes the primary ocean sinks but excludes the effects of land or ocean biota and any effects of climate change on the carbon cycle (Maier-Reimer and Hasselmann, 1987).

Climate Change

The change in global annual-mean surface temperature $\Delta T(t)$ is simulated using a box-diffusion model assuming thermal equilibrium between the atmosphere and upper ocean

$$C_E \frac{d\Delta T(t)}{dt} = S_C \ln \left[1 + \frac{\Delta C(t)}{C_0} \right] - S_T \, \Delta T(t) \tag{2}$$

(Siegenthaler and Oeschger, 1984; cf. Hoffert et al., 1980; Cess and Goldenberg, 1981; Schneider and Thompson, 1981; Wigley and Schlesinger, 1985). C_E = 12.9 w-yr/m²-°C is the specific heat of the upper ocean layer (Schneider and Thompson, 1981), S_C = 6.4 w/m² is the sensitivity of earth's radiation flux to increases in CO_2 (Siegenthaler and Oeschger, 1984), and C_0 = 280 ppmv is the pre-industrial concentration of CO_2 (Houghton et al., 1990). The climate sensitivity is given by

$$\Delta T_{2x} = \frac{S_C}{S_T} \ln(2) \; . \tag{3}$$

S_T includes the effects of both heat flux from the upper to the deep oceans (primarily through downwelling; Schneider and Thompson, 1981) and the atmospheric temperature sensitivity to radiative flux (governed by clouds and other feedback mechanisms). Reflecting current uncertainty about these factors, we consider three values for S_T = 1.77, 2.96, and 0.99 W/K m², so that Eq. (3) reproduces the IPCC "best guess" and alternative estimates of ΔT_{2x} = 2.5°, 1.5°, and 4.5° C, respectively (Houghton et al., 1990).

Greenhouse-Gas Emissions and Abatement Costs

Our simulation of anthropogenic GHG emissions and the costs of reducing them uses a heuristic model based on the primacy of the energy sector as emission source. Global emissions $F(t)$ are modeled as

$$F(t) = B(t) \, I(t) \, E(t) \; , \tag{4}$$

where $B(t)$ is a baseline emission trajectory, the IPCC business-as-usual case under which CO_2 emissions grow nearly linearly and reach twice their 1990 value by about 2040 (Houghton et al., 1990). The factors $I(t)$ and $E(t)$ represent policy-induced shifts in relative energy use (primary energy consumption per unit economic product) and emissions (GHG emissions per unit energy). In the base case, $I(t) = E(t) = 1$ for all t, and abatement costs (measured relative to this base) are zero.

We consider two stylized technological options for reducing GHG emissions, "energy conservation" and "fuel switching," that differ in cost-effectiveness (the marginal cost per unit emission reduction) and speed of implementation. Energy conservation represents a low-cost, quickly implemented policy; if adopted, energy intensity $I(t)$ falls from 1.0 to 0.7 over 20 years. The decline follows a logistic function which is intended to capture the characteristic pattern of technological diffusion—slow at first, accelerating, then slowing again as the new technology reaches a saturation level (Fisher and Pry, 1970; Hafele, 1981). Consistent with recent claims that substantial efficiency improvements are available at low or even negative marginal cost, we assume this 43 percent efficiency improvement (1/0.7) can be achieved at zero cost relative to the base case (National Academy of Sciences, 1991; Office of Technology Assessment, 1991; Fickett et al., 1990; Mills et al., 1991).

Fuel switching represents a set of high-cost, slowly adopted emission-reduction technologies. Its cost and effectiveness are simulated assuming that all emissions are produced by electric generating stations (representing long-lived emission sources) with 10-year construction periods and 30-year operating lifetimes. Two generating technologies are available: "conventional" (fossil fuel) and "non-emitting" (e.g., nuclear, solar). Incremental costs of the conventional plants are zero; non-emitting-plant costs are allocated uniformly over plant-construction and operating periods and equal $200/ton avoided carbon emissions, consistent with recent estimates, e.g., Manne and Richels (1991b), Nordhaus (1991a). At present, we assume no future reduction in the real cost of reducing GHG emissions that might accompany technological innovation.

Future energy demand is assumed to be known with certainty. Sufficient plants enter construction each year to replace plants that will retire a decade later and to satisfy increased demand. In the base case, only conventional plants are built. The policy decisions are whether to begin substituting construction of non-emitting for conventional plants, and if so, the rate at which to substitute, specified as the transition half-life or the time until half the plants then operating are non-emitting. The transition to non-emitting plants follows a logistic function to reflect the usual pattern of technological diffusion.

Illustrative climate trajectories are presented in Fig. 2. In the base case (no intervention), ΔT increases monotonically, although its rate of increase peaks and then begins to fall as the annual incremental forcing declines (greenhouse forcing is proportional to the log of the nearly linearly increasing atmospheric concentration). If the conservation and fuel-switching options are adopted, the rate of warming falls quickly, and ΔT peaks at some value ΔT_{max} depending on the climate sensitivity and rate of fuel switching. The illustrated rates of fuel switching are characteristic of historical energy-sector

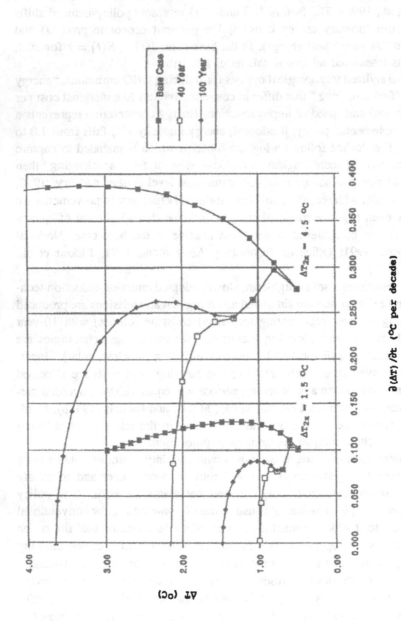

Fig. 2. Illustrative climate trajectories for three emission trajectories: base case, conservation plus fuel switching with 100-year transition half-life, conservation plus fuel switching with 40-year transition half-life, and two climate sensitivities: $\Delta T_{2x} = 4.5°C$ and $\Delta T_{2x} = 1.5°C$.

transitions: the 100-year half-life represents the rate at which first coal then oil technologies were adopted as the Western world industrialized, the 40-year rate is slightly slower than France's rapid (30-year half-life) transition to widespread nuclear electric generation (Ausubel, 1991).

RESULTS

We consider two stylized policies, "precautionary" and "wait-and-see." Under the precautionary policy, both conservation and fuel-switching options are adopted beginning in 1992, the latter with a 40-year transition half-life. Because of the growing base, this policy holds 2010 emissions only to their 1990 level, consistent with the projected effect of a 20 percent reduction by industrialized nations some groups have proposed (Morrisette and Plantinga, 1990). Under the wait-and-see policy, only the conservation option is taken in the first period (with a similar effect on 2010 emissions). In either case, at the beginning of the second period (2002) the values of ΔT_{2x} and ΔT^* are assumed to become known and fuel switching is introduced (in the wait-and-see case) or its rate is adjusted (in the precautionary case) so that ΔT_{max} will equal the revealed target ΔT^* given the revealed sensitivity ΔT_{2x}.

Two cases are illustrated in Fig. 3. In both cases, $\Delta T_{2x} = 2.5°C$, but $\Delta T^* = 1.1°$ or $2°C$. Following the precautionary policy, the fuel-switching rate is increased to 25 years if the climate target equals 1.1, and is slowed to 120 years if $\Delta T^* = 2°C$. Following the wait-and-see policy, fuel-switching is introduced with an 18 or 85 year rate, respectively. By assumption, the environmental damages associated with either trajectory leading to the same climate target are equal, so we compare the annualized abatement costs of those trajectories, i.e., the constant annual payment with the same present value as the simulated abatement costs, calculated over the period 1992–2100 using a 5 percent discount rate. For $\Delta T^* = 1.1°C$, the annualized cost of the precautionary policy is $330 billion, slightly smaller than that of the wait-and-see policy, $340 billion. For $\Delta T^* = 2°C$, the situation is reversed with costs of $55 billion and $35 billion for the precautionary and wait-and-see policies, respectively.

The annualized costs of precautionary and wait-and-see policies as a function of ΔT^* and ΔT_{2x} are illustrated in Fig. 4. For the IPCC "best-guess" climate sensitivity ($\Delta T_{2x} = A2.5°C$), the present value abatement costs are smaller under the precautionary policy if and only if $\Delta T^* < 1.2°C$. For the range of climate sensitivities we consider, the precautionary policy is always less expensive for $\Delta T^* < 0.9°C$ and always more expensive for $\Delta T^* > 1.9°C$. For intermediate climate targets, the less expensive policy depends on the climate sensitivity. As summarized by Fig. 5, the policy that is less expensive is more sensitive to ΔT^* than it is to ΔT_{2x}, over the ranges of these variables we consider. However, for given ΔT^* the abatement costs depend nonlinearly on ΔT_{2x}.

These results are only modestly sensitive to alternative values of the discount rate, the time before climate uncertainties are substantially resolved, and the future cost of non-emitting energy technologies. Using a 1 percent discount rate, the range of climate targets

448

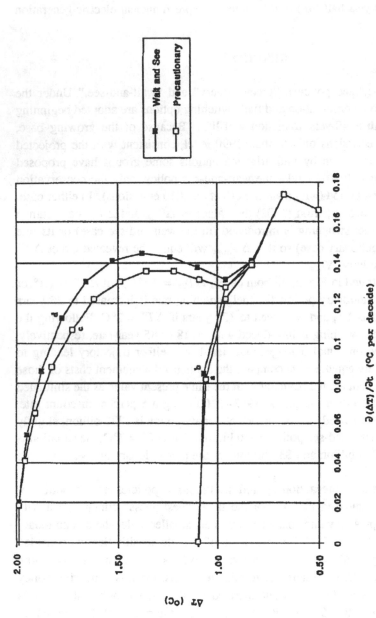

Fig. 3.   Illustrative climate trajectories for recautionary and wait-and-see policies, $\Delta T^* = 1.1°$ or $2°C$, and $\Delta T_{2x} = 2.5°C$.

[a] Precautionary policy and $\Delta T^* = 1.1°C$. Fuel switching adopted in 1992 with 40-year half-life, and amended to 25-year half-life in 2002.

[b] Wait-and-see policy and $\Delta T^* = 1.1°C$. Fuel switching adopted in 2002 with 18-year half-life.

[c] Precautionary policy and $\Delta T^* = 2°C$. Fuel switching adopted in 1992 with 40-year half-life, and amended to 120-year half-life in 2002.

[d] Wait-and-see policy and $\Delta T^* = 2°C$. Fuel switching adopted in 2002 with 85-year half-life.

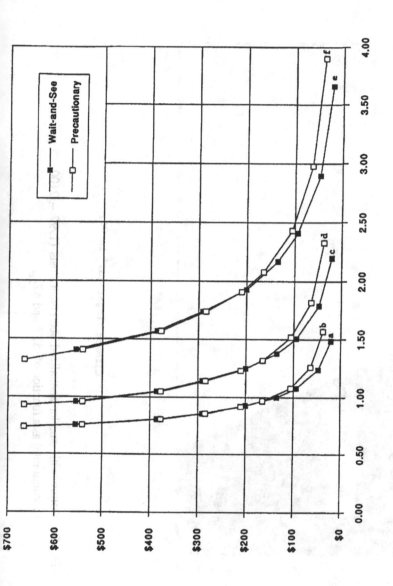

Fig. 4. Annualized costs (1992 – 2100, 5 percent discount rate) of precautionary and wait-and-see policies as a function of ΔT^* and ΔT_{2x}.

[a] Wait-and-see policy and $\Delta T_{2x} = 1.5$°C. [b] Precautionary policy and $\Delta T_{2x} = 1.5$°C.
[c] Wait-and-see policy and $\Delta T_{2x} = 2.5$°C. [d] Precautionary policy and $\Delta T_{2x} = 2.5$°C.
[e] Wait-and-see policy and $\Delta T_{2x} = 4.5$°C. [f] Precautionary policy and $\Delta T_{2x} = 4.5$°C.

450

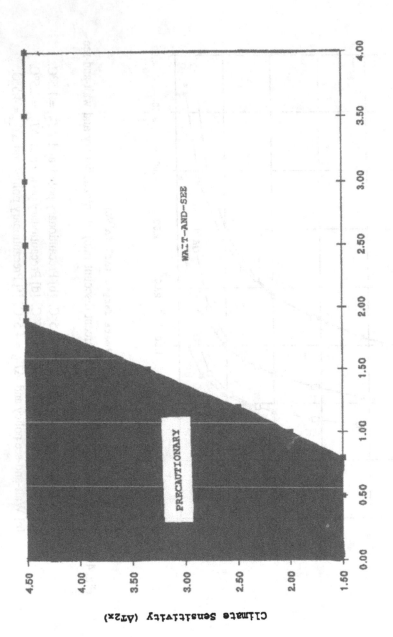

Fig. 5. Policy with smaller annualized abatement costs (1992 – 2100, 5 percent discount rate) as a function of ΔT^* and ΔT_{2x}.

for which the precautionary policy is favored shifts upwards as expected. For the values of ΔT_{2x} we consider, the precautionary policy is always less expensive for $\Delta T^* < 1.2°$ and more expensive for $\Delta T^* > 2.8°$C. Using a 10 percent discount rate, the precautionary policy is always less or more expensive for $\Delta T^* < 0.8°$ or $> 1.6°$C, respectively. Varying the duration of the first period between 5 and 20 years has little effect on the values of the climate target for which the precautionary policy is favored, although increasing the duration magnifies the cost difference between the less and more favored policies for any given ΔT^*. Reducing the incremental cost of non-emitting energy plants by 2 percent per year (to reflect technological innovation) does not significantly affect the values of ΔT^* for which the precautionary policy is favored, although it does reduce the annualized costs of achieving any target by about half, and increases the cost saving associated with the wait-and-see policy for ΔT^* greater than about 2°C.

As shown by Fig. 4, abatement costs are more sensitive to the climate target and climate sensitivity than to the choice between the two policies. Abatement costs are higher for smaller climate targets and for larger climate sensitivities. Because of the long atmospheric residence time of past GHG emissions and constraints on the rate of emission reductions, certain values of ΔT^* (depending on the climate sensitivity) cannot be achieved. For $\Delta T_{2x} = 2.5°$C, for example, this warming commitment is 0.8°C.

The climate target is implicitly defined as the value of $\Delta T_{\max}$ for which the marginal damages (costs of climate change) equal the marginal abatement costs. In our model, the marginal abatement costs strongly depend on the climate target but are not sensitive to the choice between precautionary and wait-and-see policies. For $\Delta T_{2x} = 2.5°$C, annualized marginal abatement costs are about \$2 trillion/degree C at $\Delta T^* = 1°$C, \$60 billion/degree C at 2°C (the slopes of the abatement-cost curves in Fig. 4). Marginal abatement costs vary with climate sensitivity. For $\Delta T_{2x} = 1.5°$C, annualized marginal abatement costs are smaller, \$500 billion and \$40 billion/degree C for $\Delta T^* = 1°$ and 2°C, respectively. For $\Delta T_{2x} = 4.5°$C, the marginal abatement cost is \$300 billion/degree C at $\Delta T^* = 2°$C and is undefined at $\Delta T^* = 1°$C because the warming commitment exceeds that level.

Few authors have ventured comprehensive estimates of the total damages of climate change, and none, to our knowledge, have estimated the marginal damages. Based on its judgement of the perturbation at which substantial adverse effects might occur, an international working group proposed values of 1° or 2°C for ΔT^* (combined with a target maximum rate of change of 0.1°C/decade), although it did not attempt to value the associated damages (Rijsberman and Swart, 1990). For a 3°C warming, Nordhaus (1991b) estimates steady-state damages of about 0.25 percent or more of GNP for the U.S. economy (assuming its current sectoral composition is maintained) and proposes an upper bound estimate for global damages of 2 percent of Gross Global Product. Nordhaus (1991b) and others have noted that the costs to nations that are less industrialized or less climatically diverse than the U.S. may be higher relative to GNP (Parry, 1986; Lave and Vickland, 1989). It is possible to extract crude estimates of marginal damages from Nordhaus' figures (see Appendix for details) which, combined with our estimates of marginal abatement cost, suggest that values of ΔT^* of about 1.5 or 2°C are in the ballpark of an optimal target, but that $\Delta T^* = 1$°C is too stringent.

CONCLUSIONS

By posing the problem of climate-change policy as one of sequential decision making, we attempt to highlight the need to address the appropriate balance between taking preventive action despite substantial uncertainties about the rate, magnitude, and consequences of climate change, and postponing abatement actions in anticipation of better understanding and possible innovation in response options. We also provide a framework for assessing this balance.

Although our quantitative results must necessarily be viewed as illustrative, our analysis suggests some useful insights. First, early abatement action is more appropriate the larger the damages that may accompany a specified level of change, and the more sensitive the climate is to anthropogenic GHG emissions. Our calculations suggest that costly abatement measures (beyond low-cost measures like energy conservation) are not advisable in the near term unless the damages associated with mean global warming are large enough to justify limiting overall warming to about 1.5°C.

Second, the rate at which emissions can be reduced may be more important than the date at which reductions begin; in Fig. 3, for example, a 10-year delay in beginning fuel switching can be overcome by an apparently modest increase in the rate of adoption (e.g., a decrease from 120 to 85 years in the transition half-life for $\Delta T^* = 2$°C). For more stringent targets, the delay cannot be offset as readily; e.g., at $\Delta T^* = 1.1$°C, a ten year delay requires accelerating fuel switching from a 25-year to an 18-year half life. Schlesinger and Xiang (1991) also conclude that a 10-year delay may have little effect on the ensuing climate trajectory. This suggests that substantial attention should be directed toward identifying and evaluating the factors that may slow a transition to emission-reducing technologies, and developing expertise, markets, regulations, and institutions to accelerate the transition.

Third, although improved scientific understanding of both climate sensitivity and damages is important, better information about damages appears more important for making near-term policy choices (Fig. 5). In part, this conclusion reflects the much greater progress that has been made in evaluating the likely magnitude of climate change.

The apparent desirability of postponing costly abatement options for climate targets of about 1.5°C and larger is contingent on the feasibility of rapid global transition to low-emitting technologies. Although the French shift to nuclear power was comparably fast, the relevance of this experience to a global transition is limited. General and country-specific factors that influence the rate of technological diffusion and the effectiveness of policy in speeding transition warrant attention.

Our approach presupposes that climate changes will be set in motion at a rate that is slow, compared with progress in understanding the underlying phenomena and ability to limit future consequences. A policy of postponing stringent abatement until it can be better targeted is susceptible to the possibility that dramatic adverse effects may be set in motion before they can be prevented. A high priority should be assigned to identifying mechanisms that may produce such effects, e.g., out-gassing from the tundra (Melillo et al., 1990), changes in oceanic circulation (Broecker, 1987; Stocker and Wright, 1991) and to detecting incipient changes through climate monitoring and other means.

APPENDIX

The climate target ΔT^* should be set to equate the marginal damages of further change with the marginal costs of more stringent abatement. To compare our estimates of what marginal damages must be to support particular targets with Nordhaus' (1991b) estimates of total damages, we must extrapolate from marginal annualized dollar costs to steady-state damages as a share of Gross Global Product. Such an extrapolation is necessarily crude. It requires translating: (a) from marginal to total damages, (b) from dollars to percent of gross product, (c) from Nordhaus' benchmark $\Delta T = 3°C$ to smaller values, and (d) from annualized costs 1992–2100 to steady-state annual damages at some unspecified future time.

The relationship between the marginal and total damages of a climate trajectory depends on the shape of the function relating damages to ΔT. As described in the text, this function is likely to be nonlinear and may be discontinuous; however, we have no data that would allow us to incorporate these features. Because the relationship is likely to be convex, we use a power function for the annualized total damages of a climate trajectory as a function of ΔT_{max},

$$D = \beta(\Delta T_{max})^\gamma , \qquad (A.1)$$

with $\gamma \geq 1$ chosen to reflect the degree of convexity and β calibrated to estimated total or marginal damages. Differentiating (A.1), solving for β, and substituting the result in (A.1) yields

$$D = MD\ \Delta T_{max}/\gamma , \qquad (A.2)$$

where MD is the annualized marginal damage associated with ΔT_{max}. Assuming $\gamma = 3$ for illustration, the annualized total damages corresponding to marginal damages = marginal abatement costs are about $700 billion for $\Delta T_{max} = 1°C$ and $40 billion for $\Delta T_{max} = 2°C$ (assuming $\Delta T_{2x} = 2.5°C$), respectively about 0.9 and 0.05 percent of projected Gross Global Product in 2040.*

Based on Nordhaus' work, we take 0.25 percent and 2 percent of GGP as "low" and "high" estimates of global damages accompanying a trajectory with $\Delta T_{max} = 3°C$. Using

---

* We assume 2040 GGP of about $75 trillion 1990 dollars, based on 1989 GGP of $20.7 trillion (World Bank, 1991) and the mean growth rates from Hammitt et al. (1987). See Quinn et al. (1986) for a review of long-range GGP projections.

Eq. (A.1) with $\gamma = 3$ to adjust these to $\Delta T_{max} = 1°$ or $2°C$ yields damages as a fraction of GGP of 0.01 percent (low) and 0.07 percent (high) for $\Delta T^* = 1°C$, 0.07 percent (low) and 0.6 percent (high) for $\Delta T^* = 2°C$. Accounting for the delay between abatement costs and damages would decrease the relative value of damages. Comparison with our estimated marginal abatement costs suggests that a value of ΔT^* of about 1.5 or $2°C$ appears to be in the ballpark of an "optimal" target. A target of $1°C$ appears too stringent, since estimated marginal damages there are an order of magnitude or more smaller than estimated marginal abatement costs.

REFERENCES

Ausubel, J. H., "Does Climate Still Matter?" *Nature* **350**, 649–652 (1991).

Broecker, W. S., "Unpleasant Surprises in the Greenhouse?" *Nature* **328**, 123–126 (1987).

Cess, R. D., and S. D. Goldenberg, "The Effect of Ocean Heat Capacity upon Global Warming due to Increasing Atmospheric Carbon Dioxide," *Journal of Geophysical Research* **86** (C1), 498–502 (1981).

Fickett, A. P., C. W. Gellings, and A. B. Lovins, "Efficient Use of Electricity," *Scientific American* **263**, 64–74 (1990).

Fisher, J. C., and R. H. Pry: 1970, *Simple Substitution Model of Technological Change*, General Electric Company, Research and Development Center Report 70-C-215, Schnectady, New York.

Hafele, W. (ed.): 1981, *Energy in a Finite World: A Global Systems Analysis*, Ballinger Publishing Co., Cambridge, MA.

Hammitt, J. K., and J. A. K. Cave: 1991, *Research Planning for Food Safety: A Value-of-Information Approach*, RAND R-3946-ASPE/NCTR, Santa Monica, CA.

Hammitt, J. K.: 1990, *Probability is All We Have: Uncertainties, Delays, and Environmental Policy Making*, Garland Publishing, Inc., New York and London.

Hammitt, J. K., F. Camm, P. S. Connell, W. E. Mooz, K. Wolf, D. J. Wuebbles, and A. Bamezai, "Future Emission Scenarios for Chemicals that May Deplete Stratospheric Ozone," *Nature* **330**, 711–716 (1987).

Hoffert, M. I., A. J. Callegari, C. T. Hsieh, "The Role of Deep Sea Heat Storage in the Secular Response to Climatic Forcing," *Journal of Geophysical Research* **85** (C11), 6667–6679 (1980).

Houghton, J. T., G. J. Jenkins, and J. J. Ephraums,(eds.): 1990, *Climate Change: The IPCC Scientific Assessment,* Cambridge University Press, Cambridge.

Lave, L., and K. H. Vickland, "Adjusting to Greenhouse Effects: The Demise of Traditional Cultures and the Cost to the USA," *Risk Analysis* **9,** 283–291 (1989).

Maier-Reimer, E., and K. Hasselmann, "Transport and Storage of CO_2 in the Ocean— An Inorganic Ocean-Circulation Carbon Cycle Model," *Climate Dynamics* **2,** 63–90 (1987).

Manne, A. S., and R. G. Richels, "Buying Greenhouse Insurance," *Energy Policy* **19,** 543–552 (1991a).

Manne, A. S., and R. G. Richels, "Global CO_2 Emission Reductions—The Impacts of Rising Energy Costs," *Energy Journal* **12,** 87–108 (1991b).

Melillo, J. M., T. V. Callaghan, F. I. Woodward, E. Salati, and S. K. Sinha,: 1990, "Effects on Ecosystems," in *Climate Change: The IPCC Scientific Assessment,* Houghton, J. T., Jenkins, G. J., and Ephraums, J. J. (eds.), Cambridge University Press, Cambridge.

Mills, E., D. Wilson, and T. B. Johansson, "Getting Started: No-Regrets Strategies for Reducing Greenhouse Gas Emissions," *Energy Policy* **19,** 526–542 (1991).

Morrisette, P. M., and A. J. Plantinga,: 1990, *How the CO_2 Issue is Viewed in Different Countries,* Resources for the Future, Washington, D.C.

National Academy of Sciences: 1991, *Policy Implications of Greenhouse Warming,* National Academy Press, Washington, D.C.

Nordhaus, W. D., "The Cost of Slowing Climate Change: A Survey," *Energy Journal* **12,** 37–65 (1991a).

Nordhaus, W. D. "To Slow or Not To Slow: The Economics of the Greenhouse Effect," *Economic Journal* **101,** 920–937 (1991b).

Office of Technology Assessment, U.S. Congress: 1991, *Changing By Degrees: Steps to Reduce Greenhouse Gases,* OTA-O-482, Washington, D.C.

456

Parry, M. L.: 1986, "Some Implications of Climate Change for Human Development," in *Sustainable Development of the Biosphere*, Clark, W. C. and Munn, R. E. (ed.), Cambridge University Press, Cambridge.

Quinn, T. H., K. Wolf, W. E. Mooz, J. K. Hammitt, T. W. Chesnutt, and S. Sarma,: 1986, *Projected Use, Emissions, and Banks of Potential Ozone-Depleting Substances*, RAND N-2282-EPA, Santa Monica, CA.

Rijsberman, F. R., and R. J. Swart,: 1990, *Targets and Indicators of Climatic Change*, Stockholm Environment Institute, Stockholm.

Schlesinger, M. E., and X. Jiang, "Revised Projection of Future Greenhouse Warming," *Nature* **350**, 219–221 (1991).

Schneider, S. H., and S. L. Thompson, "Atmospheric CO_2 and Climate: Importance of the Transient Response," *Journal of Geophysical Research* **86** (C4), 3135–3147 (1981).

Siegenthaler, U., and H. Oeschger, "Transient Temperature Changes due to Increasing CO_2 Using Simple Models," *Annals of Glaciology* **5**, 153–159 (1984).

Stocker, T. F., and D. G. Wright, "Rapid Transitions of the Ocean's Deep Circulation Induced by Changes in Surface Water Fluxes," *Nature* **351**, 729–732 (1991).

Wigley, T. M. L., and M. E. Schlesinger, "Analytical Solution for the Effect of Increasing CO_2 on Global Mean Temperature," *Nature* **315**, 619–652 (1985).

World Bank: 1991, *World Development Report 1991*, Oxford University Press, Oxford.

The Greenhouse Effect: Political Decision Making and the Application of Upwelling/Diffusion Models

Peter Laut

Engineering Academy of Denmark, DK-2800 Lyngby, Denmark

Jesper Gundermann

Danish Energy Agency, Landemærket 11, DK-1119 Copenhagen, Denmark

ABSTRACT

This presentation is an attempt to build a bridge from science to political decision making. In the worldwide process of preparing and following up the United Nations Conference on Environment and Development (UNCED), which will be held in Brazil in June 1992, it is likely that the urge to find a convincing basis for decision making in order to counter climatic change will become ever more acute. Even though most governments may agree that something should be done, many disagree over the extent, the pace, and the acceptable costs of action. Is it, for example, sufficient to follow the recommendation of the 1988 conference in Toronto on "The Changing Atmosphere" and reduce global CO_2 emissions by 20% until year 2005, or is it vital for the future of the World to reduce emissions at a much quicker pace? And how do we compare reductions of different greenhouse gases by different amounts, implemented over different time spans?

The present paper seeks to contribute to the creation of a more convincing and operational basis for decision making. The authors recommend the application of calculated profiles of realized global warming or avoided global warming as a direct and comprehensible way of comparing different options for reducing emissions. Computed results for two selected global and two national scenarios are presented.

INTRODUCTION

The objective in this brief presentation is to focus upon the situation of a political decision maker who has been convinced by the scientific community that action should be taken in order to counteract global warming, and who now is searching for a strategy to reduce energy consumption and the emission of greenhouse gases that he can present to the public with convincing arguments.

Where can he turn foroptions? He could, for example, have a close look at the recommendation of the Brundtland Report,[1] which operates with a 50% reduction of energy consumption per capita in industrialized countries, allowing for a 30% increase in developing countries. He can also consider basing his energy policy upon the conference statement from the international conference: "The Changing Atmosphere: Implications for Global Security" in Toronto, 1988. Here the recommendations included a reduction of global CO_2 emissions by 20% before year 2005, which, of course, would require much more radical reductions in the industrialized countries because of the expected increase in population in developing countries.

He also can try to find inspiration in studying national energy plans, as, for example, the Danish plan "Energy 2000," which has as a policy goal the reduction of Danish emissions by 20% before the year 2005.

In this process he may be surprised to find that none of the above mentioned references contain any attempt whatsoever to assess the climatic consequences of achieving these reductions, even though the main driving force behind all these plans is to limit and slow down future climate change. Thus, no attempts are made to assess the reward for restricting emissions in terms of reduced global warming.

OUR CALCULATIONS

We found this situation unsatisfactory, so we determined to do our own calculations of realized global warming and avoided global warming for a number of proposed emission scenarios. Since the IPCC report[2] in many countries has attained the status of a standard reference, we decided—in this first approach—to apply the same mathematical model as the IPCC for carrying out this task, without trying to improve or update the model. A diagram of the model for CO_2 is shown in Fig. 1.

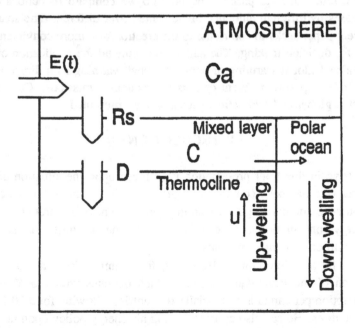

Fig. 1. Upwelling/Diffusion Model for CO_2.

Only a rough outline of the mathematics involved is described here. A detailed documentation of our method was presented at an international symposium on "Energy and Environment" in Finland in August 1991.

On the diagram, the manmade CO_2 emission per year, $E(t)$, is represented by an arrow entering the atmosphere. Part of it remains there [$= C_a \cdot dp/dt$, where C_a is a constant and dp/dt is the increase per year in the partial pressure of CO_2], while the rest is taken up by the oceans ($= f \cdot (p - p_1)/Rs$, where f is the fraction of the earth's surface that is covered by the oceans, p_1 the partial pressure of CO_2 in the mixed layer of the oceans, and R_s the surface resistance of the mixed layer). The emission is corrected for the so-called "missing sink," which is the part of the total CO_2 that is not taken up by the oceans and not, as yet, accounted for. This "missing sink" in the IPCC-model may, for example, turn out to be a significant uptake by the continental biosphere. A part of the CO_2, taken up by the ocean, will remain in the mixed layer ($= C \cdot f \cdot d \cdot dp_1/dt$, where C is the marginal increase in concentration of total carbon per unit of partial pressure and d is the thickness of the mixed layer) and the rest ($= f \cdot \Phi$) will go into the thermocline. This transport is parameterized and treated like a combination of a diffusion process and an upwelling of bottom water ($\Phi = -D \cdot \partial c/\partial z - u \cdot c$, where D is an effective diffusion constant, z the depth below the mixed layer, u the upwelling velocity and c the concentration of total carbon, exceeding the preindustrial equilibrium level, i.e., the perturbation). The polar regions there are assumed to be downwelling areas.

The model is not fully documented, neither in the IPCC report itself, nor in the scientific articles referred to. This is a deficiency that also can be observed in connection with other model descriptions and seems to become increasingly widespread in the scientific literature. So to check the model calculations required quite some effort and cumbersome, indirect analysis. But then we succeeded in establishing the differential equations corresponding to the applied upwelling/diffusion model (see Fig. 2), including the applied numerical values of the parameters, and finding the mathematical solutions to the problem.

We chose to apply the technique of Laplace-transform to the problem, which turned out to reduce the requirements for computing power to such an extent that all scenario calculations could be performed on personal computers. As an example of the type of mathematics involved, Fig. 3 shows the Laplace-transform LR(s) of the pulse response function for CO_2, with s being the complex variable and c denoting the ratio between the surface concentrations of total carbon in downwelling and upwelling regions.

Figure 3 also shows the pulse response function $R(t)$ itself, expressed in terms of its Laplace-transform. Analogous calculations must be performed to determine the pulse response function for heat. Then the CO_2 concentration and the realized global warming for an arbitrary scenario can be calculated simply as folding integrals (Fig. 4). In this context the IPCC Business-as-Usual scenario has been chosen as the reference scenario, and only differences between BaU and other scenarios have been calculated. Therefore the above-mentioned "missing sink," which is assumed to be the same for the different scenarios, drops out of our calculations.

$$E(t) = C_a \cdot \frac{dp}{dt} + \frac{f}{R_s} \cdot \left[p - p_1 \right]$$

$$\frac{f}{R_s} \cdot \left[p - p_1 \right] = C \cdot f \cdot d \cdot \frac{dp_1}{dt} + f \cdot \Phi(0, t)$$

$$\Phi(z, t) = -D \cdot \frac{dc(z, t)}{dz} - u \cdot c(z, t)$$

$$dc(z, t) = C \cdot dp(z, t)_1$$

$$\frac{d}{dt} c(z, t) = -\frac{d}{dz} \Phi(z, t)$$

Fig. 2. Upwelling/Diffusion Model for CO_2, Differential Equations.

$$LR(s) = \cfrac{1}{C_a \cdot s + \cfrac{1}{\cfrac{R_s}{f} + \cfrac{1}{C \cdot f \left[d \cdot s - u \cdot \left[\frac{1}{2} - \pi \right] + \frac{1}{2} \cdot \sqrt{u^2 + 4 \cdot D \cdot s} \right]}}}$$

$$R(t) = \text{residue} \cdot e^{-r_p \cdot t} - \frac{1}{\pi} \cdot \int_0^\infty \text{Im}(LR(-z)) \cdot e^{-z \cdot t} \, dz$$

Fig. 3. Upwelling/Diffusion Model for CO_2, Pulse Response Function.

$$p(t) = \int_0^t R \left[t - \tau \, | \cdot E \left[\tau \right] \, d\tau \right.$$

Fig. 4. Folding Integral of Emissions and Pulse Response Functions.

Figures 5 and 6 show a comparison between our results for future CO_2 concentrations and realized global warming performed for IPCC scenarios B, C and D. The lines show our results, obtained by applying our calculated pulse response functions for CO_2 and heat. The crosses indicate the IPCC values. Hereafter we concluded that our method gave results that were fully compatible with the IPCC scenario calculations.

TWO "BRUNDTLAND" SCENARIOS

Figure 7 shows our results for a kind of "Brundtland" emission scenario: The first 40 years correspond to the above-mentioned 50% reduction of CO_2 emissions per capita in industrialized countries and an increase by 30% in developing countries. Because of population growth, the total emission is likely to increase by 10–15% during that period. After that we extended the emission scenario in two ways: constant emissions until year 2100 gave a "moderate Brundtland" scenario, and an annual reduction by 1% gave a "radical Brundtland" scenario. The upper graph shows the assumed emission scenarios, the middle graph the corresponding CO_2 concentrations, and the lower graph the corresponding realized global warmings. We can read from the curves that following the "radical Brundtland" scenario instead of BaU, the realized global warming in the year 2040 will be reduced from 1.4°C to 1.2°C, which may seem somewhat disheartening. In the year 2100 the reduction will be from 3.2°C to about 2°C. In this context we have assumed proportionality between energy consumption and CO_2 emissions, which is correct if the present fuel mix is retained. The reason for this simplification in the present context is that we did not want to obscure our points by embedding them in complicated assessments of future developments in the field of energy technology.

THE DECISION-MAKER'S DILEMMA

In our opinion these results seem to indicate that the political decision-maker will find himself confronted with a dilemma: In the short and medium term the reward for pursuing aggressive reduction policies will be very small. This may create considerable difficulties for political leaders, who try to convince different countries, different political parties, and the general public to accept economical sacrifices in order to slow down climate change.

At the same time, however, there can be expected an increasing public concern about the long-term consequences of continuing the massive pollution of the global atmosphere with greenhouse gases. Here the argument could be that it is very difficult to imagine that mankind can go on increasing the content of heat-insulating gases in the atmosphere without—sooner or later—provoking a significant global warming. And this concern is not likely to be calmed by references to the uncertainty of present climate models.

462

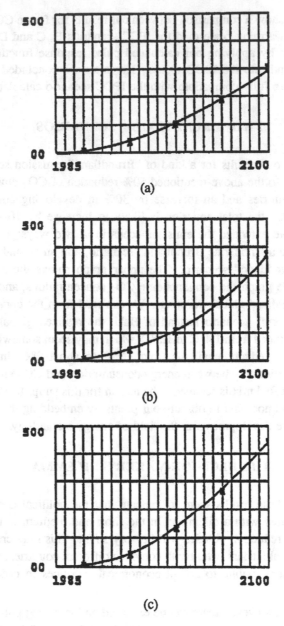

(a)

(b)

(c)

Fig. 5. Comparison: Our Results with IPCC. Avoided CO_2 concentrations for switch from (a) scenario BaU to B; (b) scenario BaU to C; and (c) scenario BaU to D (units, ppmv).

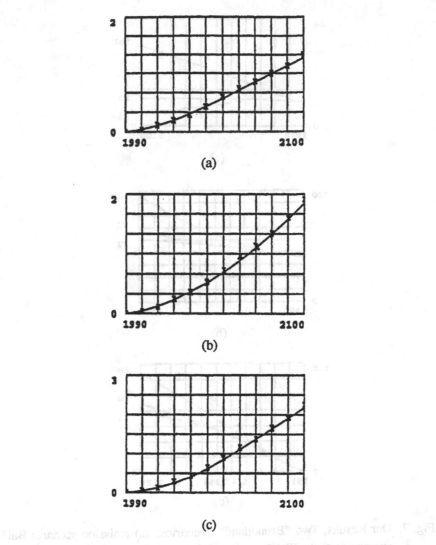

Fig. 6. Comparison: Our Results with IPCC. Avoided global warming for switch from (a) scenario BaU to B; (b) scenario BaU to C; and (c) scenario BaU to D (units, K).

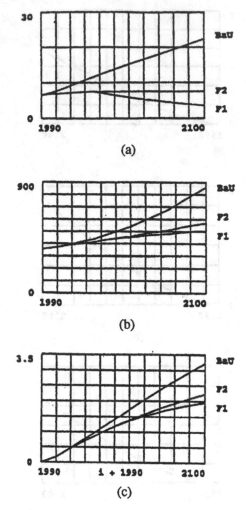

Fig. 7. Our Results, Two "Brundtland" Scenarios: (a) emission scenario BaU average from the IPCC report and two types of "Brundtland" scenarios, a radical (F1) and a more moderate (F2) (unit, Gt C/yr); (b) atmospheric concentrations of CO_2, corresponding to scenarios BaU, F1, and F2 (unit, ppmv); and (c) realized global warming, corresponding to scenarios BaU, F1, and F2 (unit, degrees Celsius).

TWO DANISH EMISSION SCENARIOS

Figure 8 gives another example of our calculations. The upper graph shows two CO_2 emission scenarios. Scenario F0, which assumes our current, stabilized national emissions to be extended to the year 2100, and scenario F5, which reduces our Danish CO_2 emissions by 20% relative to $F0$ until the year 2005 (which is the policy goal of the national energy plan) and thereafter keeps the emissions constant. The middle graph shows the amount of concentration of CO_2, and the lower graph shows the amount of realized global warming, which is avoided by Denmark's implementing these reductions. It should be stressed that this calculation only shows the effect of this single step. It does not, for example, show the impact of the successful stabilization of CO_2 emissions, which Denmark already has achieved. This type of national results should by no means be used in order to show that what a single small country can achieve does not make any difference to the future of the Earth. Rather, the impact in terms of avoided global warming should be scaled up in order to show what can be achieved if all other countries decided to follow the good example. Here one must keep in mind that Denmark has one-thousandth part of the global population and 1/350 of the global energy consumption. But the selection of an appropriate scaling factor should be done with care and will presumably lead to different results for different greenhouse gases.

Figure 9 shows the implications of a purely speculative reduction of present Danish methane emissions by 50% compared with keeping them constant.

CONCLUSIONS

We find that the application of calculated profiles of avoided global warming can be recommended as a direct and comprehensible way of comparing different options for reducing greenhouse gas emissions. So our advice to the political decision maker in search of a convincing policy option is the following: He should ask his scientific advisors to calculate the realized global warming profiles for each of the strategies for emission reduction among which he must choose. These calculations should be based upon the state-of-the-art of climate modeling. After that all further deliberations concerning policy options should be based upon these results.

We recommend this method in spite of the fact, that the absolute values of several important climatological parameters, e.g., the climate sensitivity, are still very uncertain. Therefore the numerical results should never be taken literally, but rather as an indication of the relative merits of different options and as a tool for understanding the complicated interplay between greenhouse gas emissions, concentrations, committed and realized global warming, and the long time delays involved.

However, grave difficulties are likely to arise for the political decision maker, especially because of the long time delay between the investment now of national resources in order to counter climatic change, and the resulting benefits, which may first materialize in the distant future in the form of a somewhat more moderate rise of global temperatures.

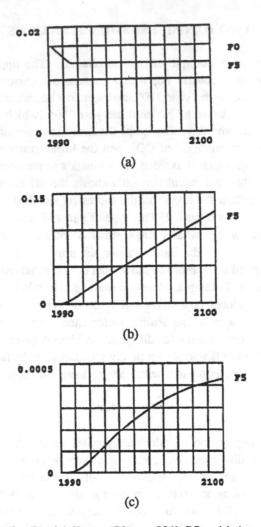

Fig. 8. Our Results, Danish Energy Plan, -20% CO_2. (a) An emission scenario (F5) for Denmark, corresponding to a reduction of CO_2 emissions by 20% until year 2005 and then keeping emissions constant. Reference scenario (F0): CO_2 emissions constant at 1990 level (unit, Gt C/yr). (b) "Avoided Concentration Profile," corresponding to a reduction of Denmark's CO_2 emissions by 20% until year 2005 and then keeping emissions constant. Reference scenario: CO_2 emissions constant at 1990 level (unit, ppmv). (c) "Avoided Global Warming Profile," corresponding to a reduction of Denmark's CO_2 emissions by 20% until year 2005 and then keeping emissions constant (scenario F5). Reference scenario: CO_2 emissions constant at 1990 level (unit, degrees Celsius).

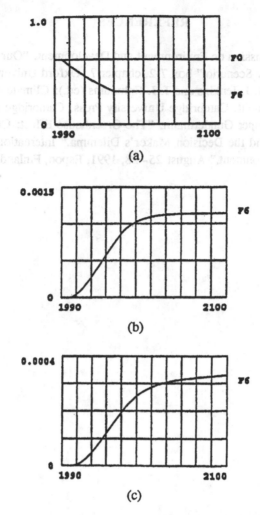

Fig. 9. Our Results, Danish Methane Reduction, 50%. (a) An emission scenario (F6) for Denmark, corresponding to a reduction of anthropogenic methane emissions to 50% of 1990 level from year 1995–2030 and then keeping emissions constant (F6). Reference scenario (F0): CH_4 emissions constant at 1990 level (unit, Mt CH_4/yr). (b) "Avoided Concentration of Methane," corresponding to a reduction of Denmark's anthropogenic methane emissions to 50% of 1990 level from year 1995–2030 and then keeping emissions constant (F6). Reference scenario: CH_4 emissions constant at 1990 level (unit, ppmv). (c) "Avoided Global Warming," corresponding to a reduction of Denmark's anthropogenic methane emissions by 50% until year 2030 and then keeping emissions constant. Includes contributions from CO_2 from methane oxidation. Reference scenario: CH_4 emissions constant at 1990 level (unit, degrees Celsius).

REFERENCES

1. The World Commission on Environment and Development. "Our Common Future." The "Case B-Low Scenario," box 7-2, chapter 7. Oxford University Press (1987).
2. J. T. Houghton, G. J. Jenkins and J. J. Ephraums (ed.): Climate change. The IPCC Scientific Assessment. Cambridge University Press. Cambridge (1990).
3. Peter Laut and Jesper Gundermann: "The Greenhouse Effect: Current Knowledge, Future Options and the Decision Maker's Dilemma." International Symposium on Energy and Environment." August 25–28, 1991, Espoo, Finland.

Energy Policy in the Light of Global Climate-Change Uncertainty

Alan T. Crane
Office of Technology Assessment
U.S. Congress
Washington, DC 20510

Congressional policy-making is an uncertain and unpredictable process. Congress receives information, advice, and influence from a great variety of sources, including the White House, industry, environmental groups, labor unions, constituents, etc. Much legislation that is finally enacted represents a compromise of many interests.

The Office of Technology Assessment was created in part to help rationalize this process. Objective, comprehensive analyses of the major aspects of technology policy issues provide guidance for Congress to understand what their choices are and what the impacts will be. In an ideal world, OTA reports would provide a road map for policy makers, who would then choose the path in the best national interest. In this far from ideal world, accurate, thorough policy analysis is extremely difficult and, even if done well, is only one source of information for policy makers. Nevertheless, policy analysis is needed and used in the decision-making process, though this fact is not always readily apparent from legislation.

Energy policy is particularly uncertain because there is no national consensus on the direction to take and no immediate crisis compelling action. Congress reflects the national diversity of opinions on energy goals. Some want to increase energy supply, in particular by allowing drilling for petroleum in the Alaskan National Wildlife Refuge (ANWR) or by expediting nuclear reactor licensing. Others want to focus on making energy use more efficient by raising standards for automobile fuel mileage (the Corporate Average Fuel Economy [CAFE] standards) and other means. Most consider it a low priority issue as long as energy supplies are adequate and prices low.

Energy policy in response to potential global climate change may represent the epitome of uncertainty. Scientific uncertainty over the potential severity of the problem precludes assurance that the correct responses are chosen. In fact, the White House argues that no significant action is yet justified because of the lack of proof.

Nevertheless, many in Congress feel that action is necessary because the consequences of climate change could be so serious. Many bills have been introduced to deal with one part or another of the issue. Senator Wirth has proposed an omnibus bill that, among other things, would promote energy efficiency and renewable energy sources. This bill may not pass in the near future, but it helps to educate Congress and the American public on the problem. As scientific certainty and congressional support gradually grow, this bill may evolve into national policy.

The Senate Energy And Natural Resource Committee took another approach to energy policy in S-1220, its version of the National Energy Strategy. This bill primarily stressed energy security rather than responding to global warming, but it tried to combine measures to improve supply and efficiency which would have been relevant. However, opposition to several provisions, especially opening ANWR, led to a major setback recently. Since President Bush has announced that he will veto any major energy bill that does not open ANWR, the prospects of any bills getting enacted in the near future are doubtful.

OTA released three reports this year that are being used in the debate. "Changing by Degrees,"[1] "Energy Technology Choices,"[2] and "Improving Automobile Fuel Economy."[3] The first is aimed at policy responses to global warming, particularly energy efficiency, to reduce CO_2 emissions. It has been the subject of a series of congressional hearings by several committees, indicating a keen congressional interest in the subject. The second takes a broader look at energy policy, reviewing options according to their contribution to the basic national goals of environmental well-being (including global warming), economic health, and national security. Many options conflict with one or two of the three goals, compounding the difficulty of achieving consensus considering the variety of value judgments operating in the debate. The third report has been particularly controversial because it concluded that, contrary to industry protestations, technology is available to substantially increase automobile fuel economy.

These reports illustrate the difficulty of policy-making in controversial situations. They are controversial precisely because different interests are involved, none of which can prove that its position is correct or its opponent's is wrong. It is rare for anyone at a congressional hearing or an OTA workshop to tell outright lies, but each side stresses the facts that support its position. OTA's mission is to sort through the arguments and conclude, where possible, which arguments are stronger.

In order to design a coherent, consistent energy strategy to respond to climate change, we need to know where it would fit in the overall strategy to resolve the climate change issue. The activities that are essential to deciding what to do about global climate change can be grouped under four headings:

1. *Understanding Effects*—identify effects as a function of emissions
   - resolve scientific uncertainties
   - improve models
2. *Remedial Actions*—costs of reducing emissions
   - energy role—CO_2, methane
   - other gases
3. *Adaption*—impact of climate change
   - environmental impact of change
   - social costs of living with a changed environment
   - ways to lessen negative environmental and social impact
4. *Policy*—balancing costs and benefits of reducing emissions.

Ideally, atmospheric scientists and modelers could accurately predict the rate of global and regional climate change as a function of CO_2 and other emissions; other researchers could predict the economic and social costs of adapting to these changes; and energy analysts could determine the costs and other impacts of following different emission trajectories. If such accuracy were possible, a set of scenarios could be prepared that would provide complete cost/benefit analyses identifying the costs and problems of controlling emissions at various levels, the resultant rate of climate change and regional effects, and the associated effects on people and the environment. Policy makers then could simply choose policies to implement the most desirable scenario.

Obviously, we are nowhere near such an ideal situation. All three sets of analyses still suffer from major uncertainties. Projections of average global temperature increase for a doubling of greenhouse gas concentration, which could easily occur in the next century, range from $1.5°C$ to $4.5°C$. This factor-of-three temperature uncertainty is compounded by environmental impact uncertainties for any given temperature change, particularly on a regional basis. The differences in resultant environmental impact between the low and the high estimates could be huge; both the rate of change and the magnitude of the total change are important to know. Research into the costs of these changes and adaptation measures is in its infancy. Even though energy has been studied intensively for almost twenty years, there is no consensus over the feasibility of greatly improving efficiency or on the prospects of renewable and nuclear technologies. Therefore we have only rough estimates of the difficulty of reducing emissions of CO_2 and other greenhouse gases.

While some of the most alarming scenarios that have been suggested now seem less probable, it is still quite possible that in fifty years the world would be much better off if CO_2 emissions were significantly reduced from business-as-usual projections. The key questions, given the great uncertainties involved, are how much reduction is warranted — and by when.

Global CO_2 emissions have increased at about 3.3%/year over the past century.[4] Reducing that increase to 1–2% would not be easy, but could be accomplished with the cooperation of all countries that are major producers of CO_2 or are growing rapidly. Stabilizing global emissions at current levels is also possible but would call for significantly more strenuous efforts. Significant reductions, say by 20% of current emissions over the next couple of decades, is technically possible but likely to be extremely difficult. Some analysts think that reductions on the order of 50% may be necessary to avoid severe problems. The greater the reduction required, the sooner action must start and the more stringent it must be.

Whatever reduction goal may be chosen eventually for the world, the United States will have to bear a disproportionate share, if it is to be met, because it is responsible for more CO_2 per capita than most other industrial countries (and about 25% of global emissions), and developing countries can't be expected to limit their growth significantly. OTA's analysis predicted U.S emissions would grow almost 50% by 2015, but could be held to about 15% with moderate, cost-effective, measures. A reduction of about 30% was predicted to be extremely difficult though technically achievable.[5] Even the moderate scenario, which would reduce U.S. emissions from the base case rate of growth of 1.4%

to 0.5%, would require implementation of a variety of measures over the next several years.

If the risk of traumatic change is sufficiently high that the goal should be major reductions in emissions levels, then action should be started essentially immediately. Energy systems, for both supply and consumption, often are built to last for decades. Changing existing facilities and equipment is much more difficult than improving new ones. Fossil power plants are expected to last 20–50 years without major overhaul. Worldwide growth of electricity could easily average 5%, resulting in 62% more generation in 10 years. If the decision to reduce fossil fuel use is deferred for 10 years, the problem for the electric sector will be 62% worse (assuming fossil fuel stays at the same proportion of total generation). Replacing these plants would be so expensive that it is likely they will continue producing CO_2 for much of their expected lifetimes. Similarly, buildings that are constructed over the next ten years with conventional energy-using systems could not be retrofitted to the same standards as new efficient buildings, even though they may have a life expectancy of 50–100 years. Of all sectors, only transportation has a major turnover of equipment in less than two decades.

If decisions to reduce emissions are delayed for a decade under a "wait-and-see" policy, then only modest reductions in the global rate of growth will be possible by 2010. Thus if the consensus of climate scientists is that CO_2 emissions are unlikely to cause major problems by say 2050, a wait-and-see policy may be a reasonable choice. Even if that consensus does evolve, ten years of uncontrolled growth will cause greater problems for policy reversal should that prove necessary.

How then do we supply the necessary information for policy makers to decide what to do? Obviously, all three *analytical* efforts must continue working simultaneously toward a convergence on acceptable scenarios. However, since we don't have unlimited resources, we must have priorities. The greatest uncertainties appear to be with our understanding of the rate of climate change and the detailed regional impact. Until we can better bound the range of uncertainties over climate change, the adaptation and energy analyses will necessarily be so broad as to be of limited use for policy making. Therefore, the climate change prediction efforts should have the highest priority. Second priority might go to the energy analysis, on the grounds that we probably don't want to adapt unless we have to. Therefore, reducing the problem is more important than minimizing the effects of the problem. This should not suggest that climate scientists and modelers should get all the funding devoted to climate-change planning; rather that their research funding needs should be given the benefit of the doubt. In fact, adaptation research has been getting less than 10% of the total budget devoted to global warming, and could productively use an increase in funding.

A vigorous analytical program should be able to reduce significantly the uncertainties within a decade. As discussed above, waiting that decade for better (but still imperfect) answers may or may not be costly depending on the urgency of the problem. Conclusions on the appropriate path to take now depend on value judgments (e.g., the level of risk to accept or the acceptability of further interference in the market) as well as assessments of the seriousness of the problem.

The White House apparently has decided that the risk at present does not warrant any significant remedial steps. Efficiency measures in early versions of the National Energy Strategy reportedly were deleted by the White House despite their attractive cost-effectiveness, on the grounds that they interfere with the market. These measures could have been part of a "no regrets policy," i.e., do those things that also make sense for other reasons, so that if the problem recedes we won't have wasted money.

The rejection of such an insurance policy, which would have been economically beneficial to the nation, on narrow ideological grounds stands in sharp contrast to congressional efforts. Legislation entailing major increases in expenditures or public sacrifices are unlikely to pass without strong presidential support. Even the very modest gasoline tax considered earlier this year died in the face of public opposition, which might have been countered if President Bush has supported it, as President Reagan had in the case of a previous proposal. Nevertheless, Congress has been laying the groundwork for future action with serious consideration of all the major elements of a CO_2 emissions reduction policy—efficiency, renewables and nuclear power. With increasing scientific evidence and a changing White House calculus, a real global climate-change policy may well emerge.

REFERENCES

1. "Changing by Degrees; Steps to Reduce Greenhouse Gases," U.S. Congress, Office of Technology Assessment, OTA-O-482, U.S. Government Printing Office, Washington, DC, February 1991

2. "Energy Technology Choices; Shaping Our Future," U.S. Congress, Office of Technology Assessment, OTA-E-493, U.S.G.P.O., Washington, DC, July 1991.

3. "Improving Automobile Fuel Economy; New Standards, New Approaches," U.S. Congress, Office of Technology Assessment, OTA-E-504, U.S.G.P.O., Washington DC, October 1991.

4. Derived from "Climate Change, The IPCC Scientific Assessment," Intergovernmental Panel on Climate Change, p. 10, Cambridge University Press, New York, 1990.

5. OTA, "Changing by Degrees," op. cit.

Environmental Factor in International Aid Programmes and Technology Transfer

Tatyana G. Kazakova

Attache, Directorate for International
Scientific and Technological Cooperation,
Ministry of Foreign Affairs
Moscow, Russia

The message of our Conference is clearly formulated in its title and all the speakers are unanimous in the conclusion that:

- climate change is evidently a global environmental problem No. 1, which requires a common response;
- the improved production and use of energy rank among first priority factors, capable of mitigating climate change;
- energy policy becomes a substantial factor of international security, taking the strategic importance of oil and gas as the principal sources of energy in a foreseeable future and the main contribution of energy pollutants to climate change processes; and
- the international community needs a clear-cut set of recommendations aimed at tackling climate change, supported by specific economic, juridical and organizational measures. This framework could well be provided by the global convention on climate change, under preparation now by a special intergovernmental body to be eventually tabled for approval at the 1992 United Nations Conference on Environment and Development.

In stressing equal responsibility of all countries for solving global environmental problems, including those pertaining to climate change, it is evident, however, that national capabilities are different and strongly dependent on economic potential. While some developed countries are ready for radical emission cuts right now—EC and Scandinavian countries, Switzerland, and Japan have already voiced themselves in favour of stabilizing their CO_2 emissions by the year 2000 and to 1990 level—the developing countries and states with economies in transition, including the USSR, have a long way to go. This applies to other environmental problems as well.

One of the critical issues of international environmental cooperation, which is also expected to find serious reflection in the future climatic convention, pertains to financial aid and technology transfer as material means to achieve the desired goals. No one today contests the necessity of external aid for the following reasons:

- because developing countries, which have two-thirds of the world population, already account for about 45% of the annual human contribution to climate change, and their subsequent industrialization without proper account of environment will bring even greater problems for the whole world;

- because poverty is correctly considered the worst polluter; and
- because developed and developing countries are actors in the same economic infrastructure, which entails mutual influence in all spheres of interaction, including environment.

It is equally important to understand that the process of East-West, North-South concerted environmental cooperation should be devoid of both greedy arrogance and parasitic attitudes in terms of financial aid and technology transfer.

The past few years brought about a number of encouraging signs to this effect. Guided by a commonly accepted perception that sound economic development can be achieved only on the basis of environmental sustainability, international economic and development institutions are becoming more and more environmentally conscious. Since 1980, under the United Nations aegis, there exists a Committee of International Development Institutions for Environment, which promotes environmental impact assessment of all the projects under consideration. About half of the loans granted currently by the World Bank have an environmental component.[1] Rich in ecological considerations are the projects of UNDP, UNIDO, FAO, UNCTAD, International Fund for Agricultural Development, ILO, ISO, HABITAT, Organization for Economic Cooperation and Development, European Community, and International Chamber of Commerce.

Environmental components are not only incorporated in the existing programs of economic aid, but also spur the appearance of special environmental aid arrangements.

In relation to climate change, may I recall the decisions taken by the countries (parties) to the Montreal Protocol on Ozone-depleting Substances in 1990, permitting the countries to make their contributions to the Protocol fund, besides hard currency, in an "in kind" form (goods and services), and in national currencies. The input from bilateral aid programs will also be counted. In addition to that, an up-to-240-million dollar ad hoc fund was created jointly by the World Bank, United Nations Development Program (UNDP) and United Nations Environment Program (UNEP) to assist developing countries in complying with the provisions of the Protocol. Similar arrangements could very well be foreseen in all global environmental agreements, including a framework convention on climate change.

Another unprecedented step by the World Bank, UNDP and UNEP was the setting up in 1990 of a Global Environmental Facility, which is a three-year pilot program providing grants and low- interest loans to developing countries to relieve pressure on global ecosystems, primarily in four domains: reducing emissions of greenhouse gases which cause global warming, preserving the Earth's biological diversity, arresting the pollution of international waters, and protecting the ozone layer. This body supplements official development assistance. In terms of climate change, areas for action include the adoption of cleaner fossil fuels and renewable energy technologies in power generation, agriculture, mining and industry, as well as management of existing forests and reforestation.

No less important is a recent decision of UNEP to establish the International Environmental Technology Centre. The Japan-located Centre would be dedicated to the transfer of environmentally sound technologies to developing countries and countries with economies in transition by means of providing training and consulting services, carrying out research, and accumulating and disseminating related information. The intention is to create similar centers in different parts of the world.[2]

May I also put on this list a concurrent UNEP decision to create, at the beginning of next year, on an experimental basis, a UN Centre for urgent environmental assistance—to the great satisfaction of the Soviet Union, which initiated this proposal in the United Nations 3 years ago.[3] Undoubtedly the new body, which would monitor prompt international responses to environmental emergencies worldwide, is yet another form of international environmental aid.

Among instruments to alleviate poverty for the sake of environment one can count debt-for-nature swaps. These transactions by now have resulted in swapping up to 100 million dollars of debt of the developing and some East European countries. Conducive to transfer of environmentally sound technologies are international environmental offset investment programs, arrangements relaxing limitations on the repatriation of income derived from environmentally benign technologies used in developing countries; tax disincentives and tariff barriers to restrain the use of, and trade in, environmentally hazardous technologies; and bilateral official development assistance programs. Thus, environmental aspects are intrinsically inscribed in ODA strategies of the FRG, Japan, Sweden, Norway, France, Great Britain, Canada, Denmark, the Netherlands, etc. The USA has a very good record in this respect as well. One of the goals of the United States Agency for International Development is to ensure the environmental soundness and long-term sustainability of its assistance programs and projects. Of great interest are some latest initiatives of American scientists and politicians, aimed at building a North-South partnership for economic progress and environmental protection, including the creation of a multilateral authority to reduce developing countries' external debt while promoting sustainable development, the establishment of "sustainable development facilities" within the multilateral development banks, setting up a center for training and research on energy efficiency and renewable energy in each of the major developing countries,[1] the establishment of an international solar energy agency, and enhancing capacities of low- and middle-income developing countries to shift to sustainable development paths.[2]

A new and substantial factor of international environmental security is connected with radical economic and political transformations in East European countries and the USSR, revealing disastrous environmental situations in that part of the globe. Some estimates show that East European countries and the USSR are losing annually up to 20% of their GNP because of environmental degradation, thus hampering their own chances for a speedy economic rehabilitation, so much in need, and putting at risk the environmental well-being of other countries. According to the IBRD studies, the restructuring of East European industry, on the basis of Western environmental standards, will cost 200 billion dollars and require long-term, all-around aid strategies, a so-called ecological "Marshall Plan" for Eastern Europe.

In the past few years the West manifested its real interest in contributing to alleviating the environmental troubles of economies in transition. In 1989, the United States Environmental Protection Agency (EPA) allocated 10 million dollars for funding environmental activities in East European countries. More than seriously environmental assistance to eastern neighbors is taken in Western Europe. Major European bodies, such as the European Community, Council of Europe, European Investment Bank, and the recently created European Bank of Reconstruction and Development, actively incorporate ecological elements in their assistance programs to East European economies and the USSR. For instance, 90% of all EC aid to Czechoslovakia is directed to environmental projects. By an EC aid program to Poland and Hungary (PHARE), it is planned to allocate 35.5 million dollars for environmental purposes in those countries.[5] In the summer of 1990, the European Parliament approved the program of aid to the countries of Central and Eastern Europe, called GREEN, aimed at assisting those countries in obtaining environmental know-how. A number of Western countries, including the USA, funded the creation of the Regional East-European Center on Environment Protection, which opened a year ago in Budapest. One may also recall the initiative of northern countries to pool resources for improving the environmental situation on the Kola peninsular in the USSR; FRG strives to cure pollution-stricken East Germany; and energy reconstruction projects between Austria, FRG and Czechoslovakia, etc.

The EEC aid experience has led to a very interesting conclusion — the level of environmental pollution in the countries of Eastern Europe could be reduced at least by half by mere improvements in industrial management, proper operation and maintenance of the existing equipment, relevant education of the personnel, acquisition of necessary environmental skills, all of which may be much more effective than large-scale, expensive, and sometimes politically sensitive, technology-transfer campaigns.

I believe it is 100% true for the Soviet Union, when we consider its role of a benefactor from foreign environmental assistance. With vast natural resources and a huge but obsolete industrial basis, all branches of our economy, of course, need restructuring, and the energy sector in this respect should be addressed as a priority. This understanding is shared by prospective donors. The plan of assistance to the USSR, adopted at the latest G-7 summit in July 1991, contained a provision on modernization of the Country's oil and gas complex, and investments in bettering the USSR energy and transportation sector are regarded by businessmen worldwide as very promising. This is the more so true because the energy branch has bigger chances than any other industry to retain a centralized all-Union management in the painful process of present political transformations in the Country. In any form the new Union will require a single, or at least a closely coordinated energy policy, and even the most ambitious leaders of Union sovereign states do not deny it.

We are living through a difficult time these years, when the old Union exists no longer, but a new one is just being born in pains. Gradually and abruptly, through ups and downs, but firmly and consistently, we are paving the way to a more stable, more secure, more prosperous future. I am deeply convinced that such a future for my country can be achieved not on the basis of confrontation and disintegration of nations, but through the constructive addition and interaction of their resources, economies, wills, and intellects. I believe this is equally true for the whole world. Let there be appreciation and success for those who pledge to help the others on this thorny, but righteous, way.

It would be senseless to make empty promises and set unfeasible targets for our industry in conditions of economic chaos and disarray and paralysis of power which we are facing now. Well-thought-out steps on climate change and related matters require a stable political decision- making mechanism, necessarily envisaging full-fledged constructive cooperation between the centre and sovereign Union states (ex-republics) yet to be formed. That is why we favor inventive combinations of all possible mitigating measures, giving any country a possibility for optional approaches and a space for maneuvering. One of the innovative proposals to combat climate change currently debated by our scientists is a so-called "zero balance" concept. It means that a country may contribute to solving climate change problem not only by reducing emissions of greenhouse gases, but with the same effect, by increasing 'sinks' of these gases primarily through intensive reforestation, thus making a zero balance between the two. It seems that the present technological situation and natural conditions of the USSR make a 'sinks' strategy a more preferable option for us right now. We could, as well, consider a possibility to subsidize planting of forests in other countries or buying absorption shares from the countries that have an excess of 'sinks' rates over emission ones.

REFERENCES

1. The World Bank and the Environment, First Annual Report Fiscal 1990, The World Bank, Washington, D.C., p. 54.
2. UNEP Governing Council Decision 16/34 of 31 May 1991.
3. UNEP Governing Council decision 16/9 of 31 May 1991.
4. Reference to testimony presented by Mr. James Gustave Speth, President of the World Resources Institute and Chairman of the Environmental and Energy Study Institute Task Force on International Development and Environmental Security before the U.S. Senate Committee on Foreign Relations on May 15, 1991. Source: UNCED Network News, July 1991, No. 2, pp. 10–11.
5. *Nature*, December 6, 1990, p. 472.

Low-Cost Energy Production: The Responsibility of the Developed Countries Toward the Less Developed

V. M. Oversby

Lawrence Livermore National Laboratory

Livermore, CA 94550

In recent years, the world has begun to develop an awareness of the environmental costs of economic expansion and development. The most widely recognized problems include depletion of the ozone layer, the effects of toxic chemicals on soils, groundwater, and wildlife, and the danger of global climate change through the accumulation of "greenhouse" gases in the atmosphere. In one area, that of ozone depletion, the nations of the world were able to reach quickly a consensus and take action to try to slow, if not stop, the depletion.

In the area of global climate change, the ability to reach a consensus is less likely. This is because the production of greenhouse gases is intimately linked to the production of energy, and increases in energy production in the developing nations of the world will be needed if they are to develop their economies and the standard of living for their peoples. It is also highly unlikely that the nations of the developed world will be willing to lower their standard of living in order to allow that of the less developed countries to be raised. In my opinion, this dilemma presents a formidable challenge and a great opportunity for both the technical community and the policy makers in the developed nations.

It is no longer sufficient to formulate a "national" energy strategy. The effects of one nation's energy policy on the global climate can negate, through increased emissions of greenhouse gases, all of the attempts of other nations to limit the increase of these gases in the atmosphere. We cannot, however, adopt the arrogant attitude that somehow the developed world can control the actions of the developing nations in the area of energy production. This type of attitude is expressed when we ask, "How are we going to allow the developing world to reach a higher standard of living?" I believe that we must assume that the less-developed countries will eventually succeed in raising their standard of living and this will imply an increase in their use of energy. In this case, our challenge is to develop the means of producing energy in a way that will limit the production of greenhouse gases and which will also limit any other adverse effects on the environment or the health and safety of the peoples of the world. This is by no means an easy task!

Although I have argued that a "national" energy strategy is impossible in the sense that the actions of each country will affect all other countries, it is still necessary to develop an energy strategy that addresses national needs and interests. Perhaps this is best thought of as the regional or national component of a global energy strategy. Each region, such as Europe or North America, would develop plans for energy management that would include the achievement of national and regional economic and environmental objectives. If these plans are properly formulated, with due regard for the fact that we all share the same planet and atmosphere, they would include the development of low-cost, non-polluting energy sources to replace at least some of the production methods presently in use. At this point the opportunity to influence and assist the developing nations would be greatest. Here we would have the opportunity to develop affordable energy options for the developing world.

We need to recognize that affordable is a relative term that covers many dimensions. The "rich" nations of the world may think that they can afford expensive—in terms of resources and currency—solutions to reducing chemical pollution of the oceans or atmosphere. If these nations fail to provide the means for the developing countries to afford their solutions, they may find that they have no "solutions" at all—merely a delay in the onset of the adverse effects they were expending resources to avoid.

The low-cost, affordable solution is the only path that will produce a real solution to our global energy and environmental issues. Let us define affordable as it applies to this problem. For a single developing nation, affordable means that the solution does not require the expenditure of a lot of hard currency. This translates into the requirement that the solution must not require the continual importation of a resource that is not available within the political boundaries of the country. This is also going to be a policy driver in developed countries; consider, for example, the development of nuclear electric power in France.

Balance of payments problems and the distribution of natural resources must be given central consideration in the development of energy options. Because of this, I believe that we cannot afford to "throw away" any energy production method. At present, the United States has "thrown away" nuclear energy in the sense that the government will no longer support in a meaningful way the development of this method of power generation. Even if the US does not intend to expand its own commercial nuclear power sector, I believe the government should support research in this area so that we can help those developing nations that choose a nuclear option to do so safely and with due attention to the safeguarding of nuclear materials.

So far, my arguments that we must develop new, affordable, environmentally friendly energy technologies have been posed from a fairly abstract, global position. This point of view is frequently difficult to use in convincing people to take action when they have a short-term view of the issues. It is basically a view that says, "We owe it to the world." A more compelling argument might be made that we also, "Owe it to ourselves." This tactic is likely to have more success in convincing the general public and the commercial sector to act in ways that achieve our long-term aims. For example, the Pacific Gas and Electric Company is presently researching methods of local energy production and distribution for northern California. It does not require a great leap of faith to see these developments as a model of Third World power systems.

The use of the profit motive is a powerful tool in providing incentive to industry. Government/industry partnerships, perhaps along the lines used in France to develop their nuclear industry, might be used to develop a wide array of alternative power generation methods suitable for both the developed and the developing world. Attention should also be given to the issues associated with power distribution. Localized power generation will help reduce losses in transmission of power. This favors the use of small, modular plants for power generation. We must, however, address the issue of power sharing among adjacent areas if we are to meet peak power demand and avoid having generation capacity that is idle except at peak demand times.

As a final comment, I would like to raise the suggestion of using technology development in the energy sector as a means of foreign aid. There are technologies, such as nuclear power, that may not be acceptable from a point of view of social acceptance or economic risk in the present US regulatory climate. This technology will, I believe, be needed by the US in future years if we are to continue to enjoy our present standard of living and to do our part in limiting atmospheric emissions of polluting gases. The Department of Energy should develop an aggressive research and development program in the area of small, modular, intrinsically-safe nuclear power for use in those countries where nuclear power is the option of choice. This will produce a better result, I am sure, than waiting for those countries to build their own nuclear plants using copies of the present generation of fission reactor plants. If the US and other developed countries were involved in developing nuclear power for the less developed world, it would allow us to address "up front" the issue of nonproliferation of nuclear weapons and special nuclear materials. If we wait for these countries to develop nuclear power on their own, which the examples of Iraq and North Korea show is likely, our task in limiting proliferation of nuclear weapons will be much harder.

The developing world will continue to develop, whether we allow it to or not. This development will put an increasing pressure on the environment and on the global climate. Perhaps our best argument for assisting the developing nations in their efforts comes from the data discussed by our keynote speaker, Dr. Kapitsa. There is strong evidence to suggest that an improved standard of living will lead to a lowering of population growth. Since the ultimate stability of the planet may depend on the limitation of population, we have a responsibility to try to achieve a stable, sustainable population level. If that goal can be achieved through raising the standard of living of the world's people at the same time, so much the better.

This work was performed under the auspices of the U.S. Department of Energy by the Lawrence Livermore National Laboratory under contract No. W-7405-Eng-48.

The British Coal Approach to Global Climatic Change

Ian S. C. Hughes

British Coal Corporation

Coal Research Establishment, Stoke Orchard

Cheltenham, Gloucestershire, GL52 4RZ, United Kingdom

INTRODUCTION

The possibility of climatic change resulting from an enhancement of the greenhouse effect by Man's activities is an issue of prime importance to politicians, industry and the world at large. It is widely accepted that there are major uncertainties about the magnitude of any changes and about their significance to the planet. Despite this it behooves us all to behave in a responsible way to ensure that as far as possible we fully understand the science underlying any potential changes and take action to minimise any damage to the well being of mankind and the ecosystems that surround us.

The objective of this paper is to outline the approach that British Coal is taking towards this vital issue.

PREDICTED CHANGES IN THE CLIMATE

As the concentration of greenhouse gases increases in the atmosphere, there is a consensus that some additional warming will occur. The crucial questions are: How large will this warming be? Over what timescale will it operate? Will it be different in various regions of the world?

The first Working Group of the Intergovernmental Panel on Climate Change (IPCC) provided a summary of scientific thinking on this issue.

The executive summary of the working group's report presents its views succinctly. It is structured to lead from points of certainty to levels of increasing doubt. The group is certain of the basic physics of the greenhouse gases and of a link between emissions and atmospheric concentrations.

But what the group says it cannot state with certainty or confidence is perhaps of even greater significance. It does not make any confident statements about the magnitude of any effects, either on temperature or on climate, or about whether increasing concentrations of greenhouse gases will lead to significant problems for the human race.

At the next level of uncertainty, the group makes predictions in its report based on the results of current climate models.

The group emphasises the amount of uncertainty involved in its own predictions, pointing out that "there are many uncertainties in our predictions, particularly with regard to the timing, magnitude, and regional pattern of climate change, due to our incomplete understanding." It then goes on to outline several areas which are presently poorly understood, including the role of both oceans and clouds in climate change, and the sources and absorbing "sinks" of greenhouse gases.

The IPCC report appears eminently sensible, but it does not appear to be a firm basis from which politicians could launch Draconian measures against greenhouse gas emissions with all the economic dislocation that could result. Such measures would present severe problems not only in industrial nations, but also in developing countries. In the latter, growth in energy and food production are crucial to the alleviation of poverty and to the general improvement of standards of living. We are faced with two potential disasters on a global scale. The first is that the worst predictions about an enhanced greenhouse effect come to fruition. The second is that we massively overreact against a threat that proves to be benign. In either of these extreme cases suffering and hardship would result for a significant proportion of the world population.

It is basically these considerations which demand a progressive approach in our response to the threat of global warming.

A RATIONAL APPROACH TO GLOBAL WARMING

It is clear that much science remains to be resolved before we can be clear about the eventual impact and significance of an enhancement to the greenhouse effect. It is equally clear that the issue is of such potential importance that immediate action is justified while the uncertainties are unravelled. These actions should minimise the buildup of greenhouse gases and so limit the scale of any potential temperature increase. The only decisions that need to be made concern how far through the suite of policy options the world should progress given present scientific understanding, and how much effort should be directed towards resolving the scientific issues.

The British Coal Industry advocates further research into the science underlying the greenhouse issue in order to generate a better understanding of the greenhouse effect and to produce improved predictions of the magnitude and timing of any climate change. It is only when this research shows results that governments will be able to make better-informed policy decisions.

At the same time, the industry supports the immediate implementation of measures of 'least regret' as a rational approach to the threat of global warming.

These are the actions which would cost least and which may bring with them other considerable benefits. Some would result in cost savings rather than expense.

Among the measures which would fall into the 'least regret' category are:

- Increasing technology transfer to less developed countries
- Increasing efficiency of energy supply
  - for power generation
  - by implementing combined heat and power (CHP) schemes
- Increasing public transport

- Phasing out CFCs
  - these also destroy the ozone layer
  - replacements should have low greenhouse potency
- Stopping deforestation
  - improved agricultural practices required
  - saves unique habitats
- Increasing efficiency of energy end use
  - better insulation
  - better education
  - more efficient appliances, light bulbs, etc.
  - for transport (there is great scope for improved private car efficiency, and short vehicle lifespan brings rapid results)

Increased energy efficiency is widely regarded as an essential first step towards combating potential global warming. For power generation in the industrialised world, efficiencies of 35% in the conversion of heat to electricity are readily attainable. But this means 65% of the energy is lost. The use of combined heat and power systems could raise the overall efficiency to 80% or more by using otherwise wasted energy in industrial processes or for heating houses. The application of such systems requires considerable forward planning and changes in infrastructure.

In less-developed countries, the average level of thermal efficiency for power generation may be as low as 25%. As a result, existing commercially available coal-fired technologies could achieve of the order of a 30% reduction in CO_2 emissions on a one-for-one replacement basis against existing plant. Money spent to bring Eastern European and developing countries up to the energy efficiencies prevalent in the more developed nations would be particularly effective. There is an urgent need to tackle the problems of technology transfer to these countries.

In transport, there is scope for considerable improvements to vehicle efficiencies. Also important would be a move away from private vehicles in city centre and urban areas and to increase substantially urban mass transport systems such as improved rail, underground, and bus networks. In terms of greenhouse gas emissions per passenger mile, these are far more efficient than individuals travelling in their own cars. The need for action on transport is made increasingly urgent by predictions of huge increases in motor transport emissions in the next century.

Chlorofluorocarbons (CFCs) appear to be a double-edged sword, they contribute to the Greenhouse Effect and they are a prime cause of stratospheric ozone depletion. Action is underway on the latter issue but there is a clear need for CFC replacements to be reassessed so their environmental impact in terms of global warming is reduced to an absolute minimum.

Deforestation occurs largely in the developing world and is linked mainly to population growth and the need for agriculture and fuel. It is estimated that as much as 46 000 square miles of tropical forest is lost each year. Actions to preserve present forest resources and to promote reforestation should be taken as a matter of urgency. Such measures should include the promotion of appropriate agricultural practices and assistance to developing nations to help them manage their forests on a sustainable basis. Action should be taken to ensure that declining forests are maintained and, wherever possible, restored.

There is much that can be done by individuals to reduce greenhouse gas emissions by using energy-efficient appliances and lighting. Although they are more expensive to buy, they last longer and the reduction in electricity bills over their life more than compensates for the increased outlay.

Energy conservation is another important area where individuals can contribute to reducing the emissions of greenhouse gases. Better insulation in houses and offices and more advanced heating control systems reduce energy consumption, increase comfort and, in many cases, save money. In the UK, the main problem appears to be lack of awareness of the potential benefits. It is therefore most important that funding is provided centrally to convey the message to the public via advertising campaigns or incentive schemes.

BRITISH COAL'S RESEARCH RESPONSE TO THE GREENHOUSE EFFECT

British Coal is active in research aimed at increasing understanding of the Greenhouse Effect, at increasing the efficiency of coal-fired power generation, and on developing technologies that could be deployed if future understanding of the Greenhouse Effect dictates dramatic action.

On greenhouse science, British Coal is funding research into natural mechanisms for removing CO_2 from the atmosphere. Contracts have been placed with a variety of expert organisations in the UK to investigate CO_2 uptake in forests and soils, and in the oceans. These projects, although modest in scale, could increase awareness of ways in which nature can be encouraged to work a little harder in our favour.

On forests and soils, for example, model simulations of plant and tree growth and of changes in soil carbon will be linked. This will provide a tool to investigate possible changes in forestry management with the aim of maximising the amount of 'locked up' carbon in the system. Possibilities include species selection or changes in silviculture or fertilisation practices.

As well as carbon dioxide, emissions of methane, other hydrocarbons, and oxides of nitrogen play a role in any enhancement to the greenhouse effect. Further information on the sources and sinks of these emissions is necessary if the role of coal is to be placed adequately in context alongside other fossil fuels and natural sources. For this reason, British Coal is coordinating an international project, partly funded by the European Community, to quantify such emissions from coal-fired plant in Europe. Sulphur dioxide from fossil-fuel burning may act as a 'negative' feedback on the greenhouse effect because of its propensity to increase the quantity and albedo of clouds. A British Coal-funded study aims to quantify the potential significance of these emissions and the implications for other environmental issues.

In addition to studies on emissions of greenhouse gases, British Coal has supported work to investigate whether the UK has experienced changes in abnormal weather events or day-to-night temperature changes as greenhouse gases have built up in the atmosphere. Results so far suggest there has been no increase in the prevalence of storms as CO_2 has increased and that much of any warming may be due to increased nighttime temperatures. If this is the case it could be that, for the UK, the effects of global climatic change would be less severe than predicted by some.

It has always made sense to maximise the efficiency of coal-fired appliances. Programmes at British Coal's Coal Research Establishment (CRE) have produced improved designs for industrial, commercial and domestic-scale coal combustion appliances. These are now in use, playing an active role in reducing emissions of CO_2 and other pollutants. Similarly, British Coal has supported and encouraged the installation of combined heat and power systems with substantial overall thermal efficiency benefits.

Further advances in efficiency are also possible, particularly with respect to power-generation options. In the UK, nearly 80% of coal is used to produce electricity and British Coal is currently leading a consortium in developing an advanced coal-fired system known as the Topping Cycle. The system links together gasification and circulating fluidised bed technologies to produce electricity via combined cycle operation. The system offers the prospect of increasing generating efficiencies to 45% with known technologies and there is the potential for further increases as the design of gas and steam turbines improve. At an efficiency of 45% a Topping Cycle power station would reduce CO_2 emissions by 20% compared to conventional pulverised fuel firing. Other significant advantages include low emissions of both sulphur dioxide and oxides of nitrogen and the prospect of a substantial drop in generating costs. The development of this is already under way and it is planned that commercial deployment will be by the year 2000.

All of this research is aimed at either improving the scientific base of knowledge or at expanding the range of 'least regret' measures available for use. If the worst predictions about global warming prove true it is clear that more dramatic action would be justified. It is for this reason that British Coal has launched an initiative targeted at the development of fossil fuel fired systems with minimal emissions of greenhouse gases.

POWER GENERATION WITH CO_2 REMOVAL

A British Coal initiative has lead to a major International Energy Agency programme aimed at developing fossil fuel-fired heat and power generating systems, with minimal gaseous emissions of greenhouse gases.

British Coal has been nominated to act as Operating Agent assessing the technology options.

A British Coal Global Warming programme forms part of the UK contribution to the IEA Programme. Preliminary assessments of technology options are underway. It is proposed to undertake laboratory-scale work on the most promising option for generation with CO_2 removal, and to investigate means of utilising, or storing, the CO_2. The aim is to maximise the efficiency of the overall system (with CO_2 removal) and ensure coal maintains its competitive position. There are three key elements to this work:

- to demonstrate the technical feasibility of CO_2 removal and storage;
- to estimate the economics of such systems; and
- to investigate the environmental acceptability of the suite of utilisation or storage options.

It is too early to determine which technologies are best, but candidate technologies should aim to remove about 90% of the CO_2 and 99% of the sulphur. Possible technologies for CO_2 removal include the use of chemical scrubbing to remove the CO_2 from the flue gases of conventional power stations with subsequent concentration and liquefaction of the CO_2 or the development of more advanced systems which could produce concentrated CO_2 streams in their own right. In this latter category, studies are presently concentrating on prospects for the use of Integrated Gasification Combined Cycles (IGCC) or on the development of combustion systems which use pure oxygen in conjunction with recycled CO_2 as the oxidising gas in the combustor. Such systems could avoid the costly CO_2 separation step but would involve additional costs in, for example, a requirement for air separation plant.

With regard to the possible utilisation or storage for CO_2 removed from power stations, there are many potential options which justify further research effort. The capacity of industry to use captured CO_2 is small and, as a result, most of the research effort will be aimed at investigating possible storage media.

Rock formations have stored oil and gas safely over geological time. As such, exhausted oil and gas wells are attractive options for CO_2 storage. A preliminary study has indicated that the available capacity in the North Sea as reserves of oil and gas become exhausted would be sufficient to store around 40 years of UK CO_2 from power stations at 1990 emission values.

On a longer time span, storage of CO_2 in the oceans may prove necessary. The potential capacity of the oceans is enormous; the total man-made emissions of carbon are about 6 GT annum^{-1}, and the oceans already hold about 34 000 GT. Initial research suggests that the most promising approach may be to discharge liquid CO_2 at depths of below about 3000 m. At this depth the characteristics of the liquid CO_2 are such that it would settle on the ocean floor forming a liquid lake. Such ideas have obvious economic and environmental implications and it is essential that appropriate studies increase our understanding of these issues as technology development continues.

It is clear that much work needs to be done, but this research offers the possibility of a fallback option, in the form of a technology that might be required, if the most pessimistic scenarios occur and a new tranche of response strategies become necessary.

CONCLUSION

Even though there are major uncertainties over the extent of the Greenhouse Effect and whether or not it is of significance to mankind, British Coal advocates the adoption of 'least regret' measures in conjunction with appropriate research and development.

British Coal recognises that it has an important role to play in responding to the Greenhouse Effect and intends to play that role to the full. Through the development and deployment of advanced clean coal systems, and via the funding of appropriate global warming studies, British Coal will continue to take a responsible and rational stance to ensure that environmental and economic objectives are fulfilled.

Canada's Responses to the Climate-Change Issue

Patrick G. Finlay

Environment Canada, Ottawa, Canada, K1A OH3

SUMMARY

After extensive consultations with stakeholders, Canada issued its Green Plan in December, 1990. The Green Plan contains more than 100 initiatives, targets and schedules that deal comprehensively with the environment. The Green Plan, and the National Action Strategy on Global Warming, which was developed with the Canadian Council of Ministers of the Environment (CCME) and federal and provincial Energy Ministers, calls for action at home and abroad on the climate-change issue. The strategies are:

- to limit greenhouse-gas emissions
- to improve our understanding of global warming
- to anticipate and prepare for global warming
- to stimulate international action on global warming solutions.

Canada's goal is to stabilize national emissions of CO_2 carbon dioxide (CO_2) and other greenhouse gases at 1990 levels by the year 2000. (This commitment does not include gases controlled by the Montreal Protocol on ozone layer depleting substances). "First steps" to limit emissions are based on improving energy efficiency, promoting alternative energy, and improving forestry and agricultural practices. Federal/Provincial/Territorial Agreements will codify actions by various governments to limit emissions. Actions "beyond first steps" will include a variety of measures that may be less economically attractive, but that may be required to respond to the climate-change issue as scientific understanding develops. The Green Plan tries to strike a balance between scientific certainties and policy implications.

The "global village" has witnessed increasing international action on environmental issues. Responses to acid rain, ground level ozone (smog), stratospheric ozone layer depletion, and global warming, have all successively involved more countries, and shortened times between the identification of the issue by scientists and responses by policy makers. The climate-change issue has broadened our perception of global security, and is linked to the broader concept of sustainable development. The work of the Intergovernmental Panel on Climate Change (IPCC) and the United Nations Conference on Environment and Development (UNCED) will provide unprecedented global responses to environmental issues.

A Brief Overview of the Netherlands Policy with Regard to the Climate-Change Issue

J. Remko Ybema

Netherlands Energy Research Foundation (ECN)

ESC-Energy Studies

1. INTRODUCTION

With regard to the issue of climate change, the Netherlands government formulated explicit policy starting in the year 1989. In order to give participants at the conference on Global Climate Change in Los Alamos an impression of the current greenhouse effect policy in one of the West European countries, the author was asked to give a brief presentation on the Dutch climate change policy. The author wished to place the disclaimer that he is not an official representative of the Dutch government and so this paper must be regarded as a personal impression of the Dutch policy.

In Sec. 2 of this overview governmental plans which show interaction with the climate change issue will be presented. In Sec. 3 a brief summary is given of the Toronto analysis for the Netherlands. This analysis is an investigation for the opportunities to cut CO_2 emissions with 20% by the year 2005.

2. GOVERNMENTAL PLANS WITH REGARD TO CLIMATE CHANGE

As in many other countries, public concern about the consequences of climate change increased rapidly in the late 1980s. Public concern with regard to climate change and sea level rise might be somewhat stronger in the Netherlands than in other countries since approximately one half of the country's area is below sea level. The following governmental plans were launched recently starting from 1989:

May 1989: National Environmental Plan: Stabilization of the national CO_2 emissions is set as an initial goal.

June 1990: National Environmental Plan Plus: The goal for CO_2 emission is a reduction by the year 2000 of 3 to 5% compared with 1988 emissions.

June 1990: Energy Conservation Plan: This contained specific goals for savings on the use of energy in all major sectors, referring to the emission reduction goal of the National Environmental Policy Plan. For each sector an energy use reduction percentage relative to the expected growth in energy use was set. Further this plan contained explicit objectives for combined production of heat and power and for renewables, e.g., the wind-turbine capacity for the year 2000 is intended to be 1000 MW. The additional annual investments related with the Energy Conservation Plan are estimated to be Dfl 2 billion.

June 1990: Transportation Plan: As in most other countries, the amount of passenger kilometers traveled is increasing. The transportation volume in the Netherlands is expected to grow if no additional measures are taken. The aim for CO_2 is a stabilization of CO_2 emissions from road transport by the year 2000 and a 10% decrease in the year 2010 relative to 1988 emissions. Such should mainly be achieved by energy savings but also by a change in driving style.

June–December 1990: Environmental Action Plans of Utilities: The utilities have got an important role in the execution of energy conservation. Actions undertaken include public information campaigns promotion of combined heat and power, extending subsidies for efficient appliances (light bulbs, condensing gas boilers, building insulation) and stimulating investments in renewable energy.

September 1991: Climate Change Document from national government: In this document the existing Dutch plans with respect to climate change have been summarized. The Dutch government will support a policy aimed at a stabilization for the end of the next century of the concentrations of greenhouse gases in the atmosphere at a level which is well below a doubling compared with pre-industrial concentrations. The role of the Netherlands in international negotiations is expressed: trying to play a catalyst role. New in this document are the goals with respect to the emissions of non-CO_2 greenhouse gases: 10% emission reduction of CH_4 in 2000 relative to 1990, stabilization of N_2O in 2000 relative to 1990, and 55% respectively 60% reduction of nitrogen oxides and volatile organic compounds, both precursors of ozone, in 2000 relative to 1988.

3. A TORONTO ANALYSIS FOR THE NETHERLANDS: 20% REDUCTION IN 2005

The governmental Climate Change Document contained also an analysis of the feasibility of reducing the CO_2 emissions with 20% by the year 2005. This Toronto analysis has been performed by the Netherlands Energy Research Foundation between January and July 1991. The study was guided by a taskgroup (Taskgroup Energy and Climate). Other taskgroups submitted information on transportation and waste. Contributions of economic sectors were only very limited, mainly passed on by related ministries. A matrix of options and sectors covering the current disaggregation of policy fields was given by the Ministry of the Environment (see Table I). Two sets of measures have been studied: a technological set of measures (TOP) and a set which also includes volume changes (VOP). The emission reduction according the technical set of measures of each option and of each sector is presented in Table I. A large effort was needed to construct a base case for the year 2005. This base case was built on current policy until the year 2000 and an annual GDP growth of 3% between 1990 and 2005. The CO_2 projection for 2005 is a stabilization compared to the year 1988; 20% CO_2 reduction (Toronto recommendation) is equivalent to 32-Mton CO_2. For the Toronto analysis existing information on the potential for energy conservation and renewables was used (as accepted by the Ministry of Economic Affairs). A simple spreadsheet model of the energy system has been used to avoid double counting in CO_2 emission reduction. The individual options were plugged into this model one after another. The order in which the measures were

TABLE I. Contribution to CO_2 Emission Reduction in 2005 of Options and Sectors According to Technical Set of Measures (in Mton CO_2).

	Residential	Industry	Transport	Others	Waste	Energy	Total
Volume measures	0.0	0.0	0.0	0.0			0.0
Carbon management					3.1		3.1
Conservation	6.8	3.8	2.7	4.4	0.8	0.3	18.8
Renewables	0.1		0.3	0.1	0.4	0.6	1.5
Fuel switch	0.2		0.4	0.0		1.7	2.3
CO_2 removal		1.8		0.0		4.7	6.5
Total	7.1	5.6	3.4	4.5	4.3	7.3	32.2

being plugged in has often an impact on the achieved emission reduction of the measure, so a ranking of options and sectors had been developed; measures which are closest to the final activities which lead to CO_2 emissions were plugged in first. (e.g., First the CO_2 emission reduction of increased penetration of efficient appliances in dwellings was calculated before efficiency improvements in electricity generation, wind energy, coal-to-gas fuel switch or CO_2 removal were allowed.)

Energy conservation is the most important option to reduce CO_2 emissions in the year 2005 in the technical set of measures. Furthermore carbon management (re-use of plastics instead of burning) and CO_2 removal give significant contributions to CO_2 emission reduction. In the set of measures that also includes volume measures (e.g., moving of aluminum industry to countries with unused potentials for clean electricity production, like Iceland; moving greenhouse horticulture to countries with a more favorable climate, like Spain) the same total CO_2 emission reduction is achieved.

Cost estimates have only been made for the TOP. To achieve 90% of the the reductions in Fig. 1, annual investments are needed which are more than Dfl 5 billion. These investment are supplemented to the investments for the Energy Conservation Plan. To estimate the average costs for emission reduction, the investments have been leveled and saved fuel has been subtracted (using current fuel prices). Average emission reduction costs were calculated to be around 110 Dfl/ton CO_2 (220 US$/ton of carbon).

The options in both sets of measures ask for structural changes in production and consumption. Some of the options cannot be achieved by individual action of the Netherlands. Individual efforts of the Netherlands seem difficult and of limited purpose.

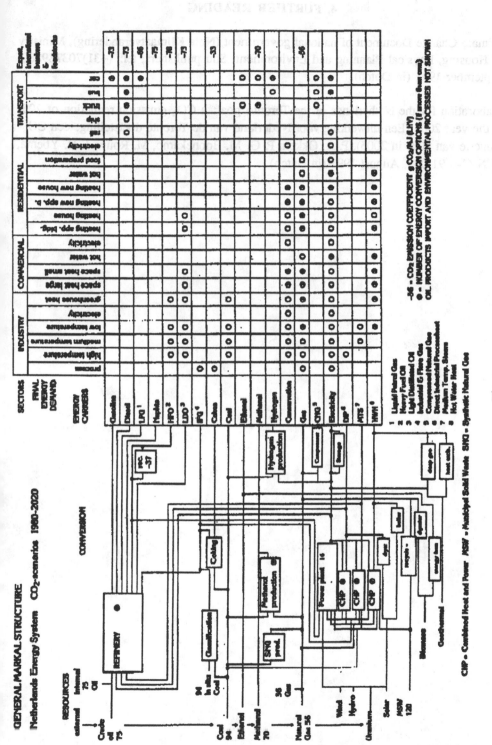

Fig. 1.

4. FURTHER READING

Climate Change Document of national government (Nota Klimaatverandering), Ministry of Housing, Physical Planning and Environment, Sdu publishers, tel: (+31)703789880, September 1991 (in Dutch).

Elaboration for the Netherlands of the Toronto goal: a CO_2 emission reduction of 20% in the year 2005 (Een uitwerking voor Nederland van de Toronto doelstelling: een CO_2-reductie van 20% in 2005), P. A. Okken, P. G. M. Boonekamp, M. Rouw, J. R. Ybema, ECN-C—91-045, August 1991 (in Dutch).

The Role of the University in Addressing Environmental Concerns

McAllister Hull
Department of Physics and Astronomy
University of New Mexico

In performing their traditional roles of research and scholarship, teaching and public service, in combination, universities are, perhaps, uniquely prepared to participate in the efforts of the broader scientific/research community in seeking to understand the environment (including, specifically, global warming resulting from anthropogenic activities), and to warn society of any impending disasters that may be forecast as a result of the understanding.

In areas where the interest of society is sufficiently clear, such as weather forecasting and national defense, governments have, especially in the last 45 years, supported focussed research efforts outside the academy. The *raison d'être* of these institutions is public service. The complexity of the applied missions of these institutions quickly led them into the basic research behind the applications, so that today no university can field the expert personnel and research equipment to match NCAR or the national laboratories—or some of the private environmental research institutions, for that matter. I make no complaint about that, and, in fact, spent significant effort as a university administrator developing interactions between some of these laboratories and appropriate faculty within the universities I have served. The increasing recognition in the national laboratories that national security depends on energy and the environment has, if anything, made the university/laboratory interactions easier.

What remains unique to the universities, however, is their role as teachers of the emerging generation of citizens: to transmit the knowledge and understanding of the world currently available, and to educate its members in ways to think about what they learn. For some students, the university will prepare them to join in adding to the world's knowledge/understanding base and its applications to solving the world's problems. It is especially important, however, that universities recognize their obligation to educate the general student about the anthropogenic problems of the environment, and appropriate citizen actions to promote their amelioration. The role of policy adviser, which some members of the research community, in and outside the universities, must assume, should be easier if there is an informed citizenry to receive their advice.

Centers and laboratories include among their personnel experts from a number of the academic disciplines into which we usually divide our pursuit of knowledge. But by their nature and mandate, they will not ordinarily include all the disciplines any well-ordered university does. The intent of the environmental research community is to inform and influence public policy on the basis of the results of its research, and this ultimate goal requires an understanding of people and governments to a depth that an atmospheric scientist, say, may not have reached. To the extent that such understanding exists, however, it is in the universities. Multidisciplinary research is not common among the disciplinary scholars of a university faculty, but it has been achieved and, in association with members of centers and laboratories, the broader approach is easier. I believe that both to educate the public and effectively to reach its leaders, universities have resources that are not generally found elsewhere in the environmental research community. Of course, we also have valuable contributions to continue to make in the basic research areas as well.

We have another asset (not unique, surely, among those concerned with the environment, but longstanding) to offer the effort. Scientists in general and university scientists in particular are traditionally independent from the political and economic pressures that focus on policies that may be advised on the basis of environmental research results. This gives us an opportunity for maintaining a general credibility that is not available to other groups wishing to influence policy. By training and instinct, most of us are incapable of polemical stands or distorted treatment of research results in application. We may not see clearly, we may lack wisdom, we may even be wrong (or working from incorrect results), but we operate from a base of professional integrity that should have an impact in societal affairs — and as a consequence, are obligated never intentionally to erode that base. Unfortunately, we are also human, and sometimes fall from grace — but, compared to some other professions, there is at least a state of grace for us to fall from!

When we step out of our laboratories or offices to offer policy advice, we encounter at least two dangers: media attention may flatter us into going farther or being more definite than the information warrants, and we may become the target of political, ill informed, polemical attacks for which the behavior of our university colleagues — as vicious as it sometimes can be — has not prepared us. Oppenheimer, for all his occasional wrong-headedness, is perhaps the classic case of encountering both dangers. But some of us, in my view, are obligated to inform the appropriate instruments of society of the policy implications of our research in any field, especially if the implication is apocalyptic, as in the case of global warming. We may encounter an uninformed President and his willfully ignorant chief-of-staff, but we cannot all of us ignore the societal impact of our work. To some extent, many of our present environmental problems might have been ameliorated had we spoken up earlier. Many of the pressing questions had been raised in the universities three or four decades ago (I've been talking about them for most of that time, and I'm not even in the forefront of the work!). Definitive solutions were not, of course, available then (or now), and there is among responsible scientists a reluctance to offer advice with incomplete results. We are quite capable of pillorying colleagues. Witness the bashing that Sagan and Ehrlich got from some members of the scientific community over so called "nuclear winter," despite the fact that their published papers pointed out the limits of their models and incompleteness of their results.

There is in this example a lesson that those of us who risk the dangers of advising Society need to learn. Within the academy, we are accustomed to dealing with the tentativeness of our conclusions, but, unfortunately with some assistance from us, the public thinks results of the scientific enterprise are fixed and proven. Thus any uncertainty in our pronouncements puzzles some of our audience and provides others with a perceived weakness to attack if the advice we're giving does not support their positions. We need to explain, much better than we have, that scientific information is not immutable, and, in the complex cases of environmental problems, is frequently incomplete. Only Popes are infallible—and then only when speaking on faith and morals. We scientists haven't even so small an area for which we can claim infallibility, and we need both to educate the public on the point, and to remain tolerant of the fallibility of our colleagues when they venture into the minefields of advice on public policy. Very important, of course, is the need to educate the public on the point that we should be listened to even when we differ among ourselves on fine points of the interpretation of our results.

A canonical example of what I'm saying is afforded by the central topic of our conference. The global environment—in particular, the biosphere—is so complex a system that we cannot, with the means at hand, provide even unchallengeable model answers to our questions. To what extent—and are the effects linear?—does increasing the CO_2 content of the atmosphere by, say 30%, increase the average temperature of the earth? How much, if any, is the warming counteracted by naturally occurring volcanic dust? Will the effect of induced El Niños occur earlier than significant warming anyway? Given the phenomenon of initial condition sensitivity, already known to be exhibited by some weather systems, is there a danger of uncontrollable development of an inimical process if some parameter regime of the atmosphere is entered? If so, we should note that the experiment may be tried only once, and one observes with sadness the inconsistency of a public policy that removes from use a carcinogen of which an ordinary human being cannot consume enough in a lifetime to induce harmful results, while it refuses to increase the required gasoline efficiency on automobiles in order to save a few percent of the CO_2 burden—and perhaps the onset of major climatic changes inimical to our continued existence.

The fact that there is an economic difference between banning a chemical (for which there may be a ready replacement anyway) and increasing the cost of automobiles to make them as fuel-efficient as current technology allows, is the kind of point that environmental policy analysts cannot afford to ignore. Much of the resistance to sound environmental policies stems from their negative economic impact—at least as these are judged by industries and politicians whose horizons are the next quarterly report or the next election. But while we may be able to restrain our sympathy for the several-million-dollar-a-year industry executive whose job seems threatened, we cannot ignore the plight of the working stiffs who would be the victims of economic displacement. Adding the appropriate university scholars to the policy analysis teams can provide comprehensive planning strength to allow a total package of advice to society that includes not only the means of saving the environment but of the kind of industrial redirection and shifts that can even enhance the overall economic climate while environmentally negative practices are halted. Providing a rational and workable plan does not guarantee its acceptance, of course, so that a part of environmental/economic planning must include proposals for convincing those responsible—in industry and government—to undertake the program. We are not terribly skilful at this, and our instinctive inability to lie professionally puts us at a disadvantage with respect to Washington lobbyists; but an open, candid, balanced, approach that utilizes all the truth (as we know it), including uncertainties and disagreements, might be so unusual that it would be welcomed. Be that as it may, I believe we have no other reasonable choice, given the complexity of the object of our study (the total earthly environment), our inability at this stage to define the problem completely to our satisfaction, the long-time scales we believe are important (with no demonstration example, like ozone depletion, to bolster our analysis in timely fashion), and the massive shifts in economic development we believe are necessary to stave off the inimical effects of unrestrained energy growth in response to unrestrained population growth in the world. In the face of impacts that require time to become measurable with confidence, we need to recall to the public (and ourselves) Berlioz's comment: "Time is a great teacher, but unfortunately it kills all its pupils." We need to assist time in teaching us to respond to distant thunder before the storm strikes.

The consequence of the failure of the scientific (and especially the university) community to inform the public fully and to plan comprehensively for the introduction of a new technology continues to be revealed by response to fission energy. In our enthusiasm to provide humankind with a new energy source, we failed to take full account of costs, failed to warn fully of the possibility of improbable but costly accidents, failed to educate on the range of questions about radioactivity (where there is still uncertainty at the very lowest end of exposure). I do not believe that the fact that there has been no fatal accident in commercial fission energy except Chernobyl is luck. We know how to build environmentally safe reactors, even in the face of incredible operator stupidity (Three Mile Island), and we know how Chernobyl's disaster could have been prevented—in the face of even greater stupidity—if optimum design had been utilized. The problems are real, and continued improvements in design—especially the inclusion of natural safety features—is necessary. But if the general public is instructed principally by news of disasters or near disasters and the negative reports of investigative teams looking at what appears to be long term misfeasance in managing some of the government's nuclear installation, we can hardly wonder if it is susceptible to the misinformation offered by many of the single-interest anti-nuclear groups that are moved by their own emotionally induced blindness. The performance is disgraceful, but a properly run nuclear power plant, as all of us here know, releases much less radioactivity into the environment than does a coal-fueled power plant of the same capacity. Industry must accept modular standardization in the design and manufacture of fission reactors, and may have to accept Pu-breeding non-power reactors in place of finicky breeder/power plants, but these requirements need not prevent useful competition any more than standardization of consumer electronic equipment (audio systems, TV, etc.) does. The problems are societal, political and economic, not technical, and we have failed to make that clear enough to those who make policy—and regulations. The bargain need not be Faustian—as one of our colleagues has suggested. I believe the university has a greater obligation than it has so far taken to make that clear: the bargain need not be Faustian, but it may be if we do not delineate all its components.

And so with the broader environmental concerns that this conference is deliberating. The universities should not only be a partner in research and policy analysis, but must take the lead in educating the general public on our work—warts and all.

There are some specific activities for universities that I recommend. First is the formation of multidisciplinary groups, with members selected for their ability to look comprehensively at the complex problems of the environment, policies that may be advisable, and the means to get the policy advice effectively to those who can act on it. Such groups may well be ideally suited to work with research groups in centers and laboratories. I will not comment on changes in university polices necessary to accommodate such groups: They have been solved at universities where I have introduced programs of that type. Second is the development of Legislative Information Assistance Programs, where legislators (at any level) may request of the program information on any subject before him or her. The programs need to be able to tap any faculty member for quick response, and while many faculty will be reluctant to provide the incomplete information that may be available, I have found that for inquiring legislators, who must act whatever the state of knowledge, a briefing paper with analysis and information is carefully considered and effectively figured into their legislation. Of course, only a few take advantage of the offered service, and I suppose these are the ones who are prepared to benefit from objectively presented information. This would add to the policy analysis and dissemination efforts of the broader environmental science community.

Finally, we should offer courses to the general student that treat the complex problems of the environment comprehensively—with several instructors in some cases, or with one willing to inform her/himself broadly enough to deal with the relevant materials from many disciplines. I offer such a course (and have for several years) that takes its content partially from the newspapers, newsmagazines, and general interest journals like Science and NSF reports. Thus about 40 students a year leave the course with (I trust) a clear idea of the limitations as well as capabilities of the sciences to address the problems of Society; of the systemic interaction among economics, policy, politics, science and technology, that characterizes all such problems, and of their roles as citizens: first, to inform themselves, and, second, to act on their information responsibly in the political process.

These programs work. They do not change the world overnight, but they are consistent with the great responsibility the universities have in a complex and dangerous milieu—a responsibility shared with others, but abrogated by the universities at our collective peril.

Socioeconomic Benefits of a Global Warming Event

R. H. Lessard, A. L. Archibeque, and E. A. Martinez
New Mexico Highlands University
Las Vegas, NM 87701

ABSTRACT

The world community is expressing a great concern about the potential detrimental effects associated with an impending period of global warming. Hominids have been injecting large quantities of waste greenhouse gases into the atmosphere since the beginning of the industrial revolution. Scientists are assuming that this activity will cause a significant rise in worldwide temperatures over a short period of time. They are predicting that the temperature increases will: change existing agricultural patterns, cause a rise in sea level that will devastate coastal cultural features, and result in the extinction/extermination of many organisms that will not be able to adapt to this change. The only alternative to global warming is global cooling. The Earth, prior to the industrial revolution, was on the verge of entering into a period of extensive glaciation. The data that support the idea that the planet is about to experience a cooling event include: changing orbital parameters that will result in reduced seasonal contrast, the atmosphere's inability to prevent the recent Little Ice Age, a possible reduction in atmospheric carbon dioxide concentrations since the early part of the present interglacial, and the worldwide expansion of deserts since the climatic optimum. The most important socioeconomic benefit that will be derived from global warming is that it could potentially prevent the inception of a global cooling event. It will: allow the world to continue to grow sufficient food to sustain its burgeoning population, prevent world governments from having to deal with mass migrations of people who will be driven from their lands by plummeting temperatures, avert the need for individuals and businesses to have to contend with problems related to building on permafrost, possibly cause a cessation of the desertification process that is in progress in many parts of the world, and eliminate the need for world governments to spend billions of dollars in order either to eliminate totally or abate the flow of waste greenhouse gases from machines and activities that may generate them.

The full paper is available from the author.

Part Four:

Panel Discussions and Conclusions on Integrating Climate-Change and Energy Policy

Panel One:
Climate-Change Scientists' Perspectives *

Lennart Bengtsson: If you look upon the scientific issue, I think I can say that, in spite of the lack of knowledge of details of this [global warming] problem, the scientists are by and large in agreement. There are always, in a case like this, different views. These may have been exacerbated by the media. There has been an antagonizing effect which has, perhaps, gotten too much press. But I think, by and large, in spite of the limitations of the models that we have been trying to explain here, and the problem of the relative smallness of the signal and the natural variability, the scientific community is aware that this is a problem.

So how can we advise the policy makers? The potential problems are certainly very large. We heard about Dr. Hammitt's using temperature increase as one factor. I think this is a mistake. One can concentrate too much on temperature. There are lots of other quantities that are more important—precipitation is one of them—and also it is quite obvious that a global effect will by necessity (for statistical reasons) lead to the fact that there will be some areas where the effect will be much larger. We have unfortunately not yet reached the state where we can make any reasonably safe scenario projections of the regional effect. The next generation of climate models, I think, will be able to do that much better. There is, secondly, the problem that the climate system may not be as stable as we believe. An example is the circulation in the North Atlantic. Some numerical experiments have indicated that even rather modest emissions of the fresh water in the North Atlantic can actually trigger a change in that circulation, which would have regional effects, at least in Northern Europe and at high latitudes around the North Atlantic.

Even if we make a decision now, it will take a very long time before the CO_2 values are going to change and that is one reason why we should not delay a decision unnecessarily. However, I think that we have to see the CO_2 problem in a wider context. There is also the question of the relation between the different greenhouse gases themselves. It has been mentioned here in the discussion by the energy producers that natural gas is something that could replace coal and oil. That would, of course, lead to an enhancement in the emissions of methane, nitric oxide, and certainly Freon, which have the additional effect, in spite of being much stronger greenhouse gases, of interacting with stratospheric ozone—which is, of course, an even more serious problem. So I think perhaps we should set some priorities about which of the greenhouse gases we should try to reduce sooner than the others.

---

* Panel discussions in this section are derived from audio transcripts of conference sessions.

Of course, it is very sensible to reduce energy costs for other reasons. Firstly, coal and oil are not infinite; they can be used in a much more intelligent way than to burn them up. Furthermore, countries that are short of these commodities will, perhaps, see an advantage to not being unnecessarily dependent on other nations. My own country, Sweden, actually introduced a CO_2 tax, in spite of the fact that a liter of petrol in Sweden is more expensive than a gallon of petrol in the United States. You can, of course, ask yourself is it sensible, because I don't think the Swedish population would be very unhappy about a slight increase in temperature! Sweden did it and I think it was a very conscious decision. The fact was that they were trying to see the global context because, even if you are very egoistic about things, there are serious environmental problems in other parts of the world, that are going to affect you anyhow. You can't avoid this because we are living in one world and we have to face the problems together. The European Community has taken a similar approach. In conclusion, I think that in spite of the fact that we have a very complex problem, this should not lead to inaction. After all, most of the decisions we are taking in our personal lives are not based upon certainty. In my view, in spite of the lack of firm evidence, there are very good reasons to make some sensible decisions as soon as possible.

Kenneth Friedman: I'd like to approach this from the standpoint of a policy person. I'm not a scientist, but I am immersed in the issue within the Department of Energy at a very practical level, being in conservation/renewable energy at the Department of Energy. Some observations: One is that the targets and goals and political commitments, whether it be from Sweden or other countries, do not guarantee an impact. We have to be very sensitive to the reality of the political statements that are being made versus the ability of countries to have the policy mechanisms at hand and the ability to implement them. Some of the things that Mark Levine talked about in terms of developing nations and the absence of an infrastructure apply to us as well. The United States requires some areas of technical expertise and needs to educate a generation to respond to these problems.

A second observation is that perceptions determine reality. In Japan the perception of energy security is very different from the perception in the United States. What they see as "vulnerability" is not necessarily what we see as "vulnerability." We have changed our perceptions of vulnerability over time, and I think we will change our perceptions of climate over time as the scientific data get better, but I can't believe that there is a model out there that climate-change scientists are going to develop that will persuade politicians that action is required. I think it's more likely that there will be political implications from destabilizing energy and environmental events — namely, dry weather in the Midwest or breakdowns in the electrical system in different parts of the country, or, as somebody said yesterday, "Summer mid-latitude continental dryness" — that will stimulate action.

Question from Audience: Once you say that it's going to take that kind of level of impacts to shift the political scene, then aren't you saying it is already essentially too late to worry about any mitigation?

Kenneth Friedman: Actually I don't think so. I'm very optimistic because we are beginning to take these steps already. We don't call it climate. We're now calling it energy efficiency. Energy efficiency is related to productivity and competitiveness. It's related to the Clean Air Act. We are starting, I think, the U.S. Government is already starting to take these steps. I think we are going to see, in national energy strategy legislation, many of the same things that you would see if you were to have, initially, a low-cost climate policy. The scientific community has only a few spokesmen whose literature I read—and it's very articulate—but you need more spokesmen and not as much noise in terms of the scientific debate. I think some more clarity would probably be useful. The final thing I'd like to say is I think there are a whole host of climate-related issues, which are also policy-related issues, that are not being discussed openly, at least in the United States.

I think other parts of the world are much more sensitive than is the United States to questions of terms of trade and technology transfer, the question of equity and additionality of aid to developing nations. The question of lifestyle is something that is not addressed in the United States; the Japanese, who have an incredibly efficient lifestyle, given their cold temperatures in the winter and warm temperatures in the summer, are trying to raise this. This is not being discussed. The whole issue of the nuclear option, which has been discussed at this meeting, for the most part has not been discussed forcefully enough as a long-term issue. It hasn't captured the imagination. Another item that is obviously not on the agenda in the United States, is the whole issue of population. Finally, we need much more data, hard data, from different countries' perspectives to allow meaningful energy comparisons of, say, the Danish refrigerator to the U.S. refrigerator. Europe is now going through this because of moves to establish the European Community.

Question: The question still arises, though, whether you really have time for all of these thoughts and decisions and comparisons or whether this problem has not overwhelmed you already.

Friedman: If it has overwhelmed us already, then we are in big trouble, because the Congress, at least, has seemingly moved away from this issue as a central theme, unless you have a political leader that's going to raise it at least in the United States.

Comment from Panel: That's the trouble. Until Iowa turns into a desert the Congressman from Iowa will not pay attention.

Friedman: I think there's a great sensitivity to this issue, but in my opinion the Congress feels that the energy-efficiency types of questions and the national energy strategy questions in the United States are a surrogate for addressing climate. Climate will come back quite quickly, especially after some political determinations are made.

George Golitsyn: There is another problem which hasn't been discussed much here, involving emissions from the energy sector that do not relate directly to the greenhouse gases but, nevertheless, influence somewhat the whole climate and environment debate. This is carbon monoxide, which is one of the main emissions of all the energy sectors. Carbon monoxide changes the hydroxyl radical, which is the main cleaner of the atmosphere. Indirect estimates show that during the last hundred years it decreased by about 30 percent, or maybe even more. It seems likely that this decrease will continue and, if so, there will be much greater pollution problems. Also, as part of the same process, atmospheric ozone (which is a greenhouse gas) is decreasing. This is just one illustration of the linkages between all kinds of environmental questions.

Another problem I see, at least in my Country and almost all others, concerns the education of the public and government. About two and a half years ago, I was in a session of my Government at which questions of global climate change were discussed. I think the then Premier understood the issues, but we need to educate the new governmental and parliamentary officials that this is something that might not be of immediate concern, but that it will eventually require their full attention. The same point relates to educating the public. How the public perceives these issues will determine the direction in which it moves.

Charles Keller: First of all, it seemed to me that the central question that we were being asked to address is: "what do we as climate scientists have to say to energy policy makers right now, not in five years, ten years, or twenty years from now?" What can we tell them that they can use next month, because that's what they want to know. I don't think that just making a massive attempt to convince you that there will be climate warming is particularly helpful. At the same time climate-change scientists need to hear what directions energy policy makers are taking. One of the reasons for getting this conference together was to see if we could open that dialogue. It seems to me that the climate-change scientists can help with the details. People are going off with big plans to do things which, if you look at the details, may require some modifications. I only need to mention the word "methanol," which was formerly described as the solution to all our gasoline problems. I think that the biggest thing we can offer the energy people is to read their writing, to comment on it, and to talk with them about it.

Education is also extremely important. Lennart [Bengtsson] was talking about the fact that it took a long time for acid rain to have a political impact and a shorter time for the ozone-depletion problem to be realized. My simplistic answer to that is that everybody waited till they saw "something" happen. It took longer from the time we saw acid rain coming till we had a lot of effects. But the minute people really noticed the effects they began to do things. The sort of explosion on the scene of the ozone hole brought people upright and they took action right away. As many people said here, the insidious thing about climate warming is that by the time you unambiguously see the effect, you may be a hundred years into a problem that you can't get out of. The public are getting more and more sensitized to these issues. One of the things I think is very interesting, and why we may have a chance to deal with global warming, is that the public no longer really distinguishes between global warming and global change. They see rain forests being chopped down, ozone, and Los Angeles smog, and they are beginning to have an integrated response to this. They are willing to look at the problems. The other thing scientists can do is launch a more articulate campaign for education. Without overstating the problem, we need to keep these vital issues on the agenda.

Jerry Mahlman: I'd like to repeat the question: What uncertainties drive policy makers the most? I think we need a psychologist perhaps to understand that, but I'd like to summarize what I said yesterday in one sentence: Do not forget that we know much, but relative to the detailed quantification of the problem, there's much we don't know, and the learning time scale is on the order of a decade or more. So there's no place to hide—but there's no place to be comfortable either. I think that's useful to know.

Some time ago I had proposed a thought experiment for policy makers. Suppose that climate models got infinitely smart tomorrow afternoon, what are you policy makers going to do about it? Do you have your act together? I'll let you work through the thought experiment in your own private way, but I think it makes a point immediately that we're all in this together and we can't hide behind somebody else not quite having their act together before we start to get our own together.

I think, to better understand what the policy makers wish to know, we should take another look at the CFC-ozone problem. I will just remind all of you of a couple of facts: One is that the decision to make a fifty-percent cut from the Montreal Protocol preceded the discovery of the ozone hole over Antarctica; the cut was based upon a theoretical projection. Yet many people are saying that the difference is that the uncertainty in the CFC-ozone problem was much less than that in the greenhouse problem. As one of the few people on the planet to have worked on both problems I say that ain't so. If you take the numbers that people wish to work on, ΔT (surface) for greenhouse warming and Δ percentage ozone depletion, global average, both of them were fundamentally uncertain to about a factor of three. And yet, the CFC policy makers immediately acted on the basis of the information they had while, in the greenhouse warming debate, the reaction has been very different. I would submit that the answer to the question is fundamentally non-scientific. In the case of the CFC problem, there were on the order of twelve companies who were the culprits, and all of them were nimble-footed enough to recognize the problem before the political system did. For the greenhouse problem there are arguably four billion culprits, and so it's a little bit harder to find the bad guy. I think that drives policy makers a lot.

The final thing I would like to point out—and I have had a lot of experience with this in the last couple of years—is the distorting aspects of the media. I don't think the press has a left-wing bias. I don't think the press has a right-wing bias. I think the only bias the press has is to find sexy news. I was quite astonished that the IPCC Report was a non-event for the press. The press, in its effort to sell news and to find the news, is always hustling the extremes of opinion. When I go to give Congressional testimony, I find that the Congress people are using the press to bolster their preconceived political notions about things. When we scientists talk about uncertainty, it's really quite amazing; they don't want to hear because their minds are made up one way or another: the sky is falling, doom is coming, or ain't nothing going to happen. So, as an educational tool, the press has been dismal. In terms of reporting of thinking and so forth, it's been very good. I think policy makers are, unfortunately, extremely influenced by this. And it's not because the press is hustling one point of view, it's hustling both points of view and the Congress people can pick off whatever they want, and they do.

Michael Schlesinger: How you view this global-warming issue fundamentally depends on your assumptions concerning the sensitivity of the climate. There is currently a factor of ten uncertainty between some assumptions concerning climate sensitivity. And that makes it extremely difficult to know what to do. So we have to narrow the uncertainty of the sensitivity in the climate system. We cannot do this with climate models alone, because we can change the treatment of the physical processes and climate models that we cannot explicitly resolve in a variety of ways to produce a variety of answers, all the while still simulating the present climate about as well as we are currently doing. In other words, simulating the present climate doesn't tell us that the model is sufficiently well formulated that we can believe its sensitivity. I think we are going to have to try and use the models in comparison with observations to pull this sensitivity out. One way of doing this is to compare the results from relatively simple models, not in the future but in the past, to try to determine what the sensitivity of the climate system is.

We need to be able to calculate the regional climate change. Here the problem is extremely difficult. Jerry Mahlman talked about this yesterday and basically said that we're going to get higher-resolution models and I think that's true. As you know, increasing the resolution by a factor of ten requires a computer a thousand times faster than what we presently have, and we'll probably have such computers in five to ten years. That's necessary, but it may not be sufficient, to do the job well. I'm not too optimistic about reducing the uncertainty of the projections of the regional climate changes and that's what we really need, of course.

At this meeting, I've been a bit surprised that we haven't had anybody to bridge the gap between projections of climate change and "what should we do about it?" The climate changes in and of themselves are not the only issue. There hasn't been anybody here to talk about impacts: so what if the climate changes? It seems to me that we need to put together a team or maybe five teams, around the planet, that are not allowed to talk to each other, except maybe quadrennially. The teams would consist of people who do climate modeling and people who do analyses of the impacts of climate change, so that they can take the simulation data and work with it. Currently, when the impacts-people get the climate data they say, "thank you very much," and then go away. You hear nothing and when you see the results it's often garbage, I'm afraid. I think we need a team of climate scientists, people who handle impacts and policy analysts. The team can't be too big or else the work won't get done. It should be on the order of twenty people that meet twice to four times a year to try to work out the whole picture.

Panel Two:
Energy Researchers' and Policy Makers' Perspectives

Louis Rosen: First of all, let me say if the world is tired of energy research, the world is doomed. I really mean that seriously and I hope that is not the case. Several hours ago I received a statement from U.S. Senator Pete Domenici. He was supposed to be here at this conference, but the Senate is in session. He just wanted me to convey to you this very brief statement, because it has to do with what we want from the global-climate-change community: "I welcome this distinguished body of scientists to New Mexico. The subject matter before you is one that has become of increasing concern to policy makers in this country and countries throughout the world. Your overall objective to determine how best to communicate and work with policy makers is a critical one. A subject of this magnitude will require that lawmakers and energy-policy planners are fully apprised of the most recent scientific information that has become available. Any decisions made on actions to be taken in the future must be based on the best possible information, given the enormity and complexity of this problem. I look forward to learning the results of this and future conferences."

How do we get to these policy makers the best possible information? As I sat and listened and learned through these discussions, I was struck by the fact that the situation with respect to data—almost raw data—about the basic physical and chemical inputs to the models appears to be at the same stage as was neutron cross-section information just after the Second World War. It was a mess. What was going on then? People were busy. They knew it was an important problem. You couldn't build reactors without these neutron cross sections, and they had to be done right. The theoretical people needed, as the modelers do right now, reasonably accurate information, the uncertainties of which they knew. It wasn't available. The data were being accumulated in the Soviet Union and as well in Europe, England, the United States, and all parts of the world. But there was no coherence.

Well, we invented a mechanism for bringing order to this chaos. The mechanism worked. It involved organizing ourselves so that the information generated would come to a central repository where it would be evaluated, where discrepancies between measurement at different laboratories could be resolved. When necessary, senior persons would travel from one laboratory to another, and it was by this means that we were able to build a base of credible, reliable, reasonably accurate information, which the reactor designers could then use with some confidence. I submit to you that we should think in terms of creating such a mechanism now. It would have been better five years ago, but it's got to be done, because without that I don't think the climate- change community and the modelers will ever be able to attain the credibility that is necessary to persuade the Congress that it must take seriously the recommendations and the results of these calculations and models and act upon them.

George Hidy: Let me just make a few comments on where the electricity industry is coming from and what it would like to see from the world of climate-change science. Basically the industry is faced with certain undesirable economic conditions at the present time. It is also faced with providing the American people reliable electricity at the lowest possible cost, including environmental cost. The socio-political environment is dictating rather stringent conditions on the utility industry as it goes into the nineties and the next century. Within that context, each issue that comes up to the electricity industry that involves, either separately or together, environmental policy, energy policy, and economic policy (let alone population increase) is questioned by the industry and its policy makers. It's questioned in a "relative risk" framework. So what is necessary for the scientific world to be convincing in terms of its predictions for the future is to couch the arguments not in terms of climate change or climate alteration alone, but in terms of the economic risk that that poses to the United States. The risks have to be traded off with one another and with other relative risks that exist within our society. I won't go into the technology and mitigation options that are available to the industry, but the short term strategy is to buy time, and that means efficiency improvements. In the long term, it is very uncertain what to do, but certainly, as we see it today, carbon is not out of the picture by any means. So we have to manage carbon with regard to our electricity requirements in the future.

The last dimension of the electricity industry approach concerns the time it takes to change to a new fuel. A forty- to fifty-year time period was mentioned earlier, based upon some of our previous experience. But whatever we do that requires extensive and disruptive change will take that fifty years scale to implement.

Finally, the industry views its situation here as one of a Jekyll/Hyde character: partly being culpable and targetable and partly one of perceiving a major opportunity. Within that context, we need in the science community information that can be used in a so-called "integrating risk assessment" framework. This means we need to link the earth sciences with environmental science. Peoples' minds on this issue will be changed by economic issues: will the economic significance of climate alteration tear us away from economic development across the world? In order to answer that question, we need not only the climate science information we've talked about here but the impact of the effects information, and what significance that has in economic terms. We don't have any kind of information on that today. All we know from zero integrated assessment analyses is that whatever that function is—loss in gross world product versus some measure of climate change—has to be very non-linear to make a compelling argument to do anything other than adapt to whatever the future holds for us. So I propose, then, that effects are going to be a very important adjunct in the climate science area that is largely untouched today in terms of quantitative information that we desperately need. Finally, socio-economic analysis is going to be very critical. We tend to look at these problems in the United States from our own perspective. We need to understand what the values and the pressures are on those societies that are quite distinct from ours, that would drive environmental protection considerations against economic projections. This means that there is a tremendous investment needed in socio-economic research that also must accompany these other elements. So I would challenge you as the climate- change community to try to find ways of putting that integrated system framework together. The U.S. billion-dollar investment in climate science should include much more effort in the socio-economic area, so that policy makers can make sense out of this extremely complex puzzle that has, thus far, surfaced in a very fragmented way.

Serguei Kapitsa: I would like to begin with a comment made by Alvin [Weinberg] about what we can expect in the middle of the next century. I remind you that world population is expected to reach the ten-billion mark. That means it will be twice what we have today. If we want those people to live as decently as the advanced and developed part of our world is now living, we would have to have from seven to maybe twelve times more energy than we have today. I think that has to be kept in mind. I'm not specifying what sort of energies, but we have to seek a non-carbon alternative on a very major scale. I think it would be worth working out in greater detail the demography with the energy projections for the future.

What I think is important, and here I will go back to the remarks of my colleague Dr. Golitsyn, is to assist in educating the public, the politicians, and the media on these matters. I can tell you of a conversation that I had in 1978 with Prime Minister Palme [Sweden], who had just lost his premiership over the nuclear issue. I asked him, rather bluntly, "How did it happen?" He said it happened because he didn't give any concern to educating the Swedish public on matters of nuclear policy. That is how one of the most remarkable politicians of this century lost his position. Things haven't changed for the better since those days in many parts of the world. Chernobyl was not so much a technological disaster as a social and psychological disaster. The human component was more important then the technical component of this accident. In a certain sense I think it was like a Greek tragedy: We were predestined for this to happen sooner or later. So the whole question of education — the human component of the energy policy — is something that has to be taken very seriously. It's really a question of the extent to which the human population at large can face the issues of modern technology. We have to explain these things as a serious international global educational effort.

Virginia Oversby: I'd like to raise an anecdote from my graduate education involving a model calculation that we were forced to read as graduate students. It was published in about 1964. I won't name the person but this eminent geophysicist did a calculation that "proved" beyond a shadow of a doubt that continental drift could not occur. This was about two years before the plate-tectonics revolution. And so, I developed at that time a dislike of models with too many adjustable parameters. My rule of thumb is, "with three I can prove the world is flat beyond a shadow of a doubt." I don't mean to say you shouldn't do modeling, but that you should distinguish clearly between numerical simulations and the adjustable parameters in the sensitivities of your results and the things that I would call "true" experiments, where you predict in advance something that will happen. I think experiments are only really valid if you start with a hypothesis that includes the modeling. The climate modelers can most help the policy people in identifying where testable hypotheses are. We need to better understand the physics and chemistry and tie down these adjustable parameters so that we can narrow these large ranges of uncertainties concerning the climate response. The other point I'd like to raise is: what is it that really affects the climate? Is it global warming or is it something that is more detailed than that? There should be more emphasis on natural variability: the effects of volcanic eruptions, aerosol injection, dust injection. This natural variability is something that was alluded to several times but I feel it's something that deserves a lot more research emphasis.

One of the things that I got from listening to people is that the cost of intervention in this business is very high. But the cost of doing nothing is also a means of intervening. So with both positive and negative intervention, the cost economically is high and the impact, if we get it wrong, is very high. We really need to pay attention to this.

I get a feeling that we're putting emphasis on changing from coal to renewable energy technologies or non-carbon energy production technologies and using global warming as an excuse for doing this. I would like to see us instead argue that it makes sense to replace the burning of a non-renewable resource with other, more sensible, renewable ways of doing things. This should be a "first principles" argument. We should separate the climatic effects, which are important to discuss, from this more fundamental issue. Hidden agendas are something I have a great dislike for.

Ruth Reck: I think it's very important to realize that most of the decision makers in the automotive business think very seriously about not just global change, but also many different environmental areas. We really think about the environment in the global-change sense, in that we have to look at all of the impacts, whether it be regional, local or something we're putting out that becomes ubiquitous, like CO_2. The concept is global and it is global change with a strong component of global warming, but not exclusively that.

In the seventies we were called upon to define a means for countering air pollution. We learned a lot of lessons there because what resulted wasn't particularly efficient. There was a great push to put catalysts on the automobile. If you lived through those times, you'll remember that it was an adversarial situation between the government and industry. I think it's absolutely mandatory that that doesn't happen with global warming. This problem is not just an urban problem. It's not something that's fixed by just one industry. We're all in it together and it's incremental for all of us. I would like to see government, non-governmental organizations — everyone — exchanging information and ideas. I really buy into this idea of the "home planet." You know we only have one Earth, and so the adversarial approach has to go. We welcome any and all suggestions from everybody and we really do see that we're in this together.

Now you've all mentioned education. I heartily recommend that too. I think that's very important, as is dealing with the media. I think we could do a lot more to improve the way we do it. Legislation is written and passed in the Congress. Individual Congressmen don't write it, their staff people do. It is, therefore, very important that climate people have a good dialogue and bring the best science and the best ideas to those people who write the legislation. This is a major responsibility.

The people at the top of General Motors and elsewhere in the automobile business are engineers; they talk engineering. When you describe something it really should, therefore, be in engineering terms. When I talk to them, I keep this in mind. We must appreciate these different perspectives.

Dwaine Spencer: I would say our most important need is to understand the implications — the plausible damage consequences — of CO_2 increases (and increases of any other greenhouse gas). There are two primary reasons that this is the single most important objective. The first and foremost is that in the context of trying to deal with the carbon dioxide mitigation issue, we're looking at a cost of tens to hundreds of trillions of dollars. These are significant amounts of money that will have many implications in the context of economic development, and the displacement of people like coal miners and the whole coal industry in this country. We were able to absorb the acid rain problem within a financial structure that was large but with a relatively small impact on our industry. I'm very dubious about the ability of the U.S. utility industry to absorb the CO_2 issue, and I'm even more dubious about the prospects on a world-wide basis.

Although I cannot properly represent the interests of the developing countries, I believe we delude ourselves in believing that the developed world will define their economic fate. We have established our powerful economies based on developing both our domestic and international resources to meet our national interests without external intervention. Can we really expect other sovereign nations to adopt a philosophy other than one that enhances or maximizes the benefit to their citizenry? They wish to use least-cost approaches to enhance their standard of living. They do not have sophisticated buyers or operators or in many cases not even sophisticated controlling entities in the governments that manage them. I have heard Tony Churchill of the World Bank say the first issue is how to alter institutional structures in these countries before we can infuse technology or new philosophy. I believe the developing world will primarily focus on the development of domestic resources, and these tend to be primarily coal. I personally believe coal will reemerge as the primary fuel type in the first half of the 21st century. Although we should do everything possible to make coal's use as efficient as possible, CO_2 emissions will inevitability increase during the next fifty years. So that's again why we need to know the associated consequences. We need to obtain realistic projected rates of change of greenhouse gases as well as maximum estimated acceptable atmospheric concentration changes as estimated by individual countries. This necessitates an international collaborative program to estimate emission sources based on individual country projections updated, perhaps, every five years. This will provide us both a firm planning basis as well as a meaningful measure of the effectiveness of conservation and energy-efficiency improvements on a global basis. Perhaps this would be a good mission for the IEA. We need to know if there are bounding economic scenarios, globally, which will inherently limit growth. I believe these should be based on historical post-World War II growth among the developed countries, as well as capital limitations being faced by developing countries. The U.S. economy has been the focus of world markets for the last four decades. How could our economy have supported so much development on a global basis over that period? I think that's a good question for the economists to address. I believe we should continue to seek specific approaches for achieving energy growth needs over a range of uncertainties of conservation and energy-efficiency impacts, in order to produce potential supply-side requirements. Again these should be carried out on a nation-by-nation basis and iterated on a five-year basis. We clearly need more careful assessments of the potentials of making carbon a renewable resource. I believe that we need to look at both terrestrial and marine biomass systems, and these should be considered in the context not only of sequestering carbon, but also in terms of providing energy displacement potential, food production potential, specialty chemicals potential and the ability to, perhaps, feed ten to fifteen billion people in this period of the mid-twenty-first century. In summary, we should be evaluating systems that not only have a CO_2 stabilization counterpoint, but also a broader sustaining role for the world's growing population, and one that will allow the developing world to attain, sometime in the twenty-first century, the standard of living that we are so blessed with today.

Robert Stokes: While it's certainly true that the budgets that the Department of Energy directs to renewable energy are only a fraction of what they were during the peak years of the Carter Administration, we're still fairly upbeat about it because the budgets for renewable energy have increased by about two-thirds in the last two years. So at least the trend is in the right direction. I personally agree that a transition to energy-efficient economies and maximizing the use of renewables makes sense regardless of the eventual answer to the CO_2 issue and the global-climate issue. I personally believe that there will come a time when we will regret the fact that we used so much of our limited petroleum resources for energy content, rather than saving it for the use of future generations as a chemical feedstock. Certainly for those of us who work in the area of energy technology one of the most critical inputs that we require from the climate-modeling community is some better sense of the time scale and the urgency attached to the transition that we must make happen. Does it need to happen in twenty-five years, fifty years, or a hundred years? We heard a little bit earlier today that a two-degree net warming is about the limit we can accommodate, and that we could have anywhere from eighty-five to one hundred twenty years to adopt policies to arrest warming at that level. The kinds of priorities you set and the allocations of resources you make for research and development depend very much on whether you are looking at a one-hundred-year problem or if you need to make a transition in twenty years, as some people have suggested. In the transportation sector, for example, we are currently giving very high priority to technologies to convert cellulose from biomass to ethanol for use more-or-less in conventional internal combustion engines. Much of the driver for that transition is the Clean Air Act, rather than concerns about global climate change. However, the result is the same. If we felt we had a longer time to deal with the transportation issue, we might invest far more than we are currently investing to develop far-out technologies, like using photovoltaics to produce hydrogen for fuel for internal combustion engines or for other energy applications. We don't see any possibility of getting that technology developed and integrated into our transportation infrastructure within twenty or twenty-five years however, but we do see a very real possibility of substantial shifts to alcohols within twenty years.

On the electric front, I certainly agree with Dwaine Spencer that the key issue for electricity production world-wide is: how much of our abundant coal resources can we permit ourselves to use over the next fifty years or so? Certainly if the nuclear and renewable agendas fail or are less able to accomplish their goals, coal will be the default fuel for the United States, China, and many other countries. You must remember that there is a very long lead time associated with the production of adequate photovoltaics to make an impact on the world's energy mix. I should also remind you that current world wide manufacturing capability for photovoltaics is only fifty megawatts — a tiny number in terms of world energy needs. It takes a long time to get the capital investment and to build up the manufacturing capacity to begin to think of gigawatt levels of production. And that's, of course, also true for conventional nuclear power and certainly for advanced, inherently safe, reactor designs. We're not going to see a big nuclear impact in the U.S. in the next twenty years. If we make some changes very quickly we might begin to see some impact after twenty years.

Finally, a better time-scale estimate would also permit us to begin to make intelligent choices relating to a variety of CO_2 mitigation technologies. For example, rather than channeling carbon dioxide recovered from coal processing to deep ocean pools (as was discussed earlier today) it might make more sense to develop and to invest in the R & D associated with solar-driven photo-conversion technology processes, which certainly can convert carbon dioxide scavenged from stack gases at power plants to high- value chemical commodities like methanol. The technology exists to do it but we're a long way from being able to introduce it in an economical way.

General Discussion:
Climate-Change and Energy Policy

Patrick Finlay: I would like to echo the unease that has been expressed by a number of participants at this conference concerning the possibility that scientific uncertainty will lead to lack of action. We may be comforted by projections that show it doesn't make any difference whether we act now or a decade hence. But I think we must put this in perspective; this is crystal ball gazing by macro-economists. I would like to go back just a little on environmental issues. Let us look at contaminated ground water, let us look at acid rain, let us look at smog. We have taken action on these issues; they have been remedial actions. The interesting thing about this global-warming issue is that we are getting an early warning from most scientists that something big may be going on. There are a lot of people here from the nuclear business who are used to dealing with risk. I would suggest to you that what we're dealing with here is a medium probability outcome with probably pretty high impact. I don't think that we should "cry wolf" and on the other hand we don't want to give too much comfort to policy makers that everything is OK, and that we will carry on just doing the science. I think the pragmatic point has been made that there is a time lag in our democratic political system, between when policies are identified and when we get action. In the seventies it was oil security concerns and now perhaps it's global warming. To summarize, I would like to echo the concern that a number of participants have shown here; I don't think we should be too comfortable about data shown here that it doesn't make any difference whether we start to act now or a decade hence, when we're sure. Because we'll never be really sure until after it's happened.

Comment from the Audience: We have to find a mechanism for keeping up a degree of pressure on policy makers so that they do not lose interest in this issue. Dr. Michael Schlesinger's research suggests that maybe it doesn't matter whether we start doing things now or ten years from now. I am increasingly uneasy over that conclusion, because it is really an artifact of the mathematics involved. When you move from a high-growth curve to a lower-growth curve you insert an additional fixed amount, which looks increasingly small as you go up. And that is true now, it's true ten years from now, it'll be true fifty years from now. If you look at it in a different way, if we need to get the CO_2 emissions down, then that fixed amount will be huge. This is a beautiful excuse for policy makers not to do anything. It's not that the conclusion is wrong; I just think we have to be careful about it.

Michael Schlesinger: If you will examine the RAND Corporation study presented earlier [Hamitt and Lempert] in which the cost of the "wait and see" approach was evaluated relative to the "precautionary" approach, you'll find that it concluded that the "wait and see" approach is less expensive.

Rejoinder: But that study was rather simple. I really think that if you look into it in more detail you would come to the opposite conclusion.

Michael Schlesinger: Let's put it this way. This [the RAND study] is the first analysis [of this kind] and it indicates that what you just stated may not be correct. Now you may ultimately be correct, but you are first going to have to do some homework.

William Zagotta: I guess the most striking thing that I didn't hear at this meeting was that nobody said we should tell our government to send more money. I think it's perfectly clear that there's a lot more data that's needed and a lot more modeling that needs to be done. We need not only U.S. money but international money devoted to this problem. Secondly, it seems clear that greenhouse gases must not be permitted to increase to infinity. That's one issue we should all be able to agree on. While we may not know what to do about it, because all the research isn't done and we don't have all the energy technology, we should at least identify who the enemy is. Thirdly, we should have spent more time on the ethical dimensions of these issues.

Louis Rosen: One of the uses for these proceedings is to help get all of you additional resources. You can't just go and ask for money and then say, "trust me, give us the money and you won't be sorry." It used to work that way; it doesn't anymore. What we must also do is show that we are productive. We should try to organize ourselves. We should ask, "How can we bring some coherence to this great variety of enterprises that are going forth world-wide to obtain data to do modeling?" How can we get to the point where what comes out of these models is believed not only by the scientific community, but also by the people who need to understand the meaning of the conclusions we reach?

Charles Keller: You'll notice that this conference is about global change, not about global warming. I think that one of the things we can do to help is to build on the fact that everywhere the public looks it sees environmental change — negative changes. There are a lot of different aspects of global change. We should be trying to address the broad spectrum, not just the narrow issue of whether it's going to be a two degree change or a three degree change. Certainly if it were going to be five degrees, I'm sure that would overwhelm all other threats. But in the absence of that, I think we should look at things more broadly — in terms of environmental change.

John Laurmann: I think the only way of making a strong argument for doing something quickly is by a completely different approach that has no economics in it at all. It is simply to look at climate projections and ask, "How do they compare with what we've known about climate change in the past?" And the critical feature here is the rate of change of climate. The argument is not that there will be big costs when you reach the rate of change of climate that matches what we had in the past, it is that we don't know what's going to happen if it gets bigger than what's happened in the past. All we can do is assume that evolution has taken care of — at least — ecosystems' abilities to respond within that temperature rate of change range, so we'll be all right.

Comment from Audience: There's been a lot of talk about getting the energy community and the climate community to combine efforts. I'm not aware of an energy community. What I see are many separate groups like a conservation community, a solar community, a fossil-fuel community, and a nuclear community. Neither at this meeting nor anywhere else is there very much communication among them, although I think there's a little communication between the solar and conservation people. I don't think that just thinking about an "energy community" covers the great complexity. I think as an early logical step, people interested in energy should reach more of a consensus and understand what they agree on, disagree on, and why. You cannot wait for people interested in energy to talk to people interested in climate because you'll end up waiting until the year 2100.

David Bodansky: I am not aware that there exists an independent U.S. organization that looks into various aspects of energy in a non-partisan way. I think such an entity would be quite useful, with a permanent base of operation. I was aware of one, the Institute for Energy Analysis at Oak Ridge. It served an extremely useful function. There is no replacement for it.

Alvin Weinberg: France has been the most successful country in terms of reducing CO_2 output—although it had no intention of doing this. Why was it so successful? Because, for a variety of reasons, the French were able to accept nuclear energy. Given the uncertainty, we should do those things that make enormous good sense whether or not you're interested in CO_2, but which will be necessary if you wish to respond to the CO_2 problem. It is worth many billions of dollars to develop nuclear power systems which are acceptable to the public and economical. My main point is that our basic priorities in R & D are looney. We are prepared to send people to a space station at costs of the order of what is necessary to make these non-CO_2-producing energy technologies economically viable, and that would allow us to avoid the necessity for heroic decisions in order to deal with this problem.

Alan Crane: Alvin is probably right that we are making some very screwy national-priority decisions. We make them on an ad hoc basis—we don't compare space stations to breeder reactors to AIDs cures. We don't take a bloc of money and say, "What is the best way to allocate this money to help the nation?" We don't have a mechanism for doing that. And that is something that could start to evolve from the scientific community as a whole if, in fact, the scientific community can resolve its micro-self-interests.

Question from Audience: All the discussions have been about trying to control the use of fuels—tax them to discourage people from using fuels. If I were a dictator of the world, it would be a quite simple problem. I would put a number up on the board, then say, "That's all the oil you've got, World, for a year. That's all the coal you've got, and I'm going to look at these numbers in ten years, and I might mark a zero at the end if I don't think it's good enough." Then these amounts should be apportioned among the countries; those countries that were clever and have money could use their energy wisely, the other countries would either have to learn to use their energy wisely or come to other countries and get help. And that's one way you could essentially encourage efficiency and keep total control over the greenhouse effect. Now, in many ways, the United States has drawn a line around the world concerning which Third World people can have nuclear bombs. I just wondered if things could get extreme enough that such similar action might not have to be considered in relation to the consumption of fossil fuels.

Ted Taylor: Well, I certainly hope not. What you are describing, which sounds similar to what people are aiming for, is "top-down" with a vengeance. I think we have to avoid putting ourselves into a situation where it all has to be top-down for, among other reasons, who at the top is going to decide how much oil and how much of this and how much of that? It smacks of energy management without knowing anything about how the management is going to be done. So I think that some of the solutions we search for will work, and some of them won't. If there's one key word to search for as a guiding principle in what we do, that word is "flexibility," the ability to act in diverse ways depending on what turns out to be important. We don't know enough now.

The market does not work in terms of providing the for-profit motive for doing the R & D to respond to this problem. But our government administration says that if a solution is all that good, the market will see to it that we have solar-electric cells or inherently safe reactors—that's nonsense. It has to be done by government; the private enterprise system won't do it, at least in the United States.

Mariano Bauer: I was a little disturbed by the comment made earlier about the revision of the climate sensitivity that could bring down the expected warming from 5 degrees Centigrade to projection to 2 degrees. This could be perceived to give countries that have a large population growth and income problems—essentially developing countries—an excuse not to participate or do their share in this global climate problem. I think this audience should be reminded that these countries have very pressing problems. They don't need that excuse to make the reasonable choice of not immediately adopting what you would like them to adopt in response to climate change. After all, the effects of ozone depletion and climate warming may be felt in five or ten years or further in the future, but if you don't have food or you catch an infectious disease and you don't have any medicine, you are sure to be dead in two weeks. Any projections of the timing and severity of climate change is not going to be taken by developing countries as an excuse for inaction. There are very valid reasons for developing countries not to accept the urgency you might see, and this should be taken into account and brought into the discussion.

Impressions of Workshops on
Energy Technologies/Environmental Impacts

Theodore B. Taylor

Many aspects of the environmental impacts of alternative energy technologies that could significantly reduce the release of greenhouse gases were presented in some 20 papers in two concurrent workshops the morning of October 22. Since this is a vast subject, and I could be present at only one of the workshops, I cannot even attempt to summarize the results in a few minutes now. I shall therefore confine myself to some selected impressions of this subject as explored in this conference.

My sense of urgency about global action to suppress releases of greenhouse gases has increased during this conference. My main concerns center more on the increases in climatic instabilities caused by changes in regional energy flows than on effects of gradual increases in global average atmospheric temperatures during the next several decades.

Realistic predictions of the character, extent, and effects of these instabilities are likely to remain impossible for many decades, if ever, since the number of types of significantly different effects are uncountable. It is folly, I am convinced, to delay actions to suppress these instabilities until they become clearly apparent.

I therefore call for unprecedented global action to develop specific regional strategies for rapid abatement of the release of greenhouse gases. Assessment of alternative components of these strategies, as appropriate for a wide variety of natural, economic, and political environments of the human habitat, must be completed with the greatest sense of urgency. I propose that these strategies be developed and ready for widespread implementation before the end of this century.

Meeting this goal will be costly, but likely to cost much less eventually than waiting another decade or so before the goal is achieved.

Present total annual costs of energy in the United States are roughly $500 billion per year. Global annual energy costs are now in the vicinity of $2 trillion. Cumulative world energy costs between now and the end of 1999 may approach $20 trillion, allowing for moderate increases in energy consumption in the developing world, while significantly reducing energy consumption in the industrialized countries.

I therefore argue that it would be reasonable to allocate at least several hundred billion dollars for thorough assessments, by the year 2000, of global alternatives, region by region, for sharp reductions in greenhouse-gas emissions within the first decade of the 21st Century. In some cases actual implementation of promising alternatives could proceed significantly before the end of the 1990s.

These alternatives include increases in the efficiency of use of energy, as well as displacement of fossil fuels by other sources of energy that do not release greenhouse gases.

In either case major new research, development, and demonstration projects must be carried out to assure that the information needed for assessing alternatives is realistic. I include the cost of such projects in the allocation of total costs of determining the best strategies for sharp reductions in release of greenhouse gases, without incurring other major environmental costs.

The strategy development efforts must also include complete system analyses of the following, for each type of alternative:

1. Projected inputs, inventories, and outputs of all forms of energy, significant materials, and money per unit product, for each major step in the production process, from acquisition of needed natural resources through ultimate disposal of waste energy and materials.

2. Comparative assessment of important environmental impacts of each step in the production process, and of the process as a whole, including not only the release of greenhouse gases, but of other potential pollutants, including waste heat. The assessed impacts should include safety of workers and the public, and possibilities for purposeful abuse of the technology for destructive purposes.

3. Overall assessments of the economic and political practicality of each type of system in a variety of types of economic and political environments.

4. Use of comparable measures for assessment of alternatives by different groups of people, to help avoid conflicting results of assessments of similar alternatives in similar settings.

In the exploration of alternatives, emphasis needs to be put on substantial reduction of emissions that can have significant environmental impacts, rather than on ways to manage impacts of large quantities of emitted materials that often cause extremely complicated environmental changes.

To help assure that a wide diversity of alternatives will be examined, and that they can be compared in meaningful ways, strategies for specific regions should be developed at various levels of social and business organization, rather than only from the top down (i.e., only by national governments or multilateral government coalitions, or by large corporate establishments).

To help assure that information used for proposing and assessing alternatives is realistic, and also widely available to many types of non-governmental enterprises and grass-roots organizations, government or industrial secrecy concerning possibilities should be minimal and perhaps abolished altogether.

Transportation and the Global Environment

Sumner Barr

Los Alamos National Laboratory

1. INTRODUCTION

Transportation accounts for about one-fourth of the energy used globally, so any discussion of the relationships between energy use and climate must take into account the transportation sector. The special requirements of transportation (e.g., liquid fuel with a high energy-to-weight ratio) also account for a fuel profile that is comprised almost completely of petroleum products. Hence, although only one-fourth of the world's energy is used in transporting people and goods, nearly two-thirds of the petroleum goes toward transportation. Emissions control presents a very different problem for mobile, distributed sources than for the centralized facilities commonly associated with power production.

In the course of planning for and participation in the *Conference on Global Climate Change: Its Mitigation Through Improved Production and Use of Energy*, we formulated a series of issues and discussion questions involving the relationships between the energy usage for transportation and global environmental change. Many of the ideas were explored in conversations that took place in the corridors and in other informal settings. However, they were part of the conference and have been recorded in this note in order to help establish a common basis for energy use and global impact of the transportation and power sectors.

Figure 1 demonstrates the perspective that we applied to the relationship between climate change and transportation energy use. It depicts the goal of transportation as providing mobility to people and distributing goods and services as widely as practical. The industry attempts to meet those goals subject to a series of constraints including economics, regulations, safety, and impact on the environment. The present discussion focuses on the last of these, in particular the emissions of greenhouse gases (mostly CO_2 as the final product of fuel combustion).

2. PROFILES OF TRANSPORTATION ENERGY USE

According to Davis (1990), approximately one-fourth of the energy used worldwide is invested in transporting people and goods. Statistics compiled by General Electric (1984) show that the U.S. energy-use profile is similarly partitioned, with about 20 quads out of a total U.S. annual consumption of 70 quads going for transportation. A quad is a quadrillion Btu, or 10^{15} Btu, approximately the amount of energy used per year in a major metropolitan area. Another convention for fossil-fuel usage is one million barrels per day crude oil equivalent. A good, simple working approximation is that one

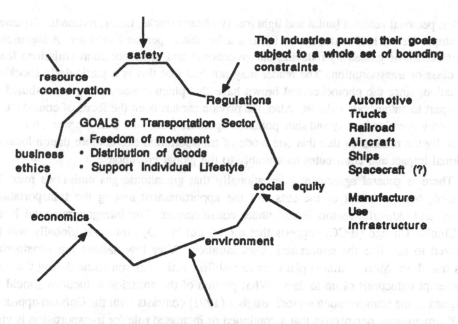

Fig. 1. The Transportation Sector Goals and Constraints.

million barrels per day is equivalent to 2 quads per year. We will also make the working hypothesis that the greenhouse-gas emissions are proportional to the energy usage.

With as much as 25% of the potential greenhouse perturbation attributable to transportation, it makes good sense to examine transportation practices and infrastructure with an eye toward conservation, alternatives, and the design of future transportation systems that will produce less CO_2. Table I, extracted from General Electric (1984), shows several measures of energy expenditure distributed among the 6 largest transportation

TABLE I. Transportation Energy Use, 1980 Statistics on Selected Modes of Transportation.

	Autos	Trucks	Buses	Rail	Air	Ships	Total	
Numbers, millions	104	36	0.5	1.7	0.2	0.04		
Vehicle-miles	82.70%	15.40%	0.40%	0.10%	0.60%		1.3	trillion
Passenger-miles	73.50%	13.20%	4.30%	0.80%	7.70%		2.8	trillion
Freight, ton-miles		19%		30.40%	0.10%	27.90%	3.0	trillion
Energy use, quads	46%	26%	0.70%	3.30%	7.80%	8.80%	20.05	quads

modes, personal vehicles (autos and light trucks), heavy trucks, buses, railroads, airplanes, and ships. Nearly half of the energy use is attributable to personal vehicles. A significant benefit could be gained by reducing the greenhouse and other pollution emissions from this class of transportation. The reader may not feel that this is a particularly shocking revelation, since the photochemical brown haze that plagues many cities is attributed in large part to automobile exhaust. Also, the private car has been the focus of considerable regulatory attention by city and state pollution agencies across the U.S. We were, however, struck by the realization that this one mode of transportation exerts more than a local or regional impact and contributes measurably to the global greenhouse.

There is general agreement internationally that greenhouse-gas emissions must be reduced, but the extent of the cuts and the apportionment among the transportation, power, and industrial sectors is still under consideration. The Intergovernmental Panel on Climate Change (IPCC) suggests that a 60% cut in CO_2 emissions globally will be required to stabilize the atmospheric concentration. Very few nations are considering cuts that deep. More common plans are to stabilize emissions sometime during the '90s or attempt reductions of up to 25%. What portion of the emission reductions should be assigned to the transportation sector? Hughes (1991) contrasts with the German approach a UK government perspective that a continued or increased role for transportation is vital to economic growth. The Germans have a goal of 25% reduction but with deep cuts in some sectors and smaller cuts elsewhere. The transportation cuts are planned at only 9%; an acknowledgment of the economic importance of transportation.

A modest amount of reflection on our approach to transporting both people and goods suggests that there is plenty of room for conservation without seriously reducing our essential mobility or negatively impacting our economy. Savings in the range of 10% to 20% could be achieved by simply educating ourselves on more efficient driving habits and vehicle maintenance procedures. Much larger reductions are available to those states or nations willing to take an integrated view of their transportation needs and options and to develop incentives to promote fuel conserving behavior.

3. A START ON A CONSERVATION STRATEGY

Figure 2 is a very preliminary approach to a cost-benefit diagram for transportation energy savings in the U.S. following the conceptual formulation of Fickett et al. (1990). The abscissa is energy savings in quads. The ordinate is the cost of implementing the particular savings strategy. Because of the serious limitations of the author in economic analysis, the ordinate has been scaled to equivalent quads of energy used. In Fig. 2 an energy strategy that introduces a benefit at an equivalent cost in some other portion of the energy budget (alternative fuels, electric vehicles operated in the same manner as current gasoline vehicles) would appear along the 45-degree diagonal. The portion of the diagram below that diagonal represents fuel conservation. Based on the data from Table I and some fairly crude estimates of conservation and efficiency goals, we have a

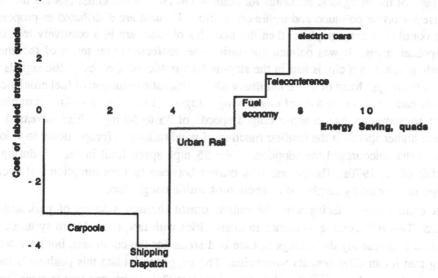

Fig. 2. Potential Benefits and Estimated Costs of Selected Energy Conservation
Strategies. Areas below the zero ordinate represent negative cost, e.g.,
savings.

proposed conservation strategy that is self-starting in a manner similar to that proposed
by Lovins for the power sector.

We recognize the dominance of urban commuting traffic in transportation energy use.
The average passenger load in the U.S. is about two per vehicle. If, in the commuting
traffic, this could be raised to three per vehicle we could save something approaching
3 quads of energy per year! The infrastructure cost would be negligible so the net cost
would be negative 3 quads (a net savings). A similar, but smaller net savings at little or no
cost would derive from dispatching cargo by the most energy-efficient routes and modes.
The 1-quad estimate represents a 10% shift from road to rail cargo. The costs saved in
the first two strategies could be applied to improved urban passenger rail service, saving
another 3 quads of fuel currently burned on commuter highways. Other innovations
such as improved fuel economy, and communications systems that promote business-at-
a-distance will generally save transportation fuel at some equivalent cost. These relatively
few modifications to our transportation infrastructure will result in a savings of one-half
of the fuel currently used (and CO_2 currently emitted).

Of course, it will take innovative legislation and considerable urban planning to
convert the savings from one conservation measure to the implementation of another. Also
it will require difficult departures from conventional urban infrastructure approaches to
others that promote energy efficient behavior. We currently choose to expand our highway
systems to relieve congestion. This approach only forestalls the inevitable worsening of
congestion and the associated air quality and greenhouse-gas accumulations.

Many of the mitigation measures for automotive greenhouse emissions are the same as those for urban pollution and traffic congestion. If autos are distributed in proportion to the population demographics, then the majority of them are in a relatively few large metropolitan areas. It was pointed out during the conference that much of the energy (as well as time) in a city is lost in the stop-and-go traffic affected by traffic signals and clogged freeways. Most of us are familiar with the dramatic reduction of fuel efficiency of our own cars when we do a lot of city driving. Hughes (1991) shows a distinct minimum in fuel consumption rate at trip-averaged speeds of 35 to 50 mph. The increased fuel usage at higher speeds is the familiar function of wind resistance (proportional to velocity squared) that encouraged the adoption of the 55 mph speed limit in the US during the oil crisis of the 1970s. The inverse relationship between fuel consumption and average trip speed reported by Hughes is a function of traffic congestion.

A control meas rfacing with the public transit through a series of park-and-ride stations. This is becoming a practice in many cities with fast, clean Metro systems. The incentive is frequently the savings in time and stress for the commuter, but now we are seeing that it can offer benefits worldwide. The trigger to initiate this positive behavior pattern is, in our observation, a public transit system that works; one that is clean, quick, offers good coverage and schedules, and is welcoming to the infrequent rider. These have generally been the responsibility of the particular city to design, build, and finance, quite reasonable when we view the issue as one of local congestion or pollution. Now that we observe that urban traffic is a significant contributor to the global greenhouse, we must acknowledge that we all have a stake in the congested traffic of a few metropolitan areas. This suggests that national or international funds may be appropriately applied to encourage motor fuel conservation, and, in turn, we should expect a level of responsible behavior from population centers of all sizes.

4. CONCLUSIONS

Private individual automobiles offer the ultimate in personal mobility, fulfill deeper needs of self-identification and, as such, represent one of the most appealing consumer products in almost every culture able to afford this luxury. The global fleet of road vehicles has grown from 50 million in 1950 to 500 million today and is projected to double again by 2030 (Boyle, 1990). The present mode of use of private vehicles for commuting in large metropolitan areas contributes significantly to local, regional, and global air quality issues. Our current infrastructure of suburban highway networks encourages this use of vehicles which, in addition to polluting the air, contributes to traffic congestion and time-wasting commutes. We have suggested here a self-starting strategy for reducing several harmful effects of the overuse of private road vehicles in urban complexes. Savings from simple conservation practices can pay for an improved transportation infrastructure that will encourage more efficient practices.

5. REFERENCES

Bleviss, D. L., and Walzer, P., 1990: Energy for Motor Vehicles, *Scientific American*, September, 1990, 103–109.

Boyle, S., 1990: Transport and Energy Policies-Only Connect, *Energy Policy*, January/February 1990, 34–41.

Davis, G. R., Energy for Planet Earth, *Scientific American*, September, 1990, 55–62.

Fickett, A. P., Gellings, C. W., and Lovins, A. B., 1990: Efficient Use of Electricity, *Scientific American*, September 1990, 65–74.

General Electric, 1984: *United States Energy Book, 1984*, GE, 72 pp.

Hughes, P., 1991: The Role of Passenger Transport in CO_2 Reduction Strategies, *Energy Policy*, March 1991, 149–160.

Concluding Observations:
Integrating Climate-Change and Energy Policy

Charles Keller and Robert Glasser

In his book *The End of Nature*, Bill McKibben points out that at the turn of the last century writers such as John Muir, who worried about the destruction of our natural environment due to economic and population growth, consoled themselves with the thought that "At least we can't do anything to the atmosphere." Well, it certainly appears as though they were wrong.

One impression of this meeting is that it seems that a consensus is emerging that the sensitivity of the climate to a doubling of greenhouse gases is probably not four to five degrees by the middle of the next century—which would give us a tremendous sense of urgency—but more like two degrees by the end of the next century. To some people this suggests that there is no problem, that we have plenty of time, and that, therefore, we should turn our attention to other more pressing threats. Fortunately, few at this conference share this rather complacent view.

The task before us is to determine how we can improve the dialogue between the climate scientists and the people responsible for contributing to the science of energy production and transportation and the energy-policy makers. First of all, we have heard here that, in the past, problems have arisen when climate scientists have tried to pass on data to the energy-policy makers. Michael Schlesinger and Jerry Mahlman have said that their data either weren't used very much, or were used incorrectly. This suggests that it is not nearly so good for energy researchers to use sterile data from a climate-change data base as it is for them to use it in coordination with the people who generated the data. There needs to be a common ground—a research dialogue. Moreover, as George Hidy pointed out, this dialogue must include other groups, such as specialists on the effects of global climate change (ecologists). Climate-change research must become better integrated, and one way to do this may be to set up a multidisciplinary, institutional home for research on environmental change.

One suggestion was made to put together teams comprised of about twenty people representing different disciplines (climate science, energy research, energy policy, ecology, etc.) which would meet regularly, work together, and be able to feed information in some integrated way to a central organization. Some discussion centered around the need to examine why existing organizations, such as the IPCC and IIASA, have failed to take or implement this approach. For example, what is the optimum number of people to involve in the process? How do we avoid imposing policy-driven deadlines on the participants? Why have previous integrated approaches failed? Is it because, in the words of one person, "There isn't enough in it for the participants," either money or, if not money, intellectual stimulation or career prospects? How do we avoid losing the invaluable institutional memory that tends to evolve with these team approaches? Once the multidisciplinary teams disband, it is hard to get them back together and the institutional memory is lost—as seems to have been the case with NAPAB.

We need a place for this integrated approach that will acquire and nurture an institutional memory. Such an institution would be a strong source of intellectual activity. One approach worthy of our attention is the "center without walls," in which you might have more than one existing institution involved, but with a mandate to integrate less formally (and hopefully some funding to prod the process along).

We have also had discussion about the problem of maintaining the public's interest in the potentially serious climate-change issue. People get tired of hearing prophecies of doom over and over again. We may only have a few years left to raise the public consciousness before people become burned out on the subject. It was suggested that we need to instill a bit of fear in people. Some have tried this approach, but it can often be counterproductive, as the story of the boy who cried wolf suggests. Yet, some way must be found of alerting the public to more distant dangers. The question is, "How should we do this?" One key means of raising public consciousness may be to describe these and other environmental challenges in terms of national and international security; these are potentially serious threats to our national well-being that require immediate global and concerted responses.

We need to improve communication and interactions across disciplines. We need to change our thinking. Integration should take place at all levels: one-on-one, in teams, and between institutions. And most importantly, it seems clear that we should give serious and urgent consideration to establishing an institutional home for interdisciplinary research on global environmental change.

Summary Talk and Concluding Remarks

Louis Rosen

This Conference has examined a wealth of information and ideas. The scope and depth and complexity of the papers and discussions, formal and informal, which have taken place during the past three and a half days make a 30-minute summary not very useful. I shall therefore confine my remarks to the general impressions I have garnered and their impact on a non-expert in the field.

When we started contemplating this conference, more than a year ago, we saw it as a timely follow-on to the 1989 Santa Fe conference: *Technology-based Confidence Building: Energy and Environment.* That conference focused on global technological cooperation towards solving energy and environmental problems as one way to improve international relations. (It appears that we were reasonably successful. We did have some help from Presidents Gorbachev and Bush.) That is still a valid goal, even if less urgent. This week we addressed global climate change and energy for strictly their own sake. Improved international relations now become the vehicle for cooperation on environment and energy rather than the other way around. They take their place as important elements of national security and international stability (Fig. 1).

Even six months ago we could not have imagined that world events would conspire to make this conference so timely, especially from the standpoint of increased availability of human and material resources to deal with the problems — technical and political, which climate change and energy production and use embody. Major research laboratories and additional highly-skilled individuals, world-wide, may now be able to devote substantially larger fractions of their resources and energies to problems of climate and energy, adding substantially to the efforts already in progress. Relaxation of the Cold War is catalyzing increased attention to the Warm War.

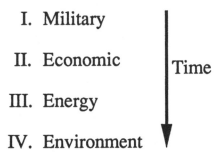

Dimensions of National Security

I. Military

II. Economic   Time

III. Energy

IV. Environment

Fig. 1. Source: 1990 Pugwash Conference on North-South Technology Transfer.

Given the nature of people and the human condition, the world is, as always, preoccupied with immediate economic problems and with good reason. Economic problems probably cause more pain and suffering than any other category. They have long been exacerbated by disparity between food production and population in many areas and the increasing scarcity of raw materials in all areas. But now another dragon is rearing its head in the form of rising pollution in air and water, which is increasingly injurious to health and comfort, but also threatens to force change in where and how we live in the not too distant future. Is global climate change a reality on a scale which can produce major physical and economic dislocations? If so, on what time scale and what would be the cost of avoidance?

Commercial views are now coalescing with academic views that global climate change is a sufficiently serious threat that we should take some level of defensive action to avoid an S&L-type catastrophe, but far more devastating, in the environmental arena.

There is not yet a consensus on the severity of the problems or on the remedial or preventive or adaptive measures we should take. There appears to be agreement that efficiency in energy production and use, where it is economically feasible, is an absolute good. It is also clear that remedial actions, once we agree on them, will not have immediate impact, because of inertia in the system and because they may require that we change our culture and our value system to reflect responsibility to future generations.

We heard disagreements of a factor of more than 2.5 on the cost of presently achievable, intrinsically safe, nuclear-energy generating capacity; and similar disparities in energy efficiencies achievable. These must be resolved.

In order to prevent, or even mitigate, the impact of the by-products of our lifestyles, it is essential that we stimulate public understanding of the problems we face, especially if solutions entail a perceived decrease in our present standard of living, based on our current value systems. So how does one go about changing the value systems of 5 billion people? If it is to be through education in a free society, the first step must be to identify and quantify the problems we face and to make these understandable to the public. But without consensus, at least in the scientific community, about the nature and severity of the problems and the costs of the solutions, this task is hopeless. Even scientific consensus is not enough. A way must be found to continually integrate science with policy or the science will get lost in the politics. To achieve consensus we must talk to each other and that, in part, is what this conference is all about.

This conference has probed for consensus on what these problems are and how we must proceed to better quantify them and devise solutions and predict the impact of these solutions. From the discussions it is clear to me that the problems cannot be solved in piecemeal fashion. It is not possible to separate variables because they are not independent. The solution of one problem can exacerbate or even make unsolvable another equally serious one. A coherent approach is called for. Basic guiding principles must be agreed upon. For example, can we agree that global population and global resources must be kept in balance? This can be characterized as planet sustainability. We still do not have an agreed-upon, peer reviewed, data base on which to structure physical and economic calculations and computer simulations that will indicate options towards sustainable development.

But what does a sustainable planet mean? We have a sustainable planet at some level of population and quality of life. But at what level of population and at what quality of life? We appear to be approaching the time when, for a given set of life styles and energy and raw-material-utilization scenarios, the product of life quality and population will be a constant, unless technological and sociological fixes can be rapidly implemented. The sociological fixes include restructuring our definitions of quality of life towards the use of fewer non-renewable resources and the generation of less environmental burden.

It appears that sociological problems and technological opportunities are becoming more intertwined. New ways of thinking about cultural and other sociological parameters are becoming essential if we are to curtail demands on the environment and other natural resources without suffering a decline in the comforts and enjoyment of life.

It has been conventional wisdom that new ways of thinking require generations to develop and implement on a large scale. But this may not be true, as evidenced by the dramatic transmutations recently witnessed in the political structures of large parts of the world and in the attitudes towards military power as the cornerstone of national security and national well-being.

We are seeing, just now, the restructuring of the military security aspects of our society. This gives some real hope that other aspects, including our value systems, can also be restructured towards a world that gives much higher priority, than is now the case, to the future, even if that entails less material affluence in the present.

We have become quite accustomed to delegating to our governments the resources and authorities to assure against military disaster, and also, but to a lesser extent, other catastrophes, including economic ones. We must now facilitate governmental efforts to make the long-term investments, support the acquisition of relevant knowledge and take the physical steps necessary to prevent and/or mitigate environmental hazards, and, when these cannot be avoided, to adapt to them in a timely fashion. We need to develop physical markers that will give us as much time as possible to avert environmental catastrophe.

For example, if significant climate change is inevitable, it will be necessary to put in place plans and programs for dealing with changing local availability of food and water, as well as to deal with physical challenges that changing patterns of precipitation will impose on demographics. Insurance-type investments in physical infrastructure may not only be a prudent precaution, but may also provide jobs and stimulate technological innovations that could be helpful in other areas of production of goods and services. Proper management of strategies may also be critical. For example, does it make sense to separate responsibilities for the management of agricultural land, forests, and water resources, as we now do?

Research activities need to be coordinated globally if they are to be maximally cost-effective in avoiding duplication of effort and in dissemination to a broad audience, so as to increase the probability of collective action.

International research centers can help to catalyze understanding and acceptance of research results. Of all issues facing humanity, none require consideration in a global context more than do environmental ones. They require human organization on a massive scale. Such organization, even if achieved, may remain extremely fragile, notwithstanding the robustness of life, even under the most desperate circumstances.

Even the robustness of life may not be a match for the dramatic changes in extreme environmental conditions that may be triggered by small changes in average temperatures.

The problems that have commanded the major attention of the superpowers during the past 46 years are fast receding, leaving them economically damaged but with their spirit and intellectual and creative powers intact. This is a major blessing in the present circumstances because it reflects on what may be our most difficult problem—the problem of education.

It is in the realm of education that I believe the scientific communities (physical and social) have been most remiss. I think we must aggressively organize ourselves to do better, individually and collectively. This I say to you not on my behalf, but on behalf of my great grandson for whose future I feel a major responsibility.

Serguei Kapitsa coauthored, some years ago, an innovative book that dealt with the insanity and futility and danger of the arms race. In it was argued the thesis that we must learn to think differently and behave differently. This, I gather from our conference, must now be done with respect to environment and energy. Old habits, old assumptions, and old value systems must be reexamined. Some must be abandoned, others altered. This is terribly difficult. But the peoples of the USSR and Eastern Europe are showing us that it is not impossible to recast one's vision of the world, and in a relatively short time.

Sir Crispin Tickell, former British Ambassador to the United Nations, in a recent address to the British Association for the Advancement of Science says: "To change our way of thinking we first need to recast our vocabulary. Words are the building blocks of thought. The instruments of economic analysis are blunt and rusty. Terms such as 'growth,' 'development,' 'cost-benefit analysis' and even 'GNP' have come to be misleading. They are more than ripe for redefinition." What Sir Crispin says does not contradict the lessons of this conference.

Finally, I leave here convinced of the necessity for intensified, on-going dialogues that catalyze communication and constructive criticism among the different scientific communities, bring to the fore the most recent knowledge and technical capabilities and explore the requirements for new research initiatives as well as action plans for safeguarding and enhancing the core quality of life for this and future generations. Improvements to energy efficiency must be pursued; renewable energy resources should be cultivated; less damaging uses of fossil fuels are to be encouraged; but it remains my perception that a combination of large-scale use of coal and nuclear fission are absolutely essential for our long-term well-being. We need to make their use as benign and inexpensive as possible, and we must certainly take those steps to mitigate greenhouse gases that are economically feasible and can be justified for reasons other than global climate change. I did not hear

534

discussed the potential impact of high-temperature superconductivity. This could save as much energy as some of the renewables will produce.

I am sure I speak for the entire Organizing Committee when I thank you all for participating in this Conference. If you have not turned in your papers, please do so within the next three weeks so that they can be included in the Proceedings which will be published and disseminated by AIP. Each of you will receive a copy. If we have them in time, these Proceedings could be helpful to next year's UN conference in Brazil.

Many people share the credit for whatever success this conference will achieve. The participants, of course, but also the people who made the multitude of arrangements, Patrick Romero, Patricia Fierro, Jean Carol Stark, and our friends in travel, the cafeteria, and security. But most of all I must recognize Robert Glasser, who really did the lion's share of the work.

Thank you.

The End

APPENDIX A

Conference on Climate-Change and Energy Policy: Executive Summary

During October 21–24, 1991, Los Alamos National Laboratory hosted and co-sponsored with Lawrence Livermore National Laboratory, the Electric Power Research Institute, and the University of California at Irvine, an international conference entitled, "Global Climate Change: Its Mitigation Through Improved Production and Use of Energy."

More than 150 distinguished physical and social scientists from over 14 countries including Mexico, Canada, the USSR, Finland, Italy, France, Denmark, Germany, and Ireland, participated in this gathering. A majority of the papers, however, were presented by US attendees from across the country.

A primary goal of the conference was to encourage discussions and forge links among climate-change scientists, energy researchers, social scientists, and policy makers. Another goal was to probe for consensus on the key issues involved in integrating climate-change research with energy production and use.

Our interpretation of the sense of the large majority of attendees is that before human society pays huge amounts to reduce greenhouse gases, the anthropogenic link to climate change must be well established. While the evidence for significant human-induced alteration is not conclusive, the potential for climate alteration exists, and the consequent environmental stress could be disruptive, even severe. The response time to return to a "natural" condition is very long (centuries), once climate alteration has occurred. Therefore, timely and judicious global efforts to establish options for mitigation and adaptation to potential change should be made now, and precautionary measures need to be considered—especially those that would be neutral or beneficial to the economy, conserve non-renewable resources, and improve the protection of our environment from harmful contaminants. It was, moreover, felt that enhanced research efforts are warranted to improve our ability to predict global and localized long-range climate change.

Action can be taken to improve the economic and physical quality of life while reducing the probability of adverse or undesirable effects of climate alteration. They include the use of minimally polluting renewable resources such as non-organic fuels and cleaner fossil fuels, increased energy efficiency in production and utilization, and reduction of the rate of population growth. No one at the conference quarreled with the assumption that population expansion may, by itself, defeat all efforts to achieve sustainable development. Indeed, this may be the fundamental challenge facing our planet. Population growth and energy-use patterns suggest that any response to climate change must be global in scope, with strategies of the developing countries playing a pivotal role in the future.

There were, of course, extreme views. A few argued that money can be saved, sustainable development achieved, and greenhouse gases reduced to necessary levels solely by conservation and the use of solar energy, wind power, water power, biomass, and through reforestation. A few others argued that significant warming — at least in the foreseeable future — is unconfirmed and, therefore, that nothing out of the ordinary need be done for at least the next decade.

However, most of those presenting papers did not identify with either of the above positions. On the contrary there appeared to be a consensus among those present that climate alteration from atmospheric greenhouse gas concentrations could occur in the next century. These alterations could lead to effects worldwide that would be adverse to human and natural eco-systems, even though there may be some local benefits of such change. Thus, a global "insurance policy" approach is warranted, whereby research is devoted to studying the effects of, and possibilities for, adaptation to the hypothesized warming.

One disturbing revelation was that large discrepancies exist in estimates of how long it will take to provide a means for suppression of greenhouse-gas emissions given known predictions of energy use and production. Additional uncertainty surrounds estimates of the costs of a new generation of inherently safe nuclear power plants (not to mention the cost-effectiveness of major changes in the design of nuclear plants) and the extent to which efficiencies in the production and use of energy can be achieved at a socially acceptable price. It is hoped and expected that the Conference participants will take a leading role in addressing these fundamental issues.

Finally, there appeared to be considerable enthusiasm for establishing an institutional home as part of an integrated research strategy on global warming, encompassing climate-change scientists, energy researchers, ecologists, energy-policy makers, and social scientists. Such an approach would lead to a peer-reviewed data base including basic information upon which different groups develop their climate simulation models and also their calculations of the costs, efficiencies, and by-products — physical and economic — of various energy production and use scenarios. This integrated approach would also provide valuable information to government science and policy advisors. Some of the participants intend to give this idea further thought and to develop concrete recommendations.

APPENDIX B

Global Climate Change:
Its Mitigation Through Improved Production and Use of Energy
Conference Agenda

Monday, October 21, 1991

Opening Session: 8:30–9:15 a.m.

Chairman: John Hopkins, Associate Director-at-Large, Los Alamos National Laboratory; Director, Center for National Security Studies, Los Alamos National Laboratory

Welcome: Siegfried Hecker, Director, Los Alamos National Laboratory

Conference Goals: John Whetten, Associate Director, Energy and Technology, Los Alamos National Laboratory

Final Agenda: Robert Glasser, Center for National Security Studies, Los Alamos National Laboratory

Introduction of Keynote Speaker: Siegfried Hecker, Director, Los Alamos National Laboratory

Keynote Speaker: Serguei Kapitsa, Director, Institute for Accelerator Design at the P. Kapitsa Institute for Physical Problems, Moscow, Russia

Break: 9:15–9:45 a.m.

Plenary Session 1: Assessing Global Climate Change: 9:45 a.m.–12:30 p.m.

Chairman: Charles F. Keller, Jr., Director, Institute of Geophysics & Planetary Physics, Los Alamos National Laboratory

Talk I: "Model Simulations of Greenhouse Warming—Any Observational Evidence?" Lennart O. Bengtsson, Director, Max-Planck Institut fur Meteorologie, Hamburg Germany

Talk II: "Observed Global Warming: Are We Sure?" Thomas R. Karl, Chief of Global Climate Laboratory, National Climatic Data Center, Asheville, North Carolina

Talk III: "Reliability of the Models: Their Match with Observations," Warren M. Washington, Senior Scientist and Director of Climate and Global Dynamics Division, National Center for Atmospheric Research, Boulder, Colorado

Talk IV: "When Will We Have Better Evidence?" Jerry D. Mahlman, Director of Geophysical Fluid Dynamics Laboratory, National Oceanic and Atmospheric Administration, Princeton University, Princeton, New Jersey

Talk V: George S. Golitsyn, Director, Institute of Atmospheric Physics, Academy of Sciences, Moscow, Russia

Lunch: 12:30–2:00 p.m.

Plenary Session 2: Energy Technologies and Economics: 2:00–6:15 p.m.

Chairman: John Weyant, Professor, Department of Engineering/Economic Systems, Stanford University, Stanford, California

Talk I: "The Problem: The Role of Energy Production in Greenhouse Gas Emissions," Irving Mintzer, Coordinator, Stockholm Environment Institute, Climate Programme

Talk II: "The Role of Energy Conservation," Mark Levine, Head Energy Analysis Group, Lawrence Berkeley Laboratory, Berkeley, California

Talk III: "Advanced Techniques for Efficient Energy Use," Amory B. Lovins, Director, Rocky Mountain Institute, Snowmass, Colorado

Break: 4:00–4:30 p.m.

Talk IV: "The ATW Concept for Radioactive Waste Destruction and Energy Production," Edward Arthur, Program Director for ATW, Los Alamos National Laboratory

Talk V: "The Technology and Economics of Energy Alternatives," Dwain F. Spencer, Vice President of Commercialization and Business Development, Electric Power Research Institute, Palo Alto, California

Talk VI: John Weyant, Professor, Department of Engineering/Economic Systems, Stanford University, Stanford, California

Director's Hosted Reception: 6:30–7:30 p.m.

Tuesday, October 22, 1991

Concurrent Workshops: 8:30 a.m.–12:30 p.m.

Workshop IA: Energy Technologies/Environmental Impacts

Chairman: William Fulkerson, Associate Director for Advanced Energy Systems, Oak Ridge National Laboratory, Oak Ridge, Tennessee

Introduction: Robert Schock, Energy Program Leader, Lawrence Livermore National Laboratory, Livermore, California

Ten Minute Talks:

"U.S. Efforts to Reduce Waste in Electronics Industry," Richard Benson, Program Manager Industrial Waste Production, Los Alamos National Laboratory

"Implementing the Nuclear Option," David Bodansky, Professor of Physics, University of Washington, Seattle, Washington

"Hot Dry Rock," David V. Duchane, Program Manager, Los Alamos National Laboratory

Break: 10:00–10:30 a.m.

"Adaptive Policies for Abating Climate Change," James K. Hammitt, Mathematician, RAND, Santa Monica, California

"Energy Conservation and Renewables in Global Warming," John A. Laurmann, Department of Civil Engineering, Stanford University, Stanford, California

"Extension of Nuclear Fuel Supplies," Duane R. Pendergast, Licensing Engineer, AECL Canada, Ontario, Canada

"The Energy Technology Clearing House—An SED/OECD Effort," Paolo F. Ricci, Principal Administrator, International Energy Agency, Paris, France

"Integrated Assessment of Energy Technologies for CO_2 Reduction in the Netherlands," Remko Ybema, Junior Scientist, Netherlands Energy Research Foundation, Petten, The Netherlands

Workshop IB: Energy Technologies/Environmental Impacts

Chairman: Serguei Kapitsa, Director, Institute for Accelerator Design at the P. Kapitsa Institute for Physical Problems, Moscow, Russia

Ten Minute Talks:

"Resource Cost Analysis of Energy Management and Energy Supply Technologies," Steven G. Diener, President, Steven G. Diener & Associates Ltd., Ontario, Canada

"Calculating the Net Greenhouse Warming Effect of Renewable Energy Resources: Methanol from Biomass," Rex T. Ellington, Faculty Associate, Science and Public Policy Program, University of Oklahoma, Norman, Oklahoma

"Waste to Energy, Institutional-Municipal," Alex E. Green, Graduate Research Professor of Engineering, University of Florida, Gainesville, Florida

Break: 10:00–10:30 a.m.

"The British Coal Approach to Global Climate Change," Ian S. Hughes, Team Leader, Environmental Sciences, Coal Research Est., Cheltenham, United Kingdom

"The Modular HTGR: Its Possible Role in the Use of Safe and Benign Nuclear Power," Massoud T. Simnad, Professor/Consultant to General Atomics, Center for Energy and Combustion Research, University of California, San Diego, La Jolla, California

"Future Contributions of Renewable Energy Technologies and Their Impact on CO_2 Emissions," Robert Stokes, Deputy Director of National Renewable Energy, Solar Energy Research Institute, Golden, Colorado

"CFCs and Halon Restrictions — Impacts on Energy Technologies," Robert E. Tapscott, Director, Center for Global Environmental Technologies, University of New Mexico, Albuquerque, New Mexico

"Environmental Impacts of Renewable Energy Technologies," Theodore B. Taylor, President, Southern Tier Environmental Protection Society, Inc., West Clarksville, New York

"Making Better Use of Carbon: The Co-Production of Iron and Liquid Fuels," John H. Walsh, Energy Advisor Consultant, Ontario, Canada

Lunch: 12:30–2:00 p.m.

Plenary Session 3:
Integrating Climate Change and Energy Policy—A Panel Discussion: 2:00–5:00 p.m.

Introduction of the Panels: Robert Glasser, Center for National Security Studies, Los Alamos National Laboratory

Introductory Talk: "The Human Dimensions of Climate Change," Charles N. Herrick, Senior Analyst, National Oceanic and Atmospheric Administration, Washington, DC

Panel I: "How We Think We Can Help the Energy Policy-Making Community— *"Climate-Change Scientists' Perspectives"*

> *Chairman:* Fredrick Zachariasen, Professor, California Institute of Technology, Pasadena, California

>> Lennart O. Bengtsson, Director, Max-Planck Institut fur Meteorologie, Hamburg, Germany

>> Kenneth Friedman, Special Assistant to the Deputy Assistant Secretary for Industrial Technology, Office of Conservation and Renewable Energy, Washington, DC

>> George S. Golitsyn, Director, Institute of Atmospheric Physics, Academy of Sciences, Moscow, Russia

>> Charles F. Keller, Jr., Director, Institute of Geophysics and Planetary Physics, Los Alamos National Laboratory

>> Jerry D. Mahlman, Director of Geophysical Fluid Dynamics Laboratory, National Oceanic and Atmospheric Administration, Princeton University, Princeton, New Jersey

>> Aristedes Patrinos, Acting Director, Environmental Sciences Division, Department of Energy, Washington, DC

>> Warren M. Washington, Senior Scientist and Director of Climate and Global Dynamics Division, National Center for Atmospheric Research, Boulder, Colorado

Break: 3:15–3:45 p.m.

Panel II: "What We Need from the Climate Change Community—Energy Production and Transportation Sectors' Perspectives"

> *Chairman:* Alvin M. Weinberg, Distinguished Fellow, Oak Ridge Associated University, Oak Ridge, Tennessee

>> George M. Hidy, Vice President, Environment Division, Electric Power Research Institute, Palo Alto, California

>> Serguei Kapitsa, Director, Institute for Accelerator Design at the P. Kapitsa Institute for Physical Problems, Moscow, Russia

>> Jose Roberto Moreira, Secretary Advisor of Science and Technology, Brazilian Government, San Paolo, Brazil

>> Virginia M. Oversby, Associate Director Staff of Chemistry and Material Science, Lawrence Livermore National Laboratory, Livermore, California

>> Ruth A. Reck, Staff Research Scientist, General Motors Research Laboratories, Wannen, Michigan

Louis Rosen, Senior Fellow Emeritus, Los Alamos National Laboratory

Dwain F. Spencer, Vice President of Commercialization and Business Development, Electric Power Research Institute, Palo Alto, California

Wednesday, October 23, 1991

Concurrent Workshops: 8:30 a.m.–5:00 p.m.

Workshop II: Climate Change and the Electric Utilities Sector

Co-Chairman: George M. Hidy, Vice President, Environment Division, Electric Power Research Institute, Palo Alto, California

Co-Chairman: John O. Roads, Research Meteorologist, University of California, San Diego, Scripps Institute of Oceanography, La Jolla, California

Ten Minute Talks:

"General Circulation Model Predictions and their Consequences for Renewable Energy Technologies," A. David Corbus, Environmental Analyst, Solar Energy Research Institute, Golden, Colorado

"Climate Change: Outlook for Energy Legislation," Alan Crane, Senior Associate, Office of Technology Assessment, U.S. Congress, Washington, DC

"Climate Impact of the Gulf's Fires," Alexander S. Ginzburg, Institute of Atmospheric Physics, Moscow, Russia

"Environmental Factor in International Aid Programmes and Technology Transfer," Tatyana G. Kazakova, Attaché, Ministry of Foreign Affairs, Moscow, Russia

"The Greenhouse Effect: Political Decision Making and the Application of Upwelling/Diffusion Models," Peter Laut, Professor, Engineering Academy of Denmark, Lyngby, Denmark

"Socioeconomic Benefits of a Global Warming Event," Robert H. Lessard, Geology Professor, New Mexico Highlands University, Las Vegas, New Mexico

"Snow Hydrology," Susan Marshall, Staff Member, Los Alamos National Laboratory

"Revised Projection of Future Greenhouse Warming," Michael E. Schlesinger, Professor, Department of Atmospheric Sciences, University of Illinois at Urbana-Champaign, Urbana, Illinois

Break: 10:30–10:45 a.m.

"Atmospheric Modeling of Greenhouse Trace Gases," C.-Y. Jim Kao, Staff Member, Los Alamos National Laboratory

"Low-Cost Energy Production: The Responsibility of the Developed Countries Toward the Less-Developed," Virginia M. Oversby, Associate Director Staff of Chemistry & Material Science, Lawrence Livermore National Laboratory, Livermore, California

"Regional Climatology," John O. Roads, Research Meteorologist, University of California, San Diego, Scripps Institute of Oceanography, La Jolla, California

"Coal or Natural Gas: Calculations of the Global CO_2/Methane Tradeoff," XueXi Tie, Institute of Geophysics and Planetary Physics, Los Alamos National Laboratory

Lunch: 11:45 a.m.–1:00 p.m.

Discussion: 1:00–5:30 p.m.

Workshop III: Climate Change and the Transportation Sector

  Co-Chairperson: Ruth A. Reck, Staff Research Scientist, General Motors Research Laboratories, Warren, Michigan

  Co-Chairperson: Sumner Barr, Group Leader, Earth and Environmental Science, Los Alamos National Laboratory, Coordinator, Los Alamos Global Change Program, Los Alamos National Laboratory

Ten Minute Talks:

  "Global Simulations of Smoke from Kuwaiti Oil Fires and Possible Affects on Climate," Gary Glatzmaier, Institute of Geophysics and Planetary Physics, Los Alamos National Laboratory

  "Solid Fuel Conversion for the Transportation Sector," Alex E. Green, Graduate Research Professor of Engineering, University of Florida, Gainesville, Florida

  "Evaluation of CO_2 Emission Control Measures in Several Nations," Douglas Hill, Consulting Engineer, Huntington, New York

  "Meso- and Large-Scale Ocean Dynamics Model," Darryl D. Holm, Fellow, Los Alamos National Laboratory

  "The Role of the University in Addressing Environmental Concerns," McAllister H. Hull, Provost Emeritus, Professor of Physics, Emeritus, The University of New Mexico, Albuquerque, New Mexico

Break: 10:00–10:15 a.m.

"What the Greenhouse Skeptics are Saying," William W. Kellogg, Senior Scientist NCAR (Retired), Boulder, Colorado

"Implication of Anthropogenic Atmospheric Sulfate for the Sensitivity of the Climate System," Michael E. Schlesinger, Professor, Department of Atmospheric Sciences, University of Illinois at Urbana-Champaign, Urbana, Illinois

"Toward Regional Climate Change Scenarios: How Far Can We Go?" Peter H. Whetton, Division of Atmospheric Research CSIRO, Victoria, Australia

Lunch: 11:30 a.m.–12:45 p.m.

Discussion: 12:45–5:30 p.m.

Wednesday Evening

5:30 p.m. Transportation from Conference Site to Santa Fe for Dinner at Selected Restaurants

9:30 p.m. Transportation from Santa Fe (La Fonda Hotel) to Los Alamos

Thursday, October 24, 1991

Plenary Session IV: Reports from Workshops/Conference Summary

Chairman: William W. Kellogg, Senior Scientist, Boulder, Colorado

8:45–9:15 a.m. "Climate Change and Energy Policy: Canada's Approach," Patrick G. Finlay, Deputy Director, Climate Change Response, Environment, Canada

9:15–9:45 a.m. "Climate Change and Energy Policy: Netherland's Approach," Remko Ybema, Netherlands Energy Research Foundation

Reports From Workshops and Concluding Discussions and Remarks

9:45–10:15 a.m. "Report from the Energy Workshops," Theodore B. Taylor, Southern Tier Environmental Protection Society, Inc.

10:15–10:45 a.m. Break

10:45–11:15 a.m. "Report from the Workshops on Integrating Climate Change and Energy Policy," Charles F. Keller, Director, Los Alamos Branch, Institute of Geophysics and Planetary Physics

11:15–11:45 a.m. Integrating Climate Change and Energy Policy: Follow-on Interactions Between the Two Disciplines

11:45–12:15 p.m. Conference Summary/Concluding Remarks

Louis Rosen, Senior Fellow Emeritus, Los Alamos National Laboratory

12:15–2:00 p.m. Private Discussions

2:00 and 5:00 p.m. Buses to Hotel

APPENDIX C

Global Climate Change: Its Mitigation Through Improved Production and Use of Energy

October 21–24, 1991

CONFERENCE PARTICIPANTS

Anita Archibeque
New Mexico Highlands University
Environmental Science Department
Las Vegas, NM 87701

Eugene A. Aronson
Sandia National Laboratories
P.O. Box 5800, Div. 6601
Albuquerque, NM 87185

Edward Arthur
Los Alamos National Laboratory
T-DO, MS B210
Los Alamos NM 87545

George A. Baker
Los Alamos National Laboratory
T-11, MS B262
Los Alamos NM 87545

Fairley J. Barnes
Los Alamos National Laboratory
EES-15, MS J495
Los Alamos NM 87545

Sumner Barr
Los Alamos National Laboratory
EES-5, MS F665
Los Alamos NM 87545

Mariano E. Bauer
Universidad Nacional Autonoma de Mexico
Programa de Energia
Apartado Postal 70-172
Mexico D.F. 04510

Lennart O. Bengtsson
Max-Planck-Institut fur Meteorologie
Bundesstrasse, 55
D-2000 Hamburg 53 GERMANY

Charles Bensinger
Advanced Energy Systems
P.O. Box 2685
Santa Fe, NM 87504

Richard A. Benson
Los Alamos National Laboratory
ET-EET, MS F643
Los Alamos NM 87545

Michael E. Berger
Los Alamos National Laboratory
ET-EET Program Office, MS F643
Los Alamos NM 87545

Christopher Bernabo
Science & Policy Associates, Inc.
1333 H Street, NW
West Tower, Suite 400
Washington, DC 20005

Leslie G. Black
U.S. Senate Energy Committee
364 Dirksen Bldg.
Washington, DC 20510-6150

David Bodansky
University of Washington
Department of Physics, FM-15
Seattle, WA 98195

James E. Bossert
Los Alamos National Laboratory
MS D446
Los Alamos NM 87545

James N. Bradbury
Los Alamos National Laboratory
MP-DO, MS H844
Los Alamos NM 87545

Karl Braithwaite
Los Alamos National Laboratory
DIR-ESD, MS A103
Los Alamos NM 87545

Richard J. Burick
Los Alamos National Laboratory
ADET, MS D455
Los Alamos NM 87545

John M. Callaway
RCG/Hagler Bailly Inc.
P.O. Drawer O
Boulder, CO 80306-1906

David C. Camp
Lawrence Livermore National Laboratory
Nuclear Chemistry Division
P.O. Box 808
Livermore, CA 94550

Robert K. Carr
Los Alamos National Laboratory
CNSS, MS A112
Los Alamos NM 87545

David C. Cartwright
Los Alamos National Laboratory
ADR, MS A114
Los Alamos NM 87545

Shi Yi Chen
Los Alamos National Laboratory
CNLS, MS B258
Los Alamos NM 87545

John F. Clarke
Battelle, PNL
370 L'Enfant Plaza SW, Suite 900
Washington, DC 20024

William E. Clements
Los Alamos National Laboratory
EES-5, MS D466
Los Alamos NM 87545

Donald D. Cobb
Los Alamos National Laboratory
SST-DO, MS D455
Los Alamos NM 87545

A. David Corbus
Solar Energy Research Institute
1617 Cole Blvd., Bldg. 17-3
Golden, CO 80401

Kurt Covey
Lawrence Livermore National Laboratory
P.O. Box 808, L-264
Livermore, CA 94550

George A. Cowan
Los Alamos National Laboratory
ADR, MS D434
Los Alamos NM 87545

Alan T. Crane
Office of Technology Assessment
U.S. Congress
Washington, DC 20510

Hassan A. Dayem
Los Alamos National Laboratory
C-DO, MS B260
Los Alamos NM 87545

Ruth B. Demuth
Los Alamos National Laboratory
EES-5, MS F665
Los Alamos NM 87545

Raffaele di Menza
ENEA of Italy
818 18th Street NW
Washington, DC 20006

Steven G. Diener
Steven G. Diener & Associates, Ltd.
123-21 Vaughan Road, Suite 123
Toronto, Ontario, Canada M6G 2N2

David J. Dodd
Ontario Hydro
800 Kipling Avenue
Toronto, Ontario, Canada M8Z 584

David V. Duchane
Los Alamos National Laboratory
MST-7, MS E549
Los Alamos NM 87545

John K. Dukowicz
Los Alamos National Laboratory
T-3, MS B216
Los Alamos NM 87545

Donna M. Edwards
Sandia National Laboratories
Center for Computational Engineering
P.O. Box 969
Livermore, CA 94551-0969

Rex T. Ellington
University of Oklahoma
Energy Center
100 East Boyd Street
Norman, OK 73019-0628

Patricia Fierro
Los Alamos National Laboratory
IGPP, MS K305
Los Alamos NM 87545

Patrick G. Finlay
Environment Canada
373, Sussex Drive
Ottawa, Canada K1A-OH3

Gwen B. Fleener
P.O. Box 926
Boulder, CO 80306-0926

Ann M. Florini
University of California, Los Angeles
Center for International & Strategic
  Affairs
405 Hilgard Ave., 11381 Bunche Hall
Los Angeles, CA 90024-1486

Kenneth Friedman
U.S. Department of Energy
Office of Conservation & Renewable
  Energy
Washington, DC 20585

William Fulkerson
Oak Ridge National Laboratory
P.O. Box 2008
Oak Ridge, TN 37831-6247

Carlos E. Garcia
Los Alamos National Laboratory
ET-EET, MS F643
Los Alamos NM 87545

W. Lawrence Gates
Lawrence Livermore National Laboratory
P.O. Box 808, L-264
Livermore, CA 9455

Richard W. Getzinger
American Assoc. for the Advancement
  of Science
1333 H Street, NW
Washington, DC 20005

Alexander S. Ginzburg
Institute of Atmospheric Physics
3, Pyzheusky
Moscow 109017 Russia

Robert Glasser
Los Alamos National Laboratory
CNSS, MS K306
Los Alamos NM 87545

Gary A. Glatzmaier
Los Alamos National Laboratory
EES-5, MS F665
Los Alamos NM 87545

George S. Golitsyn
Institute of Atmospheric Physics
3 Pyzheuscky
Moscow 109017 Russia

548

Rex Gram
Albuquerque Journal
P.O. Drawer J
Albuquerque, NM 87103

Alex E. Green
University of Florida/ICAAS
311 SSRB
Gainesville, FL 32611

Ludwig A. Gritzo
Los Alamos National Laboratory
ADO, MS A120
Los Alamos NM 87545

Francisco Guzman
Instituto Mexicano del Petroleo
Eje Central Lazaro Cardenas 152
Mexico D.F. 07730

Kjell G. Håkansson
Studsvik AB
Rossenkallav 27
5-61136 Nykoping, Sweden

James K. Hammitt
The RAND Corporation
1700 Main Street, P.O. Box 2138
Santa Monica, CA 90407-2138

Michal Harthill
Department of the Interior
MS 4103, 18th & C Streets, NW
Washington, DC 20035

Sarah M. Hayes
Los Alamos National Laboratory
MS P368
Los Alamos NM 87545

Siegfried Hecker
Los Alamos National Laboratory
DIR OFF, MS A100
Los Alamos NM 87545

Charles N. Herrick
NOAA
1825 Connecticut Avenue, NW, #518
Washington, DC 20235

George M. Hidy
Electric Power Research Institute
3412 Hillview Avenue
Palo Alto, CA 94303

Douglas Hill
15 Anthony Court
Huntington, NY 11743

Darryl D. Holm
Los Alamos National Laboratory
T-7, MS B284
Los Alamos NM 87545

David B. Holtkamp
Los Alamos National Laboratory
P-3, MS D449
Los Alamos NM 87545

John C. Hopkins
Los Alamos National Laboratory
ADAL, MS A112
Los Alamos NM 87545

John W. Hopson
Los Alamos National Laboratory
T-3, MS B216
Los Alamos NM 87545

I.S.C. Hughes
British Coal
Coal Research Est.
Stoke Orchard
Cheltenham Glos. GL52 4RZ, England

McAllister H. Hull
University of New Mexico
Department of Physics & Astronomy
Albuquerque, NM 87131

O'Dean P. Judd
Los Alamos National Laboratory
ADDRA, MS A110
Los Alamos NM 87545

C. Y. Jim Kao
Los Alamos National Laboratory
EES-5, MS D446
Los Alamos NM 87545

Serguei P. Kapitsa
Institute for Physical Problems
Academy of Science
2 Kosygina Street
Moscow 117334 Russia

Thomas R. Karl
National Climatic Data Center
Federal Building
Asheville, NC 28801

Tatyana G. Kazakova
Ministry of Foreign Affairs
32/34 Smolenskaya
Sennaya pl
121200 Moscow, Russia

Ann C. Keller
Los Alamos National Laboratory
CNSS, MS K306
Los Alamos NM 87545

Charles F. Keller
Los Alamos National Laboratory
ET-IGPP, MS K305
Los Alamos NM 87545

William W. Kellogg
445 College Avenue
Boulder, CO 80302

Ronald P. Koopman
Lawrence Livermore National Laboratory
P.O. Box 808, L-209
Livermore, CA 94550

Richard H. Kropschot
Los Alamos National Laboratory
MS A103
Los Alamos NM 87545

Burgess Laird
Los Alamos National Laboratory
DIR-ESD, MS A103
Los Alamos, NM 87545

John A. Laurmann
University of California, Santa Barbara
Marine Science Institute
Santa Barbara, CA 93106

Peter Laut
Engineering Academy of Denmark
DK-2800 Lyngby, Denmark

Gerald G. Leigh
New Mexico Engineering Research Inst.
The University of New Mexico
Albuquerque, NM 87131

Robert J. Lempert
RAND
1700 Main Street
Santa Monica, CA 90407

Robert H. Lessard
New Mexico Highlands University
Geology Department
Las Vegas, NM 87701

Mark D. Levine
Energy Analysis Program
Lawrence Berkeley Laboratory
MS 90-4000
Berkeley, CA 94720

Arthur E. Lewis
Lawrence Livermore National Laboratory
P.O. Box 808, L-209
Livermore, CA 94550

Robert A. Lott
Gas Research Institute
8600 West Bryn Mawr Avenue
Chicago, IL 60631

Jean-Francois Louis
Atmospheric and Envir. Research, Inc.
840 Memorial Drive
Cambridge, MA 02139

Amory B. Lovins
Rocky Mountain Institute
1739 Snowmass Creek Road
Snowmass, CO 81654-9199

Jerry D. Mahlman
NOAA/Geophysical Fluid Dynamics Lab.
Princeton University
P.O. Box 308
Princeton, NJ 08542

Robert C. Malone
Los Alamos National Laboratory
C-3, MS B265
Los Alamos NM 87545

Ms. Kim Manley
4691 Ridgeway Drive
Los Alamos, NM 87544

Len G. Margolin
Los Alamos National Laboratory
P-15, MS D406
Los Alamos NM 87545

Susan Marshall
Los Alamos National Laboratory
EES-5, MS K401
Los Alamos NM 87545

Edward A. Martinez
New Mexico Highlands University
Environmental Science Deprtment
Las Vegas, NM 87701

Eric Martinot
University of California, Berkeley
Energy & Resources Group, Bldg. T-4
Berkeley, CA 94720

Robert A. McClatchey
USAF Phillips Lab/Geophysics Directorate
Hanscom AFB, MA 01731

Tom McManus
Department of Energy
25 Clare Street
Dublin 2, Ireland

Mark Meo
University of Oklahoma
Energy Center
100 East Boyd Street
Norman, OK 73019-0628

Nicholas C. Metropolis
Los Alamos National Laboratory
MS B210
Los Alamos NM 87545

Arthur J. Miller
SIO-UCSD, A-024
La Jolla, CA 92093

Warren F. Miller
University of California, Berkeley
4161 Etcheverry Hall
Berkeley, CA 94720

Irving M. Mintzer
Stockholm Environment Institute
9514 Garwood Street
Silver Spring, MD 20901

Jose Roberto Moreira
Brazilian Government
Esplanada dos Ministerios, Bloco E
Brasilia DF 70062 Brazil

Pablo Mulas
Instituto de Investigaciones Electricas
Interior Internado Palmire
Cuernavaca, Morelos, Mexico

Joseph P. Mutschlecner
Los Alamos National Laboratory
EES-5, MS F665
Los Alamos NM 87545

Hector Nava-Jaimes
Instituto Mexicano del Petroleo
Eje Central Lazaro Cardenas 152
Mexico D.F. 07730

Derek M. Oaks
Los Alamos National Laboratory
CNSS, MS K306
Los Alamos NM 87545

Dana B. Orwick
The Aspen Institute
6004 Winnebago Road
Bethesda, MD 20816

Virginia M. Oversby
Lawrence Livermore National Laboratory
P.O. Box 818, L-353
Livermore, CA 94550

Duane R. Pendergast
AECL
Sheridan Park Research Community
Mississauga, Ontario, Canada L5K 1B2

John L. Petersen
The Arlington Institute for National Strategy
2101 Crystal Plaza Arcade, Suite 136
Arlington, VA 22202

Rick S. Piltz
Committee on Science, Space,
  & Technology
U.S. House of Representatives
388 Ford House Office Building
Washington, DC 20515

James C. Porter
Los Alamos National Laboratory
NWD-NDR, MS B218
Los Alamos NM 87545

M. Juan Quintanilla
Universidad Nacional Autonoma de Mexico
Programa Universitario de Energia
A.P. 70-172
Mexico D.F. 04510

Ruth A. Reck
General Motors Research Laboratories
Environmental Sciences Department
30500 Mound Road
Warren, MI 48090-9055

Paolo F. Ricci
International Energy Agency
2 rue Andre Pascal
Paris, France Cedex 16 75016

John O. Roads
UCSD/Scripps Inst. of Oceanography, 0224
La Jolla, CA 92093

Michael Rodemeyer
House Committee on Science, Space,
  & Technology
Rm. 2320 Rayburn House Office Building
Washington, DC 20515

Patrick Romero
Los Alamos National Laboratory
IGPP, MS K305
Los Alamos NM 87545

Louis Rosen
Los Alamos National Laboratory
CNSS, MS K306
Los Alamos NM 87545

Julian Sanchez
Inst. Nactional de Investigaciones
  Nucleares
Sierra Mojada #447
Mexico, D.F. C.P. 11010 Mexico

Michael E. Schlesinger
Department of Atmospheric Sciences
University of Illinois at
  Urbana-Champaign
105 South Gregory Avenue
Urbana, IL 61801

Robert N. Schock
Lawrence Livermore National Laboratory
P.O. Box 808, MS L-209
Livermore, CA 94550

Stanley O. Schriber
Los Alamos National Laboratory
AT-DO, MS H811
Los Alamos NM 87545

Stephen T. Shankland
Los Alamos National Laboratory
CNSS, MS K306
Los Alamos NM 87545

James P. Shipley
Los Alamos National Laboratory
SST-DO, MS D460
Los Alamos NM 87545

L. M. Simmons
Santa Fe Institute
1660 Old Pecos Trail, Suite A
Santa Fe, NM 87501

Massoud T. Simnad
University of California, San Diego
Center for Energy & Combustion Research
Mail Code R-011
La Jolla, CA 92093

Raji Sinha
State of New Mexico/DOE
Environmental Improvement Dept.
Santa Fe, NM 87501

Richard C. Slansky
Los Alamos National Laboratory
T-DO, MS B210
Los Alamos NM 87545

Richard D. Smith
Los Alamos National Laboratory
IT-3, MS B216
Los Alamos NM 87545

Dwain F. Spencer
Electric Power Research Inst.
3412 Hillview Avenue
Palo Alto, CA 94304

Larry Spohn
Albuquerque Tribune
P.O. Drawer T
Albuquerque, NM 87103

Sampat Sridhar
Atomic Energy of Canada Research
344 Slater Street
Ottawa, Ontario, Canada K1A 0S4

Robert A. Stokes
National Renewable Energy Laboratory
1617 Cole Blvd.
Golden, CO 80401

Gerald E. Streit
Los Alamos National Laboratory
A-4, MS B299
Los Alamos NM 87545

William G. Sutcliffe
Lawrence Livermore National Laboratory
P.O. Box 818, L-19
Livermore, CA 94550

Kevin C. Tan
Los Alamos National Laboratory
CNSS, MS K306
Los Alamos NM 87545

Robert E. Tapscott
Ctr. for Global Environmental
    Technologies
NM Engineering Research Inst.
The University of New Mexico
Albuquerque, NM 87131-1376

Theodore B. Taylor
Southern Tier Environ. Protection
    Society, Inc.
P.O. Box 40
W. Clarksville, NY 14786

Charles Teclaw
Los Alamos National Laboratory
A-4, MS F611
Los Alamos NM 87545

N. Anne Tellier
Los Alamos National Laboratory
ET, MS F643
Los Alamos NM 87545

Les E. Thode
Los Alamos National Laboratory
ADDRA, MS A110
Los Alamos NM 87545

XueXi Tie
Los Alamos National Laboratory
IGPP, MS K305
Los Alamos NM 87545

Linda Trocki
Los Alamos National Laboratory
A-4, MS B299
Los Alamos NM 87545

Carlos Velez
Inst. Nacional de Investigaciones
    Nucleares
Sierra Mojada #447
Mexico, D.F. C.P. 11010 Mexico

John Vitko, Jr.
Sandia National Laboratories
P.O. Box 969
Livermore, CA 94551-0969

Robert S. Vrooman
Los Alamos National Laboratory
ADO-ISEC, MS G733
Los Alamos NM 87545

John H. Walsh
19 Lambton Avenue
Ottawa, Ontario, Canada K1M 0Z6

Mrs. Barbara Walsh
19 Lambton Avenue
Ottawa, Ontario, Canada K1M 0Z6

Ronald A. Walters
Los Alamos National Laboratory
ADR, MS A114
Los Alamos NM 87545

Warren M. Washington
National Center for Atmospheric Research
P.O. Box 3000
Boulder, CO 80307-3000

Bryan C. Weare
University of California
Dept. of LAWR
Davis, CA 95616

Alvin M. Weinberg
Oak Ridge Associated Universities
P.O. Box 117
Oak Ridge, TN 37830

John P. Weyant
Terman Engineering Building, Rm 406
Stanford University
Stanford, CA 94305

John Whetten
Los Alamos National Laboratory
ADET, A107
Los Alamos NM 87545

Peter H. Whetton
CSIRO - Division of Atmospheric Research
Private Bag No. 1
Mordialloc 3195, Victoria Australia

Michael D. Williams
Los Alamos National Laboratory
A-4, MS B299
Los Alamos NM 87545

John D. Wirth
North American Institute
128 Grant Ave., Suite 221
Santa Fe, NM 87501

Remko Ybema
Netherland Energy Research Foundation
Westereduinweg 3, P.O. Box 1
Petten, The Netherlands 1755Z6

Leland W. Younker
Lawrence Livermore National Laboratory
P.O. Box 808, L-203
Livermore, CA 94550

Fredrik Zachariasen
California Institute of Technology
452-48 Caltech
Pasadena, CA 91125

William E. Zagotta
Lawrence Livermore National Laboratory
P.O. Box 808
Livermore, CA 94550

John Zinn
Los Alamos National Laboratory
SST-7, MS D466
Los Alamos NM 87545

COLOPHON

These Proceedings are set entirely in the classic typeface Times Roman, designed for the *London Times* ca. 1890, rarely equaled for beauty and readibility.

Our typesetter software is TeXtures, $T_{\!E}X$ for the Macintosh, from Blue Sky Research. We have used the latest version, Lightning TeXtures, which typesets as one types, another surprising development for minicomputers.

Other softwares we have used include Microsoft *Word*, Adobe *Illustrator*, and Claris *MacDraw II*.

We used the AppleScanner and AppleOneScanner with softwares *AppleScan* and Light Source *Ofoto*.

Our computers are the Apple Macintosh IIcx and IIci.

The editors at Los Alamos were for the most part delighted with the care that authors took with their papers, especially with graphics. There was minimal need to cope with scanning problems, crooked figures and excessive tabular information. However, in the interest of rapid publication we accepted some graphics that do not measure up either to the standards of the American Institute of Physics or those of the Los Alamos National Laboratory.

We believe that data interpretation and arguments were not compromised by our approach, which did result in the desired early publication.

Finis

Printed in the United States
By Bookmasters